Lecture Notes in Physics

Springer-Verlag Berlin Heidelberg GmbH

The Editorial Policy for Proceedings

The series Lecture Notes in Physics reports new developments in physical research and teaching – quickly, informally, and at a high level. The proceedings to be considered for publication in this series should be limited to only a few areas of research, and these should be closely related to each other. The contributions should be of a high standard and should avoid lengthy redraftings of papers already published or about to be published elsewhere. As a whole, the proceedings should aim for a balanced presentation of the theme of the conference including a description of the techniques used and enough motivation for a broad readership. It should not be assumed that the published proceedings must reflect the conference in its entirety. (A listing or abstracts of papers presented at the meeting but not included in the proceedings could be added as an appendix.)
When applying for publication in the series Lecture Notes in Physics the volume's editor(s) should submit sufficient material to enable the series editors and their referees to make a fairly accurate evaluation (e.g. a complete list of speakers and titles of papers to be presented and abstracts). If, based on this information, the proceedings are (tentatively) accepted, the volume's editor(s), whose name(s) will appear on the title pages, should select the papers suitable for publication and have them refereed (as for a journal) when appropriate. As a rule discussions will not be accepted. The series editors and Springer-Verlag will normally not interfere with the detailed editing except in fairly obvious cases or on technical matters.
Final acceptance is expressed by the series editor in charge, in consultation with Springer-Verlag only after receiving the complete manuscript. It might help to send a copy of the authors' manuscripts in advance to the editor in charge to discuss possible revisions with him. As a general rule, the series editor will confirm his tentative acceptance if the final manuscript corresponds to the original concept discussed, if the quality of the contribution meets the requirements of the series, and if the final size of the manuscript does not greatly exceed the number of pages originally agreed upon. The manuscript should be forwarded to Springer-Verlag shortly after the meeting. In cases of extreme delay (more than six months after the conference) the series editors will check once more the timeliness of the papers. Therefore, the volume's editor(s) should establish strict deadlines, or collect the articles during the conference and have them revised on the spot. If a delay is unavoidable, one should encourage the authors to update their contributions if appropriate. The editors of proceedings are strongly advised to inform contributors about these points at an early stage.
The final manuscript should contain a table of contents and an informative introduction accessible also to readers not particularly familiar with the topic of the conference. The contributions should be in English. The volume's editor(s) should check the contributions for the correct use of language. At Springer-Verlag only the prefaces will be checked by a copy-editor for language and style. Grave linguistic or technical shortcomings may lead to the rejection of contributions by the series editors. A conference report should not exceed a total of 500 pages. Keeping the size within this bound should be achieved by a stricter selection of articles and not by imposing an upper limit to the length of the individual papers. Editors receive jointly 30 complimentary copies of their book. They are entitled to purchase further copies of their book at a reduced rate. As a rule no reprints of individual contributions can be supplied. No royalty is paid on Lecture Notes in Physics volumes. Commitment to publish is made by letter of interest rather than by signing a formal contract. Springer-Verlag secures the copyright for each volume.

The Production Process

The books are hardbound, and the publisher will select quality paper appropriate to the needs of the author(s). Publication time is about ten weeks. More than twenty years of experience guarantee authors the best possible service. To reach the goal of rapid publication at a low price the technique of photographic reproduction from a camera-ready manuscript was chosen. This process shifts the main responsibility for the technical quality considerably from the publisher to the authors. We therefore urge all authors and editors of proceedings to observe very carefully the essentials for the preparation of camera-ready manuscripts, which we will supply on request. This applies especially to the quality of figures and halftones submitted for publication. In addition, it might be useful to look at some of the volumes already published. As a special service, we offer free of charge LATEX and TEX macro packages to format the text according to Springer-Verlag's quality requirements. We strongly recommend that you make use of this offer, since the result will be a book of considerably improved technical quality. To avoid mistakes and time-consuming correspondence during the production period the conference editors should request special instructions from the publisher well before the beginning of the conference. Manuscripts not meeting the technical standard of the series will have to be returned for improvement.

For further information please contact Springer-Verlag, Physics Editorial Department II, Tiergartenstrasse 17, D-69121 Heidelberg, Germany

Lorenz Ratke   Hannes Walter
Berndt Feuerbacher (Eds.)

Materials and Fluids Under Low Gravity

Proceedings of the IXth European Symposium on Gravity-Dependent Phenomena in Physical Sciences Held at Berlin, Germany, 2–5 May 1995

DARA   esa

Springer

Editors

Lorenz Ratke
Berndt Feuerbacher
Institut für Raumsimulation, DLR
D-51140 Köln, Germany

Hannes Walter
European Space Agency
8–10 rue Mario-Nikis
F-75738 Paris, France

Cataloging-in-Publication Data applied for.

Die Deutsche Bibliothek - CIP-Einheitsaufnahme

Materials and fluids under low gravity : proceedings of the IXth European Symposium on Gravity Dependent Phenomena in Physical Sciences held at Berlin, Germany, 2 - 5 May 1995 / Lorenz Ratke ... (ed.).
(Lecture notes in physics ; Vol. 464)
ISBN 978-3-662-14066-6 ISBN 978-3-540-49260-3 (eBook)
DOI 10.1007/978-3-540-49260-3

NE: Ratke, Lorenz [Hrsg.]; European Symposium on Gravity Dependent Phenomena in Physical Sciences <9, 1995, Berlin>; GT

ISBN 978-3-662-14066-6

Originally published by Springer-Verlag Berlin Heidelberg New York in 1996
Softcover reprint of the hardcover 1st edition 1996

Typesetting: Camera-ready by the authors
SPIN: 10515243 55/3142-543210 - Printed on acid-free paper

Foreword

The IXth European Symposium on Gravity Dependent Phenomena in Physical Sciences took place at the Berlin Congress Center close to the famous Alexanderplatz. This venue offered not only the opportunity to become acquainted with recent results of scientific investigations on the influence of gravity on phenomena in physics, fluid sciences and materials sciences, but also with the remarkable progress in urban building activity and evident improvements in the daily life of people in the new federal states of Germany. And, indeed, about three hundred scientists from all over the world took time by the forelock to present their research results, and whenever allowed by the tight symposium time schedule, to inspect locations of preferably historical significance in Berlin. However, the content and course of the conference also doubtlessly underlined that the meeting was suited to preparing the science community for the enormous scientific and technological challenge of using the forthcoming International Space Station. Indeed, this permanent research laboratory in space will offer more possibilities for microgravity research than ever before, even despite the fact that since the beginning of this decade more experiments world wide could have been performed than in the 15 years before.

The present book of conference papers selected by the scientific committee represents the most interesting recent results out of about two hundred presentations at the symposium. Both theoretical and experimental findings of gravity dependent phenomena are treated, the latter ones gained by using Space Shuttle and MIR missions, re-entry satellites, sounding rockets, parabolic flights, drop tubes, and even centrifuges. The scope of research fields covers metallurgical solidification processes, growth of semiconductor and protein crystals, the behaviour of liquids and gases, combustion and other physico-chemical processes. Although most of the questions involved are of a fundamental nature, the application potential of the scientific results for improving living conditions, technologies and products on Earth becomes more and more clear. With regard to various innovations the transfer of results has been impressively demonstrated, especially in the quality improvement of castings for cars and airplanes as well as the strip casting of self-lubricating bearing alloys.

I am convinced one can make today the clear-cut statement that research under space conditions has gained an increasing reputation. The scientific results of such experiments – not least documented in this publication – are quite impressive. Therefore, I would like to take the opportunity – on an occasion like this fruitful symposium organized by the European Space Agency ESA and the German Space Agency DARA together with the local organizer Technical University of Berlin - to express my appreciation and gratitude to all who have contributed to this success.

Bonn, August 1995

Jan-Baldem Mennicken
Director General
German Space Agency, DARA GmbH

Preface

The importance of gravity as an experimental parameter and as a variable in a large diversity of physical phenomena and processes was recognized some 25 years ago, when first exploratory experiments under reduced gravity conditions were carried out during the return flight of Apollo capsules from the Moon. Meanwhile numerous experiments have been conducted with Spacelab, the Russian space stations Salyut and Mir, various unmanned satellites, furthermore sounding rockets and drop towers.

An acceleration field, be it induced by gravity or inertia, strongly affects fluids, inducing well-known phenomena such as hydrostatic pressure, natural convection, sedimentation. A large number of processes used on earth to produce solid materials like, e.g., semiconductors, metals or alloys, necessarily employ the liquid state. The microstructure of these materials, with crucial influence on their physical properties, is greatly influenced by processes in the molten or liquid state (e.g. local variations of dopants in semiconductors), since convective motions transport both heat and mass. On the other hand, there are phenomena that can be investigated on earth within severe limitations only, such as critical point phenomena, where the compressibility diverges at the critical point such that fluids in this thermodynamic state are strongly compressed by the hydrostatic pressure.

A major fraction of all microgravity results obtained sofar has been obtained in the course of the past three years. These recent results form the basis of the IXth European conference on gravity-dependent phenomena in physical sciences, held in Berlin, May 1995, which convened researchers from all over the world, including the former eastern bloc countries. The truly interdisciplinary character of "microgravity research" has led to an extensive and very fruitful discussion between experts working in diverse fields such as fluid dynamics, combustion, critical point phenomena, undercooled melts, crystal growth, and basic physics.

The contributions collected in this volume of the "Lecture Notes in Physics" series reflect some of this flavour. They represent a selection from about 200 papers presented during the symposium. The papers selected do not cover all aspects of research und reduced gravity conditions, but they give the reader an overview of the main research topics investigated and the progress in understanding the physical phenomena involved, as obtained during the last few years. These lectures should therefore be a prime source of information for the expert as well as for graduate students.

Cologne and Paris,
August 1995

L.Ratke, B.Feuerbacher and H.U.Walter

Table of Contents

I Critical Point Phenomena and Adsorption

III Crystallization

Part I

Critical Point Phenomena and Adsorption

New Critical Phenomena Observed Under Weightlessness

D. Beysens

Service de Physique de l'Etat Condensé du Commissariat à l'Energie Atomique, Centre d'Etudes de Saclay F-91191 GIF-SUR-YVETTE Cedex France

Abstract

A review is given of several new phenomena observed when the effects of gravity are suppressed in fluids and liquid mixtures near a critical point. A new mechanism of heat transport, the "Piston effect", has been discovered. This mechanism is due to the expansion of a conductive thermal boundary layer and leads to a "critical speeding up" instead of the usual "critical slowing-down". In the presence of an external force (e.g. gravity) the boundary layer is affected by convection and the effect is reduced. Under certain circumstances, a jet may be formed. New results have been obtained concerning the kinetics of phase ordering. After a temperature quench, which leads to the formation of two phases (e.g. liquid and gas), the morphology and growth of the droplet pattern can be connected and leads to only two universal growth laws: growth linear with time at high volume fraction ($\Phi > 0.35$) where the droplets are interconnected, and growth as $(\text{time})^{1/3}$ for smaller volume fraction ($\Phi < 0.35$). This universality seems to be ruled only by the volume fraction and should apply to any fluid phase ordering process. The effect of an electric field leads to interesting observations, where electrostriction is able to reproduce some features of the gravity phenomena. These observations were made by using density-matched liquid mixtures and fluids onboard Texus rockets,furthermore with experiments on the space shuttle missions D-2, IML-1 and IML-2 and the MIR orbital station during the Antares and Altair missions.

1. Introduction

Fluids are supercritical when their temperature and pressure are above the critical point temperature and pressure (Fig.1). They exhibit a number of interesting properties (large density, low viscosity, large diffusivity) which make them intermediate between liquids and gases. In addition, their isothermal compressibility ($\kappa_T = (1/\rho)(\partial\rho/\partial P)_T$) can become very large when they approach the critical point.

The use of supercritical fluids under conditions of reduced gravity, e.g. for the storage of cryogenic propellants, has raised a number of fundamental questions concerning heat and mass transport phenomena when gravity-driven convection is either suppressed or at least greatly reduced. We address in the following, a number of situations where the temperature of a reservoir is changed [1, 2, 3] or a source of heat is located in the fluid [4, 5, 6]. We also consider the kinetics of phase ordering when the temperature is quenched from the supercritical, homogeneous phase to the gas-liquid coexistence region. We analyse in the light of recent theoretical studies [3, 4, 5, 6], the results of several experiments performed under near-zero gravity onboard the Texus rockets, the space shuttle missions D-2, IML-1 and IML-2 and the MIR orbital station during the Antares and Altar missions.

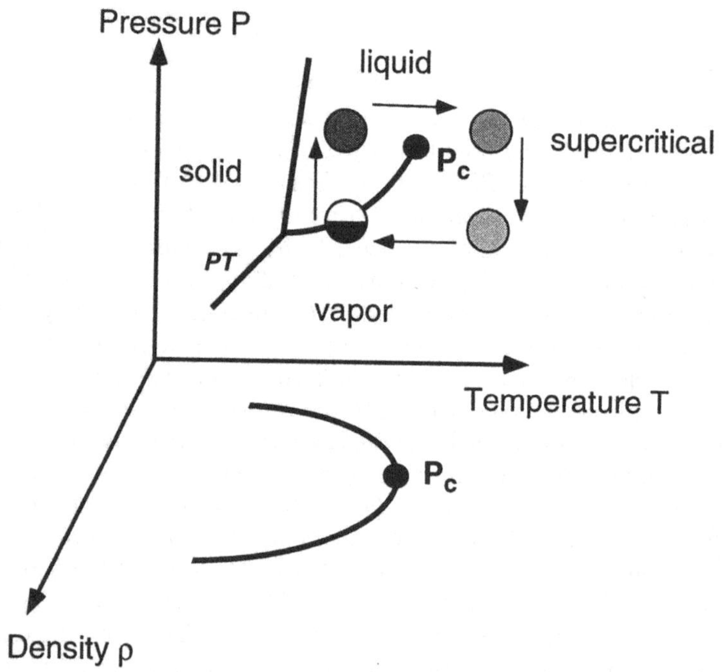

Fig. 1. Phase diagram of a pure substance. P_c: critical point; P_T: triple point.

2. The critical region

2.1 Universality and scaling laws

An important aspect of the critical region [7] is that most of the anomalies of the thermodynamic and transport properties can be expressed in the form of scaled, universal (power law) functions with respect to the critical point (CP) coordinates. Then any result obtained with one single fluid can be generalized to a whole class of systems, the "class of fluids", to which also belong liquid mixtures, including polymer melts and solutions, microemulsions, molten salts, monotectic liquid metals, etc. This scaling is of a fundamental nature. It stems from the universal behavior that asymptotically the free energy F must obey in order to fulfill the conditions of a 2nd order phase transition - the CP (Fig.2). In that sense, scaling is generic to CP phenomena.

The *order parameter* (OP) of the transition is the parameter (M) which determines the CP coordinates by being zero at temperatures above the CP, where the system is homogeneous, and unequal zero below, where two phases coexist. In fluids the order parameter is the density difference $M=\rho-\rho_c$, with ρ, (ρ_c) the (critical) density. Supercritical fluids on the critical isochore correspond to M=0, and the liquid and gas phases in the coexistence region below the CP correspond to $M=M^+$ or M^-, that is, $\rho=\rho^+$ or ρ^-.

The vicinity of a second order phase transition is characterized by the presence of *large fluctuations* of the order parameter M. Fluctuations give rise to an unusually strong scattering of light, the so-called *critical opalescence*. These fluctuations are large near the critical point because of the (universal) asymptotic form of the free energy. Excited by the thermal fluctuations in the system, the order parameter fluctuates more and more strongly as the CP is approached. This corresponds to a divergence of the susceptibility χ of the system

$$\chi = \left(\frac{\partial^2 F}{\partial M^2}\right)^{-1} \propto \epsilon^{-\gamma} \tag{1}$$

with $\epsilon = (T-T_c)/T_c$ and γ=1.24.

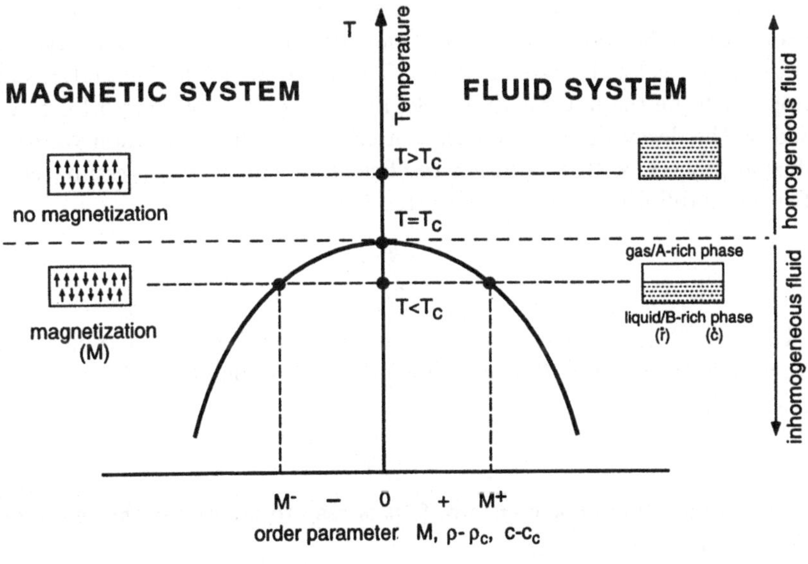

Fig. 2. Phase transition in a T,M phase diagram. ρ: density; c: concentration.

2.2 Correlation length of fluctuations as a natural lengthscale

The importance of fluctuations in the vicinity of the critical point corresponds to a space-dependent susceptibility correlation function $\chi(r)$,

$$\chi(r) \propto \langle \delta \mathrm{M}(r) \delta \mathrm{M}(0) \rangle \propto \frac{\exp(-r/\xi)}{r} \tag{2}$$

with the parameter ξ, the correlation length of the order parameter (OP) fluctuations:

$$\xi = \xi_0 \epsilon^{-\nu} \tag{3}$$

with ν=1/2 and ξ_0 a system-dependent amplitude. The correlation length is a function of temperature; it diverges at the CP and appears as the natural lengthscale of CP phenomena.

All the systems which belong to the same "universality class", as defined by the space dimensionality D and the dimensionality n of the order parameter obey the same asymptotic, universal, scaled behavior. The class of fluids is defined by D=3, n=1. In addition to the fluids (OP: density), can also be considered the partially miscible liquid mixtures (OP: concentration), including the polymer-solvent and microemulsions systems. Many other systems belong to this class, e.g. the 3D Ising model (OP: magnetization), the current representative of the class.

2.3 Critical slowing-down. Unit of time

As above for statics, it is possible to define universal dynamics scaling laws for the transport coefficients. The theory is more delicate and the universality classes are not so large as for the case of statics. The dynamics of the fluctuations of the OP appears to be the natural timescale, like the correlation length for space. It is ruled by the decay time of a fluctuation of size ξ. The fluctuation vanishes by a diffusion process, with a diffusion coefficient D which can be estimated from the Brownian diffusion of a cluster of size ξ [8]

$$D = \frac{k_\mathrm{B} T}{6\pi\eta\xi} \tag{4}$$

Here k_B is the Boltzman constant and η is the shear viscosity. A typical time τ is naturally defined as the diffusion time of a fluctuation of size ξ on the typical lengthscale ξ, i.e.

$$\tau = \frac{6\pi\eta\xi^3}{k_\mathrm{B} T} \tag{5}$$

From the temperature dependence of ξ, it is easy to determine the temperature dependence of τ, i.e.

$$\tau \approx \epsilon^{-3\nu} \tag{6}$$

which shows that the fluctuations of the OP relax more and more slowly as the system approaches its critical point. This is the well known phenomenon of *critical slowing-down*. For a fluid, defined by its density, pressure and temperature, both density (the OP) and temperature fluctuations are slowed down, since pressure equilibrates nearly instantaneously (in fact, at the velocity of sound). The diffusion coefficient can be written also as

$$D \equiv D_{\mathrm{T}} = \frac{\Lambda}{\rho C_{\mathrm{p}}} \tag{7}$$

with D_{T} the isothermal heat diffusion coefficient, Λ the heat conductivity and C_{p} the specific heat at constant pressure. To give an idea of critical slowing down, thermal equilibration of a CO_2 sample at 1 mK from T_{c} with thickness $2e$=1cm would need a time t_{D} of the order of

$$t_{\mathrm{D}} \approx \frac{e^2}{D_{\mathrm{T}}} \tag{8}$$

which corresponds to about one month to reach thermal equilibrium!

2.4 Scaling by unit of length and time

It then appears that the correlation length of fluctuations is the natural length-scale of CP phenomena, as the fluctuation lifetime is the natural timescale. Then most of the processes, once properly scaled by ξ and τ, are universal.

2.5 Pure fluids, liquid mixtures, and gravity effects

Binary liquid mixtures near a critical point of miscibility belong to the same universality class as pure fluids for static properties (the three dimensional Ising model [7]). However, their behavior in a gravity field can be markedly different. This difference stems from the order parameter $\mathrm{M} = c - c_{\mathrm{c}}$ for liquid mixtures (c_{c} is the critical concentration), leading to an osmotic compressibility, which diverges at the critical point and a mutual diffusion coefficient which goes to zero. For pure fluids, M is the density difference $\rho - \rho_{\mathrm{c}}$ and at the critical point the isothermal compressibility diverges and the thermal diffusivity goes to zero. Fluids are very sensitive to the earth gravity field especially in the vicinity of the liquid-gas critical point [9]. Very important convection flows, often turbulent, are observed [10]. In addition, experiments are complicated by the very high compressibility of the fluid which induces density gradients even in the one-phase region.

3. The "Piston effect"

3.1 Heat transport by the Piston Effect

The transport of heat in dense pure fluids classically involves the mechanisms of convection, diffusion, and radiation. Recently, the understanding of thermal equilibration of a pure fluid near its gas-liquid critical point (CP) has required a fourth mechanism, the so-called "Piston Effect" (PE). In contrast to ideal gases [11], this effect originates from the high compressibility of the critical fluid.

The numerical simulation of the Navier-Stokes equations for a one dimensional (1-D) van der Waals gas [2, 5] reveals the basic physical mechanisms giving rise to the PE (Figs.3). When a homogeneous bulk fluid enclosed in a two-wall sample cell is suddenly heated from one wall, a diffusive thermal boundary fluid layer (thickness δ) forms at the wall-fluid interface. Due to the high compressibility of the bulk, the fluid layer expands and acts as a piston generating an acoustic wave which propagates into the bulk and which is reflected at the second wall enclosing the fluid. Thermal conversion of this pressure wave is, in turn, able to heat the fluid in an adiabatic way. As a result, a spatially uniform heating of the bulk fluid can be detected only after a few acoustic cahracteristic times $t_{\mathrm{a}} = L/c$ (with L the characteristic sample dimension and c the sound velocity in the fluid). During repeated travel of the pressure wave in the fluid, the bulk temperature progressively reaches thermal equilibrium.

An additional result inferred from a recent asymptotic analysis of the PE [6] shows that the fluid velocity produced by the expansion of the hot boundary layer reaches its maximum value at the edge of the layer. This value is proportional to the heat input (per unit surface) at the wall. The true PE driving force is this velocity which induces the compression of the bulk fluid by a small transfer of matter and makes the boundary layer act as a converter which transforms thermal energy into kinetic energy. The effect of an energy source, as provided by continuous heating of the fluid, is then markedly different from that of a temperature step at the sample wall, where the heat power decreases with time as the fluid equilibrates.

As initially discussed by Onuki and Ferrell [1, 3], the first characteristic time scale for the PE is the time scale (t_0) which corresponds to the transfer from the boundary layer of the energy (E_{b}) able to adiabatically heat the remaining fluid (size $L - \delta \approx L$) to the desired temperature $\Delta T_{\mathrm{b}} \approx E_{\mathrm{b}}/LC_{\mathrm{v}}$. C_{v} is the specific heat at constant volume and we consider a sample of unit volume and unit mass. The energy transfer occurs at an acoustic time scale [2], i.e. instantaneously when compared to t_0. Equilibration is obtained when the boundary layer temperature $E(t)/\delta C_{\mathrm{p}}$ (with C_{p} the specific heat at constant pressure) reaches ΔT_{b}, i.e. when the energy $E(t)$ which has diffused in the boundary layer during t_0 has reached E_{b}. Therefore $E_{\mathrm{b}}/\delta C_{\mathrm{p}} \approx E_{\mathrm{b}}/LC_{\mathrm{v}}$ and $\delta \approx L/\gamma_0$, with $\gamma_0 = C_{\mathrm{p}}/C_{\mathrm{v}}$. A more refined treatment [5] gives $\delta = L/(\gamma_0 - 1)$. The value of t_0 can readily be written as $t_0 = \delta^2/D_{\mathrm{T}}$, where D_{T} is the thermal diffusivity of the fluid.

The second and longer characteristic time scale is the diffusion time t_{D} as a function of the lengthscale L, that is,

$$t_{\mathrm{D}} = \frac{L^2}{D_{\mathrm{T}}} \tag{9}$$

t_{D} and t_0 are then related as follows:

$$t_0 = \frac{t_{\mathrm{D}}}{(\gamma_0 - 1)^2} \tag{10}$$

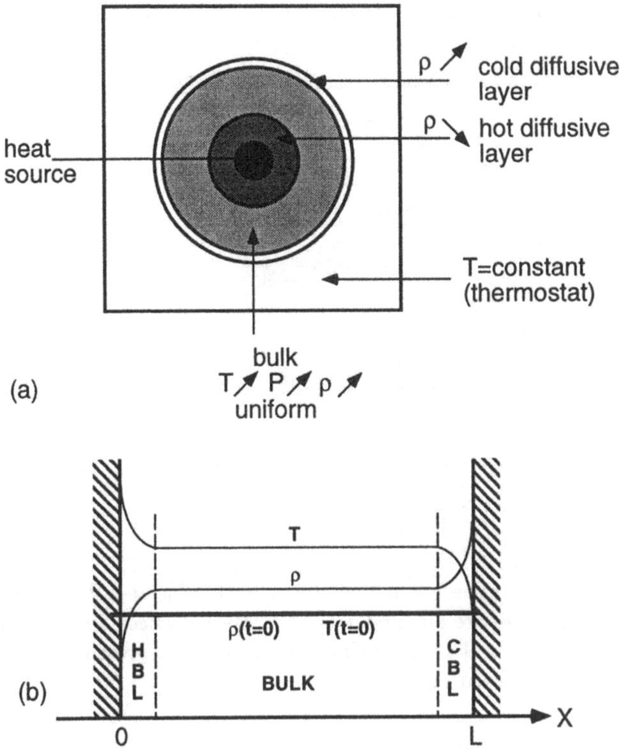

Fig. 3. Schematic of the temperature and density rise in a closed initially isothermal sample when heated in the centre. Sketch of (a) the sample and (b) the temperature T and density ρ distribution. The wall at X=0 is heated; the temperature of the wall at X=L is kept constant. A diffusive boundary layer forms, hot at X=0 (HBL) and cold at x=L (CBL). Outside these layers, in the bulk fluid, temperature and density rise homogeneously ("Piston Effect"). Pressure is uniform within an acoustic time scale.

A striking result is the critical speeding up of the PE when going closer to the CP [12]. As a matter of fact, near T_c, γ_0 diverges and both δ and t_0 go to zero, although t_D goes to infinity. This result represents an enormous reduction in the time required for temperature equilibrium.

A number of experiments in sounding rockets have demonstrated this acceleration of dynamics. Straub, et al. in Texus 6 [13], observed that the inner temperature of a cell filled with SF_6 at critical density did follow the outer temperature when crossing the critical point. The same conclusion has been drawn by Klein et al. in Texus 26 [14]. In Texus 25 [10], a thermal cycle has been performed with CO_2 around its critical point. Thermalization was found to be homogeneous and the kinetics were limited by the thermostat and the sample cell. In addition, this experiment showed that the fluid inhomogeneities still relax by diffusion, which makes the final equilibration dependent of the diffusion time scale [15].

3.2 Bulk temperature relaxation

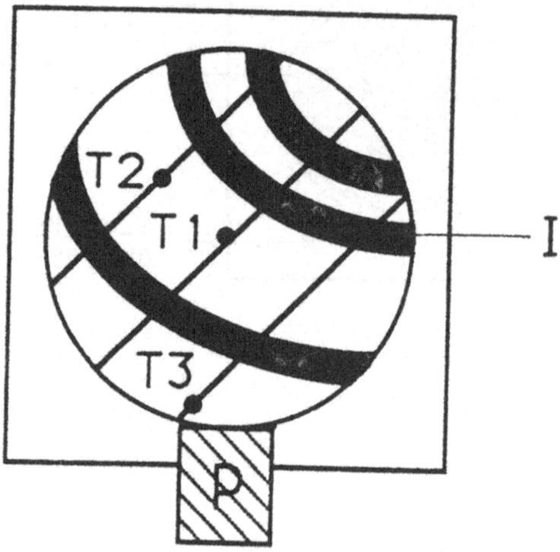

Fig. 4. Schematic of the experiment cell, filled with SF_6 at 1.27 ρ_c. The density is off-critical in order to compare on similar time scales PE effect and heat diffusion. T1: heating thermistor; T1, T2, T3: measuring thermistors. P: pressure gauge. I: interference fringes for density measurements.

In order to study the PE dynamics, one can produce a heat pulse directly in a fluid enclosed in a sample placed in a thermostat and study the density, temperature and pressure evolution [16]. This has been done on MIR during the missions ANTARES and ALTAIR, and onboard the space shuttle during the mission IML2 (Fig.4,5,6). Samples of CO_2 and SF_6 at liquid, gaseous and critical densities have been used.

Three regions were observed in the sample (Figs.3,4,5, 6). First, a hot boundary layer, centered around the heat source, with large coupled density and temperature inhomogeneities. It relaxes by a diffusive process, the density and temperature relaxation is slowed-down close to the critical point. Second, the bulk fluid, which remains uniform in temperature and density and its dynamics are governed by the PE time. At the thermostat walls a weak, cold boundary layer forms, which cools the bulk by a PE mechanism. The final equilibration in temperature and density of the fluid is governed by the diffusion time t_D, which represents the slowest mechanism.

Let us analyse the temperature evolution after the heat pulse. The data can be described by the following function, proposed by Ferrell and Hao [4] and Carlès and Zappoli [5, 6]:

$$\Delta T_B = \frac{E}{mC_v} \exp(t/t_c) \mathrm{erfc}(\sqrt{t/t_c}) \tag{11}$$

Here ΔT_B is the bulk temperature rise and E is the energy input into the sample. This function is the response of the system to a very short pulse. The general

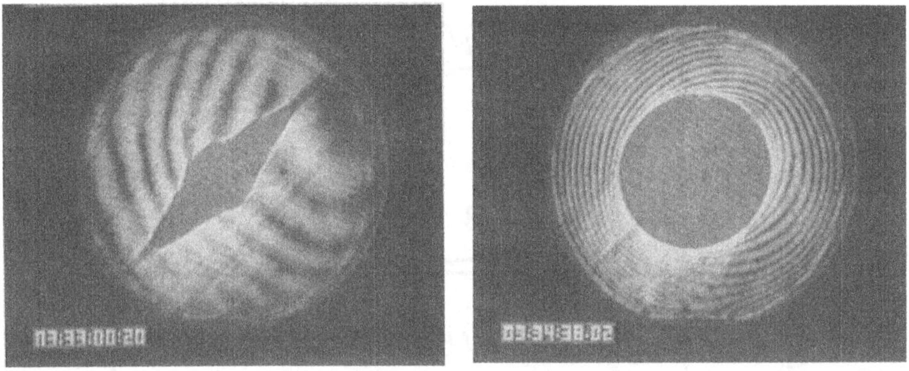

Fig. 5. (left) Aspect of the cell during a heat pulse under micro-g. The black region around T1 is the hot boundary layer, which extends along the thermistor leads. (right) Aspect of the cell after the heat pulse. The hot boundary layer vanishes due to diffusion.(from BEM3 in IML2).

case is a convolution product of the above function with the exact time profile of the heat pulse. Fiure 7 shows the data obtained at 16.8 K in CO_2 at critical density are shown, together with a fit using (11) (best fit: $E/(mC_v)$=22.6 mK; $t_c \approx 4.3$ s.).

3.3 Scaling behavior of the temperature at the hot wall

The relaxation of the temperature at the hot wall is expected to follow a diffusive relaxation. Within the approximation $T_b \approx T_i$, one expects [5, 6]

$$\Delta T^* = \frac{1}{\sqrt{\pi t^*}} \tag{12}$$

with $t^* = t/t_D$ and $\Delta T^* = mC_p \Delta T/E^*$, with $E^* = E\, S_{CW}/SW$ the energy reduced by the cold and hot wall surface areas and ΔT the temperature difference between the hot wall and the bulk. Figure 8 shows a log-log plot of ΔT^* as a function of the reduced time for different equilibrium temperatures. All data follow a single curve, which confirms the diffusive transport of the heat localised in the hot boundary layer. The exponent $\omega \approx 2/3$ differs from the expected classical value 1/2 because of transcient effects, as has been discussed by Fröhlich et al in [17].

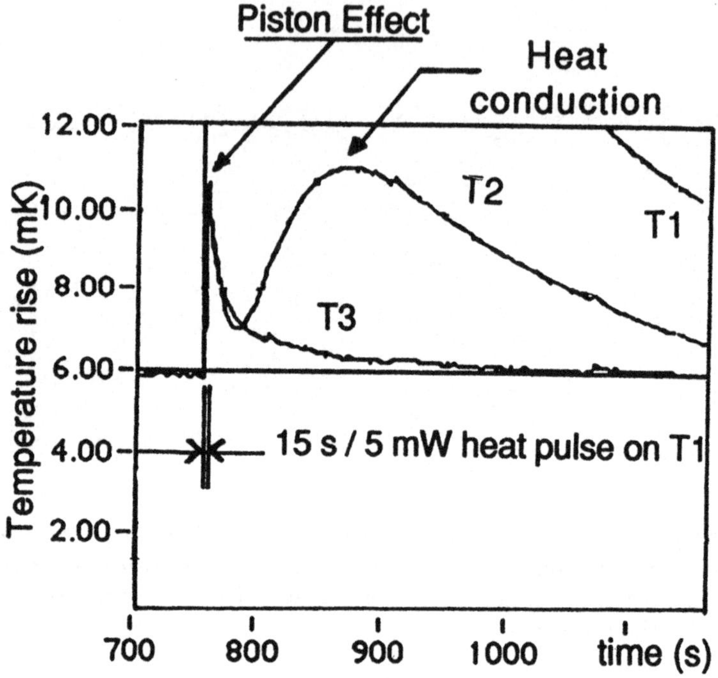

Fig. 6. Temperature evolution in T1, T2 and T3 showing the relative contributions and different timescales of PE and heat diffusion. Thermalization by the PE is fast and complete at the end of the pulse. Heat conduction gives the same temperature value (11 mK here) much later. (from BEM3 in IML2).

3.4 Measurements of thermophysical properties

When the temperature of the thermostat is changed, the temperature of the fluid varies accordingly in the bulk. The process can be considered as an adiabatic PE heating for times larger than t_c where temperature and density changes can be expressed as:

$$\frac{\Delta T}{\Delta \rho} = \left(\frac{\partial T}{\partial \rho}\right)_S \tag{13}$$

This coefficient is difficult to determine on earth and can be a test for an equation of state. Figure 9 summarizes results obtained by Michels et al. in determining this coefficient during the IML2 mission [18]. Although the variation with $T - T_c$ is the same as deduced from the equation of state proposed by Sengers, [19] there is a systematic difference of 40%.

3.5 Diffusion, convection and Piston effect

A further step is to analyse how the PE and diffusion are affected by a weak acceleration. For this purpose, a weak acceleration (30 μg), as provided by the

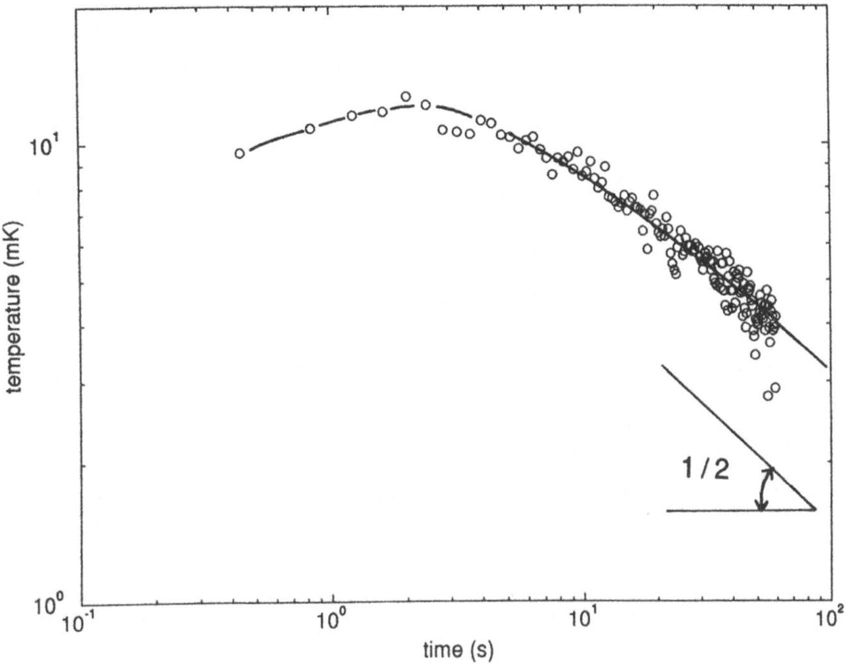

Fig. 7. Temperature evolution of the bulk temperature (log-log plot) after a heat pulse (power $\approx$ 20 mW, duration $\approx$ 2.5 s) in a cell with CO_2 at ρ_c. The cell is similar to that in Fig.4, with T1 as a heat source and T2 to measure the bulk temperature. Heating is ensured by the expansion of the hot boundary layer around the thermistor T1. Cooling is due to the cold boundary layer at the thermostat cell wall. Heating and cooling follows Eq.(11), with a slope near 1/2. (From the ALICE-Antares experiment in MIR)

rotation of the spacecraft, has been imposed on the sample during a heat pulse. The results are shown in Figs.10,11. The thermalization by the PE starts in the bulk together with the heat pulse (Fig. 11). When the sample is submitted to the acceleration, thermalization is affected during a few seconds, although the hot boundary layer is only slightly affected (Fig.10, left). It is likely that the convection of the boundary layer, although weak, disturbs the density gradients in the layer, thus resulting in a pressure reduction. Once heating has stopped, the occurence at T2 of the diffusion temperature peak is markedly accelerated by convection, as shown in Fig.10(right),11.

3.6 Jet flow

A striking phenomenon has been observed with CO_2 and SF_6 at critical density. A jet of fluid is projected from the hot boundary layer into the bulk fluid. This phenomenon is observed in a CO_2 sample with 0.9 mm diameter thermistors, at temperatures close to T_c, and in a SF_6 sample with 0.25 mm diameter thermistors, at temperatures above T_c, and also below T_c. The ejection mechanism

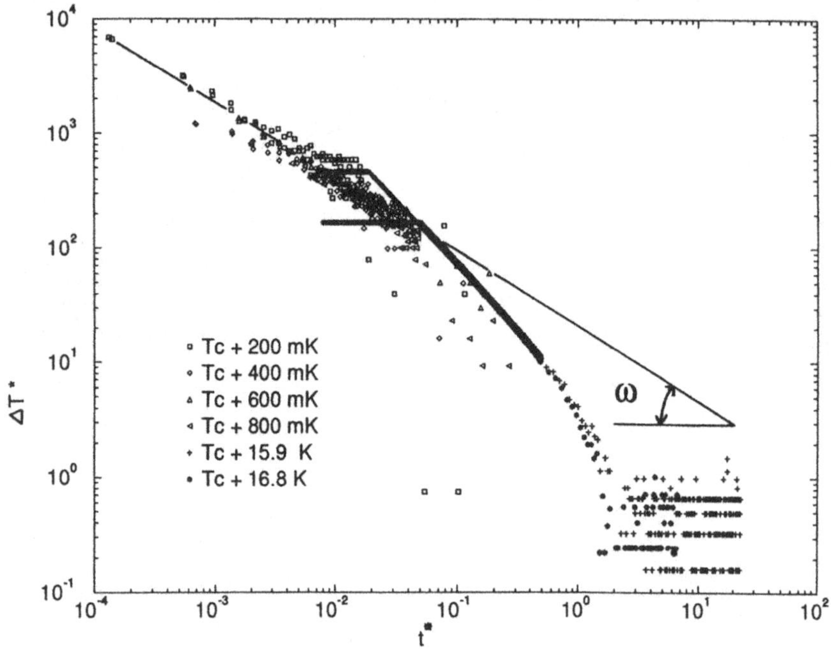

Fig. 8. Reduced temperature evolution of the hot wall (T1 in the Fig.4 arrangement) after a heat pulse. ΔT^* is scaled by the energy sent in the fluid and its specific heat at constant pressure, and time is scaled by the diffusion time. Scaling demonstrates the hot boundary layer diffusive behavior. The exponent $\omega \approx 0.6$ differs from the classical exponent 1/2 because of transient effects (see [17]). (From the ALICE-Antares experiment in MIR)

in CO_2 occurs when the fluid is closer than 1K to T_c and after a time sufficient to obtain a well developed thermal layer. The jet is randomly oriented. In contrast, when the experiment is reproduced with SF_6 and with smaller thermistors, where the heating energy per surface area is considerably larger, the jet is already present at 3.25 K from T_c and occurs always at the same position. It appears even for the shortest heating period of 40 ms (Fig.12). In all cases, the jet has a vortex shape with an initial ejection speed of a few cm/s. Gravity levels during this phenomenon were lower than a few 10^{-4}g.

The jet can be also observed on earth, however there is convection at the same time and only its early development ($t < 100$ms) can be analysed. The time dependence of the jet length $\mathcal{L}$ for different duration of the heat pulse is presented in Fig.13. No significant variations are observed. When the injected power P_0 is varied, the initial velocity $V_0 = (d\mathcal{L}/dt)_{t=0} \propto P_0$. The temperature dependence of V_0, under zero-g conditions and on earth, is presented in Fig.14. There is a systematic difference between the two sets of data, with the ground results showing the smaller velocities. The difference is probably due to gravity

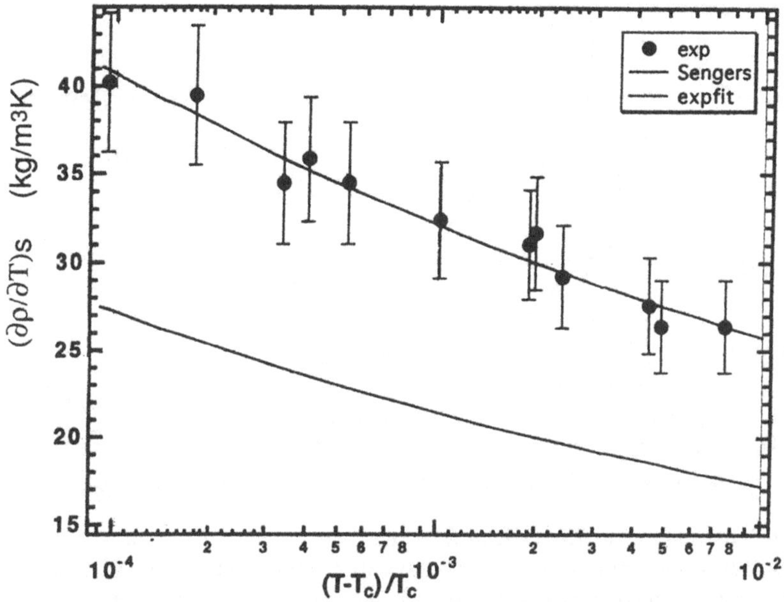

Fig. 9. The experimental derivative $(\partial\rho/\partial T)_S$ versus temperature as obtained by Michels et al. during the IML2 mission. The bottom curve is obtained from an equation of state proposed by Sengers [19]; expfit is an empirical exponential fit. (From MIM in IML2 [18])

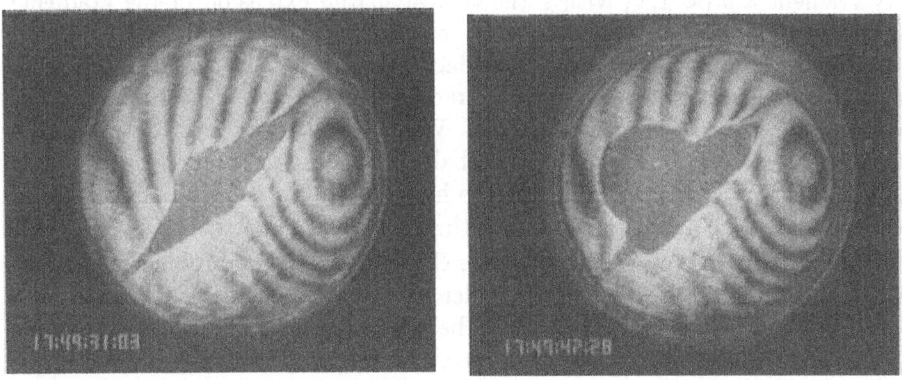

Fig. 10. Thermalization after a heat pulse under variable acceleration. The cell is that of Fig.4, filled with SF_6 at 1.27 ρ_c. (left) A weak acceleration ($\approx$ 30 μg), causing convection, is provided by a spacecraft rotation starting 7.5 s after the beginning of the heat pulse. Right: the hot boundary layer touches the thermostated wall, causing the temperature of the bulk to drop (from BEM3 in IML2).

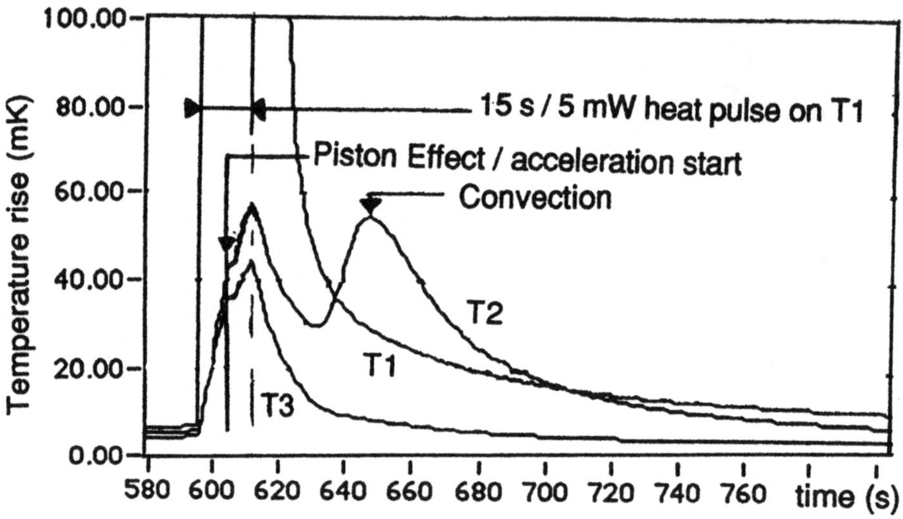

Fig. 11. Thermalization after a heat pulse under variable acceleration. The cell is that of Fig.4, filled with SF_6 at 1.27 ρ_c. Temperature measured in T1, T2 and T3. Note that thermalization by the PE is lowered when acceleration starts. Convection increases further thermalization in T1 and in the bulk.(From BEM3 in IML2)

induced convection on earth in the boundary layer, which, even for such very small times, can affect the development of the jet. The velocity slightly increases when approaching T_c.

A first possible explanation of this phenomenon could be found in the density gradient $\nabla\rho$ ($\approx 1/\zeta$, with ζ the typical spatial extension of the gradient) at the border between the thermal layer and the bulk fluid. To this gradient corresponds an out-of-equilibrium interfacial tension $\sigma \approx k_B T/\zeta^2$. The thermal layer is analogous to a hot "bubble" of radius δ, with internal pressure counterbalanced by the capillary pressure σ/δ. When the thermal layer develops, the capillary pressure can no longer counterbalance the internal pressure, the bubble breaks and a jet is formed. According to hydrodynamics, the jet velocity should scale as $V \propto 0.1\sigma/\eta \propto 1/(\eta\zeta^2)$, with η the fluid shear viscosity.

When compared to experiments, this theory leads to two problems. The first is the amplitude of the velocity. In order to obtain velocities of the order of 1 cm/s, the gradient extension ζ has to be microscopic. It is natural to assume in this case that ζ scales as the correlation length ξ, which is the ultimate lengthscale of a density inhomogeneity. Then the temperature dependence should be different to what is observed since both η and ξ increases when T is closer to T_c.

A plausible explanation is that the jet is merely due to the expansion of the hot boundary layer in the hottest region of the thermistor. According to the Zappoli and Carlès analysis [5, 6], the velocity of the fluid during the growth of the boundary layer is maximum at the border with the bulk and is proportional

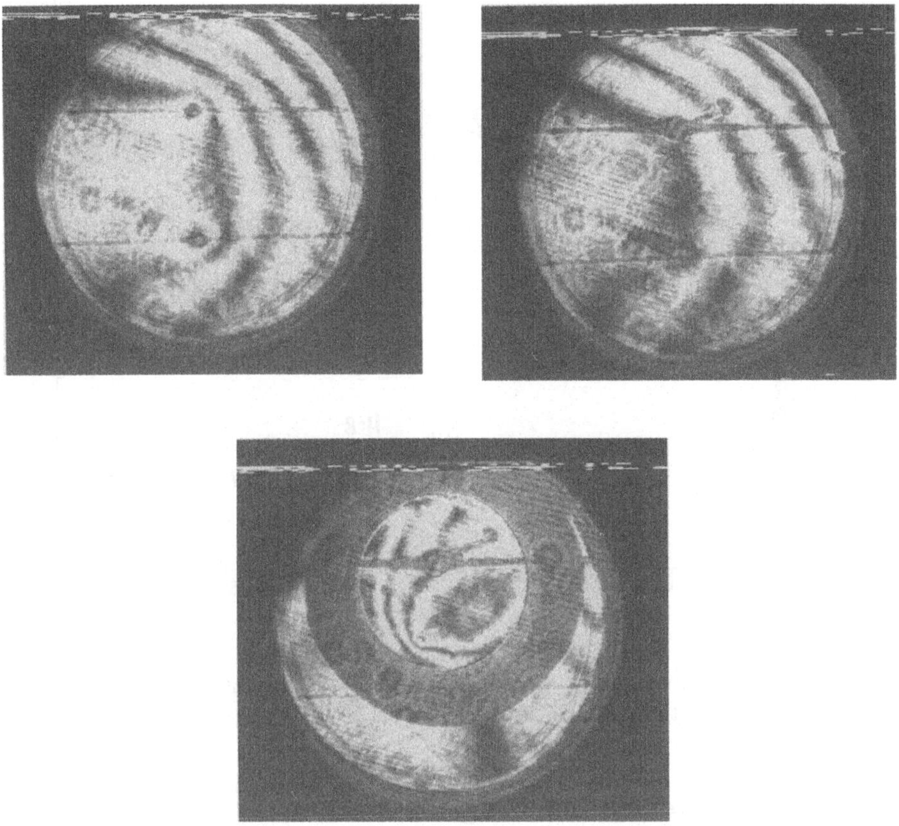

Fig. 12. Formation of a jet flow under zero-g (SF_6 sample, similar to the Fig.4). The arrow shows the jet. The residual accelerations are: X: $2 \cdot 10^{-4}$; Y:$5 \cdot 10^{-5}$; Z:$5 \cdot 10^{-5}$ (units: earth acceleration). A heat pulse (20 mW power, 40 ms duration) is sent in a sample filled with SF_6 at critical density. (top left). T=T_c+3.25K, before the pulse. (top right) same as in top left, 280 ms after the pulse. (bottom row) In the two-phase region, at T=T_c-0.1K. (From the ALICE-Altair experiment in MIR).

to the expansion coefficient at constant density $(\partial T/\partial P)_\rho$ ($\approx P_c/7T_c$ for CO_2) and the injected power P_0 per unit area S:

$$V_0 = \frac{1}{T_c}(\frac{\partial T}{\partial P})_\rho \frac{P_0}{S} \tag{14}$$

This expression accounts for the above observations and measurements; with $P_c \approx 7$ MPa, $S \approx 4 \cdot 10^{-8}\text{m}^2$, $P_0 \approx$20 mW, one finds V_0 of the order of a few cm/s, which compares well with the experimental velocity. In this view, the jet should thus be a spectacular demonstration of the existence of the "piston" in the PE.

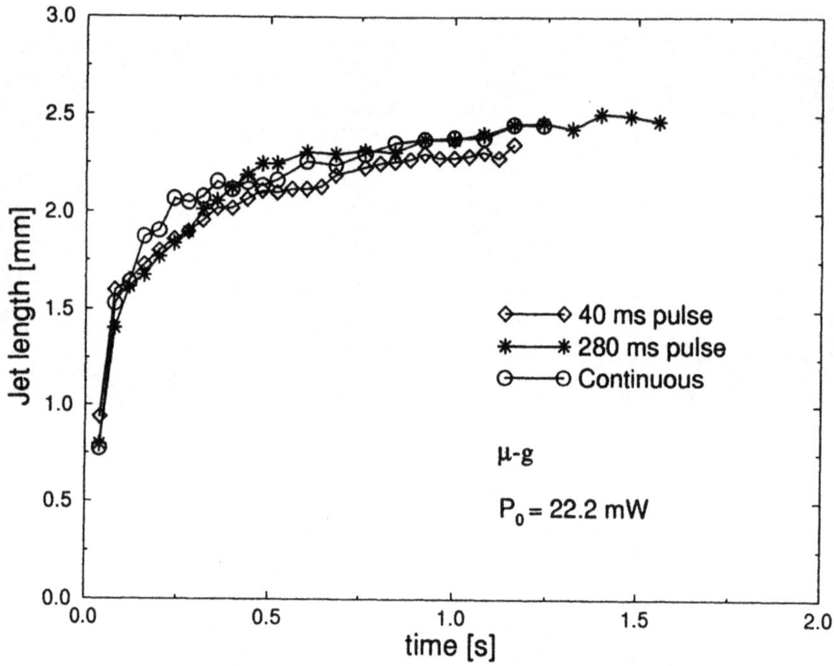

Fig. 13. Length evolution of the jet from Figs.10,11, under weightlessness, for different pulse duration.(From the ALICE-Altair experiment in MIR)

4. Kinetics of phase ordering

Although phase separation in fluids and liquid mixtures is a common process which occurs in many areas of science and technology [20], the connection between the morphology of domains and the growth laws is still unclear. Only a few experimental results are available when density or concentration is systematically varied [21, 22]. In the critical region, it is very easy to vary continuously the physical parameters (Fig.15). The critical slowing down enables a detailed investigation of the mechanisms involved in the separation process and the results can be described by universal, scaled master curves, valid for all fluids within two scale factors.

During phase separation, pure fluids and binary liquids are always sensitive to the gravity field, through convection and/or sedimentation. In order to avoid convection, a density-matched liquid mixture can be prepared by adding a small amount of deuterated cyclohexane in the cyclohexane-methanol mixture [23]. Nevertheless, it is impossible to avoid long-term sedimentation because density-matching can never be perfect.

A problem which is specific to fluids still complicates the experimental study: The thermal diffusivity goes to zero at the critical temperature T_c. However, as discussed above, the "Piston effect" speeds up thermalization (at the cost of a

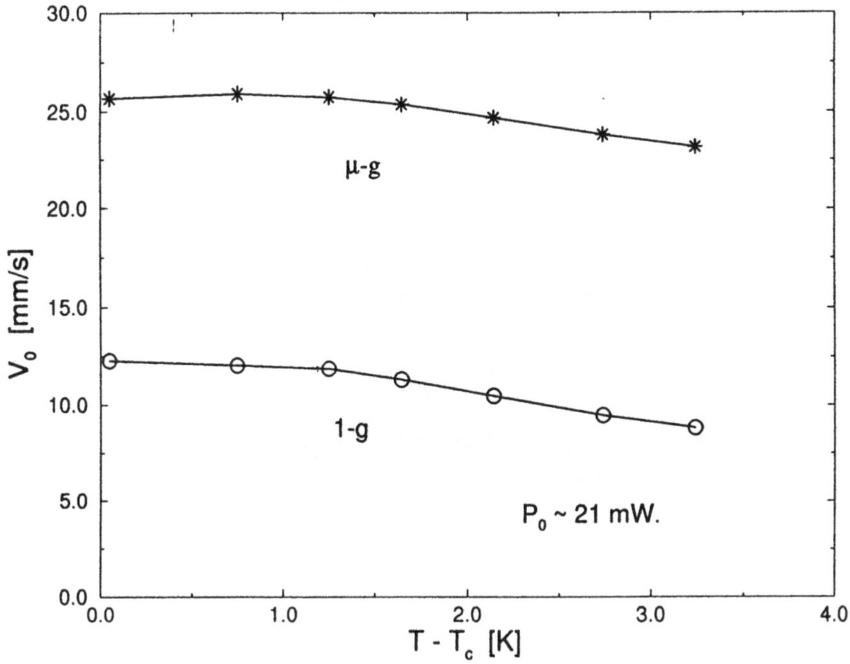

Fig. 14. Initial jet velocity V_0 versus time at different temperatures after a 40 ms heat pulse in SF_6 under 1-g and zero-g conditions. (From the ALICE-Altair experiment in MIR).

thin boundary layer) so that thermal quenches very close to T_c are limited only by the thermal response of the thermostat.

Experiments with density-matched binary liquids and fluid CO_2 under reduced gravity (Figs.15) show [24] that when M=0, that is when the volume fraction of the minority phase is $\Phi = 1/2$, an interconnected pattern of domains which coalesce continuously is formed. Later, when the equilibrium concentration has been rduced, the characteristic wavelength L_m grows linearly with time t. When expressed in the scaled units

$$K_m^* = \frac{2\pi\xi}{L_m} \tag{15}$$

and

$$t^* = \frac{t}{t_\xi} \tag{16}$$

the results obtained in all liquid mixtures and in CO_2 during the gravity-free experiment can be placed on the same master curve [24], and thus strongly support universality (Fig.16).

When the volume fraction of the domains is very small ($\Phi < 0.03$), and the gravity effects are negligible, the droplets do not coalesce and they grow by a

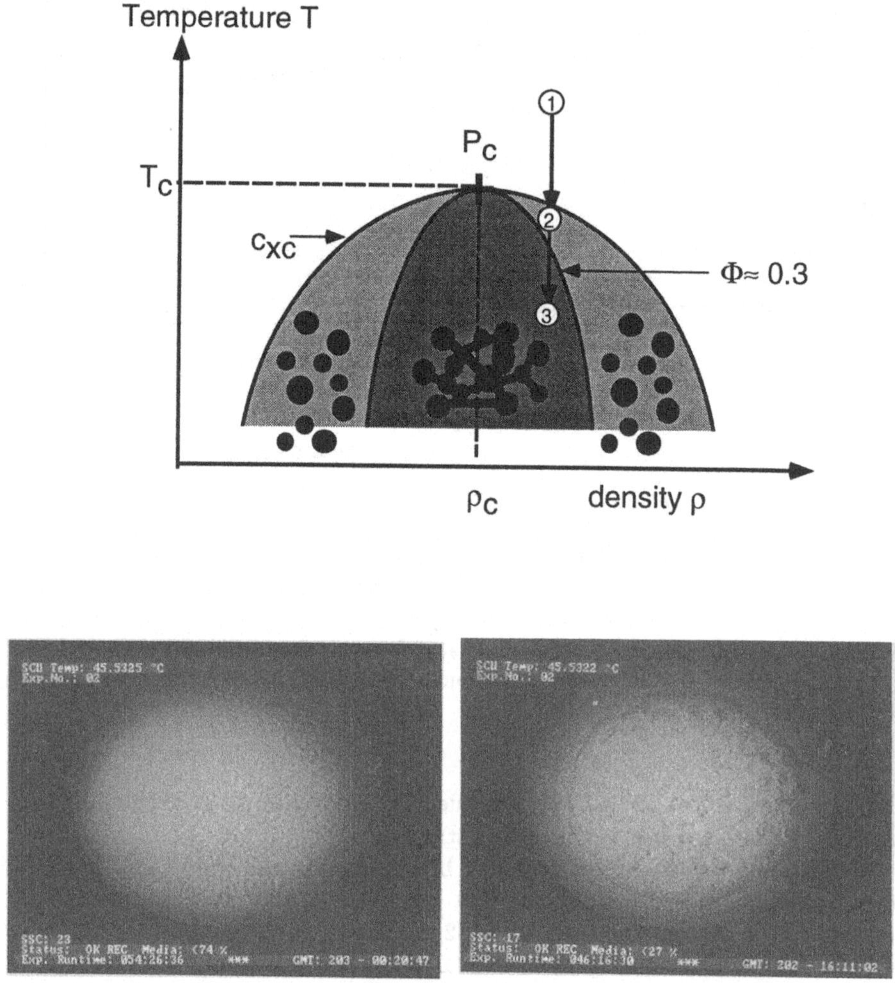

Fig. 15. Morphology and growth (from BEM in IML1 and IML2 missions). Phase diagram (schematic) of fluid SF_6. P_c: critical point; T_c: critical temperature (45.55°C); ρ_c: critical density (0.737 g cm-3); c_{xc}: coexistence curve. When the supercritical fluid (①, density off-critical by +0.5%) is quenched below the c_{xc}, droplets of vapor (volume V1) and liquid (volume V2) nucleate and grow. It is expected that the volume fraction threshold $\Phi = V1/(V1+V2) \approx 30\%$ separates a region of "slow" growth (*bottom left:* after the quench ① → ② of 50 μK below c_{xc}) where the droplets are disconnected and grow as $(\text{time})^{1/3}$, to a region of "fast" growth; (*bottom right:* after the quench ① → ③ of 600 μK below c_{xc}), where the droplets are interconnected and grow as $(\text{time})^1$.

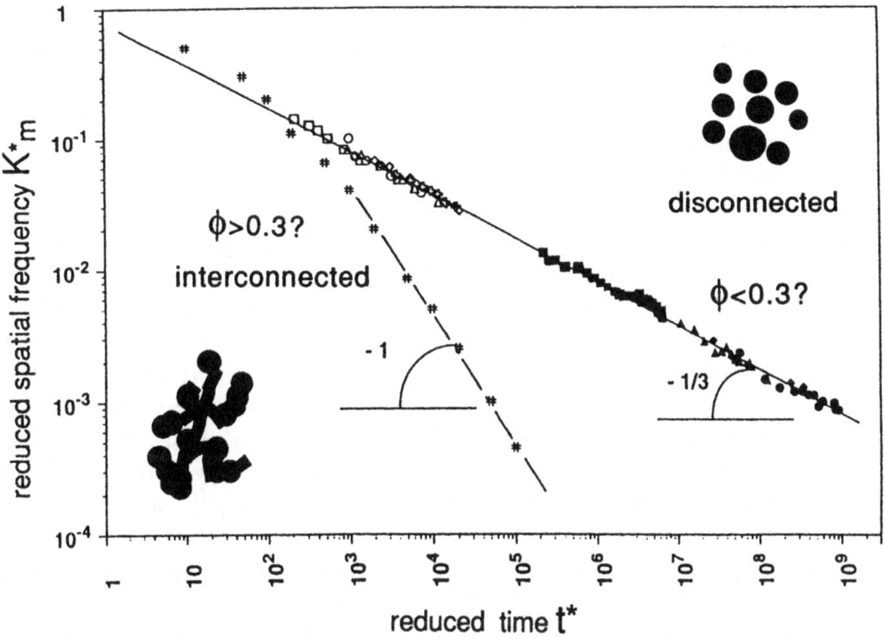

Fig. 16. Only two universal growth laws are evidenced when the size of the droplets are expressed in scaled units. The governing parameter seems to be the volume fraction Φ, which governs also the morphology of the drop pattern (interconnected, disconnected), with a sharp cross-over expected for $\Phi \approx 0.3$. (From [24, 27]).

diffusion mechanism. Experiments [25] show that, although the initial growth follows a power law with time with exponent $1/2$ or $1/3$ depending on the initial supersaturation, the late stages are always characterized by a $1/3$ growth law exponent as described by Lifshitz and Slyosov [26]. Off-critical systems at small volume fractions have already been studied in a liquid mixture by Wong and Knobler [21]. For $\Phi > 0.10$ a "slow" growth characterized by a $1/3$ growth law exponent is reported in the beginning and a faster growth later on, with the growth exponent varying from $1/3$ to 1 as a function of Φ. In the case where the densities of each component are close, gravity effects are expected to be weak in these experiments.

The idea of density matching was used in a subsequent investigations [22] where the temperature quench has been done in the presence of a controlled concentration gradient. In a single experiment, this allowed the position of the coexistence curve to be determined as well as the boundary between the "fast" and "slow" zones. However, due to intrinsic experimental difficulties, the standard error of the experiment was large and difficult to determine. Also the experiment was performed with a concentration gradient the influence of which is diffiult to estimate. Based on these results it was only possible to state that the threshold value of Φ is between 0.3 and 0.4.

In the IML-1 mission with the CPF [27], an off-critical density-matched binary mixture of methanol and partially deuterated cyclohexane at concentration $c \approx c_c - 0.01$ and an off-critical fluid of sulfure hexafluoride at $(\rho - \rho_c)/\rho_c = 0.036$ have been tested by using different quench steps. All the data obtained in liquid mixtures and in SF_6 during the gravity-free experiments, when expressed in the scaled units K_m^* and t^* can be placed on the same master curve ($K_m^* \approx 1t^{*1/3}$), making clear the universality of phase separation in fluids and liquid mixtures. Moreover, the crossover between slow and fast growth as reported by Wong and Knobler cannot be observed at Φ=0.1 when gravity effects are suppressed. It is likely that this discrepancy can be attributed to gravity effects which are difficult to evidence directly.

An experiment with an off-critical fluid (SF_6) was carried out during the IML-2 mission. The density difference $(\rho - \rho_c)$ has been measured on earth by studying the temperature dependence of the meniscus between the phases.This method is precise and gave $(\rho - \rho_c)/\rho_c = 0.005$. Under microgravity, a determination of the transition temperature T_{cx} with a precision of 50 μK has been made. Fast growth, with interconnected pattern, occured for a quench depth of 300 μK and 1000 μK while for a quench depth of 50 μK and 100 μK slow growth was observed. This shows that the volume fraction is the relevant parameter for studying the connection between morphology and growth. Nevertheless, the presence of a density gradient in the cell, related to the very large compressibility of the fluid near its critical point, complicates the interpretation of the data, in particular the estimation of the cross-over volume fraction between fast and slow growth. In addition, these experiments were hampered by the short time available. In particular, the precise determination of the transition is essential and needs much time to avoid density gradients during the experiment. This determination cannot be performed on earth within a few mK because gravity-induced gradients are present (the fluid is compressed under its own weight).

Although the experimental results are scarce, phase separation has been the object of much theoretical interest during the last two decades. The work by Siggia [28] has initiated the attempts to explain the two growth laws by choosing a hydrodynamics approach. Recently, several groups [2, 30, 31, 32, 33] performed large scale direct numerical simulations by using slightly different approaches to solve coupled equations of diffusion and hydrodynamics. Some of them have recovered a t^1 growth law [29, 30]. The others [32, 33] were not able to reach the late stages of separation and measured the transient values of the growth exponent (between 1/3 and 1). In spite of these efforts, the physical mechanism for the linear growth has not been clarified. To our knowledge, the simulations never showed two asymptotical laws which would depend on the volume fraction: the exponent is either larger then 1/3 when accounting for hydrodynamics or 1/3 for pure diffusion. Thus the simulations still do not explain the transition from one regime to one another. Recently, a new theoretical approach has been suggested by Nikolayev, Beysens and Guenoun [34] which explains the existence of the two different regimes, their relation to the pattern morphology and the absence of a crossover. Growth is due to the coalescence of droplets; depending on the volume fraction, coalescence can occur through the Brownian diffusion of droplets ($t^{1/3}$ growth law, isolated droplets) or to hydrodynamics correlation coming from the coalescence process itself, where a coalescence event induces another coalescence event (t^1 growth law, interconnected droplets). The transition between the two regimes appears at a well defined volume fraction, of order 30%.

5. Other aspects

Because the fluids become extremely compressible close to the critical point, a number of interesting situations can be induced under weightlessness. One is the effect of an electric field, investigated by Zimmerli et al. [35] during the IML2 mission. An electric field of order 200 V/mm was applied to SF_6 at critical density. Due to electrostriction, a density gradient followed, which diverges near T_c as the isothermal compressibility. Therefore it becomes possible to induce well defined density gradients in microgravity, which could reproduce some features of the gravity phenomena. In particular, axisymmetric and centrosymmetric geometries can be obtained.

6. Conclusion

As has been reported above, weightlessness prevents flows when density gradients are present in the fluids. Such gradients are unavoidable on earth near the critical point because of the divergence of the isobaric thermal expansion coefficient $(\partial r/\partial T)_p$: even minute temperature gradients lead to high density gradients.

It is therefore not surprising that new effects, usually hidden on earth, can be demonstrated under reduced gravity. The most spectacular effect is the thermo-compressible transport of heat, or Piston effect, which paradoxically leads to an acceleration of the thermalization process when approaching the critical point.

Applications of these findings are found in the supercritical storage of fluids, and especially of cryogenic fluids (O_2, H_2). The use of supercritical fluids prevents the gas-liquid interface from being monitored, which is always difficult under microgravity.

How gas-liquid interfaces can order is also a problem where weightlessness brings some enlightenment. The coarsening of a liquid-gas emulsion is seen to follow very simple universal laws, depending on the drop pattern morphology.

One may conclude that the study of supercritical fluids under weightlessness, demonstrates the very unusual hydrodynamics of these supercritical fluids, compressible, dense, and weakly viscous, which makes their behavior quite particular when compared to gases or liquids. Indeed one can speak of new hydrodynamics, or "super" hydrodynamics.

Further developments are manyfold. One may mention the effect of an acceleration field (Rayleigh-Bénard compressible problem) or oscillatory acceleration (vibration), and how chemical reactions (combustion) can proceed in such fluids.

Acknowledgments

This review has been made possible thanks to the help of many friends. The author wishes especially to thank T. Fröhlich, M. Bonetti, Y. Garrabos and F. Perrot. The support of the Centre National d'Etudes Spatiales is gratefully acknowledged.

References

1. A. Onuki, H. Hao, and R.A. Ferrel, *Phys. Rev.***A41** (1990) 2255 **A45** (1994) 4779.
2. B. Zappoli, D. Bailly, Y. Garrabos, B. Le Neindre, P. Guenoun, and D.Beysens, *Phys. Rev.* **A41** (1990)2264; B.Zappoli, A.Durand-Daubin, *Phys. Fluids* **6** (1994) 1929-1936.
3. A. Onuki and R.A. Ferrel, *Physica* **A164** (1990) 245.
4. R.A. Ferrel and H. Hao, *Physica* **A197** (1993) 23.
5. B.Zappoli, P.Carles, *Eur. J. of Mech. B/Fluids* **14** (1995) 4165.
6. P. Carlès, thesis (Institut National Polytechnique, Toulouse, 1995).
7. H. E. Stanley, *Introduction to Phase Transitions and Critical Phenomena*, Clarendon Press, Oxford 1971.
8. K. Kawasaki K.,1970, *Ann. Phys. NY* **61** (1970) 1.
9. M. R. Moldover, J. V. Sengers, R. W. Gammon and R. J. Hocken, *Rev. Mod. Phys.* **51** (1979) 79.
10. P. Guenoun, D. Beysens, Y. Garrabos, F. Kammoun, B. Le Neindre and B. Zappoli, *Phys. Rev.* **E47** (1993) 1531.
11. A.M. Radhwan and D.R. Kassoy, *J. Eng. Math.* **18** (1984) 183.
12. H. Boukari, M.E. Briggs, J.N. Shaumeyer, and R.W. Gammon, *Phys. Rev. Lett.* **65** (1990) 2654.
13. K. Nitsche and J. Straub, in: *Proc. 6th Eur. Symp. on Materials Sciences under Microgravity*, ESA SP-256, Noordwijk 1986, p. 109.
14. H. Klein, G. Schmitz, and D. Woermann, *Phys. Rev.* **A43** (1991) 4562.
15. Y. Garrabos, B. Le Neindre, P. Guenoun, F. Perrot, and D. Beysens, *Microgravity Sci. Tech.* **2** (1993) 108.
16. M. Bonetti, F. Perrot, D. Beysens, and Y. Garrabos, *Phys. Rev.* **E 49** (1994) 4779.
17. T. Frhlich, S. Bouquet, M. Bonetti, Y. Garrabos and D. Beysens, *Physica A* **2523** (1995, to appear).
18. A. C. Michels (1995, private communication)
19. A. Abbaci and J.V. Sengers, Technical report no. BN1111 (IPST, University of Maryland, College Park, 1990).
20. J. D. Gunton, M. San Miguel and P. S. Sahni , in: *Phase Transitions and Critical Phenomena*, vol 8, edited by C. Domb and J. L. Lebowitz (Academic Press, 1983), p. 269 and references therein.
21. N. C. Wong and C. M. Knobler, *Phys. Rev. A* **24** (1981) 3205.
22. Y. Jayalakshmi, B. Khalil and D. Beysens, *Phys. Rev. Lett.* **69** (1992) 3088.
23. D. Beysens, P. Guenoun and F. Perrot, *Phys. Rev. A* **38** (1988) 4173.
24. Y. Garrabos, B. Le Neindre, P. Guenoun, B. Khalil and D. Beysens, *Europhys. Lett.* **19** (1992) 491 and refs. therein.
25. A. Cumming, P. Wiltzius, F. S. Bates and J. H Rosedale, *Phys. Rev. A* **45** (1992) 885; T. Baumberger, F. Perrot and D. Beysens, *Phys. Rev. A* **45** (1992) 7636.
26. I. M Lifshitz and V. V. Slyosov, *J. Phys. Chem. Solids* **19** (1961) 35.

27. F. Perrot, P. Guenoun, T. Baumberger, D, Beysens, Y. Garrabos and B. Le Neindre, *Phys. Rev. Lett.* **73** (1994) 688.
28. E . D. Siggia, *Phys. Rev. A* **20** (1979) 595.
29. S. Puri and B.Dnweg, *Phys. Rev. A* **45** (1992) R6977.
30. F. J. Alexander, S. Chen, and D.W. Grunau, *Phys. Rev. B* **48** (1993) 634.
31. A. Shinozaki and Y. Oono. *Phys. Rev. E* **48** (1993) 2622.
32. O. T. Valls and J. E. Farrell, *Phys. Rev. E* **47** (1993) R36.
33. K. Kawasaki, T. Ohta, *Physica A* **118** (1983) 175; T. Koga, K. Kawasaki, M. Takenaka and T. Hashimoto, *Physica A* **198** (1993) 473.
34. V. S. Nikolayev, D. Beysens and P. Guenoun , preprint (1995).
35. G. Zimmerli, R. A. Wilkinson, R. A. Ferrell, H. Hao and M. R. Moldover, preprint (1995).

Numerical Solutions of Thermoacoustic and Buoyancy-Driven Transport in a Near Critical Fluid

B. Zappoli[1], *S. Amiroudine*[2], *P. Carles*[1], *J. Ouazzani*[3]

[1] CNES, 18 Av. Edouard Berlin, 31055 Toulouse Cedex, France
[2] Institut de Mécanique des Fluides, 1 rue Honnorat 13003 Marseille, France
[3] Arcofluid, IMT, Technopole de Chateau-Gombert, 13451 Marseille, France

Abstract

This paper presents the mechanisms of heat and mass transport of 1D and 2D low Mach number, unsteady, viscous, low heat diffusing, hypercompressible Navier-Stokes equations of a van der Waals gas (CO_2). The results have been focused on some striking behaviours compared to those obtained for normally compressible gases: i) heat equilibration is still achieved very fast under normal gravity conditions, as under zero-g conditions, by the Piston Effect before buoyancy convection has time to enhance heat transport; ii) mass equilibration is achieved on a much longer time scale by a quasi isothermal buoyant convection; iii) due to the very high compressibility, a stagnation point effect as that encountered in high speed flows provokes an overheating of the upper wall of a heated square cavity; iv) a significant difference with the convective single roll pattern generated under the same condition in normal CO_2 is also found: on the Piston Effect time scale, under the form of a Marangoni-like pattern due to the very thin boundary layer-localised density gradients; on the heat diffusion time scale under the form of a double roll convective structure.

1. Introduction

Heat transport in dense fluids is achieved by the basic mechanisms of convection, diffusion and radiation. In the absence of those mechanisms as it is the case in low heat diffusing near critical fluids under zero-g conditions, it has been shown recently that a fourth mechanism of heat transport named the Piston Effect (PE) is responsible for a very fast heat transport. This effect has been extensively studied and described in number of papers [1]-[10], involving one-dimensional analytical [1, 5, 8, 9] and numerical [2, 6, 7, 10] analysis, thermodynamical theory [4] and space-borne experiments [3] in low-gravity conditions where buoyant convection is strongly decreased. The PE mechanism originates from the particular properties of near-critical fluids, and more specifically from their diverging isothermal compressibility and vanishing thermal diffusivity. It can be briefly described by the following steps: (i) when a confined, near-critical fluid is heated from the boundaries, a thin thermal boundary layer forms; (ii) the fluid in this layer expands strongly due to the large compressibility, and induces an adiabatic compression of the bulk fluid; (iii) temperature is homogenised very rapidly, as a result of the adiabatic compression. The thermoacoustic nature of this effect is now well established as well as its limits since it has been shown that all the heat brought to the wall of a container filled with a near-critical pure fluid could ultimately be transferred to the bulk at the speed of sound when approaching the critical point [7, 8].

Under these conditions, the isothermal compressibility can be five or six orders magnitude higher than that of the perfect gas. It is thus necessary to have a robust and efficient algorithm in order to solve the unsteady, low heat diffusing, hypercompressible Navier-Stokes equations. We have chosen the finite volume method with its related algorithms [7] and an acoustic filtering procedure [12] which is necessary to reduce the computational time when an acoustic-wave description is not needed.

A number of authors [11, 12, 13] have studied the basic fluid science approaches of normally compressible media, and earlier, in the beginning of the seventies, a number of technical studies have been devoted to the hydrodynamic behaviour of very compressible cryogenic supercritical oxygen, hydrogen or helium, which are known to be of a great technological interest. Heinmiller [13] has in particular, and for example, studied numerically the convective flows in cryogenic oxygen tanks to obtain a description of the pressure collapse which follows a fast mechanical destratification of a microgravity-stored near critical fluid.

Although no numerical solutions have been obtained with a reliable Navier-Stokes code written for a hyper-compressible, near-critical fluid, some theoretical work has been done in an attempt to obtain stability criteria for fluid layers subjected to adverse temperature gradients [14, 15]. More recently a 1-D numerical description of the hydrodynamically stable infinite fluid layer cooled from below [16] has been given while 1-D heat transport has been studied in clearly stable conditions [17]. Rather than examining the problem of a layer heated from below (with the unknown potential for Rayleigh-Bénard instability), we choose to look at the simpler case of convection (driven by both buoyancy and thermoacoustic effects) in a box which is heated from the side. It is demonstrated that the bulk of the temperature equilibration in this situation is achieved by the PE and not by buoyant convection. It has also been found that, due to the very high compressibility, significant density gradients remain even after the thermal field has been nearly completely homogenised by the PE, leading to a quasi-isothermal convective motion in a pure fluid which is then responsible for mass equilibration. In addition, a stagnation-point overheating effect has been observed, although the flow velocity is very low. The next section presents the modeling and governing equations, then we describe the numerical method, and results are discussed finally.

2. The Model and the Governing Equations

2.1 The Model

As mentioned in the introduction, the main goal of this work is to investigate the competition of convective and thermoacoustic (PE) heat transport. We consider a square, 2-D cavity with a single, vertical, heated wall (on the left-hand side), the others being perfectly insulated (see Fig.1). Under such conditions, the PE and buoyant convection both originate at the heated boundary while the other boundaries are not thermally active. This would not have been the case if one had considered, as in [5], an isothermal right-hand boundary which would have also led to a PE, but also, to a more complex situation without providing more information regarding the problem under study.

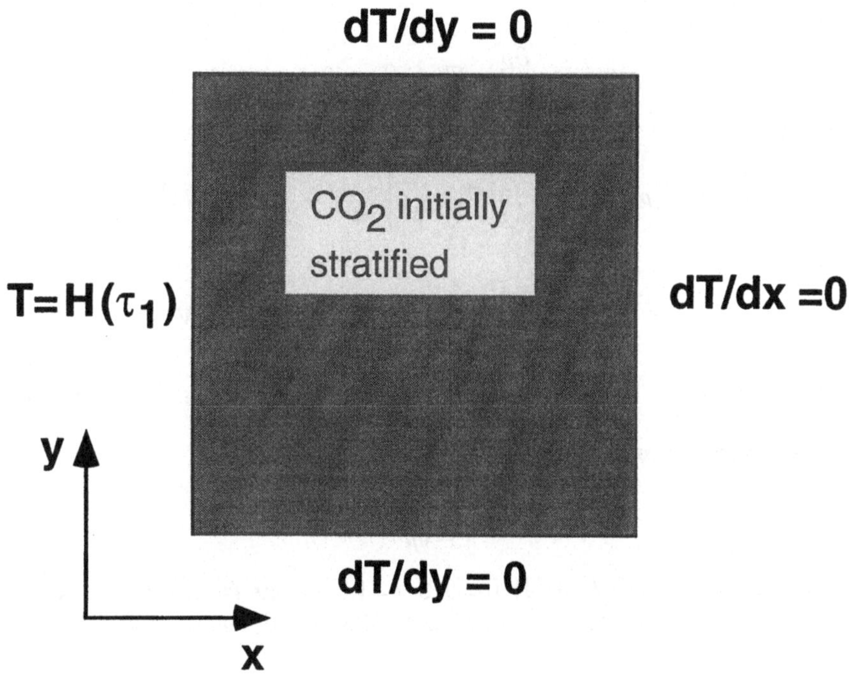

Fig. 1. The 2D Model.

The fluid is initially at rest and stratified in the thermodynamic equilibrium such that $T_0' - T_c' = \mu \ll 1$ where T_0' and T_c' are respectively the initial and critical temperature while the primes denote dimensional variables. μ defines the proximity to the critical point. The following transport coefficients are considered [18]:

$$\lambda = \frac{\lambda'}{\lambda_0'} = 1 + \Lambda\left(\frac{T' - T_c'}{T_c'}\right) \qquad C_v = \frac{C_v'}{C_{v0}'} = 1 \qquad \tilde{\mu} = \frac{\mu'}{\mu_0'} = 1$$

where λ', C'_v, μ' are respectively the thermal conductivity in which Λ =0.75, the heat capacity at constant volume and the viscosity; the subscript "0" represents the value far from the critical point. The heat capacity at constant volume and the viscosity have been considered as constant and equal to the value held by the perfect gas.

2.2 The Governing Equations

As just mentioned, the governing equations are those of a Newtonian, viscous, heat-conducting van der Waals gas initially at rest and stratified, subjected to a boundary heating. If the pressure is normalised with respect to the value it would have at the critical condition if the fluid were a perfect gas, and if the other variables are referred to their critical values, the governing equations can be written as follows:

Continuity:

$$\frac{\partial \rho}{\partial t} + \nabla \cdot (\rho \mathbf{V}) = 0 \tag{1}$$

Momentum:

$$\frac{\partial (\rho \mathbf{V})}{\partial t} + \nabla(\rho \mathbf{V}\mathbf{V}) = -\gamma^{-1}\nabla P + \epsilon \left[\nabla^2 \mathbf{V} + \frac{1}{3}\nabla(\nabla \cdot \mathbf{V})\right] + \frac{1}{F_r}\rho \mathbf{g} \tag{2}$$

Energy:

$$\frac{\partial (\rho T)}{\partial t} + \nabla \cdot (\rho \mathbf{V} T) = -(\gamma - 1)(P + a\rho^2)(\nabla \cdot \mathbf{V}) + \tag{3}$$

$$+\frac{\epsilon\gamma}{P_r}\nabla \cdot [\{1 + \Lambda(T-1)^{-0.5}\}\nabla T] + \epsilon\gamma(\gamma-1)\left[\frac{2}{3}\mathbf{V}_{i,i}\mathbf{V}_{j,j} - \mathbf{V}_{i,j}\mathbf{V}_{j,i} - \mathbf{V}_{i,j}\mathbf{V}_{i,j}\right]$$

Van der Waals equation of state:

$$P = \frac{\rho T}{1 - b\rho} - a\rho^2 \tag{4}$$

The values of a and b in (4) (a=9/8, b=1/3) come from the expression of critical coordinates as a function of a' and b' for the dimensional van der Waals equation, namely:

$$T'_c = \frac{8a'}{27b'} \qquad \rho'_c = \frac{1}{3b'} \qquad P'_c = \frac{a'}{27b'^2}$$

in which T'_c =304.13 K and ρ'_c=467.8 kg/m^3. The following non-dimensional variables have been defined:

$$\rho = \frac{\rho'}{\rho'_c} \qquad T = \frac{T'}{T'_c} \qquad \mathbf{V} = \frac{\mathbf{V}v}{C'_0} \qquad t = \frac{t'C'_0}{L} \ ,$$

where $C'_0 = \sqrt{\gamma R' T'_c}$ is the sound velocity for the perfect gas in which γ is the ratio of specific heats and R' is the perfect gas constant. One should note that the energy equation (3) embodies strong variations of the heat conduction coefficient and as shown by this equation, it goes to infinity as $\mu^{-1/2}$ when T' approaches T'_c.

In (1)-(3), ϵ is a Knudsen number-like parameter defined by $\epsilon = P_r t'_a/t'_d$, t'_d is the characteristic time of heat diffusion for the ideal gaz (IG) $t'_d = L'^2/\kappa'_0$ where κ'_0 is the thermal diffusivity of the IG at critical density and t'_a is the characteristic acoustic time $t'_a = L'/\sqrt{\gamma R' T'_c}$. In the case of CO_2 confined in a square cavity of side length of 10 mm, t'_a=35 ms and t'_d=3000 s. $P_r = \nu'_0/\kappa'_0$ is the Prandtl number (ν'_0 is the kinematic viscosity) and $F_r = C'_0/L'g'_0$ is the acoustic Froude number (g'_0=9.8 m/s^2 is the earth gravity).

Since the problem is to investigate the interaction between the PE and the buoyant convection following a boundary heating, the characteristic time scale to be chosen to this end should be the shorter of the two phenomena in order not to miss important and determining possible interactions between them. The PE time scale defined by [1] is: $\tau = \frac{\epsilon f(\mu,\Lambda)}{\mu^2} t$ where $f(\mu,\Lambda) = \mu\left(\frac{1}{\Lambda} + \frac{1}{\sqrt{\mu}}\right)$. This dimensional PE timescale is of order of tenths of seconds for CO_2 at some K from T_c. The convective time scale of this range of temperature is of same order as the PE timescale which shortens as the critical point is approached. Consequently, The PE time scale is thus taken as characteristic time while velocity is normalised with respect to the characteristic velocity on that timescale : $\mathbf{V}_\tau = \frac{\mu^2}{\epsilon f(\mu,\Lambda)} \mathbf{V}$. The dimensionless conditions can be written as:

$$\begin{aligned} \rho_0(y) &= 2 - \exp(K(y-1)) \\ T_0 &= 1 + \mu \\ P_0(y) &= P_{01} + \frac{9}{4}\mu[1 - \exp(K(y-1))] \\ U = V &= 0 \end{aligned} \tag{5}$$

where $K = 4\gamma g/(9\mu F_r)$ and $P_{01} = 3/2(1+\mu) - 9/8$. The boundary conditions for the problem defined in the previous section are:

$$\begin{aligned} U = V = 0 & \qquad \text{at all walls} \\ T(0,y,t) &= H(\tau_1) \\ \frac{\partial T}{\partial x}(1,y,t) = \frac{\partial T}{\partial x}(1,0,t) = \frac{\partial T}{\partial x}(x,1,t) &= 0 \end{aligned} \tag{6}$$

The parameter τ_1 is the boundary heating time scale, which has been chosen to be that of the PE because in most of the experiments the heating law characteristic time is of the order of one second which corresponds to the order of magnitude of the PE characteristic time.

3. Numerical Method

3.1 General Description

The numerical method used in this analysis is based on finite volume method with the SIMPLER algorithm [19, 20]. The detailed developments and comparisons of different algorithms can be found in [7, 10].

A staggered mesh has been developped where velocity components are placed on the sides of the cell and thermodynamic quantities on the cell center, to avoid pressure oscillations [20]. As we approach the critical point, the thermal boundary layer becomes much thinner and is of the order [1]: $\delta_{BL} = f(\epsilon, \mu) \approx o(\sqrt{\epsilon}\mu^{0.25})$. At acoustic times, all the physics of the problem are in this thin boundary layer and one needs a small grid size in order to have a relatively small cell-Peclet number. Thus, a variable mesh has been developped which is a "power law" type mesh [7] commonly used in finite volume methods. The numerical code we have used solves two-dimensional unsteady, viscous, compressible flows in rectangular cavities on variable meshes. It takes into account hypercompressible flows with physical properties which diverge at the critical point, as shown in (3). Our code has been carefully tested on some relevant benchmark cases. The results of such comparisons can be found in [10]. It is important to note here that, besides the fundamental role played by the physics of the problem under study, we demonstrate the feasability of simulating unsteady near-critical hypercompressible flows using finite-volume method. The numerical modelling of supercritical fluids has been developed in recent years, mostly in one-dimensional situations [2, 6, 7] and multidimensional numerical studies have not previously performed for this problem with the method of acoustic filtering described below.

3.2 Acoustic Filtering

As is the case for normally compressible flows, the acoustic filtering procedure is necessary to reduce the computational time when acoustic waves description is not needed. As a matter of fact, without such filtering, the semi-implicit character of the present algorithm would not be able to reduce the time step (compared to an explicit one) [9] since the time step reduction is governed mainly by acoustic phenomena and not by stability criteria. On the other hand, as pointed out by Paolucci [12], for small Mach numbers, implicit schemes making use of relaxation methods are relatively inefficient to time-step reduction and overall pressure changes since the relaxation of long-wavelength errors is a slow process in the low-Mach-number limit.

The thermodynamic field is expanded with respect to the small Mach (Ma) number; thus, the pressure is written for example,

$$p = p^{(0)} + Ma^2 p^{(0)} + o(Ma^2)$$

all the other variables are expanded in a similar manner. This results in the leading term of the pressure being a function of time only and the contribution of the pressure to the momentum equations comes from $O(Ma^2)$ terms in the Mach number expansion. In practice, in our numerical approach, we have splitted the pressure as follows:

$$P = p'(x, y) + \overline{P(t)}$$

where the leading term $P(t)$ is homogeneous and the perturbation term, which is of order Ma^2, only appears in the momentum equations.

The splitting of the pressure has resulted in one more unknown than the number of available equations and an extra equation is needed to determine the static pressure $\overline{P(t)}$. This pressure is determined by requireing mass conservation at each timestep: $\int_{V(t)} \rho dV(t) = \int_{V(t)} \rho_0 dV_0$ where V represents the volume of the cell and subscript "0" refers to initial values. This method of the filtering of sound is used for timescales greater than acoustic timescales. Another way to determine this static pressure [10] is to use the combined continuity, energy and equation of state and integrate over the volume of the cell.

4. Results and Discussion

In what follows, the results of the computations previously defined are analysed with a special emphasis on the heat and mass transport mechanisms leading to the new thermal and mechanical equilibrium in the cavity after the temperature of the left hand sidewall is increased by 10 mK over a period 1s. In order to examine the difference between the normal gas response and that of supercritical fluid, the same problem has been solved with the same numerical code for CO_2 assuming it behaves like a perfect gas at standard conditions, i.e T'_0=300 K and ρ'_0=1.8 kg/m^3. Although the computations are performed in terms of dimensionless variables, it is more convenient to discuss results in terms of dimensional quantities.

4.1 Temperature Homogeneisation

In Fig.2(b) it is shown that 4.7 s temperature equilibration in the cavity is almost achieved through the PE, while the effect of buoyancy is restricted to a low-density thermal plume visible at the hot wall and top-left part of the cavity. This part of the flow field will be described more in detail in section 4.3. On the contrary, the perfect gas thermal field shown in Fig.2(a) for the same time is inhomogeneous because of more rapid thermal diffusion coupled with very weak convection. The homogeneous temperature field in the bulk of the cavity for the supercritical fluid at a time much shorter than the thermal-diffusion time are the signature of the PE and shows that there is no significant interaction between the buoyant convection and the PE.

The temperature is plotted in Fig.3(a) as a function of x at y=0.5 for several times during the temperature homogeneisation period for both supercritical fluid and perfect gas. The supercritical fluid exhibits the characteristic profiles of the PE (a very thin thermal boundary layer followed by a thermally homogeneous bulk) and the perfect gas shows a diffusive profile. For the density field, we observe some very interesting features as seen in Fig.3(b), which shows profiles for both supercritical fluid and perfect gas near the heated wall at y=0.5 for two times belonging to the PE period. The first observation is that the density for the supercritical fluid is four orders of magnitude larger than that for a perfect gas owing to the very high compressibility of the supercritical fluid. The second is that the boundary layer is much thinner for the supercritical fluid than for the perfect gas.

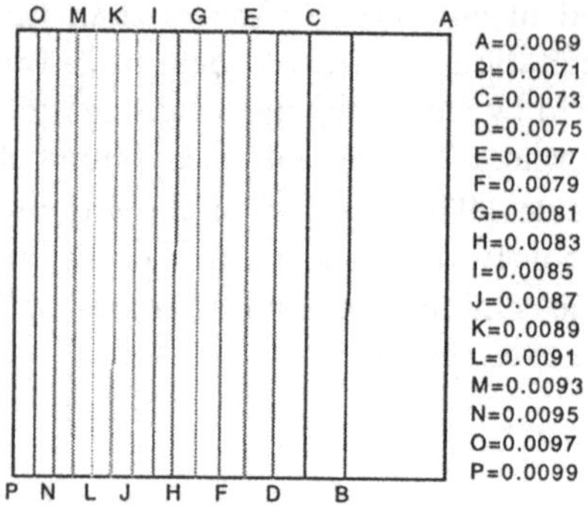

Fig. 2. (a): Temperature difference field in K at 1g and 4.5 seconds for PG.

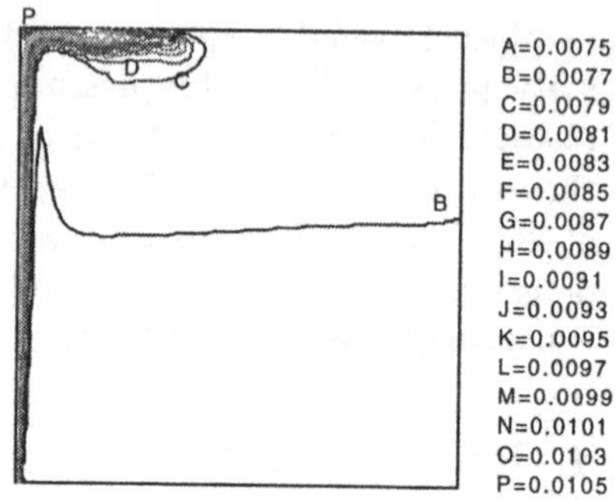

Fig. 2. (b): Temperature difference field in K at 1g and 4.5 seconds for for SCF.

The direct consequence of these two features concerning the density field is the structure of the buoyant convection field. This is confirmed on Fig.4 where velocity vectors are shown for the two cases at a time of 4.7 s. While the perfect gas exhibits a classical circular one-roll pattern, the supercritical fluid exhibits a pattern reminiscent of zero-gravity Marangoni [24] convection (in which the left-hand-side heated wall would be replaced by a free surface). It should also be emphasised that the intensity of convection for the supercritical fluid is about 300 larger for the supercritical fluid than for the perfect gas.

4.2 Density Homogeneisation

If it is now well established that temperature equilibration near the critical point is very fast compared to diffusion, the long observed equilibration time for density [21, 22, 23] is not well understood and seems paradoxal in light of the fast equilibration time for temperature. Boukari [16] has shown by numerical solution of the 1-D, convectively-stable, flow equations that the formation of the density profile following, on ground, a temperature quench of the lower boundary, could take hours whereas temperature is homogeneised very fast by PE. Owing to the very large compressibility, the remaining temperature inhomogeneity (of order $\alpha\mu$) left by the PE gives rise to a significant density inhomogeneity which relaxes very slowly together with heat diffusion and leaves time to gravity to generate a significant convection in a quasi-isothermal medium. This can be checked on Fig.5(a) which shows clearly that under increasing gravity conditions the mass depletion is the TBL is less and less. It must be also emphasized that owing to the long density homogeneisation time, the convective pattern is quasi-steady when refered to the convective time scale.

Fig.5(b) which represents temperature as a function of x at y=0.5 for different gravity levels confirms that the PE is responsible for most of the temperature equilibration: while PE is responsible for 85% of the equilibration in zero gravity, only 10% more are achieved with convection.

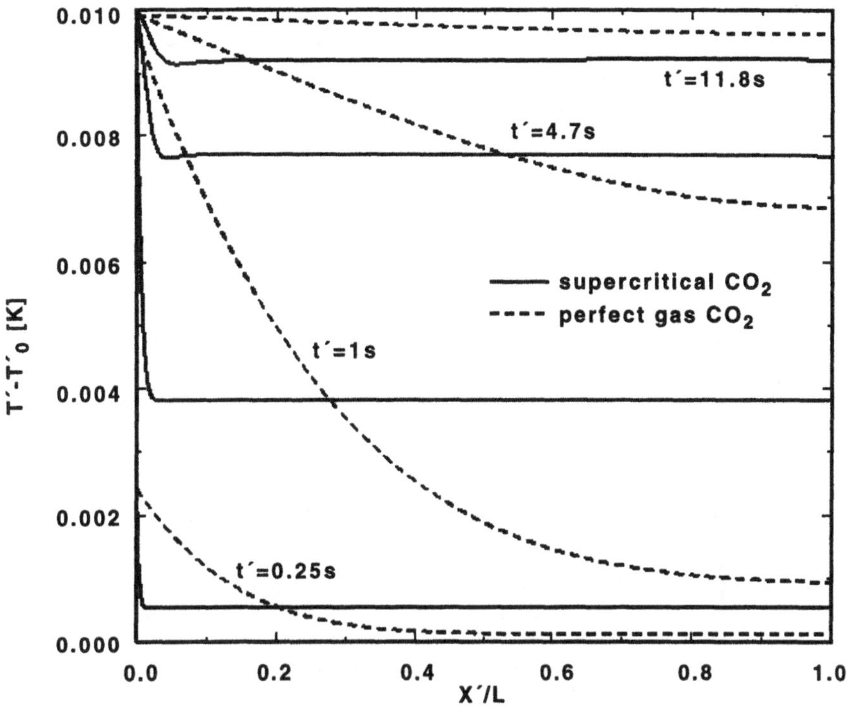

Fig. 3. (a) Temperature difference in K at y=0.5

4.3 Stagnation-Point Effect and the Two-Roll Pattern of Isothermal, Quasi-steady Critical Convection

The interaction between the thermal plume and the upper left corner of the cavity that begins to occur within the PE time scale mentioned in section 4.1, requires additional discussions. As shown on the temperature plots, this is a very a hot region, even hotter than the boundary itself. This is clearly not an artifact of the numerical procedure, since mesh refinement effects in the corresponding region as well the influence of time step on the overheating of the upper wall have been carefully checked. Therefore, the following mechanism is proposed. During the rise of the supercritical fluid particle along the heated wall, the gain in kinetic energy for a very small temperature increase at the boundary is much higher than it would be in a Perfect Gas due to the strong density decrease in the Thermal boundary layer. On the other hand, the vertical pressure gradient is quite small (because of the low Mach number of the flow; see section 3.2), so that the increase in kinetic energy results from the work of the buoyancy force. When the thermal plume reaches the upper, insulated boundary, a turning flow region forms that surrounds a stagnation point where kinetic energy turns into internal energy (thus explaining the hot spot) and into work of pressure forces to expand the fluid because of the high isothermal compressibility. In high-speed flows of normally compressible fluids, the transformation from kinetic energy into

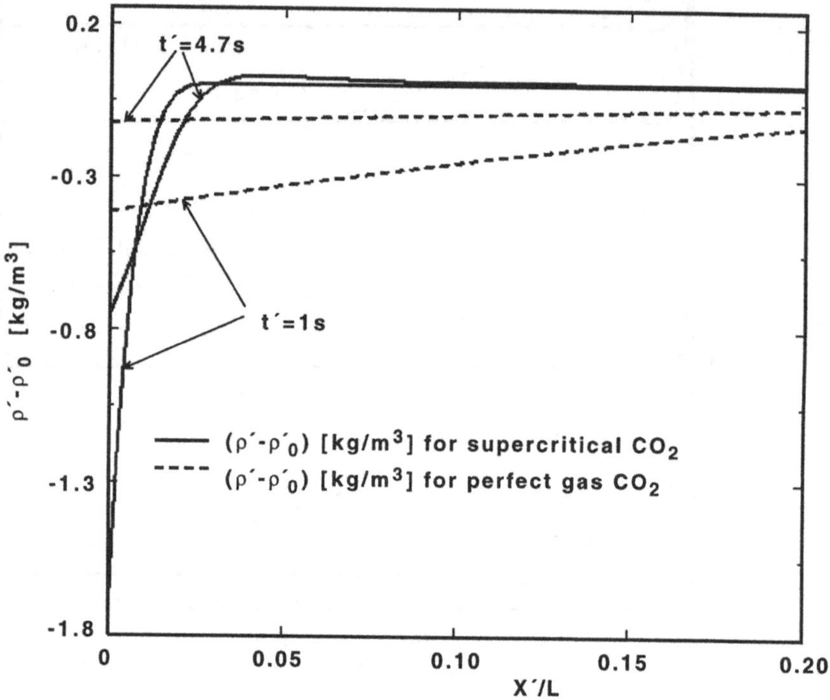

Fig. 3. (b) Density difference in kg/m^3 at y=0.5.

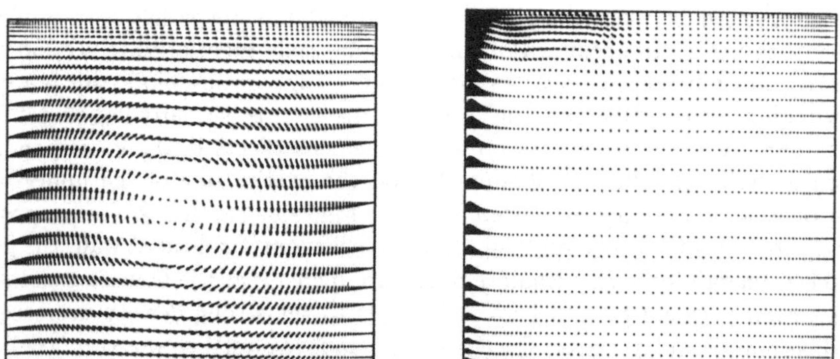

Fig. 4. Velocity field at 1g and 4.5 s (left) for the perfect gas, the maximum vector is 6.5 mm/s and (right) for supercritical fluid, the maximum vector is 2247 mm/s.

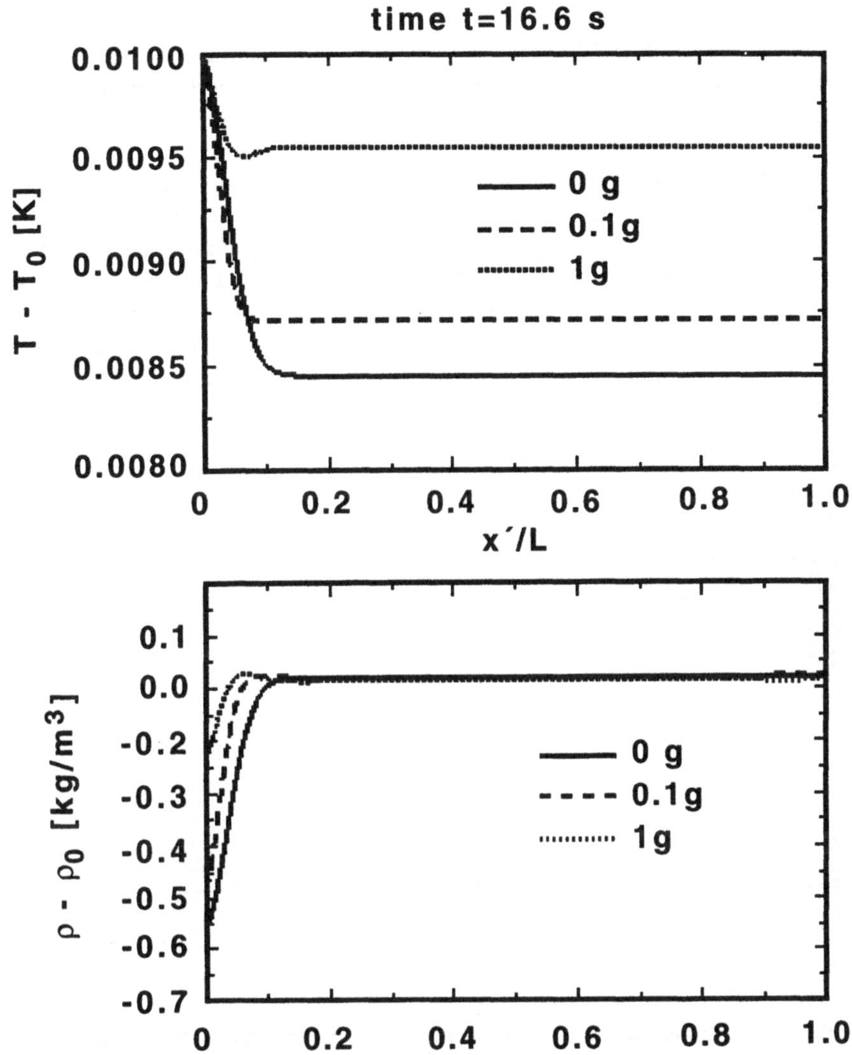

Fig. 5. (a [top]) Temperature difference in K for the supercritical fluid at t'=16.6 s and in (b [bottom]) the density difference.

internal energy comes from the dynamic effects of the fluid that cause pressure gradients according to the velocity-pressure coupling. In low-speed convective flows, the velocity-pressure linkage weakens in the low-Mach-number limit and the pressure-source term in the internal-energy equation is negligible owing to the moderate compressibility. Therefore, temperature is governed by diffusion and convection, density is coupled with temperature and it is thus unusual to find this effect in Boussinesq fluids.

In low Mach number hypercompressible flows, the pressure is still homogeneous because of the little velocity-pressure coupling as for normal gases, but due to the high thermal expansion coefficient, there is an enhancement of the density-temperature coupling. This leads to a strong velocity-temperature coupling through the strong velocity divergence term in the pressure work forces of the internal energy balance which is responsible for the overheating of the upper wall. It is likely that convection phenomena very close to the critical point exhibit new thermal behaviours both coming from the dependance of the thermodynamic coefficients on density and from dynamic effects due to the increase in the kinetic energy. This hot spot on the upper wall is responsible of a counterclockwise roll appearing in the upper left corner on Fig.6(left). It tends to disappear as time goes by, both by diffusion and by the decrease of the thermomechanical coupling. Note also that temperature is quite homogeneised (Fig.6(right)).

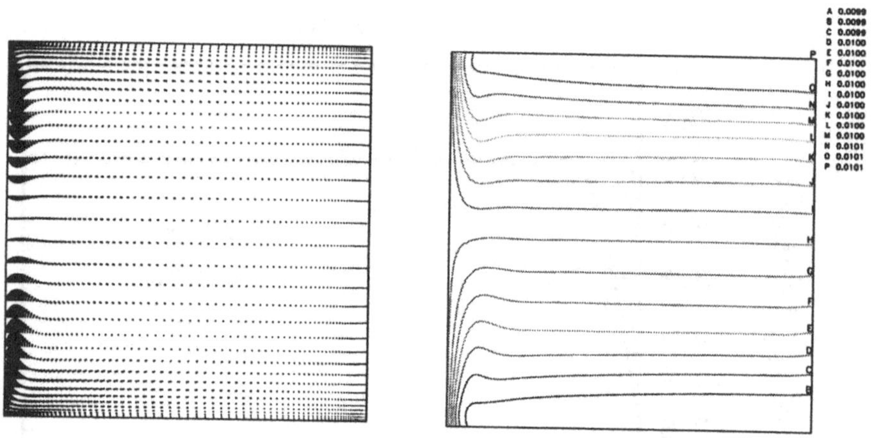

Fig. 6. For the supercritical fluid at 1g and 713 s. *Left:* Velocity field, the maximum vector is 86 mm/s; *Right:* Temperature difference field.

5. Conclusion

A 2-D, unsteady numerical code has been developed which is able to simulate hypercompressible, low-Mach-number, buoyant-convection flows. This has been applied to a particular problem involving near-critical fluids subjected, under normal gravity conditions, to heat addition at a boundary. The results of these computations have identified some important basic mechanisms of heat transport, mass equilibration and the existence of interesting convective structures. As expected, the description of convective motion in a dense liquid which can be

104 times as compressible as a perfect gas has been a numerical chalenge. The numerical solution of the Navier-Stokes equations for this case has been successfully achieved using a finite-volume method with the SIMPLER algorithm, together with acoustic filtering which proved to be the most efficient in terms of precision and computational cost. The results demonstrate that the high compressibility of near-critical fluids and their low heat diffusivity lead to a decoupling of the heat- and mass-equilibration processes. The PE initially homogenises the temperature on a very short time scale (shown to be convection-independent) compared to diffusion and identical to the microgravity PE time scale. At the end of this PE time period, the thermal field is almost completely homogenised while density differences are still significant owing to their very large initial value. These remaining density inhomogeneities then relax very slowly to equilibrium by (slow) thermal diffusion, providing time for convection to begin. This is similar to what is observed for a normal gas, but convection persists for a very long time in the case of a Supercritical Fluid in an already thermalized fluid. The structure of this phenomenon has been explored so as to emphasize its peculiarities vis vis normal gases under the same heating conditions. Whereas normal gases exhibit a classical single-roll structure, near-critical convection first exhibits a Marangoni-like structure during the fast PE period, then (owing to the presence of the overheating of the upper, insulated boundary due to a surprising stagnation point effect), a long-lasting isothermal convection with an original counter-rotating two-roll pattern. This work clearly points out the important role played by the critical behaviour of the transport coefficients in the formation of convective structures in near-critical fluids, which may have implications for other phenomena such as instability or the onset of turbulence.

Acknowledgements

The authors gratefully acknowledge support from CNES (Centre National d'Etudes Spatiales) through IMFM (Institut de Mécanique des Fluides de Marseille).

References

1. B.Zappoli, *Phys. Fuids A* **4** (1992) 1040.
2. B.Zappoli, D.Bailly, Y.Garrabos, B.Le Neindre, P.Guenoun and D.Beysens, *Phys.rev. A* **41** (1990) 2224.
3. P.Guenoun, D.Beysens, B.Khalil, Y.Garrabos, B.Kammoun, B. Le Neindre and B.Zappoli, *Phys.Rev. E* **47** (1993) 1531.
4. A.Onuki, H.Hao and R.A Ferrell, *Phys.Rev.A* **41** (1990) 2256.
5. B.Zappoli, P.Carles, *Europ.J. of Mech. B* **14**(1) (1995) 41-65.
6. B. Zappoli, A.Durand-Daubin, *Phys. Fuids* **6** (1994) 1929-36.
7. S.Amiroudine, J.Ouazzani, B.Zappoli, P.Carles, Numerical Solutions of 1D unsteady hypercompressible flows using finite volume methods, submitted to *Int.J. of Num. Meth. for Heat and Fluid Flow* (1995).
8. P.Carles, B.Zappoli, Acoustic Saturation of the critical speeding up, to appear in *Physica D.*

9. P.Carles, PhD thesis, *l'Effet Piston et les phénomènes thermoacoustiques dans les fluides supercritiques*, Institut National Polytechnique de Toulouse (France), 1995.
10. S.Amiroudine, PhD thesis, Modélisation numérique des phénomènes de transport de chaleur et de masse dans les fluides supercritiques, Institut de Mécanique des Fluides (France), 1995.
11. L.W. Spradley, S.W. Churchill, *J. Fluid Mech.* **70**(4) (1975) 705-720.
12. S.Paolucci, On the filtering of sound from the Navier-Stokes equations, SAND 82-8257 (1982).
13. P.J. Heinmiller, A numerical solution of the Navier-Stokes equations for supercritical fluid thermodynamic analysis, T.R.W. Rep.no.17618-H080-R0-00, Houston, Texas, (1970).
14. M. Gitterman and V.A. Steinberg, *High Temp.* **8** (1971) 754.
15. M. Gitterman and V.A. Steinberg, *J. Appl. Math. Mech.* **34** (1970) 305.
16. H. Boukari, R. L. Pego, R.W. Gammon, Calculation of the gravity-induced density profiles near the liquid-vapor critical point, Private Com. (1995).
17. Fang Zhong and Horst Meyer, Density equilibration near the liquid-vapor critical point of pure fluids" (1995), Pivate Com.
18. H.E Stanley, *Introduction to Phase transition and Critical Phenomena*, Clarendon, Oxford, 1971.
19. S.V Patankar and D.P Spalding, *Int.J.Heat and Mass Transfer* **15** (1972) 1787-1806.
20. S.V Patankar, Numerical Heat Transfer and Fluid Flow, Hemisphere, Washington, D.C, 1980.
21. D.Bailly, M.Ermakov, B.Zappoli, Density relaxation in near-critical pure fluid, *Phys. Rev. A*, in preparation
22. R.A Ferrel, IML-2, Spacelab mission, 1994.
23. Klein and B.Feuerbacher, *Phys. Lett. A*, **123**, (1987) 183.
24. B.M Carpenter, G.M Homsy, *High Marangoni number convection in a square cavity: Part 2*, Phys. of Fluids A, 2.

Adsorption Kinetics and Exchange of Matter at Liquid Interfaces and Microgravity

H. Fruhner[1], *K. Lunkenheimer*[2] *and R. Miller*[2]

[1] University of Potsdam, Solid State Physics, Potsdam, Germany
[2] Max-Planck-Institut für Kolloid- und Grenzflächenforschung, Berlin-Adlershof, Germany

Abstract

The oscillating bubble set-up allows accurate measurements of relaxation processes of surfactant, protein and mixed surfactant/protein adsorption layers in a frequency interval from few Hz up to several hundred Hz. Experimental dependencies of dilational elasticity as a function of oscillation frequency obtained for some surfactant solutions are in good agreement with a diffusional exchange of matter theory. The relaxation behaviour of gelatin adsorption layers show no significant dependence in the studied frequency interval as remarkable changes are expected at much lower frequencies. However, in presence of a surfactant the mixed protein/surfactant adsorption layer shows a linear dependence $E(f)$ which would be in line with a dilational viscosity effect. The instrument has the capacity to be used also for adsorption kinetics studies by enlarging the bubble surface several times of its initial area at the beginning of the experiment and then recording the pressure change with time. Moreover, transient relaxation studies can be performed setting in after generating small-amplitude step-type or square pulse changes of the bubble area.

1. Introduction

Surface active substances are able to modify significantly the properties of interfaces by adsorption. This fact is used in many processes and many new technologies are based on adsorption effects, for example froth flotation, foam generation, demulsification, emulsification. In general, these technologies work under dynamic conditions and improvement of their efficiency is possible by a controlled use of surface active material, such as surfactants, polymers and mixtures of them. To optimise the use specific knowledge of their dynamic adsorption behaviour rather than of equilibrium properties is of great importance [5].

The most frequently used parameter to characterise the dynamic properties of adsorption layers at a liquid interface is the dynamic surface and interfacial tension. Many techniques to measure dynamic tensions have different time windows varying from milliseconds to hours and days. The drop and bubble techniques (drop volume, drop shape, drop pressure methods) seem to be the most general ones, and they can usually be run, after small modification, with bubbles as well.

A second important group of experiments deals with interfacial relaxation and mechanical properties of adsorption layers and the exchange of matter between molecules in the bulk solution with those at the interface. The damping of capillary and longitudinal waves is the most frequently used technique for harmonic interfacial disturbances at medium and high frequencies [3, 6, 8, 11, 12, 18, 26, 37, 38, 39]. Transient relaxation technique is preferred to investigate predominantly slower processes [19, 23, 36]. New developments have been made, which are most of all drop and bubble methods: a modified pendent drop, drop/bubble pressure experiments, or the oscillating bubble method [42, 7]. The main differences between adsorption kinetics and relaxation studies is that the first starts at a more or less free surface while the second determines properties of adsorption layers close to equilibrium. The dynamic behaviour of adsorption layers in which molecules after adsorption undergo changes in structure and arrangement, can show different mechanisms at proceeding surface coverage.

The aim of the paper is to present a method which can be used for adsorption kinetics and exchange of matter studies at liquid/gas and liquid/liquid interfaces. The method is designed in a way that its use under microgravity conditions is possible. Such conditions are advantageous most of all due to less pronounced disturbing convections. Moreover, experiments as well as respective theories are easier to derive because of the spherical geometry of drops and bubbles under reduced gravity conditions.

2. The Experimental Set-Up

One of the more recently developed methods to investigate the surface relaxation of soluble adsorption layers due to harmonic disturbances is the oscillating bubble method. The technique involves the generation of radial oscillations of a gas bubble at the top of a capillary immersed into the solution under study. The first set-up was described by Lunkenheimer and Kretzschmar [27] and Wantke [41] followed by a new design of apparatuses using novel pressure transducer techniques to monitor the pressure changes inside bubbles or drops [1, 10, 13, 28, 29, 35, 43].

The principle of the new designed oscillating bubble experiment is shown in Fig. 1 [7]. A small air bubble is formed at the tip of a capillary which is immersed into the solution. Via a piezo-driven excitation system mounted to the temperature-controlled container the bubble volume, and consequently the bubble surface area, is excited to harmonic oscillations. A pressure sensor directly connected to the solution bulk allows to measure the pressure changes at the bubble surface due to changes in the curvature and the adsorption layer density.

From the excitation frequency dependence of the pressure response, while keeping the bubble oscillation amplitude constant, the dilational elasticity and the exchange of matter can be calculated [41, 42]. The method can be applied in a

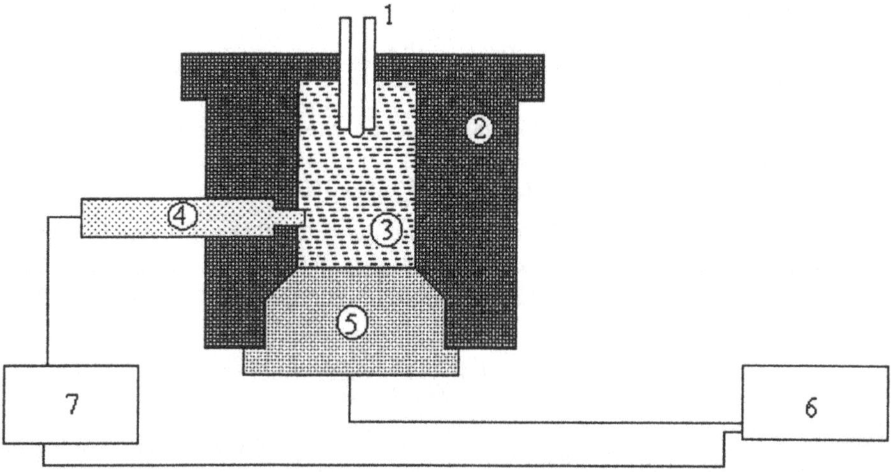

Fig. 1. Oscillating bubble set-up to measure the dilational elasticity and the exchange of matter of a surfactant solution as a function of frequency; 1 - capillary, 2 - brass block with lid, 3 - solution, 4 - piezoelectric driver, 5 - low pressure quartz transducer, 6 - amplifier and measuring instrument for signal amplitude and phase lag, 7 - frequency generator

frequency interval from less than 1 Hz up several hundred Hz. The piezo-element used to generate sinusoidal area changes of the bubble surface can also produce arbitrary area changes, such as step-type or square-pulse area changes, which are of interest in transient relaxation experiments. Moreover, a fast increase of the bubble surface area creates a surface almost free of any adsorbed molecules so that subsequently the adsorption process at the bubble surface can be studied. These additional features make the experimental set-up useful for quite a number of different interfacial studies. Even experiments at the liquid/liquid interface seem to be possible by replacing the gas bubble by a drop of a second immiscible liquid.

3. Theory for Harmonic and Transient Relaxations

If area changes are performed on a Langmuir trough, the theoretical model has to take into account the lateral transport of adsorbed molecules [25, 4, 17]. Assuming isotropic area deformations, the diffusional flux at the interface is given by

$$\frac{1}{A}\frac{d(\Gamma A)}{dt} = D\frac{\partial c}{\partial x} \quad \text{at} \quad x = 0 \tag{1}$$

where $A(t)$ is the time function of the interfacial area, Γ is the surfactant surface concentration, t the time, D the diffusion coefficient, c the surfactant bulk concentration and x the coordinate normal to the surface. The transport of molecules in the bulk by diffusion is given by Fick's diffusion law

$$\frac{\partial c}{\partial t} = D\frac{\partial^2 c}{\partial x^2} \tag{2}$$

The solution to the problem for a harmonically changing surface area has the general form [24]

$$c(x,t) = c_0 + \alpha \exp(\beta x) \exp(i\omega t) \tag{3}$$

Here c_0 is the equilibrium bulk concentration, $\omega = 2\pi f$ the circular frequency with f being the frequency, and α and β are parameters, and i is the imaginary unit. The introduction of ansatz (2) leads to the expression for $E(i\omega)$ as the final result

$$E(i\omega) = E_0 \frac{\sqrt{i\omega}}{\sqrt{i\omega} + \sqrt{2\omega_0}} \tag{4}$$

with

$$E_0 = -\left(\frac{d\gamma}{d\ln\Gamma}\right)_{\mathrm{A}} \quad \text{and} \quad \omega_0 = \left(\frac{dc}{d\Gamma}\right)^2 \frac{D}{2} \tag{5}$$

and γ is the surface tension, E_0 is the dilational elasticity modulus, $E(i\omega)$ the complex dilational elasticity, and ω_0 is the characteristic frequency.

Methods applicable with arbitrary area changes to induce relaxation processes are discussed in the literature: Langmuir trough technique [4], the elastic ring [20, 21], different stress relaxation experiments [14, 15, 16, 40], a modified pendent drop experiment [31, 32], or drop pressure methods [29, 35, 36]. By moving a barrier at the trough, changing the shape of the elastic ring, lifting the funnel or the strip, or increasing/decreasing the volume of a pendent drop, a variety of area changes can be performed, such as jumps, square waves, ramp type, trapezoidal and harmonic area changes or continuous linear and non-linear expansions. The whole theoretical treatment of the derivation of interfacial response functions was discussed recently by Miller an coworkers [30]. It was shown by Loglio [22, 23] that exchange of matter functions derived for harmonic disturbances can be applied to transient ones.

As the result for a diffusion-controlled exchange of matter , using the theory of Lucassen and van den Tempel [24], the following functions result when assuming a square pulse disturbance of the surface area,

$$\Delta\gamma_1 = E_0 \frac{\Delta A}{A_0} \exp(2\omega_0 t)\mathrm{erfc}(\sqrt{2\omega_0 t}) \quad \text{at} \quad 0 < t < t_1 \tag{6}$$

$$\Delta\gamma_2 = \Delta\gamma_1(t) - \Delta\gamma_1(t - t_1) \quad \text{at} \quad t > t_1 \tag{7}$$

with $\Delta\gamma = \gamma_0 - \gamma(t)$ the surface tension of the surfactant free system, A_0 the base value of the surface area and ΔA the area change. The time t_1 designates the duration of the square pulse. Beside the diffusion-controlled models, others exist to describe the adsorption kinetics and exchange of matter. De Feijter et al. [2] have developed a kinetic relation which allows to describe the formation of a protein adsorption layer,

$$\frac{d\Gamma}{dt} = k_{\mathrm{ad}} c_0 \left(1 - \frac{\Gamma}{\Gamma_\infty}\right)^{\mathrm{f}} - k_{\mathrm{des}} \frac{\Gamma}{\Gamma_\infty} \tag{8}$$

Here k_{ad} and k_{des} are the rate constants of adsorption and desorption, respectively, Γ_∞ is the area per adsorbed polymer segment, and f is the number of adsorption sites per adsorbing molecule. The special case f = 1 is known as the Langmuir reaction mechanism. This equation can be used to describe the relaxation process of adsorbed polymer molecules. Under the assumption of a square pulse area disturbance for example the surface tension response function reads [34]:

$$\Delta\gamma_1 = E_0 \frac{\Delta A}{A_0} \exp(-Kt) \quad \text{at} \quad 0 < t \leq t_1 \tag{9}$$

$$\Delta\gamma_2 = \Delta\gamma_1(t) - \Delta\gamma_1(t - t_2) \quad \text{at} \quad t > t_1 \tag{10}$$

with

$$K = \mathrm{f} k_{ad} \frac{c_0}{\Gamma_\infty} \left(1 - \frac{\Gamma_0}{\Gamma_\infty}\right)^{(\mathrm{f}-1)} + \frac{k_{ad}}{\Gamma_\infty} \tag{11}$$

The surface dilational modulus E_0 is related to the surface coverage by the equation

$$E_0 = R\,T\,\gamma_\infty \left((1-\mathrm{f})\frac{\Gamma_0}{\Gamma_\infty} + \mathrm{f}\frac{\frac{\Gamma_0}{\Gamma_\infty}}{1 - \frac{\Gamma_0}{\Gamma_\infty}} \right) \tag{12}$$

It was shown by Loglio et al. [22] that the most useful disturbance for interfacial relaxation experiments is the trapezoidal area change. For time regimes realised in most of the transient relaxation experiments the trapezoidal area change can be approximated adequately by a square pulse.

These theories, only summarised here (for more details look for example into the book of Dukhin et al. [5]), can serve to interpret experimental data obtained for adsorption layers of surfactants or proteins under harmonical or transient disturbances. As the result the dilational elasticity and the exchange of matter mechanism are available.

4. Experimental Results

To demonstrate the function of the instrument experimental examples of studies with aqueous solutions of surfactant, gelatin as a model protein, and gelatin/surfactant mixtures will be given. Fig.2 represents the frequency dependence of the effective surface elasticity of cetyltrimetyl ammonium bromide solutions at three concentrations below the critical micellar concentration (CMC).

The curves indicate two characteristic properties of soluble adsorption layers, the influence of the diffusion exchange of matter on the effective elasticity values and the effect of concentration on the plateau values.

The diffusion exchange leads to a steep decrease of the effective elasticity values at low frequencies. The frequency at which the elasticity reaches a constant level depends strongly on the surfactant concentration. Above this frequency the effective elasticity is equal to the dilational elasticity modulus E_0. Measurements above the CMC show that under sinusoidal oscillations the surface tension

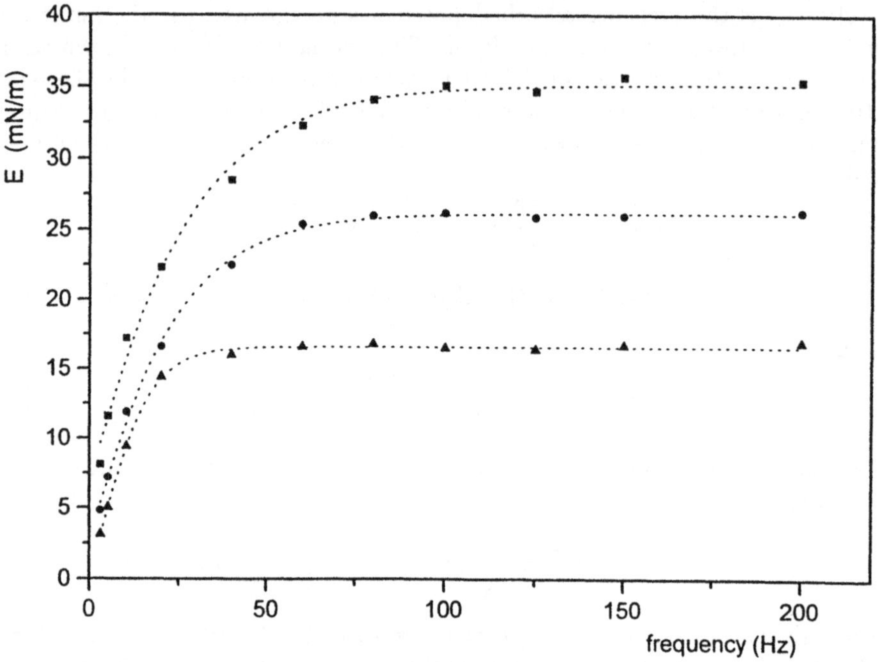

Fig. 2. Frequency dependence of the effective surface elasticity of cetyltrimethyl ammonium bromide solutions; (*) $2.5 \cdot 10^{-4}$ mol/l; (*) $5 \cdot 10^{-4}$ mol/l; (*) $7.5 \cdot 10^{-4}$ mol/l

response is also sinusoidal and reaches values at compression about 3 mN/m below the equilibrium surface tension value at the CMC. The amplitude of surface tension oscillation depends on the area change of the bubble surface.

To evaluate the exchange of matter mechanism the phase angle between the area deformation and the surface tension response has to be determined. From the determination of the phase shift it becomes evident that the mechanism is diffusion controlled. The surface tension response of two aqueous gelatin solutions is shown in Fig. 3 as effective dilational elasticity. It is evident that gelatin behaves as an almost insoluble layer. A frequency dependence is observed only at very low frequencies. The relaxation mechanism behind is not the diffusion exchange of matter but rather a mechanism like that describe by (8).

The dynamic surface tension amplitudes $\Delta\gamma$ measured for three solutions, a surfactant without gelatin, a gelatin without surfactant, and a mixed gelatin/surfactant system, in the frequency range from 3 Hz to 300 Hz are shown in Fig. 4. The values were obtained at area variations of 15.5 %. The measured surface tension variations are caused by the change in surface concentration Γ of surfactant and protein and an additional part which is proportional to the rate of deformation $d\Gamma/dt$ or frequency. The system 0.7% gelatin + $8 \cdot 10^{-4}$ M sodium dioctyl sulfosuccinate shows a remarkable form. In a frequency range from about 50 Hz to 300 Hz a linear region was obtained which cannot be explained by a

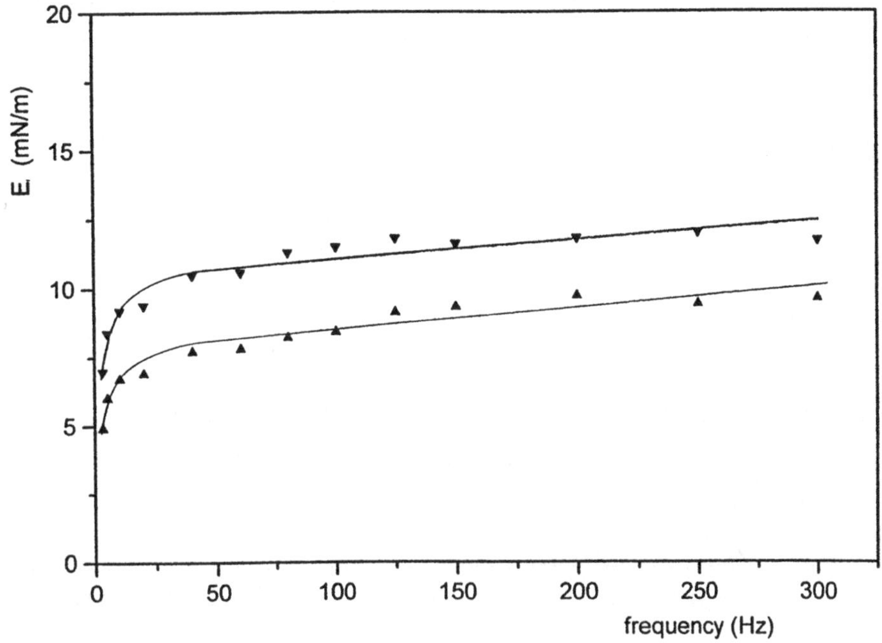

Fig. 3. Frequency dependence of the effective surface elasticity of aqueous gelatin solutions at T=296 K; (*) 0.7 % gelatin; (*) 1.0 % gelatin

diffusion exchange of matter. A linear increase of the stress amplitude ($\Delta\gamma$) with deformation rate ($d\Gamma/dt$) is a characteristic feature of an intrinsic viscosity and the slope of the line would allow to calculate the dilational viscosity. However, the mixed system under study is very complex and other relaxation mechanisms could be responsible for the observed frequency dependence, for example the exchange of surfactant molecules bound to the protein molecules and those independently adsorbed at the surface. This has to be checked in further studies.

5. Conclusions

The experimental data show that the oscillating bubble set-up allows accurate measurements of relaxation processes in adsorption layers of surfactant, protein and mixed surfactant/protein solutions. The experimentally obtained frequency dependence of the dilational elasticity of surfactant solutions is in good agreement with existing theories. The given example of a protein adsorption layer shows only small relaxation effects of the molecules in the given frequency interval. Significant changes are expected at much lower frequencies indicated at frequencies below 10 Hz.

In the presence of surfactants the mixed protein/surfactant adsorption layer exhibits very interesting features. While the pure surfactant solution shows elas-

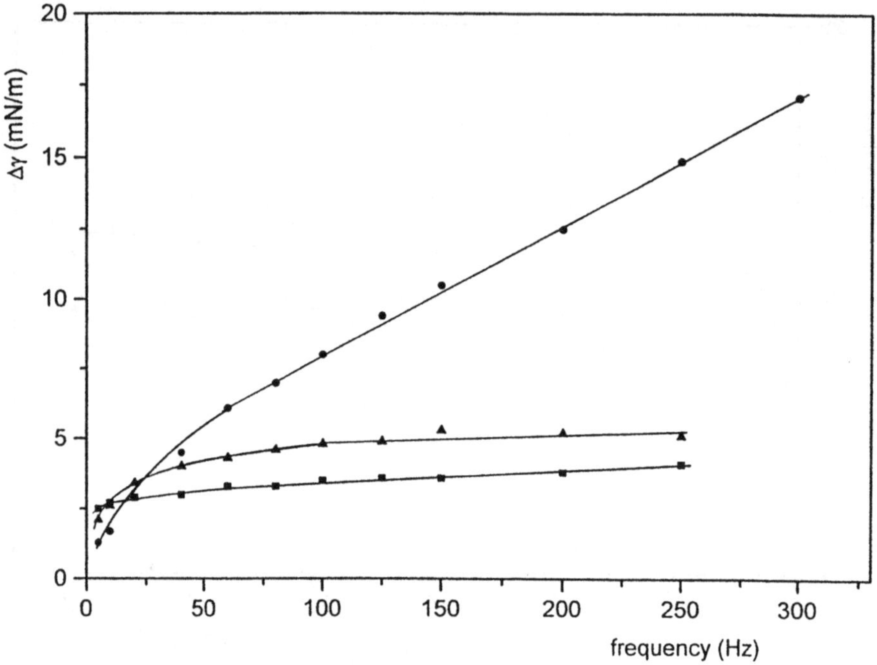

Fig. 4. Dynamic surface tension amplitudes $\Delta\gamma$ at area variations of 15.5 % as a function of frequency at T=296 K; (*) 0.7 % gelatin; (*) $8 \cdot 10^{-4}$ M sodium dioctyl sulfosuccinate; (*) 0.7 % gelatin + $8 \cdot 10^{-4}$ M sodium dioctyl sulfosuccinate

ticity values levelling off at higher frequencies, the mixed system shows a linear dependence of $\Delta\gamma(f)$ which would be in line with a dilational viscosity effect. As discussed above the instrument has the capacity to be used for adsorption kinetics and transient relaxation studies as well. This requires further development of the hardware of the present instrument as well as the software needed to control the piezo-element and to acquire and store data obtained from the pressure sensor.

6. Acknowledgement

The work was financially supported by the Fonds der Chemischen Industrie.

References

1. C.H. Chang and E.I. Franses, *J. Colloid Interface Sci.*, **164** (1994) 107
2. J. de Feijter, J. Benjamins and M. Tamboer, *Colloids Surfaces*, **27** (1987) 243
3. F, de Voeght and P. Joos, *J. Colloid Interface Sci.*, **98** (1984) 20
4. D.S. Dimitrov, I. Panaiotov, P. Richmond and L. Ter-Minassian-Saraga, *J. Colloid Interface Sci.*, **65** (1978) 483

5. D.S. Dukhin, G. Kretzschmar and R. Miller, in *Studies of Interface Science*, D.Möbius and R.Miller (Eds.), Vol. 1, Elsevier, Amsterdam, 1995
6. J.C. Earnshaw, R.C. McGivern, A.C. McLaughlin and P.J.Winch, *Langmuir*, **6** (1990) 649
7. H. Fruhner and K.-D. Wantke, "A new oscillating bubble technique for measuring surface elasticity", presented at the *Bubble & Drop 95* workshop, May 1995, Empoli
8. S. Hârd and R.D: Neuman, *J. Colloid Interface Sci.*, **115** (1987) 73
9. C. Hempt, K. Lunkenheimer and R. Miller, *Z. Phys. Chem.* (Leipzig), **266** (1985) 713
10. T. Horozov, K. Danov, P. Kralschewsky, I. Ivanov and R. Borwankar, *1st World Congress on Emulsion*, Paris, Vol. 2, (1993) 3-20-137
11. H. Hühnerfuss, P.A. Lange and W. Walter, *J. Colloid Interface Sci.*, **108** (1985) 442
12. Q. Jiang, Y.C:. Chiew and J.E. Valentini, *Langmuir*, **8** (1992) 2747
13. D.O. Johnson and K.J. Stebe, *J. Colloid Interface Sci.*, **168** (1994) 21
14. P. Joos, M. Van Uffelen and G. Serrien, *J. Colloid Interface Sci.*, **152** (1992) 521
15. P. Joos and M. Van Uffelen, *J. Colloid Interface Sci.*, **155** (1993) 271
16. P. Joos and M. Van Uffelen, *Colloids Surfaces*, **75** (1993) 273
17. G. Kretzschmar and J. König, *SAM*, **9** (1981) 203
18. G. Kretzschmar and R. Miller, *Adv. Colloid Interface Sci.* **36** (1991) 65
19. L. Liggieri, F. Ravera and A. Passerone, *J. Colloid Interface Sci.*, **140** (1990) 436
20. G. Loglio, U. Tesei and R. Cini, *Colloid Polymer Sci.*, **264** (1986) 712
21. G. Loglio, U. Tesei and R. Cini, *Rev. Sci. Instrum.*, **59** (1988) 2045
22. G. Loglio, U. Tesei, N. Degli-Innocenti, R. Miller and R. Cini, *Colloids and Surfaces*, **57** (1991) 335; **61** (1991) 219
23. G. Loglio, R. Miller, A. Stortini, N. Degli-Innocenti, U. Tesei and R. Cini, *Colloids and Surfaces*, **90** (1994) 251
24. J. Lucassen and M. van den Tempel, *Chem. Eng. Sci.*, **27** (1972) 1283
25. J. Lucassen and D. Giles, *J. Chem. Soc. Faraday Trans. 1*, **71** (1975) 217
26. E.H. Lucassen-Reynders, J. Lucassen, P.R. Garrett, D. Giles and F. Hollway, *Adv. in Chemical Series*, Ed. E.D. Goddard, 144 (1975) 272
27. Lunkenheimer, K. and Kretzschmar, G., *Z. Phys. Chem.* (Leipzig), **256** (1975) 593
28. C.A. MacLeod and C.J. Radke, *9th Intern. Symposium 'Surfactants in Solution'*, Varna, 1992, T2.A3.2
29. C.A. MacLeod and C.J. Radke, *J. Colloid Interface Sci.*, **160** (1993) 435
30. R. Miller, G. Loglio, U. Tesei and K.-H.Schano, *Adv. Colloid Interface Sci.*, **37** (1991) 73
31. R: Miller, Z. Policova, R. Sedev. and A.W. Neumann, *Colloids and Surfaces*, **76** (1993) 179
32. R. Miller, R. Sedev, K.-H. Schano, C. Ng, and A.W. Neumann, *Colloids and Surfaces*, **69** (1993) 209
33. R. Miller, P. Joos and V.B. Fainerman, *Adv. Colloid Interface Sci.*, **49** (1994) 249
34. R. Miller, G. Loglio and A.W. Neumann, *J. Colloid Interface Sci.*, (1995), submitted
35. R. Nagarajan and D.T. Wasan, *J. Colloid Interface Sci.*, **159** (1993) 164
36. A. Passerone, L. Liggieri, N. Rando, F. Ravera, and E. Ricci, *J. Colloid Interface Sci.*, **146** (1991) 152

37. C. Stenvot and D. Langevin, *Langmuir*, **4** (1988) 1179
38. S.M. Sun and M.C. Shen, *J. Math. Analysis Appl.*, **172** (1993) 533
39. M. van den Tempel and E.H. Lucassen-Reynders, *Adv. Colloid Interface Sci.*, **18** (1983) 281
40. M. Van Uffelen and P. Joos, *J. Colloid Interface Sci.*, **158** (1993) 452
41. K.D. Wantke, R. Miller and K. Lunkenheimer, *Z. Phys. Chem.* (Leipzig) **261** (1980) 1177
42. K.D. Wantke, K. Lunkenheimer and C. Hempt, *J. Colloid Interface Sci.*, **159** (1993) 28
43. D.T. Wasan, K. Koszo and R. Nagarajan, *67th Annual Colloid and Surface Science Symposium*, Toronto, 1993, 285

Critical Depletion of Pure Fluids in Colloidal Solids: Results of Experiments on EURECA and Grand Canonical Monte Carlo Simulations

M. Thommes[1], *M. Schoen*[2] *and G.H. Findenegg*[1]

[1] Iwan-N.-Stranski-Institut für Physikalische und Theoretische Chemie Technische Universität Berlin, Straße des 17. Juni 112, D-10623 Berlin
[2] Institut für Theoretische Physik, Technische Universität Berlin, Hardenbergstraße 36, D-10623 Berlin

Abstract

A microgravity experiment on the EURECA-1 [1] mission of ESA was performed to study the adsorption of a near-critical fluid (SF_6) on a finely dispersed graphitic adsorbent (Vulcan 3-G graphitized carbon). The experimental set-up was housed in a three-stage high-precision thermostat (HPT) which allows for a temperature control within 1 mK over periods of days. The adsorption (surface excess amount) Γ was measured as a function of temperature T along near-critical isochores using a volumetric technique. Five independent runs, either at the critical density ($\rho/\rho_c = 1.01$) or a slightly higher density ($\rho/\rho_c = 1.04$), were performed. The EURECA experiment confirmed a novel critical sorption phenomenon, which is due to the colloidal state of the adsorbent. At temperatures well above the critical temperature T_c the adsorption excess amount Γ increases with decreasing temperature, but closer to the critical temperature T_c exhibits a maximum and then decreases sharply for $T \to T_c$. The phenomenon was also observed for near-critical isochores of SF_6 in a mesoporous glass material. Grand canonical Monte Carlo simulations for a fluid in a slit-pore suggest that the negative critical adsorption effect is caused by depletion in the core region of the pore as T approaches T_c. This effect, which we call *critical depletion*, is believed to be driven by the proximity of the bulk fluid to its critical point. It may be of significance whenever fluids in contact with mesoporous or colloidal materials approach their critical point.

1. Introduction

The study of physical adsorption and wetting phenomena at fluid/solid interfaces near the coexistence curve and in the critical region of a pure fluid represents an active field of current research [1, 2, 3]. Physical adsorption of vapors in a temperature and pressure range close to gas/liquid coexistence depends on a

[1] EURECA is the acronym for European Retrievable Carrier, an autonomuous satellite developed by the European Space gency ESA. The satellite was a platform for experiments in fluid physics and material science. (Eds. note)

competition between the long-range attractive gas/solid interactions and the cohesive energy of the fluid. For sufficiently attractive solid substrates a multilayer adsorbed film of the fluid is formed at the solid/gas interface. This behavior, which is a signature of complete wetting prevails in a temperature range from the wetting temperature T_w up to nearly the critical temperature T_c of the fluid [1]. On approaching the critical region, the adsorbed film is believed to transform from a liquid-like layer into a diffuse layer in which the local density $\rho(z)$ changes gradually from its value at the wall ($z = 0$) to the bulk density ρ_b (at $z \to \infty$). Near the gas/liquid critical point the correlation length ξ of density fluctuations in the bulk fluid increases and thus becomes the dominating length scale both in the bulk and also for fluid interfacial phenomena. According to the scaling theory of critical adsorption [1, 3] the density profile $\rho(z)$ of a near-critical fluid next to a *flat* semi-infinite wall is a slowly decaying algebraic function of the scaled distance $x = z/\xi$, where z is the distance from the surface and ξ is again the correlation length of critical fluctuations in the bulk fluid. Asymptotically close to T_c the correlation length increases by a power law $\xi = \xi_0 t^{-\nu}$ in $t = (T-T_c)/T_c$ for $t \to 0$, where ξ_0 is a critical amplitude and ν a critical exponent ($\nu = 0.63$ for 3D Ising-like systems). Experimentally one measures Γ, the overall surface excess amount per unit surface area,

$$\Gamma = \int_0^\infty (\rho(z) - \rho_b)dz \tag{1}$$

where $\rho_b = \rho(z \to \infty)$ is the bulk density of the fluid. The scaling form of the density profile $\rho(z)$, the power laws for the correlation length and for the order parameter of the fluid imply an asymptotic power law for the surface excess $\Gamma(T)$ along the critical isochore, namely $\Gamma \propto t^{(\nu-\beta)}$. Here β is the critical exponent of the order parameter ($\beta = 0.31$ for 3D Ising-like systems). This prediction for critical adsorption was tested experimentally mainly for the adsorption from binary liquid mixtures on a solid substrate or at fluid interfaces [1, 2, 3]. On the other hand a different behavior was found for the critical sorption isochore of SF6 on a colloidal graphitic adsorbent (graphitized carbon black) [4, 5]. Well above T_c the surface excess increases weakly as to be expected, but falls off sharply close to T_c contrary to the expected behavior. In order to find out if this behavior near Tc was an artifact caused by gravitational effects on the near-critical bulk fluid, this experiment was repeated under microgravity conditions on the EURECA-1 space platform.

2. Experiments Under Microgravity and Laboratory Conditions

Surface excess isochores $\Gamma = \Gamma(T)$ for SF_6 on Vulcan 3-G graphitized carbon black were measured by a volumetric technique. Basically the apparatus consists of a sorption cell, containing the adsorbent and the fluid at the density of the experimental isochore, and a reference cell without adsorbent. The pressure difference between the two cells is monitored by a sensitive differential pressure transducer. The volumes of the sorption and the reference cell can be changed by means of piston-cylinder devices. All parts of the system containing the fluid are mounted on the inner stage of a three-stage high-precision thermostat (HPT), which controls the temperature within 1 mK over periods of days. A detailed description of the experimental set-up is given in [6]. Measurements start at a temperature T_1 ca. 15 K above T_c by adjusting the fluid density in the sorption and the reference cell. As the temperature is now lowered, a pressure difference (ΔP) between the two cells builds up due to the sorption of the fluid on the substrate and the volume of the sorption cell has to be changed to restore $\Delta P = 0$, thus restoring the original bulk density of the fluid in the sorption cell. Changes in Γ are therefore given in terms of volume changes of the sorption cell. A series of temperature steps yields a complete scan of the experimental isochore. Scans without adsorbent in the adsorption cell (blank measurements) yield a baseline correction $\Delta V_b(T)$ which accounts for the asymmetry of the apparatus. This correction has to be taken into account in the calculation of $\Gamma(T)$.

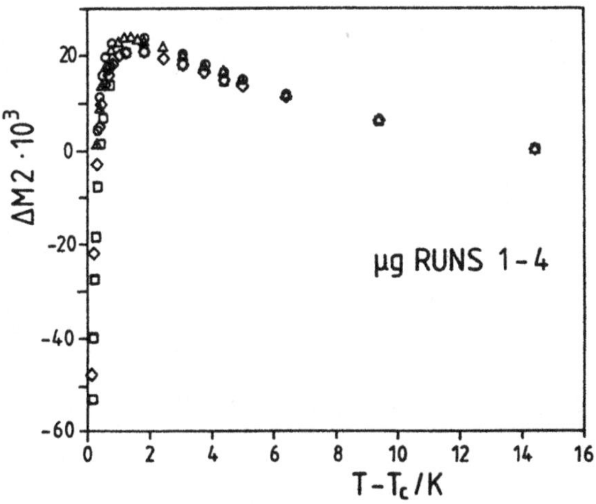

Fig. 1. Adsorption excess (raw data) of SF_6 on Vulcan 3-G graphitized carbon black along near-critical isochores measured on the EURECA I space platform $\rho/\rho_c = 1.01$: RUN 1(O) and RUN 2 (△); $\rho/\rho_c = 1.04 = 1.04$: RUN 3 (◇) and RUN 4 (□).

On the EURECA-1 space mission five independent runs (each extending over about 30 days) at near-critical densities were performed, either at $\rho/\rho_c = 1.01$ or at a slightly higher density $\rho/\rho_c = 1.04$. In addition complementary tests, in which the influence of systematic changes of the experimental conditions on the result could be assessed, were also performed by telecommands on EURECA [7]. Telecommands were found to be very useful in all EURECA runs in order to optimize the experiment's performance. More than 500 telecommands were used to adjust the experimental density, to optimize the thermal conditions of the HPT and to monitor changes of experimental parameters under microgravity.

In Fig.1 EURECA results of RUN 1 and 2 ($\rho/\rho_c = 1.01$) and RUN 3 and

4 (ρ/ρ_c = 1.04) are shown. The data are expressed in terms of motor steps M of the stepping motor which changes the volume of the sorption cell relative to the position at the initial temperature. Well above T_c the microgravity results exhibit the expected temperature dependence of adsorption as in the laboratory experiments. Closer to T_c the surface excess reaches a maximum and then decreases sharply for $T \to T_c$ in qualitative agreement with the laboratory results. In RUN 3 and 4 sorption data closer to the critical temperature (down to $T - T_c$ = 0.14 K) were obtained. At temperatures well above T_c the motor step sum of RUN 3 and 4 lies somewhat below the corresponding values of RUN 1 and 2 as to be expected for the higher bulk density of these runs. The results of RUN 5 confirm those of RUNS 1 - 4 in all aspects.

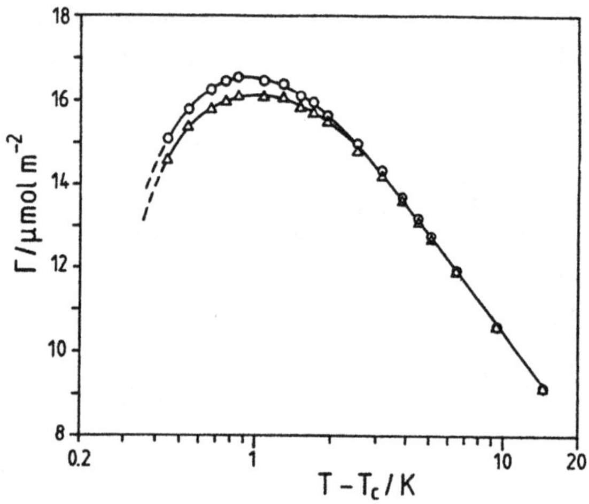

Fig. 2. Preliminary quantitative results for μg-RUN 2 (Δ) and the corresponding 1-g reference run (O).

Preliminary results of the blank measurements at the critical density (ρ/ρ_c = 1.01) have been used to calculate the surface excess amount Γ for EURECA RUN 2 and the corresponding 1-g reference run performed with the original HPT Flight Module after the mission. The two runs are shown in Fig. 2. At temperatures $T - T_c$ >3 K the adsorption excess Γ increases weakly with decreasing temperature as expected, but at $T - T_c \approx 1$ K it exhibits a maximum and then decreases sharply for $T \to T_c$. The 1-g and the μg-runs are in good quantitative agreement down to $T - T_c$ =1.5 K; closer to the critical temperature differences in Γ between the two runs are observed. The significance of these deviations cannot be assessed before the final results of the blank measurements are available.

The qualitative agreement of the microgravity results with the laboratory measurements indicates that the decrease of the adsorption excess in the near-critical region is not an artifact caused by gravity effects but represents a genuine critical sorption phenomenon. It was suggested that this effect may be due to confinement effects of the fluid in the colloidal graphite substrate [5]. The graphitized carbon black used in the EURECA experiment consists of particles with a mean-size of ca. 30 - 35 nm. In the agglomerates formed by these primary particles the distance between the exposed surfaces of neighbouring particles is distributed widely, from a few molecular diameters up to very large distances; a typical distance is estimated to be 5 - 10 nm. One expects that if the correlation length ξ of critical density fluctuations approaches half the magnitude of the wall-to-wall distance D the finite geometry may influence the sorption isochore due to suppression of density fluctuations in the pores [8]. The wall-to-wall distance between neighbouring carbon particles may therefore play a similar role as the pore size D of a porous material. In order to test this conjecture we have also measured the critical adsorption of SF_6 in a controlled-pore glass of mean-pore width ca. 30 nm (CPG 350), which unlike the colloidal graphite substrate, comprises a rigid interconnected mesoporous system. These experiments are described in detail in [9]. The result for the critical sorption isochore of SF_6 in this mesoporous glass is in qualitative agreement with the SF_6/graphite system and an even more pronounced decrease of the surface excess Γ was observed in case of the SF_6/CPG 350 system for $T \to T_c$.

3. Grand Canonical Monte Carlo Simulations

In order to elucidate microscopic details of the experimentally observed decrease of the surface Γ along near-critical isochores for $T \to T_c$ Monte Carlo simulations were performed. Because in the experiments, the fluid in the colloidal (porous) materials is in thermodynamic equilibrium with a bulk fluid reservoir at a near-critical density, temperature T, chemical potential μ and volume V are appropriate state variables to define the thermodynamic states of the fluid in the pores and the reservoir in the simulation. Accordingly, the grand canonical ensemble is most appropriate for such a computer simulation. Since the depletion phenomenon was found both with the colloidal graphite substrate and the porous glass, it was conjectured that it should not depend on details of the pore geometry. Therefore we have chosen a rather simple model system, namely a Lennard-Jones fluid in slit pores with wall-to-wall distances up to 20 molecular diameters (20σ, where σ is the molecular diameter) in virtual contact with a bulk reservoir at near-critical density. Details of the Monte Carlo simulation studies are given in [9, 10].

Figure 3 shows plots of the local density $\rho^*(z)$ as a function of position between the walls of a slit-pore with distances of 10 and 7.5 molecular diameters for a given strength of attractive fluid-solid interactions sufficiently strong to cause complete wetting of the walls by the fluid. Density profiles are shown for a reduced temperature $T^* = 3.00$ well above the critical and a near-critical temperature $T^* = 1.36 \approx T_c^*$ (T^* is given in units of ϵ_{ff}/k_B where ϵ_{ff} is the well-depth of the fluid-fluid Lennard-Jones potential and k_B the Boltzmann constant). We will focus first on Fig. 3 where the density profiles in the 10σ-pore model

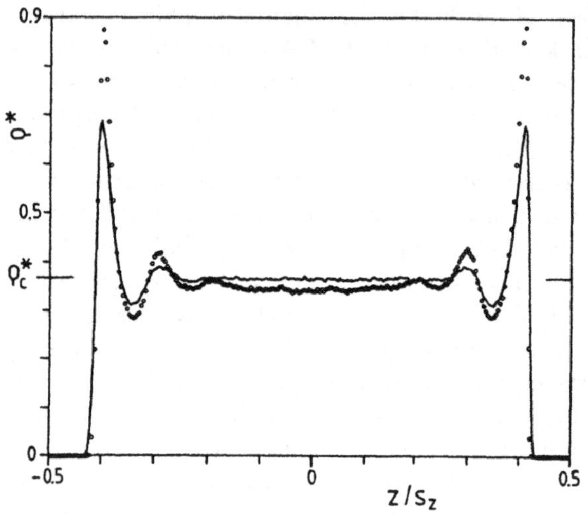

Fig. 3. Local density $\rho^*(z)$ of the pore fluid as a function of position z between lower and upper wall for a pore width of 10 molecular diameters ($s_z = 10\sigma$) and reduced temperatures $T^* = 1.36 \approx T_c^*(0)$; $T^* = 3.00(--)$.

are shown. Close to the wall, $\rho^*(z)$ is an oscillatory function of z, reflecting the formation of individual layers of adsorbed molecules. As to be expected the density near the wall increases with decreasing temperature due to adsorption. Well above the critical temperature at $T^* = 3.00$ the fluid in the *core region* between the walls appears to be homogeneous, i.e. $\rho^*(z)$ is nearly constant and equal to the near-critical bulk density $\rho_c^* \approx 0.365$ (ρ^* is given in units of $1/\sigma^3$). At the near-critical temperature $T^* = 1.36$, on the other hand, the density in the core region is depleted with respect to the bulk density. This behavior was also found in case of a slit pore with a wall to wall distance of 20 molecular diameters. Appropriate density profiles for a series of temperatures along the critical isochore are presented in [9].

Additional Monte-Carlo studies have shown that this depletion effect in the core of the pore depends on the pore size as well on the net strength of the attractive gas-solid interactions. If these parameters are chosen improperly, depletion may not occur at all. Depletion was not found for a fluid in a 7.5σ-pore as shown in Fig.4. As in the case of the 10σ-pore the density near the walls increases for $T \rightarrow T_c$, however the fluid in the core region is slightly inhomogeneous. At both temperatures ($T^* = 3.00$ and $T^* = 1.36$) $\rho(z)$ still oscillates weakly around ρ_c^* in the core region. The fluid in the core region of the 7.5σ-pore is stronger influenced by the attractive walls as compared to pores having wall-to-wall distances greater than 10 molecular diameters. Therefore it seems that depletion can only occur for a given strength of attractive solid-fluid interactions if the pore is sufficiently wide.

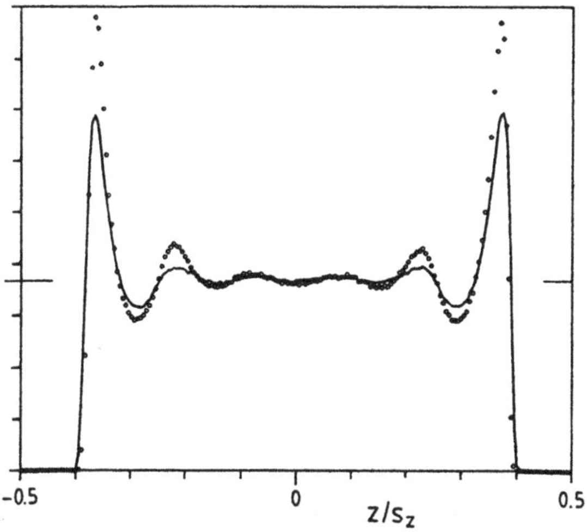

Fig. 4. The same as in Fig.3 but for a pore width of 7.5 molecular diameters ($s_z = 7.5\sigma$).

4. Discussion

To compare the results of the computer simulation with the experiments it is instructive to investigate the behavior of the overall average density ρ_p^* of the pore fluid obtained by integration of the $\rho^*(z)$ profiles. The temperature dependence of this average density may be compared with the temperature dependence of the experimentally accessible surface excess amount Γ. Figure 5 shows results of ρ_p^* for a pore width of 20 molecular diameters at several temperatures. Qualitatively, the mean pore density exhibits a strikingly similar temperature dependence compared with Γ (Figures 1 and 2). From the density profiles in Fig. 3 it is evident, that for $T \to T_c$ a tendency exists to remove molecules from the core region of the pore while at the same time the density in the vicinity of the walls increases due to adsorption. Accordingly, this effect is intrinsically different from "ordinary" drying, where the local density would decrease near the walls, but not in the core region of the pore [8, 11].

The experimentally observed depletion for SF_6 on Vulcan 3-G graphitized carbon black and in controlled-pore glass (mean pore size 31 nm) takes place rather abruptly in the true near-critical region at $(T-T_c)/T_c < 5 \cdot 10^{-3}$, whereas the temperature range over which density depletion occurs in the simulations is much wider. According to the Ornstein-Zernike theory the total correlation function h(r) [12] of homogeneous fluids becomes long-ranged near the critical point. Thus an analysis of the range of h(r) as it results from the MC simulations for the bulk fluid can tell over what temperature range one may expect to see the influence of criticality in the simulations. Results of this analysis presented in [10] show that the temperature range affected by the growing extend of fluctu-

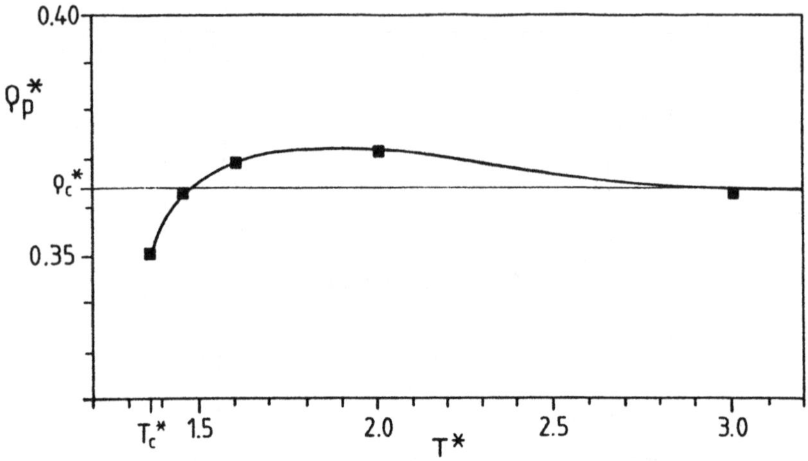

Fig. 5. The overall density ρ_p^* of the pore fluid for a pore width of 20 molecular diameters ($s_z = 20\sigma$) as a function of temperature along a near-critical isochore.

ations is much wider than in the real system. The fact that in the simulations depletion commences at much larger $T - T_c$ than in the experimental system is consistent with these findings. Therefore, we conclude that the simulations are indeed reflecting the same physical phenomenon.

The observed depletion effect is believed to be caused by the restriction of critical fluctuations (or equivalently, a lower compressibility) in the pore fluid relative to the bulk fluid. Theoretical and experimental studies [13, 14] have shown that due to the suppression of critical point fluctuations in the confined fluid pore criticality is shifted to lower temperatures compared with the critical point of the bulk fluid. The depletion effect may be due to the fact that the pore fluid is away from its critical point when the bulk fluid approaches criticality which in turn leads to a difference in compressibility between the pore fluid and the bulk fluid. In order to maintain mechanical equilibrium under this condition the density in the core region of the pore decreases [10]. In this sense the depletion phenomenon is driven by the criticality of the bulk fluid and is therefore called *critical depletion.* This intuitive picture suggests that the *critical depletion* effect may be of significance whenever fluids in contact with mesoporous or colloidal substrates approach their critical point. Such situations arise for example in extraction processes using supercritical fluids.

Acknowledgements

This work was supported by the Deutsche Agentur für Raumfahrtangelegenheiten (DARA) under Grant 50 WM 9115. We would like to thank DLR-MUSC, ESA-ESTEC and Kayser-Threde for their continuing cooperation. One of us (M.S.) thanks the Deutsche Forschungsgemeinschaft (DFG) for a Heisenberg fellowship (Scho 525/5-1). We are also grateful to the Scientific Council of

Höchstleistungsrechenzentrum (HLRZ) at Forschungszentrum Jülich for an allowance of computer time on the CRAY Y-MP/864.

References

1. S. Dietrich, in: *Phase Transition and Critical Phenomena* ,Vol. 12, C. Domb and J.L Lebowitz (Eds.), Academic Press, London, 1988; S. Dietrich, *NATO ASI Ser.* **B 267** (1991) 391.
2. M. Schick, in: ***Liquides aux Interfaces***, Les Houches 1988 Session XLVIII, Course 9, J. Charvolin, J. Joanny, J. Zinn-Justin, North-Holland, Amsterdam, 1990.
3. D. Beysens, in: ***Liquides aux Interfaces***, Les Houches 1988 Session XLVIII; Course 10, J. Charvolin, J. Joanny, J. Zinn-Justin (Eds.), North Holland, Amsterdam,1990.
4. G.H. Findenegg, M. Thommes and T. Michalski in: Proc. VIIIth Europ. Symp. on Materials and Fluid Sciences in Microgravity, **ESA SP-333** (1992) 795.
5. M. Thommes, H. Lewandowski and G.H. Findenegg, *Ber. Bunsenges. Phys. Chem.* **98** (1994) 477.
6. G.H. Findenegg, M. Thommes and K. Kemmerle, Proc. of the 44 th Congress of the International Astronautical Federation (IAF) **IAF-93-J.1.260** (1993).
7. M. Thommes and G.H. Findenegg , *Advances in Space Research* **16** (1995) 83.
8. U. Marini Bettolo Marconi, *Phys. Rev. A* **38** (1988) 6267.
9. M. Thommes, G.H.Findenegg, M. Schoen, *Langmuir* **11** (1995), in press.
10. M. Schoen, M. Thommes, *Phys. Rev. E* (1995), submitted.
11. A. Jamnick, D. Bratko, *Chem.Phys. Lett.* **203** (1992) 465; J. Jamnick , *J. Chem. Phys.* **102** (1995) 5811.
12. J.S. Rowlinson, B. Widom, *Molecular Theory of Capillarity*, Clarendon Press, Chapter 9.2, Oxford, 1982.
13. M.E. Fisher, H. Nakanishi, *J. Chem. Phys.* **75** (1981) 5857; H. Nakanishi, M.E. Fisher, *J. Chem. Phys. 78* (1983) 3279.
14. M. Thommes and G.H.Findenegg , *Langmuir* **10** (1994) 4270.

Part II
Solidification

Dendritic Growth Measurements in Microgravity

M.E. Glicksman[1], *M.B. Koss*[1], *L.T. Bushnell*[1], *J.C. LaCombe*[1] *and E.A. Winsa*[2]

[1] Materials Engineering Department
Rensselaer Polytechnic Institute Troy, NY 12180-3590
[2] Space Experiments Division NASA
Lewis Research Center Cleveland, OH 44135

Abstract

The Isothermal Dendritic Growth Experiment (IDGE) is an orbital space flight experiment, launched by NASA, in March, 1994, as part of the United States Microgravity Payload (USMP-2). The IDGE provided accurately measured dendritic growth rates, tip radii of curvature, and morphological observations of ultra-pure succinontrile in the supercoolings range between 0.05 - 2.0 K. Data were received in the form of pairs of digitized binary images telemetered to the ground from orbit in near-real-time, and as 35mm photographic film. The data taken above ca. 0.4 K supercooling provide the first comprehensive assessment of diffusion-limited dendritic growth, and those taken below this value yield new insights into the influence of convection. The IDGE also demonstrated that prior terrestrial dendritic growth data were generally corrupted by melt convection, making their comparison with diffusion-limited theories questionable. A key element in the design and operation of this experiment, which was in development for over ten years at a cost of approximately $12 million U.S., was its reliance on telescience for accomplishing in situ near-real-time monitoring and control of the experiment.

1. Introduction

The growth of dendrites is one of the commonly observed forms of solidification encountered when metals and alloys freeze under low thermal gradients, as occurs in most casting and welding processes. In engineering alloys, the details of the dendritic morphology directly relates to important material responses and properties. Of more generic interest, dendritic growth is also an archetypical problem in morphogenesis, where a complex pattern evolves from simple starting conditions. Thus, the physical understanding and mathematical description of how dendritic patterns emerge during the growth process are of interest to both scientists and engineers [1, 2].

The Isothermal Dendritic Growth Experiment (IDGE) is a basic science experiment designed to measure, for a fundamental test of theory, the kinetics and morphology of dendritic growth without complications induced by gravity-driven convection. The IDGE, a collaboration between Rensselaer Polytechnic Institute, in Troy NY, and NASA's Lewis Research Center, in Cleveland OH, was developed over ten year period from a ground-based research program into a space flight experiment. Important to the success of this flight experiment was provision of in situ near-real-time teleoperations during the spaceflight experiment.

2. Background on Dendritic Growth Theory

A number of theories of dendritic crystal growth, based on various transport mechanisms, physical assumptions, and mathematical approximations, have been developed over the last forty years. These theories attempt to predict a dendrite's tip velocity, V, and radius of curvature, R, as a function of the supercooling, ΔT (see the review by one of the authors [1]). The growth of dendrites in pure melts is known to be controlled by the transport of latent heat from the moving crystal-melt interface as it advances into its supercooled melt. Ivantsov, in 1947, provided the first mathematical solution to the dendritic heat conduction problem [3, 4], and modeled the steady-state dendrite as a paraboloidal body of revolution, growing at a constant velocity, V. The resultant thermal conduction field can be expressed exactly in paraboloidal coordinates moving with the dendritic tip. The temperature field solution is known as the Ivantsov, or "diffusion-limited" transport solution. This solution is, however, incomplete, insofar as it only specifies the dendritic tip growth Péclet number, $Pe = VR/2\alpha$, (here Pe is the growth Péclet number, and α is the thermal diffusivity of the molten phase) as a function of the initial supercooling, and not the unique dynamic operating state; V and R. The Péclet number obtained from the Ivantsov solution for each supercooling yields instead an *infinite* range of V and R values that satisfy the diffusion-limited solution at that particular value of ΔT.

In the early 70's, succinonitrile (SCN), a BCC organic plastic crystal, was developed as a model metal analog system for studying dendritic growth [5]. SCN solidifies like the cubic metals, i.e., with an atomicially "rough" solid-liquid interface, yet retains advantages because SCN displays convenient properties for solidification experiments, such as a low melting temperature, optical transparency, and accurate characterization of its thermophysical properties. The use of SCN greatly facilitated dendritic growth studies over the past twenty years, where because of its use, dendritic tip velocities could be accurately measured and used as a critical test of theory.

Theoretical efforts have concentrated on trying to discover an additional equation or length scale, which when combined with the Ivantsov conduction solution, "selects" the observed operating states (see references within [1] and [6]). Although the underlying physical mechanisms for these "theories of the second length scale" are quite different, their results are invariably expressed through a scaling constant, $\sigma^* = 2\alpha d_0/(VR^2)$, where d_0 is the capillary length scale, a materials parameter defined from the equilibrium temperature of the crystal-melt interface, the solid-liquid interface energy, and the specific and latent heats.

Although some theories predict the value of this scaling constant, in practice the scaling constant is used as an adjustable parameter to describe dendritic growth data in various materials.

Subsequent experiments with SCN showed that gravity-induced convection *dominates* dendritic growth in the lower supercooling range typical of metal alloy castings [4, 5]. Convection, unfortunately confounds any straightforward analysis of dendritic solification based on conductive heat transfer. There have been a few attempts to estimate the effects of natural or forced convection on dendritic growth [6], but these calculations are themselves based on yet unproven elements of dendritic growth theory, and, consequently, can not provide an independent test of the theory. In the higher supercooling range, where thermal convective influences diminish in comparison to thermal conduction, the morphological scale of dendrites becomes too small to be resolved optically at the high growth speeds encountered. The experimental situation prior to the microgravity experiment reported here was that there appeared to be too narrow a range of supercoolings in any crystal-melt system studied terrestrially that remains both free of convection effects, and also permits an accurate determination of the dendrite tip radius of curvature.

3. The Isothermal Dendritic Growth Experiment

The microgravity environment allowed us to broaden the range of supercoolings for which convection free measurements can be made and, therefore, permits for the first time a critical assessment of theory. The use of microgravity as a way to continue dendritic growth studies underscores several relevant criteria for pursuing a microgravity spaceflight experiment. The IDGE had clear scientific objectives, i.e., to record controlled dendritic growth phenomena in the absence of gravity-driven convection. There was clear enunciation of the anticipated scientific impact, including a definitive test of dendritic growth theory, or a data set to be used as a benchmark for future theoretical developments. The microgravity facility for spaceflight is unique. In the case of IDGE, there were attempts made to "scale" gravity out of the problem, but because of the way gravity and convection couples to heat conduction, this did not prove feasible. Furthermore, sounding rockets, originally suggested for the IDGE, do not provide a long enough duration of quality microgravity for a definitive test of theory. Finally, all the details of a proposed orbital experiment, and its anticipated scientific results, must by NASA's rules meet rigorous peer review.

Details of the development and testing of the IDGE apparatus are reported elsewhere [7, 8]. In brief, central to the flight experiment apparatus are the flight growth chamber and the optic photographic system. The stainless steel and glass flight growth chamber provides an unobstructed view of the growing dendrites through four windows, set perpendicularly to allow for stereo observations. A thin capillary injector tube, called the stinger, has thermoelectric coolers attached to one end, so dendritic growth can be initiated on command. The other end of the stinger, where the dendrite emerges, is located near the center of the chamber. The growth chamber is mounted inside a temperature controlled tank, or thermostat, and its windows from a portion of a modified shadowgraphic optical system that includes the 35 mm photographic system and the Slow Scan Television (SSTV) dendrite detection and near-real-time monitoring system (Fig. 1).

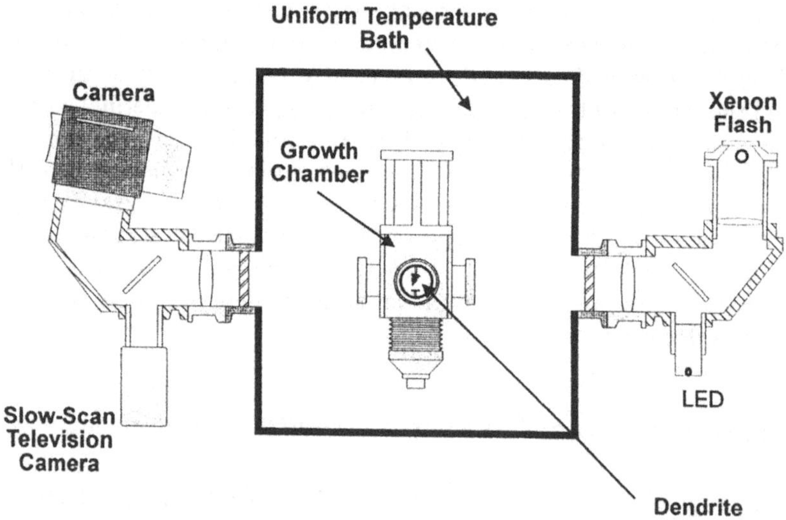

Fig. 1. Schematic of IDGE growth chamber, thermostatic bath, 35 mm cameras, and Slow Scan Television (SSTV) cameras.

During each growth cycle, the SCN is melted completely, and the bath is set to the prescribed supercooling. The system waits for the occurance of isothermal conditions, and then the thermo-electric coolers, initiating dendritic growth, turn on. Measurements of dendritic growth under terrestrial conditions showed that a dendrite usually grows from 2 to 4 mm from the tip of the stinger before achieving steady-state velocity. For the flight experiment, after the dendrite is detected, the system waits for 2.5 times a pre-programmed interval before the photography sequence commences. The objective here is to not consume any film before the dendrite achieves steady-state growth. At the same time, for each dendritic growth cycle, the IDGE records a time series of stereopairs of low-resolution SSTV binary digitized electronic images (Fig.2), and transmits data streams characterizing the dendrite's growth conditions during the cycle to the ground and to the on board memory.

The SSTV images (transmitted from the IDGE at a rate of one every 90 seconds) and the digital data stream (sent to the ground once every two seconds) were available to the experiment team in near real time at the Payload Operations Control Center (POCC) at the Marshall Space Flight Center (MSFC). The actual supercooling and the growth velocity could be calculated from the telemetry data, and experimenters could make reliable judgements of the scientific value and quality of the exposed (but unavailable) photographs. In additon, we could also measure the orientation of the observed dendrite with respect to the photographic axis, which provides information crucial for the tip radii analysis. All this information can be assembled quickly to make time-critical calculations and evaluations of the on-going space flight experiment [9]. Based on this "telescience" analysis, we uplinked changes and corrections to the programmed flight protocol which enhanced the scientific data return of the IDGE. The level of analysis that is possible from the telescience data steam exceeded the processing capability of the POCC science team. Therefore, a second science team, located at RPI, was also required. The POCC team sent data via the Internet, phone and fax to the RPI team, and the RPI team in turn supplemented the work of the POCC team by transmitting their "off-line" analysis back to the POCC by phone and fax. The rapid interchange of images, data, and analysis permitted the two experiment teams to "replan" subsequent experimental cycles and, thereby, efficiently use the limited resources of film and time in orbit.

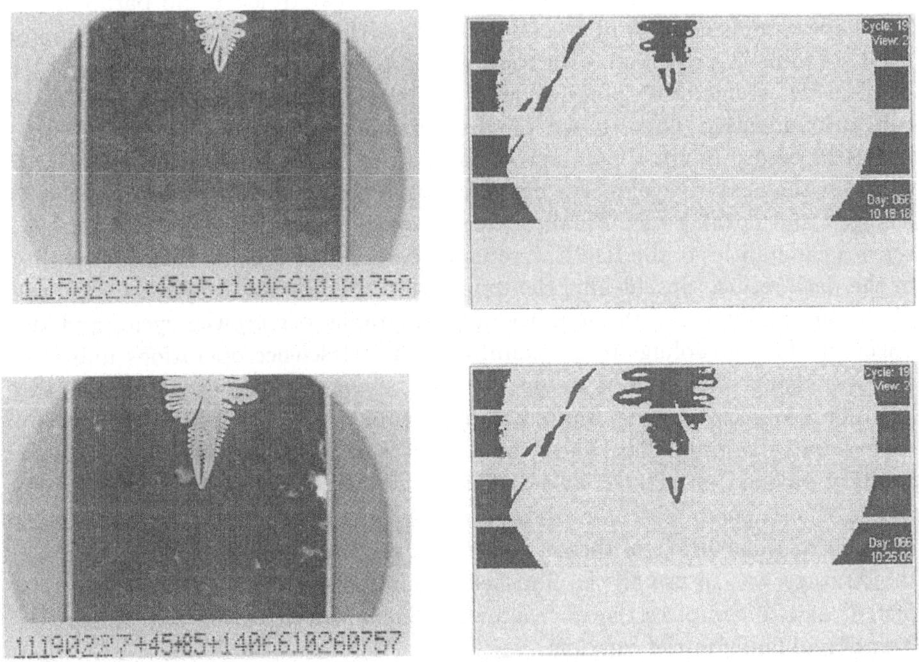

Fig. 2. Example of 35 mm photographs and Slow Scan Television (SSTV) images from the same dendritic growth cycle from the space flight experiment.

4. Flight Experiment Operations

As a default, the IDGE has an autonomous pre-programmed experimental protocol in place. The IDGE autonomous flight protocol is divided into 38 cycles. The initial warm up-cycle consists of heating the system up to the operational temperature where all system components are checked and initialized. A slow melting plateau is performed as a check of specimen purity. The next cycle provides a freezing plateau, which yields the precise liquidus temperature of the SCN (± 0.001 mK) in the chamber and serves as the calibration point from which all subsequent supercoolings are measured. The next 30 cycles provide dendritic growth data acquisition, where dendrites are grown and photographed at each of 10 programmed supercoolings taken in logarithmic steps between 0.1 and 1.0 K, and repeated three times. An additional seven growth cycles were scheduled for repeating some of the dendritic growths at the larger supercoolings, where, because of the difficulties in measuring small tips ($< 25\mu$m), redundant data would be beneficial. These extra growth cycles also provide the option to extend the supercooling range of the experiment above and below the nominal pre-programmed range. The protocol specifies that a sequence of six photographs were to be exposed for each camera view during each growth cycle. The photo intervals for each growth cycle were determined from dendritic growth theory based on the assumptions of conductive heat transfer and paraboloidal tip shape, coupled to a simple scaling rule that the velocity times the square of the tip radius is a constant with respect to supercooling.

The IDGE operating plans allowed the initial 12 pre-programmed cycles to run autonomously. Then, based on the data analysis performed during those initial 12 cycles, minimal parameter changes were made to the experiment protocol for the next 10 cycles. For example, when necessary, the photo times were changed and nothing else. Finally, after 22 cycles were completed, the full telescience capabilities of the IDGE were used to maximize the quantity and quality of the data return. Specifically, the data return was improved by modifying the photo timing intervals, the number of photographs per growth cycle, and the selection of supercoolings to be examined. The telescience operations managed all these changes to produce an optimal data return under the constraint of two limiting resources: 1) time, limited to approximately *nine* days of microgravity with priority commanding when the USMP-2 payload was designated as the primary payload of STS-62; and 2) film frames, limited to approximately *225 usable 35 mm photo frames* available on each camera. The IDGE operations are *terminated* when either of these two limiting resources is exhausted. Therefore, the strategy was to use all the film before our the microgravity time was completed, and to complete the microgravity time without leaving any unexposed frames remaining in the cameras.

On March 4, 1994, 8:53 AM EST, the space shuttle Columbia (STS-62), with the IDGE on board, lifted off from the Kennedy Space Center, Cape Canaveral Florida, into a low earth orbit of approximately 163 nautical miles. The IDGE powered up 4 hours after launch, ran through the calibrations cycles, and then commenced the dendritic growth cycles. The first growth cycle, at a supercooling of approximately 0.17 K, afforded the first opportunity to obtain by telemetry critical feedback from the experiment. The initial velocity analysis, completed around day 01- hour 08 - mission elapsed time, already showed a significant departure from theory. The velocity calculated from the SSTV in microgravity was about 3.6 μm/s, as compared to 9.8 μ m/s at terrestrial gravity, and 2.8 μm/s as predicted by theory. Based on the pre-programmed photo intervals for this growth cycle, the disparity between the predicted velocity and the observed velocity meant that 3 of the 6 programmed photographs were taken when the dendrite was beyond the camera's field of view. Later, a more detailed analysis revealed that steady-state growth on orbit ceased after only 5.5 mm of growth (Fig. 3), rather than after 10 mm of growth experienced under terrestrial conditions. This unexpected occurrence, combined with the error in the estimated velocity, resulted in only one photograph being exposed during the steady-state regime. Thus, insufficient film data were captured to determine the steady-state velocity from this growth cycle.

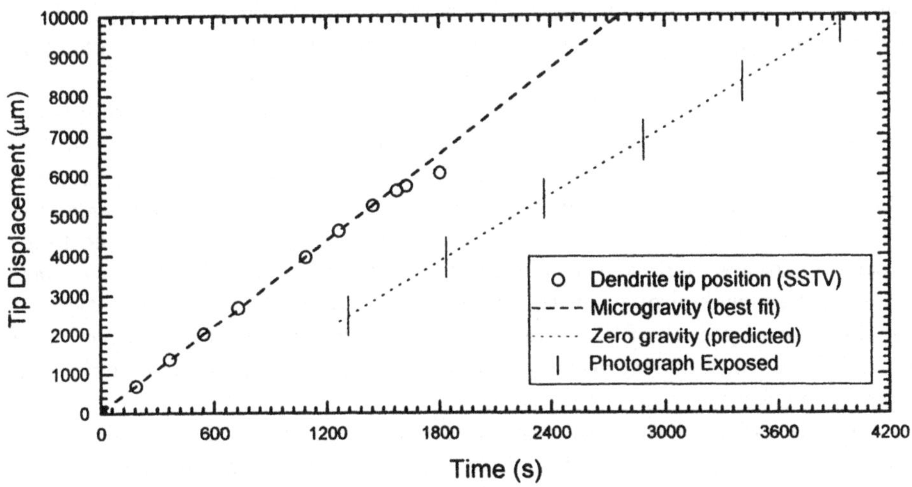

Fig. 3. Plot of the dendrite tip displacement data from the Slow Scan Television analysis and the predictied tip displacement curve from dendritic growth theory.

A similar analysis as described above was conducted for all analyzable data obtained through the first 11 dendritic growth cycles, for supercoolings between 0.1 and 1.0 K. This analysis revealed four significant trends that were used in subsequent mission replanning: 1) the measured velocity in microgravity is different from the theoretical estimates; 2) steady-state dendrite growth ceases in microgravity over a total growth distance smaller than under terrestrial conditions; 3) the total steady-state growth distance is proportional to the supercooling; and 4) the growth transient prior to achieving steady-state growth is shorter in microgravity than in terrestrial gravity, and the required displacement of the dendrite tip before steady-state growth is achieved decreases as the supercooling is increased. Many of the experimental parameters for the subsequent growth cycles were adjusted using the empirical "rules" developed from the analysis from the SSTV data of the pre-programmed growth cycles. These adjustments greatly improved both the quality and quantity of the data obtained from the flight experiment.

5. Results for IDGE on USMP-2

There are dramatic differences in the observed growth rates, particularly at the lower supercoolings, between terrestrial and microgravity conditions (Fig. 4). At about 1.7 K supercooling, there still appears to be a detectable (5%) reduction in the growth velocity under microgravity conditions as compared to terrestrial conditions. This reduction in velocity becomes more evident at 1.3 K supercooling where the velocity measured in microgravity is approximately 15% lower than the terrestrial data, increasing to almost a 200% reduction at small supercoolings. These particular results show that there are large convective effects occuring at 1 g_0, even at the highest supercoolings for which the solidification micro-structures can be resolved during *in situ* observations.

The slope of the dendritic velocity data in microgravity changes from being steeper than the terrestrial data at the higher supercoolings, to being almost parallel to it at the lower supercoolings (less then ca. 0.4 K). This observation reveals that even in microgravity, substantial convective effects reappear at the lower supercoolings. The higher scatter in the microgravity data at the lower supercoolings are ascribed (although not yet proven) to variations in the magnitude of the residual quasi-static acceleration environment, or to variations of the direction of growth of the measured dendrite with respect to the direction of the local net microgravity acceleration vector, or perhaps to some combination of the two effects.

The discussion of the radii results (Fig.5) parallels that already described for the velocity data, except that here the radius is not as strong a function of supercooling as is the velocity, so that the ratios between the terrestrial and microgravity results are smaller. An experimentally determined microgravity scaling constant, $\sigma^* = 2\alpha d_0/(VR^2)$, calculated from the microgravity measurements of V and R, at high supercoolings where convection affects are assumed minimal, yields σ^*=0.0194. Curiously, this value of σ^* is almost exactly the same as the terrestrially measured value of σ^*. This surprising agreement between the terrestrial and microgravity results for σ^* will be discussed in more detail later.

Theoretical velocity calculations using value of σ^* measured in microgravity and Ivantsov's transport theory do not describe the microgravity velocity data over *any* supercooling range. The theoretical predictions are, however, parallel to and above the microgravity data at the higher supercoolings. If instead of directly calculating σ^*, one treats σ^* as an adjustable scaling parameter (which we designate, σ_V), one is able to fit the observed microgravity velocity data at

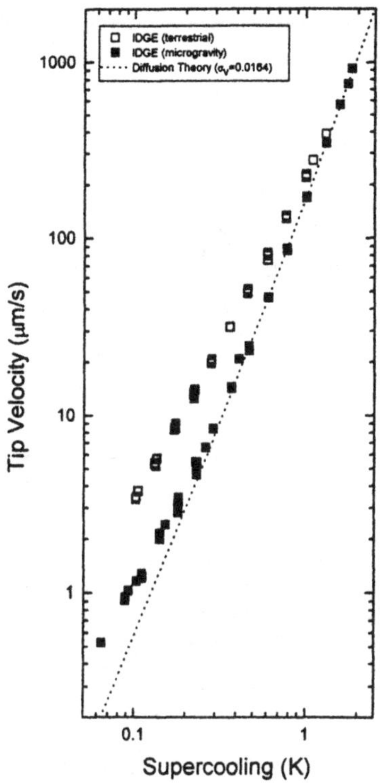

Fig. 4. Dendrite tip velocity as a function of supercooling for terrestrial and microgravity data sets.

supercoolings of 0.47 K and above with $\sigma_V = 0.0164$. The deviations between the microgravity data at the lower supercoolings and the fit at the larger supercoolings using the best value of σ_V provide quantitative evidence that even at a quasi-static residual microgravity level of approximately 0.7 μg_0, there are significant, and eventually dominant, effects on the dendritic growth. Thus, at sufficiently low supercoolings, even in microgravity, convection effects substantially alter the dendritic growth velocity from what would be expected by heat conduction alone. The excellent agreement of velocity calculations with σ_V with the microgravity data over the range of supercoolings from 0.47 K to 1.7 K show that this is the supercooling range over which one must test dendritic growth theories which employ heat conduction as the rate limiting transport process.

Theoretical radii calculations based on the microgravity measured value of σ^* and Ivantsov's transport theory do not describe the dendritic tip radii data over *any* supercooling range in microgravity either. A one-parameter fit to the radii data over the supercooling range from 0.47 K to 1.57 K yields another adjustable scaling parameter, σ_R=0.0179, which accurately describes the dendritic radii data at higher supercoolings in microgravity. Thus, there appears to be no *single* scaling parameter, whether calculated directly from the velocity and radius

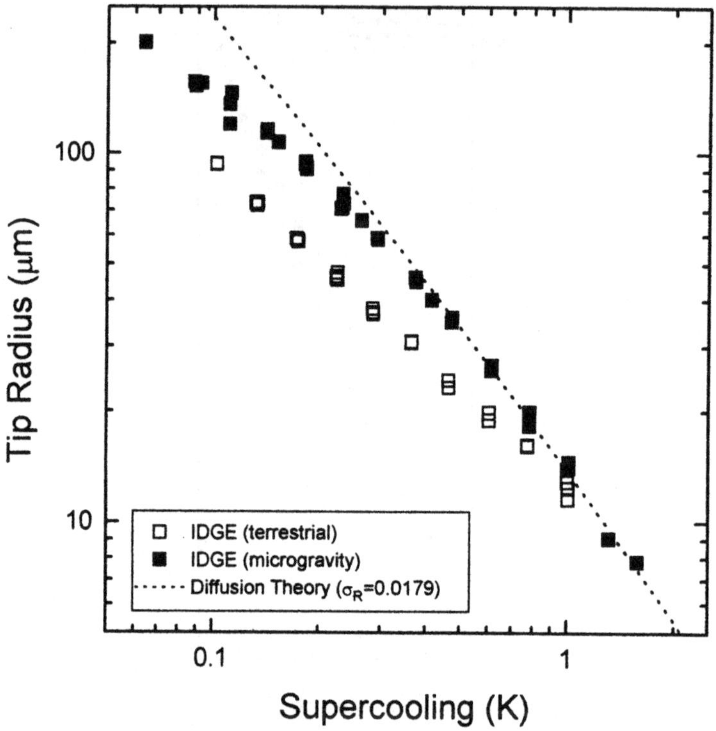

Fig. 5. Dendrite tip radius as a function of supercooling for terrestrial and microgravity data sets.

data combined, or inferred as a adjustable parameter from theory, that properly describes *both* the velocity and radius data. This result suggests strongly that the Ivantsov function combined with a *unique* scaling constant is not a robust description of the dendritic growth process. An important part of this lack of robustness in the theoretical description of dendritic growth theory is evident in the Péclet number data (Fig.6 and 7). The microgravity measured Péclet number data set is clearly separated from the terrestrial data. Most importantly, even in the diffusion-limited regime, the microgravity data are systematically lower than for paraboloidal dendrites in the Ivantsov model.

Before the IDGE, it was not possible to test separately the Ivantsov transport solution and the interface scaling hypothesis. To our knowledge, the IDGE provides the first solid evidence that Ivantsov's formulation for paraboloidal dendritic growth does not accurately describe dendritic growth in SCN. The approximate agreement achieved between the transport theory and the microgravity data indicates that dendritic growth is indeed most likely governed by the conduction of latent heat from the crystal-melt interface, but the detailed Ivantsov formulation to describe that conduction process is in need of some modification.

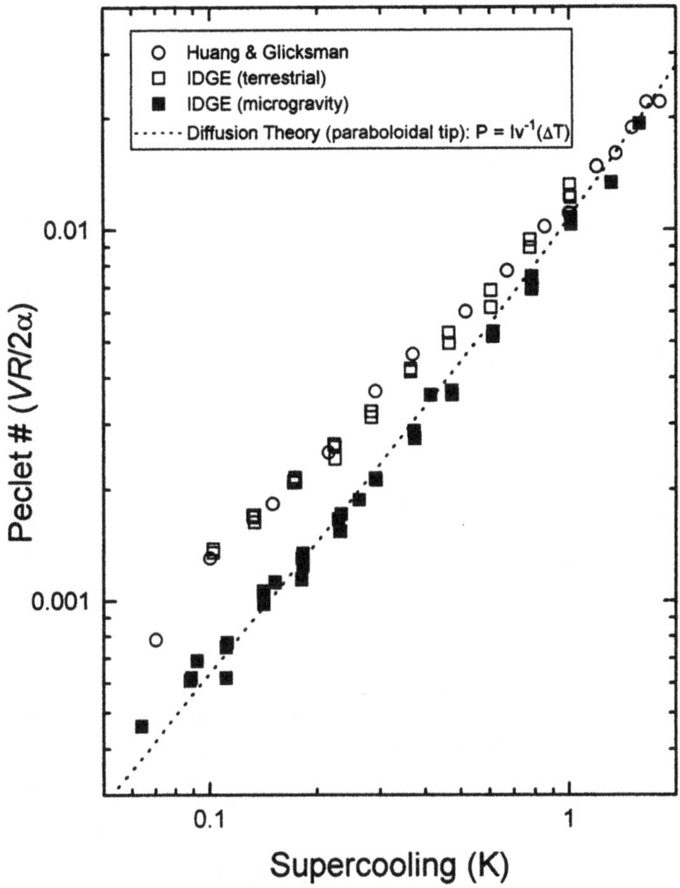

Fig. 6. Dendrite tip growth Pe number as a function of supercooling for terrestrial and microgravity data sets.

6. Summary and Conclusions

We measured the dendritic growth velocities and tip radii of curvature of succinonitrile in microgravity using the IDGE instrument flown on the USMP-2 platform in the payload bay of the space shuttle Columbia (STS-62). The on-orbit microgravity data, when compared to terrestrial dendritic growth data, demonstrate that: (1) Convective effects under terrestrial conditions cause growth speed increases of a factor of 2 at lower supercoolings ($\Delta T < 0.5$ K), and remain significant even up to values as high as $\Delta T = 1.7$ K supercooling. (2) In the supercooling range from 0.47 K to 1.7 K, the data remain virtually free of convective effects and may be used reliably for examining diffusion-limited dendritic growth theories. (3) A diffusion solution to the dendrite problem, combined with a unique (measured) scaling constant, σ^*, does not yield individual growth veloc-

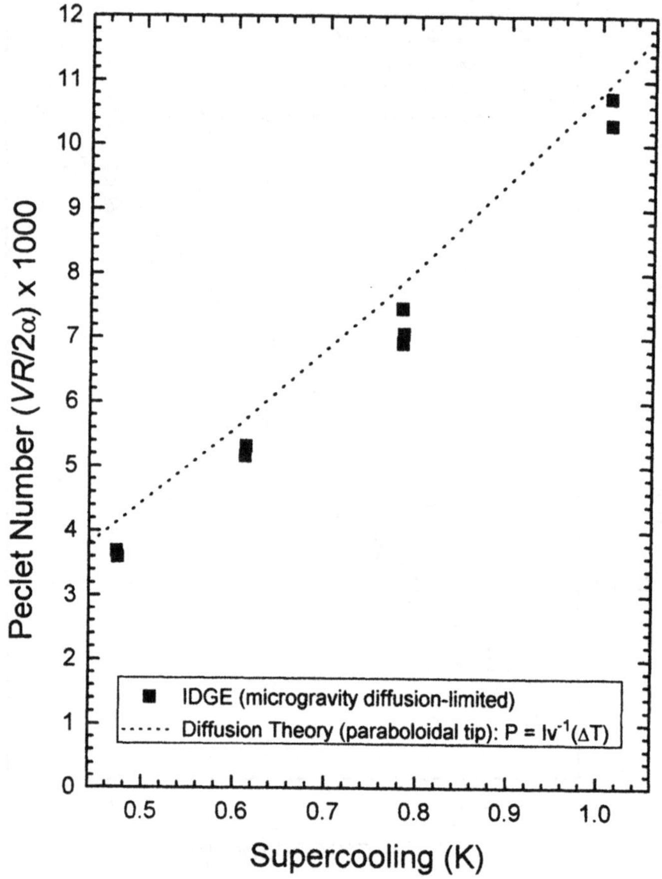

Fig. 7. Dendrite tip growth Pe number as a function of supercooling for microgravity measured data and the Ivantsov model for paraboloidal tip shape.

ity and radius predictions consistent with the observed dendritic growth velocities and radii as a function of supercooling. (4) The failure of this conventional formulation is currently attributed to small departures from the Ivantsov thermal diffusion solution, which is formulated for paraboloidal dendrites. Ivantsov's theory describes the overall dependence of Péclet number on supercooling, but predicts a value higher (5% - 15%) than the data we observed in microgravity in the diffusion-limited regime.

At completion of the IDGE flight on USMP-2, where 38 experimental cycles were originally planned, 60 cycles were accomplished. 57 of the 60 were dendritic growth cycles, from which 56 SSTV velocities were calculated. Of those, photographic sequences were attempted for 52, of which 40 useful velocity measurements were made. The supercooling range of the experiment was increased from the proposed range of 0.077 - 1.3 K, to 0.06 - 1.9 K supercooling. In addition, without IDGE teleoperations, the actual range from which usable data could have been obtained would have been limited to only 24 cycles in the range 0.22 - 1.3 K supercooling.

Based in part on the description of the experiment operations and results narrated in this paper, an experiment's operations appear to be as crucial to the scientific success as is the hardware development. Furthermore, teleoperations, the remote observation, and operation of an experiment, allows the scientist on earth to become "*the man in space*". Access to long duration, high quality microgravity environments is limited and expensive. If approached properly, rewarding scientific results may be obtained. The "*romantic*" phase of microgravity research is now over [10] and access to spaceflight demands application of the highest standards of laboratory science.

Acknowledgements

We thank, for the continuing interest, and financial support, the NASA Life and Microgravity Sciences and Application Division, (Code U), Washington, DC, under Contract NAS3-25368, with liaison provided through the Space Experiments Division at NASA Lewis Research Center, Cleveland, OH.

References

1. M.E. Glicksman and S.P. Marsh, *The Dendrite*, in Handbook of Crystal Growth, Ed. D.J.T. Hurle, (Elsevier Science Publishers B.V., Amsterdam, 1993), Vol. 1b, p.1077.
2. M.E. Glicksman, M.B. Koss, and E.A. Winsa, *Phys. Rev. Lett.*, **73** (1994) 573.
3. G.P. Ivantsov, Dokl. Akad. Nauk SSSR **58** (1947) 567.
4. S.C. Huang and M.E. Glicksman, *Acta Metall.*, **29** (1981) 701.
5. M.E. Glicksman and S.C. Huang, Convective Transport and Instability Phenomena, ed. Zierep and Ortel, Karlsruhe, (1982), 557.
6. R. Ananth and W.N. Gill, *J. Crystal Growth*, **91** (1988) 587, and **108** (1991) 173.
7. M.E. Glicksman, et al., *Met. Trans. A*, **19A** (1988) 1945.
8. M.E. Glicksman, et al., *Advances in Space Research* **11** (1991) 53.
9. M.E. Glicksman, M.B. Koss, L.T. Bushnell, J.C. LaCombe, and E.A. Winsa, 6th International Symposium on Experimental Methods for Microgravity Science, ed. Robert J. Schiffman and J. Barry Andrews, (The TMS, 1994, p. 141).
10. Leo Steg, Private communication (1987).

A Study of Morphological Stability During Directional Solidification of a Sn-Bi Alloy in Microgravity.

J.J. Favier, P. Lehmann, B. Drevet, J. P. Garandet, D. Camel[1] *and S.R. Coriell*[2]

[1] Commissariat à l'Energie Atomique, DTA/CEREM/DEM/SES, 17 rue des Martyrs, 38054 Grenoble Cedex 9, France
[2] NIST, A153 Materials, Gaithersburg, MD 20899, USA

Abstract

The morphological stability of the solid-liquid interface in Sn-0.58 at.% Bi alloys was studied during the first MEPHISTO space experiment. The transition from a planar front to a destabilized front is investigated through the in-situ measurement of the Seebeck voltage between the ends of a directionally solidified sample. This technique allows to precisely detect the instability threshold in steady state conditions as well as during transient stages. Wavelengths and segregation patterns of microstructures corresponding to well-defined distances to threshold are characterized by post mortem analyses. The experimental results are discussed within the frame of the Mullins and Sekerka's theory and 3D non linear calculations.

1. Introduction

It is known experimentally since the pioneering work of Rutter and Chalmers [1] in the early 50's that above critical growth velocity conditions, a binary alloy which is supposed to form a single phase crystal exhibits regular morphologies with segregated patterns. The formed microstructures and the induced solute distributions determine many of the mechanical or electrical properties of the product, justifying the tremendous efforts made to better understand and control the structure formation.

The behavior of the solid-liquid interface has received much recent attention in the last decade. From a theoretical viewpoint, the system shows many of the type of pattern formation that occur in well-known hydrodynamic stability problems with some specificity which makes the stability threshold determination but also the pattern selection still more complex. However, the progress made in the theoretical approaches in hydrodynamics, including non linear theories, allowed to refine previous models. In spite of these efforts, there is no rigorous criterion nowadays to predict the wavelength of the instability even though the selection process in dendritic growth is rather well established [2]. Present theories, even the most advanced, are far from accounting for the coupling between morphological instability and hydrodynamic instabilities which are always present in any ground based experiment. Only rough models predict the potential impact of convection on morphological stability of the front, essentially in terms of a stability threshold [3]. A precise test of the morphological stability theories in solidification therefore requires pure diffusive growth conditions.

The key issues in cellular growth are the dynamics of destabilization and the mode selection process. The first is related to the local conditions allowing the instability to develop on the interface. The way the critical conditions are approached can affect the occurence of the growing perturbation. One of the tasks is also to characterize if the bifurcation is normal or subcritical, which is still an open question because only a few reliable experimental data exist due to the difficulty to measure simultaneously all the relevant parameters [4]. From the knowledge gained from pure hydrodynamic studies, it is well known that pattern selection is very sensitive to boundary conditions and that modes can be inhibited if the experimental configuration does not allow them to develop [5]. In the morphological stability problem, 3D interfaces must be used to reproduce the conditions existing in real systems : the thickness of the sample should be at least an order of magnitude larger than the characteristic wavelength of the instability. In typical experiments, this means working with samples of millimeter size.

Experimentally, the problem has been considered extensively in various types of systems, including transparent "model systems". These systems present the advantage of allowing a direct observation of the interface. If from the morphological stability point of view the relevant parameters seem to be the ratio of thermal conductivities, the phase diagram and the solute diffusion coefficient, which can be found comparable in metallic alloys and in some transparent organic materials, the differences in viscosities and thermal diffusivities make the similarity inadequate for the study of coupling between interface instability and hydrodynamic instabilities. These systems used in the presence of gravity are therefore limited in representing reality of alloys when dealing with the study of convective effects in solidification.

One of the main techniques used in solidification is the Bridgman technique which allows the thermal gradient at the interface and the solidification rate to be imposed independently. However, when dealing with metallic systems, special techniques must be used to measure "in situ" the most relevant solidification parameters, including the signature of the morphology of the interface.

The MEPHISTO [1] instrument, which has been extensively described in previous papers [6, 7], has been designed to fulfill these aims. This instrument has been flown on the space shuttle for 11 days during the USMP1 [2] mission, allowing more than 40 solidification and melting cycles to be realized under 10^{-4}g or

[1] Acronym of Materiel pour l'Etude des PHénomenes Intéressant la Solidification sur Terre et an Orbite. It refers either to the CEA/CNES/NASA cooperative program on the fundamental and applied study of material solidification, or the space instrument, especially designed to study metallic solidfication.(Eds. note)

[2] USMP is the acronym of the United States Microgravity Payload which is a carrier for dedicated stand-alone experiment facilities located in the Space Shuttles cargo bay. (Eds. note)

less. The first part of the programme dealt with the planar front solidification configuration and the segregation behavior. It has been found that the gravity level encountered during the flight led to pure diffusive solidification conditions [8]. This programme enabled us also to independently measure all the thermophysical parameters of the Sn-Bi system with an accuracy of 5 to 10%. This is a prerequisite for testing and validate the different theories whose predictions often only differ by these margins.

In this paper we present the experimental data gained in real time through the various techniques used. A special attention is made to the Seebeck signal delivered by the non-planar solidification front. Then, the analysis of the different samples is given. Steady and unsteady states are characterized. Special care is taken in analyzing the information acquired during transient states. The morphological stability theories are discussed within the frame of the present results. We propose an interpretation of the apparent discrepancies between recent non-steady-state calculations and the experiment and emphasize the dynamical aspect of the transition.

2. Experimental Techniques

The MEPHISTO instrument is composed of two sets of heat diffuser and heat sink positioned along three cylindrical samples in such a way that they create a liquid bridge between two solid ends of each sample. The diffuser temperature was 600°C and the heat sinks were maintained at 50°C . The alloys of Sn-0.58 at.% Bi were prepared previously from 6N pure metals and then placed in quartz crucibles. Two solid-liquid interfaces are therefore created in each sample. One interface is movable at a given velocity through a pulling table, imposing directional solidification or melting of the corresponding parts of the samples. The velocities ranged between 2 and 26 μm/s. The first sample can be fed with an electrical current (Peltier marking of 40 A during one second) in order to mark the morphology of the interface that can be revealed by post-mortem analysis. The second sample is equipped with two thermocouples embedded in 1 mm quartz capillaries. They are passed by the moving interface during solidification and melting. Very reproducible recordings allow to measure thermal gradients in both liquid and solid phases (G_{L} and G_{S} respectively). For the considered velocities we deduced G_{L}=173 K/cm and G_{S}=89 K/cm. The ratio of thermal gradients is then:

$$\frac{G_{\mathrm{L}}}{G_{\mathrm{S}}} = 1.94 \tag{1}$$

which is slightly different from the ratio $k_{\mathrm{S}}/k_{\mathrm{L}} = 1.96$ of thermal conductivities of polycrystalline and liquid tin at its melting point given by literature [9]. This was expected since, at the low growth velocities used, the release of latent heat can be assumed negligible. In this sample, an on-line measurement of the growing interface position is performed by a resistance technique. Since liquid and solid tin have different electrical conductivities, solidification (or fusion) of any new part of the sample induces a change in the overall electrical resistance of the sample. Actually, the solid-liquid interface does not follow the movement of the furnace instantly. There is a velocity transient due to heat transfer between the furnace and the sample, which can be described to the leading order by an exponential expression:

$$R = R_0[1 - \exp(-t/\tau)] \ , \tag{2}$$

where R_0 is the pulling rate. The transient cannot be precisely followed with the present resolution of the resistance measurement but there remains a lag between the positions of the interface and the furnace which can be easily measured. It is found to be of the order of 120 s at a 2 μm/s pulling speed and of 60 s at 26 μm/s. This sample is also devoted to the quenching of the solid-liquid interface at the final solidification by moving rapidly into the heat sink.

The third sample is equipped at each end with two thermally regulated electrodes enabling the thermoelectric voltage to be measured. Based on the Seebeck effect, the solid/liquid/solid sample acts as its own thermocouple. If T_{fix} represents the temperature of the fixed interface and T_{mov} the temperature of the moving one, the measured voltage V is linearly linked to the total moving interface undercooling, according to the expression :

$$V = (\eta_L - \eta_S)(T_{\text{mov}} - T_{\text{fix}}) \tag{3}$$

where η_L and η_S are the thermoelectric powers of the liquid and solid phases. Note that this equation is valid if the interface can be considered isotherm.

3. The Seebeck Signal

Seebeck recordings in planar front solidification have been analyzed earlier [10]. A typical signal is given in Fig. 1, showing three different regimes. After an initial transient corresponding to the build-up of the solute boundary layer, a steady state occurs with a constant drift. Then, on stopping the pulling, a final transient appears. The drift seems to be related to thermoelectric effects produced by the movement of the formed solid in the gradient zone of the furnace. However, its physical origin is still not well understood. The Seebeck signal related to steady-state solidification is differentially measured at the end of pulling. It has been shown [10] that the 11 measurements of the undercooling in planar front do not depend on growth rate, enabling us to conclude that mass transfer was purely diffusive during flight. Futhermore, comparison between Seebeck signals during transients and numerical modelling allowed us to check the consistency of physical parameters independently measured in the space experiment [10].

Seebeck voltage values during the stationary regime have been recorded in a large range of growth rates (Fig. 3). These measurements allow to precisely define the transition from a planar front (where the voltage is independent of velocity) to a destabilized front (where the voltage decreases monotonously). In this last case, it has been shown that the Seebeck signal arises from a thermal contribution (the mean temperature of the interface) and an ohmic one due to thermoelectric currents generated along the front [8, 11]. Above the critical

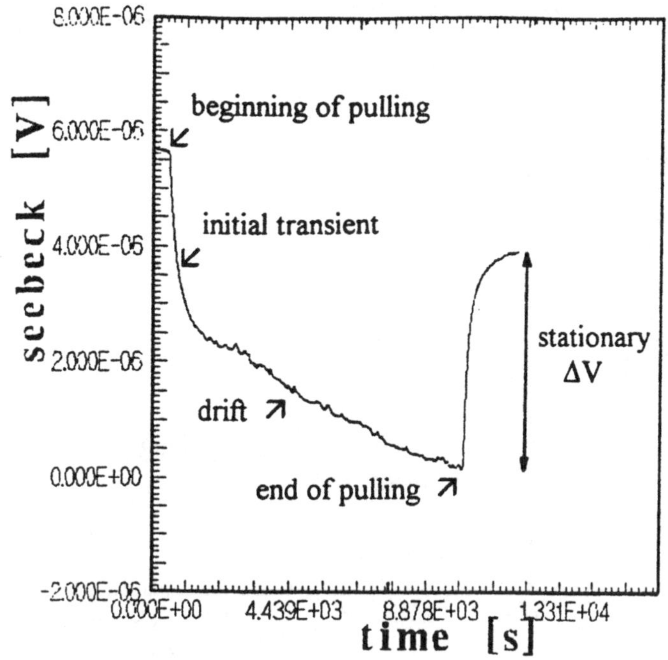

Fig. 1. Typical Seebeck signals as a function of time for planar front.

velocity of morphological instability R_C, instability occurs during the initial thermal and chemical transients. In that case, the curve of the Seebeck voltage as a function of time exhibits a particular shape, shown on Fig. 2. An initial transient is first observed, corresponding in part to a planar front solidification since the interface velocity has not reached the nominal pulling rate. It is followed by a peak, which becomes more apparent when the pulling rate increases, and a steady state regime charaterized by a drift as in the case of planar front signals. Then, a final transient occurs at the end of pulling. From the three following considerations, one can conclude that the peak is the signature of the first detectable instability:

1. First, a metallography in longitudinal section has been performed on the segment of the Seebeck sample solidified at R/R_C=4.3. It exhibits a single phase zone followed by a microsegregated one (Fig. 4). The position where the microstructure first appears is found to coincide with the peak of the corresponding Seebeck signal.
2. Second, the modelled curves of the planar front initial solute transient, accounting for heat and mass transfer phenomena, fit the experimental curves at the beginning of the solidification cycle and deviate from them in the neighbourhood of the peak, which indicates that a morphology change takes place.

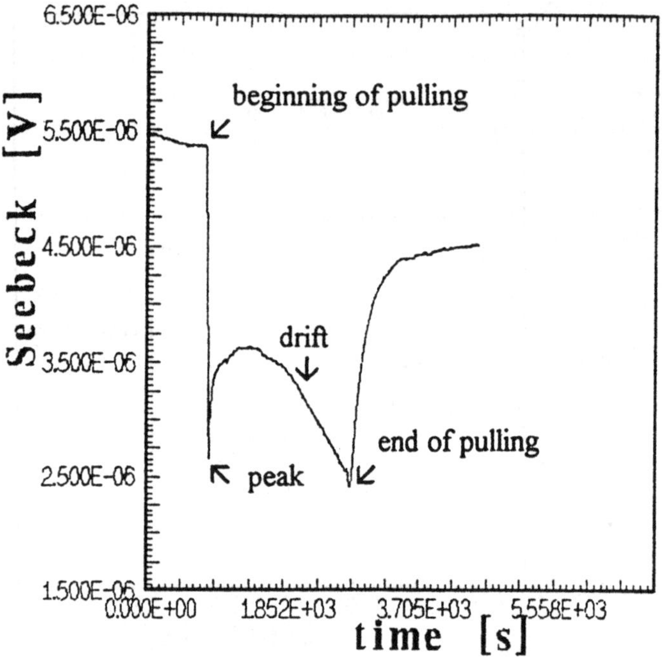

Fig. 2. Typical Seebeck signals as a function of time for cellular front solidifications.

3. Finally, the occurence of a peak has been explained through a complete model of the thermoelectric behavior of a destabilized front [11].

Indeed, it has been shown that thermoelectric currents are generated in the microsegregated solid, circulating in opposite senses in cellular and intercellular regions respectively. These currents induce an additional ohmic contribution to the Seebeck signal. After the appearance of the first instability, the amplitude of this contribution evolves, because the length of the microsegregated solid lying in the gradient zone of the furnace increases with time. This explains the decrease of the signal after the peak. Once the length of the microsegregated solid reaches that of the gradient zone, the evolution of the signal is only due to the drift previously described.

4. Sample Analysis

During the last solidification on 150 mm, five different velocities were used on 30 mm segments. For departures from instability threshold R/R_C of 1.3 and 4.3, stationary microstructures consist respectively in nodes and cells. The periodicities and microsegregations have been characterized on cross sections by image analysis and electron microprobe. In the case of nodes, the concentration field is given in Fig. 5. The concentration extrema are 1.14$\pm$ 0.08 at.% Bi in the node

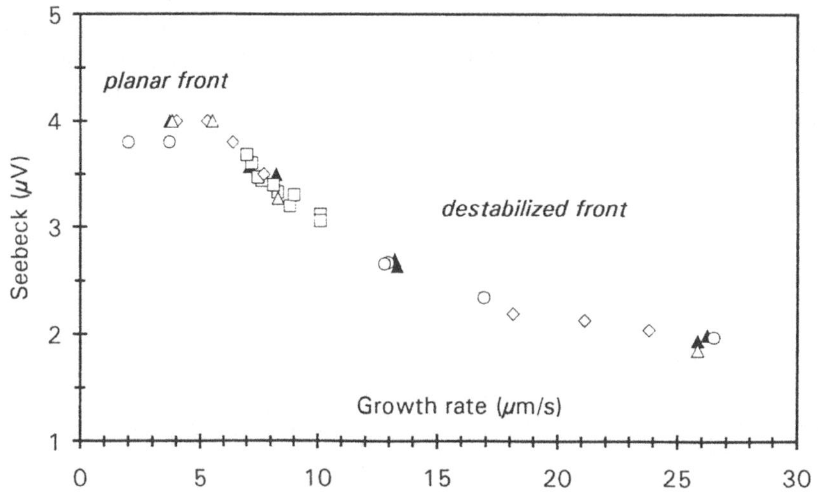

Fig. 3. Experimental values of the Seebeck voltage at steady state as a function of pulling rate.

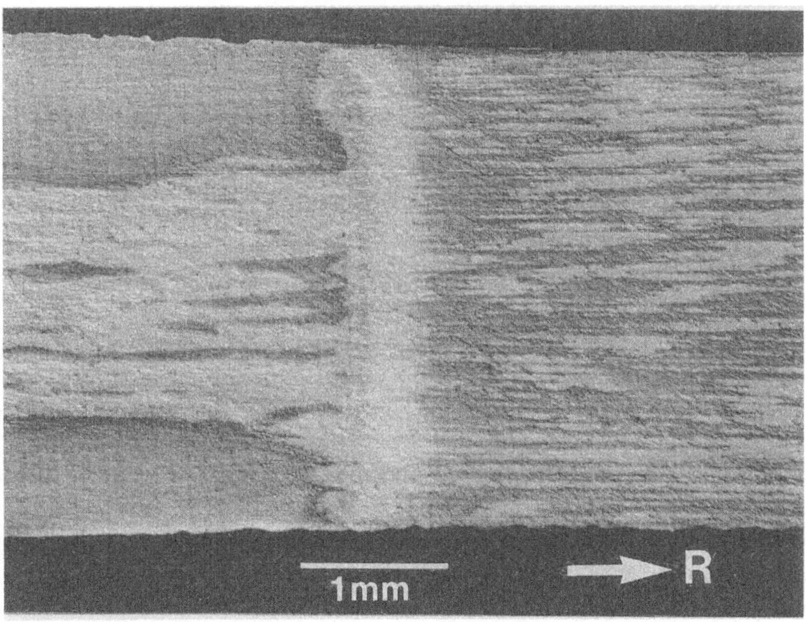

Fig. 4. Micrograph in longitudinal section of the sample solidified at 25.85 μm/s showing from left to right the unmelted zone, a light band corresponding to planar front solidification, and the cellular morphology.

centers and 0.51 ± 0.02 in the matrix. The mean spacing between node centers is $\overline{\lambda} = 88\mu m$. The regularity of the node hexagonal array is not perfect, resulting in a high value of the standard deviation (18 μm). For the cellular microstructure, a distorted hexagonal array is observed, characterized by two distinct wavelengths λ_1 and λ_2 (Fig. 7).

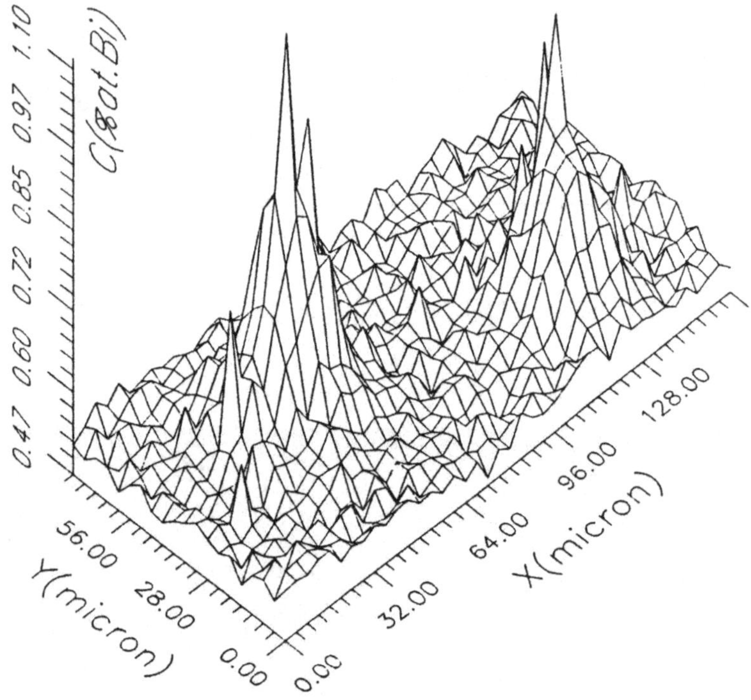

Fig. 5. Concentration distribution measured by electron microprobe in a two node region of the sample solidified at 7.8 $\mu m/s$.

As shown by the 3D histogram of spacings and angles formed by the segments joining neighbouring cells (Fig.8), λ_1 and λ_2 are close to 45 and 55 μm respectively. The concentration field in a region containing four cells is given in Fig. 6. The concentration extrema are 0.30 0.05 at.% Bi in the cell centers, 0.50 0.05 in intercellular regions located at $\lambda_1/2$ and $0.87\pm$ 0.10 at $\lambda_2/2$.

The solidified samples are polycrystalline. X-ray diffraction analyses indicate that the majority of grains grow along the [110] axis of the body-centered tetragonal structure of tin, which is the commonly observed growth direction for pure tin. X-ray Laue diffraction analyses were also performed in several grains of the sample solidified at R/R_C=4.3. Results show that the bands of cells characterized by the smallest spacing λ_1 follow the [001] axis in each grain (Fig.7).

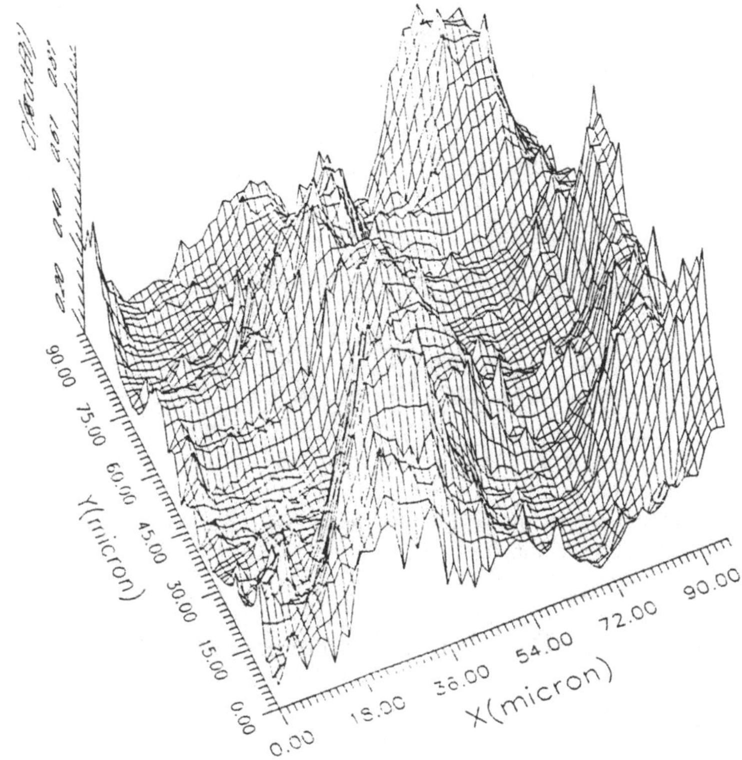

Fig. 6. Concentration distribution measured by electron microprobe in a four cell region of the sample solidified at 25.85 μ m/s.

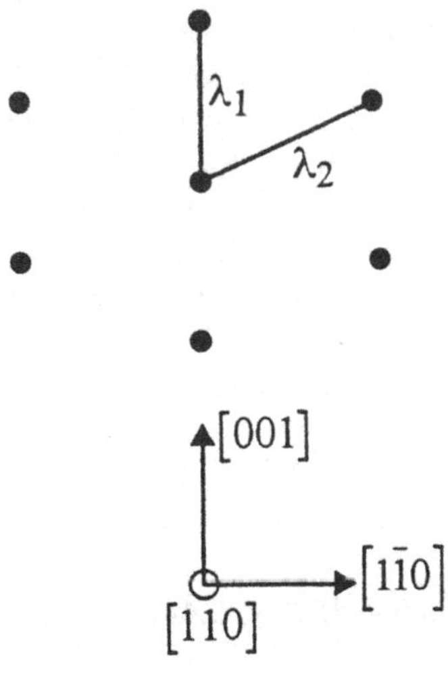

Fig. 7. Schematic of the distorted hexagonal array of cells at R=25.85 μm/s and the corresponding crystallographic directions determined by Laue diffraction.

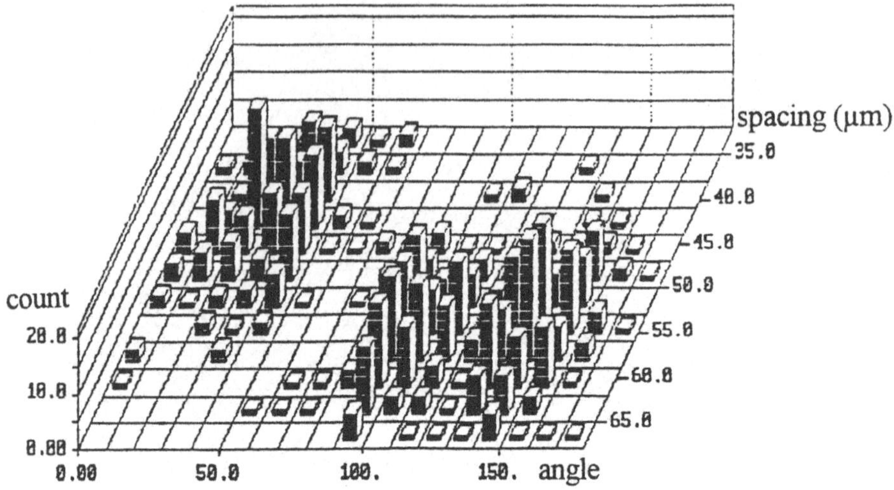

Fig. 8. Histogram of spacings and angles formed by the segments joining neighbouring cells for the structure of Fig.7.

5. Steady State Results and Interpretation

5.1 Stability Threshold

The critical velocity of morphological instability has been calculated using Mullins and Sekerka's complete dispersion relationship expression [12], with the thermophysical parameter values deduced from the planar front solidification study [10] (see Table 5.1). The result is very sensitive to the ratio of the thermal conductivities of solid and liquid. We decided to take the experimental values deduced from the thermal gradient measurements which are reproducible within 5%. The ratio k_S/k_L=1.94 is thus deduced from relation (1). Using for the diffusion coefficient D=$1.3 \cdot 10{-}9\text{m}^2/\text{s}$ and for the partition coefficient k=0.27, the critical velocity is found to be R_C=5.363 μm/s. With the error bars of one digit on each value of the above parameters (Tab.5.1), the value R_C=6.087 μm/s is obtained. We can thus admit that R_C=5.36± 0.73 (± 13%) μm/s or :

$$4.63 < R_C < 6.09\mu\text{m/s} \ .$$

The experimental critical velocity $R_C^{\text{exp}} = 6 \pm 0.4(\pm 6\%)\mu\text{m/s}$ is deduced from the measured Seebeck voltage as a function of pulling rate (Fig. 3), leading to :

$$5.6 < R_C^{\text{exp}} < 6.4\mu\text{m/s} \ .$$

It can be concluded that the experiment shows 10% overstability but is compatible with the Mullins and Sekerka's steady state predictions as the error bars on the experimental and theoretical values of the critical velocity overlap.

Partition coefficient:	k = 0.27±0.01
Bi diffusion coefficient:	$D = 1.3 \pm 0.1\,10^{-9}\mathrm{m}^2/\mathrm{s}$
Global thermoelectric power:	$\eta_L - \eta_S = 1.3\ \mu V/K$
Ratio of thermal conductivities of solid and liquid tin:	$k_S/k_L = 1.94$

Table 1. Thermophysical parameters for the Sn-0.58 at.% Bi alloy. From [10].

5.2 Wavelengths

The experiments show that "hexagonal like" structures are obtained for departures from threshold R/R_C of 1.3 and 4.3. Following McFadden et al. [13], the wave number of 3D instabilities forming an hexagonal pattern is given by

$$\omega = \frac{4\pi}{\sqrt{3}\lambda} , \tag{4}$$

where λ is the spacing between neighbouring hexagon centers. Calculations of these authors show that hexagon centers correspond to cell centers (concentration minima) for a cellular structure and to node centers (concentration maxima) for a node structure. Thus, the above definition of λ is in agreement with the spacings measured in this work (see Sec.4.). The experimental values of deduced from equation (4) are reported on the 3D calculated marginal stability diagram of Fig. 9.

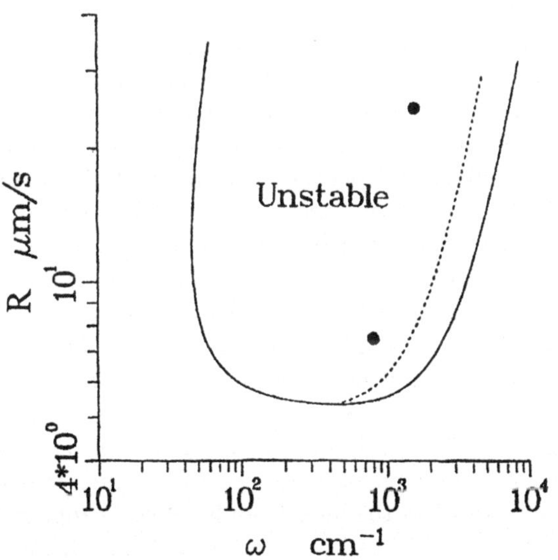

Fig. 9. 3D calculated marginal stability curve and fastest growing mode (dashed line). Full circles indicate experimental points.

Increasing the distance to instability threshold, node followed by hexagonal cell structures have been observed experimentally. No 2D structures (elongated cells) have been found. This is in agreement with 3D weakly non linear calculations performed by McFadden et al. [13] for systems with k< 1. The experimental wavelengths fall inside the marginal stability curve, indicating that the experimentally excited modes belong to the potentially unstable ones. The experimental modes differ significantly between the fastest growing mode predicted by the linear theory and that represented by the dashed line in Fig. 9.

5.3 Anisotropy of the Microstructure

The cell pattern has been found experimentally slightly anisotropic with two wavelengths. The smallest one corresponds to a well characterized crystallographic axis, i.e. [001] (Fig. 7). Considering a 30% anisotropy in the thermal conductivities of the solid between the c axis and the direction normal to c, and using the model of Coriell et al. published in [14], the $[1\bar{1}0]$ direction is predicted to be less stable than the [001] direction. At the critical velocities, the wavelength is shorter for the [001] direction. This is at least in qualitative agreement with the experiment (45 μm versus 55). However, other sources of anisotropy might have an effect such as the surface tension, but thermal conductivities have been considered here because the anisotropy of conductivities is measured for tin and reported in the literature. Further, the morphological stability model does exist for conductivities which is not the case for surface tension.

5.4 Characterization of the Bifurcation

Using the analytical expression given by Alexander et al. [15], the Landau coefficient is evaluated as $a = -320$. This means that under our experimental conditions, the bifurcation between planar and non planar front is expected to be subcritical. Experimentally, the interface should be almost less stable than predicted by the Mullins and Sekerka's theory and the selected wavelengths should not be the fastest growing ones. The second statement is true as shown by the analysis of the structures (Fig. 9). But the interface is more stable than predicted. It is thus difficult to draw any conclusion on the nature of the bifurcation. Another feature typical of subcritical behaviour is the existence of a hysteresis between the transition from planar front to nodes and that from nodes to planar front. Such a hysteresis is in principle visible in the Seebeck signal. When the pulling is stopped, the interface velocity decreases exponentially and a small peak is observed. This peak could mark the morphological re-stabilization. But the resistance measurements are not accurate enough to give a precise value of the interface velocity when the initial and final peaks appear. So no significant difference in interface velocity can be detected and the nature of the bifurcation is still an open question. A careful study of the transitions on microstructure in transient regimes is under current investigation and will allow us to draw definitive conclusions.

6. Transient Results and Interpretation

We saw in Sec.3. that the peak marks the moment where the first structure appears during the initial thermal and chemical transients. The instantaneous Sekerka number [12] is

$$S = \frac{\bar{G}}{mG_C} \,, \tag{5}$$

where G_C is the chemical gradient at the interface and $\bar{G} = (k_S G_S + k_L G_L)/k_S + k_L)$ has been calculated at the characteristic time t_p of the peak (Fig. 11), using the instantaneous interface velocity and the value of the undercooling deduced from the Seebeck value at that time. All these Sekerka numbers follow the same law, starting from an extrapolated value of 1 at the critical velocity and raising to a value of the order of 2.3 for the highest velocities used (Fig. 10). This clearly indicates that the interface is found much more stable when looking at its dynamical behavior, the higher the pulling velocity, the more stable the interface. This result is very reproducible.

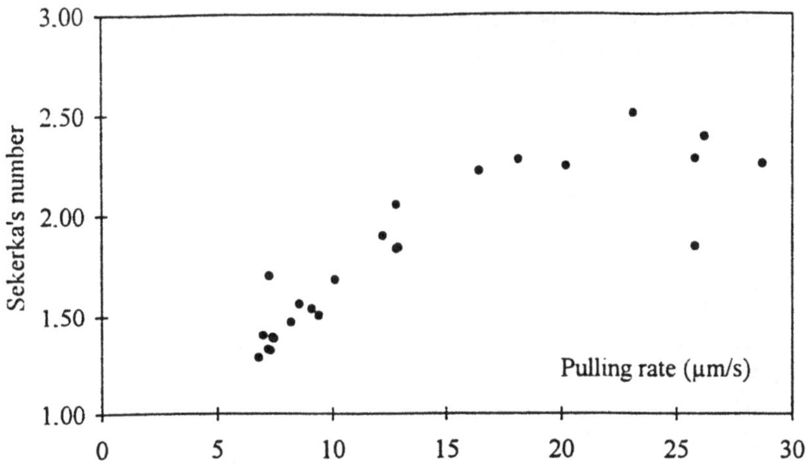

Fig. 10. Sekerka number at the peak of the Seebeck signal, deduced from instantaneous undercooling and interface velocity, as a function of pulling rate.

The time t_p at which the peak occurs in the initial transient of the Seebeck signal has been reported as a function of pulling rate in Fig. 12. The planar front initial transients have been modelled for the range of experimental pulling rates. For each time step, the Sekerka number has been calculated, allowing to estimate the time at which the modified constitutional supercooling criterion is satisfied. The calculated times show the same trend as the experimental ones for the decrease of t_p when the pulling rate increases, but shifted by a factor of about 2 to the shorter times (Fig. 12). The instability conditions have also been calculated numerically at NIST, with a completely time dependent scheme, including the thermal transfer in the sample leading to the observed instantaneous interface velocities. Some of these results are reported on the diagram of Fig. 12. They are located close to the quasi steady state Mullins and Sekerka calculations and therefore do not explain the observed first instabilities appearance.

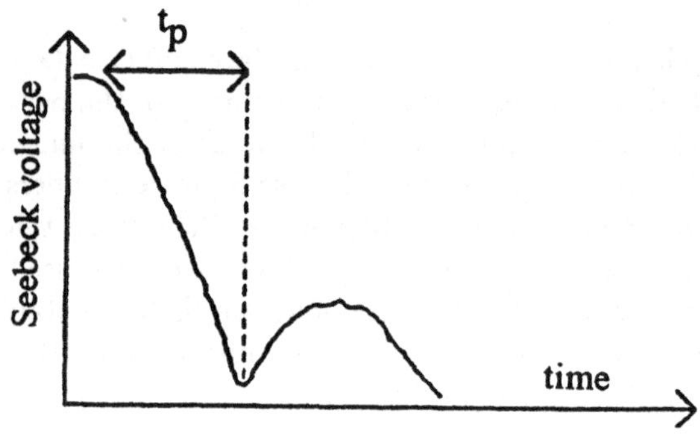

Fig. 11. Schematic of the peak of the Seebeck signal at the characteristic time t_p, featuring the onset of instability.

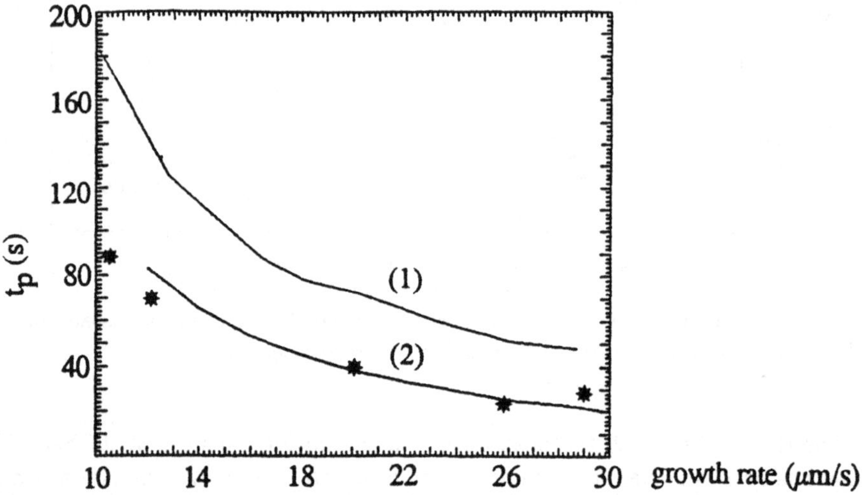

Fig. 12. Onset of instability during the initial transient: experimental (1), application of the Mullins and Sekerka criterion with a quasi permanent approximation (2), and complete non linear calculation of the evolution of an initial perturbation (stars).

7. Concluding Remarks

In this experiment, the morphological instability threshold has been extensively studied in steady state conditions as well as during initial transient stages. The microgravity environment allowed Sn-0.58 at.% Bi alloys to be solidified under pure diffusive conditions of solute transport. We are thus able to propose experimental results concerning morphological stability threshold and patterns in the frame of the hypothesis of the Mullins and Sekerka theory and subsequent linear and non-linear refinements.

The Seebeck technique has been proven to be a very accurate way to look at the first breakdown of the interface, in good agreement with the other characterization means such as microprobe analysis and interface quenching.

Many different aspects have been examined and quantified. Steady state theory correctly predicts, within the error bar, the stability threshold, even if the interface seems experimentally more stable than theoretically predicted.

Although tin is known as behaving essentially as a metal, some anisotropy in the cellular pattern has been found. The complete 3D segregation patterns have been measured. However, the existing non linear numerical models are unable to predict these patterns for departure from the threshold over a few percent. There is a need for further theoretical effort since clean reference microgravity data exist.

Another important result concerns the mode selection. It is clear from the comparison between experiment and calculation that for 3D interfaces the stable mode is definitely not the fastest growing one (Fig. 9) as supposed by many authors. The observed ones fall inside the neutral stability curve, which is consistent with linear theory. However, contrary to hydrodynamics, the minimum of the neutral stability curve is very smooth, allowing many modes to be excited. The selection criterion is not understood yet.

During the initial transient, after a planar front solidification, the peak of the Seebeck signal marks the moment when morphological instability occurs. However, the theoretical threshold is located far before (up to twice), suggesting that in dynamic condition the interface is far more stable than predicted. This systematic deviation of the experimental breakdown from the theoretical prediction leads us to reconsider the definition of instability itself. From a theoretical point of view, the front is considered unstable as an infinitesimal perturbation can find conditions to grow. By assumption the initial perturbation is supposed to be infinitesimal; therefore, even if it is allowed to grow, it takes a certain time before it can reach an amplitude making it experimentally visible.

Beside the sign of the dispersion relationship, the pre-exponential term must be considered. It is in fact the amplification factor transforming the infinitesimal perturbation into a macroscopic stable instability.

The ratio of the amplitude δ to the initial amplitude δ_0 of the perturbation has been calculated as a function of time during the initial transients. Results at R=25.8 μm/s are given in Fig. 13. In this particular case, the front is supposed to become unstable after 35 s. Experimentally, the first instability is found after 51 s, which corresponds to an amplification factor (ratio of the actual amplitude

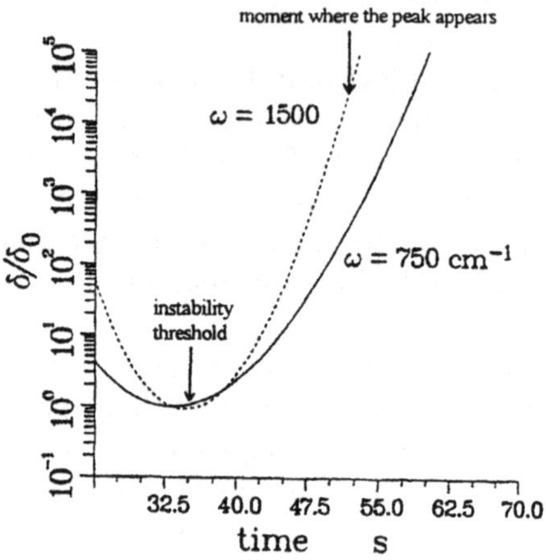

Fig. 13. 3D calculations of the ratio of the amplitude δ to the initial amplitude δ_0 of the perturbation as a function of time for the experimental mode (dashed line) and the fastest growing mode (full line) during the initial transient at R=25.8 μm/s.

of the perturbation to the amplitude at threshold) of $2 \cdot 10^4$. The same result can be drawn in a δ versus time plot parametered by δ_0 (Fig. 14). With this representation, it is clear that the amplitude of the instability becomes macroscopic at a very well defined time due to its exponential growth. The calculation has been performed for most of the velocities used. The amplification factor allowing to fit the experimental instability is found to be always in the 10^4 to 10^5 range. Correspondingly, the instantaneous Sekerka numbers have been calculated for the same times: they are found varying between 1 and 2.25 in very good agreement with the values deduced from the experiment (Fig. 10). These considerations allow us to estimate the size of the initial perturbation at threshold, which is found to fall into the nanometer range. For the first time, the study of the morphological stability during initial transients allows us to deduce an important parameter for the theoretical understanding, the amplification factor or equivalently the size of the initial perturbation at threshold. This perturbation is found infinitesimal as assumed by the theories, in the nanometer range, which is the characteristic scale where specific behaviors of the interface, such as surface tension anisotropies or faceting, can be expressed. This could explain why morphological instabilities are so sensitive to these parameters.

To sum things up, this experimental contribution presents very accurate and reproducible results on the instability threshold determination in steady state and during the dynamical process of destabilization, as well as pattern selection and 3D characterization of the structures. These results, obtained in pure diffusion without any interference from convective origin, represent a very reliable data base for theoreticians. This data base will be extended in the near future with the new flight opportunities for MEPHISTO.

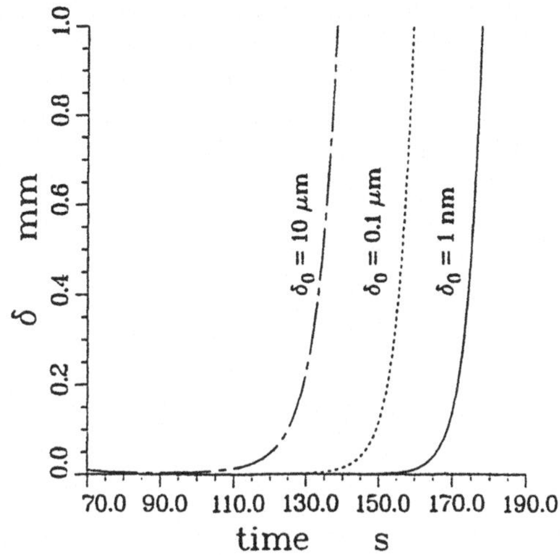

Fig. 14. 3D calculations of the amplitude δ of the fastest growing perturbation as a function of time for different initial amplitudes δ_0 during the initial transient at R=10.5 μm/s.

Acknowledgements

The authors have greatly appreciated the assistance of F. Herbillon, J. E. Mazille, I. Schuster, I. Touet and N. Valignat in the sample characterization. The present work was conducted within the frame of the GRAMME agreement between the CNES and the CEA.

References

1. J.W. Rutter, B. Chalmers, *Can. J. Phys.*, **31** (1953) 15
2. M.E. Glicksman, S. P. Marsh, "The dendrite", in: *Handbook of Crystal Growth*, vol. I, Chap. 15, D. T. J. Hurle Editor, North-Holland (1993)
3. J.J. Favier, A. Rouzaud, *J. Crystal Growth*, **64** (1983) 367
4. S. de Cheveigne, C. Guthmann, P. Kurowski, *J. Crystal Growth*, **92** (1988) 616
5. E.L. Koschmiedr, S.A. Prahl, *J. Fluid Mech.*, **215** (1990) 571
6. A. Rouzaud, J.J. Favier, D. Thevenard, *Adv. Space Res.*, **8** (1988) 49
7. A. Rouzaud, J. Comera, P. Contamin, B. Angelier, F. Herbillon, J.J. Favier, *J. Crystal Growth*, **129** (1993) 173
8. J.J. Favier, J.P. Garandet, A. Rouzaud, D. Camel, *J. Crystal Growth*, **140** (1994) 237
9. *Thermophysical Properties of Matter*, vol. 1, Ed. Y. S. Touloukian, IFI/Plenum, New York-Washington, 1970

10. P. Lehmann, J.P. Garandet, B. Drevet, F. Herbillon, J.J. Favier, to be published
11. P. Lehmann, R. Moreau, O. Laskar, D. Camel, to be published
12. W.W. Mullins, R.F. Sekerka, *J. Appl. Phys.*, **35** (1964) 444
13. G.B. McFadden, R.F. Boisvert, S.R. Coriell, *J. Crystal Growth*, **84** (1987) 371
14. S.R. Coriell, G.B. McFadden, R.F. Sekerka, *J. Crystal Growth*, **100** (1990) 459
15. J.I.D. Alexander, D.J. Wollkind, R.F. Sekerka, *J. Crystal Growth*, **79** (1986) 849

Response of Crystal Growth Experiments to Time-Dependent Residual Accelerations

J.I.D. Alexander

Center for Microgravity and Materials Research, University of Alabama in Huntsville, Huntsville, Alabama, 35899, USA

Abstract

Real-time Seebeck voltage variations across a Sn-Bi melt during directional solidification in MEPHISTO on USMP-1[1] show a distinct variation which can be correlated with thruster firings. The Seebeck voltage measurement is related to the response of the instantaneous average melt composition at the melt-solid interface. This allows a direct comparison of numerical simulations with the Seebeck signals obtained on USMP-1. The effects of such accelerations on composition for a directionally solidifying Sn-Bi alloy have been simulated numerically. USMP-1 acceleration data was used to assist in our choice of acceleration magnitude and orientation.

1. Introduction

Under terrestrial conditions, thermally and compositionally inhomogeneous fluids generally give rise to buoyancy-driven flows. In a low-earth-orbit spacecraft, the effective gravity can, in principle, be locally reduced by factors of 10^6 to 10^7 in comparison to conditions at the earth's surface. For crystal growth, this suggests that melt convection, which can lead to undesirable compositional nonuniformity, may be sufficiently reduced to allow diffusive conditions to prevail. Thus, at least for planar melt-crystal growth interfaces, this could lead to lateral or radial composition uniformity in the grown sample. However, it has been recognized for some time [1, 2, 3] that residual accelerations arising from the gravity gradient tides, atmospheric drag, vernier firings, crew motions, etc., are sufficient to cause significant deviations from "zero-gravity" conditions. Since the purpose of conducting most low gravity experiments is, generally, to eliminate or substantially reduce convective motion, this has caused some concern. Indeed, it is now apparent from sensitivity studies (for reviews see [4, 5] and references therein) that significant composition nonuniformities can arise as a result of residual acceleration or "g-jitter".

[1] The acronyms MEPHISTO and USMP-1 are explained in the contribution of Favier et al. in this volume. (Eds. note)

The awareness of the potential disruptive effect of g-jitter for crystal growth (and other experiments involving inhomogeneous fluids or fluid surfaces) has spawned a still growing body of literature that deals with the theoretical estimation of g-jitter effects on experiments (see for example [4, 5, 6, 7, 8, 9]). These estimates range in complexity from order of magnitude analyses to the numerical solution of the full nonlinear equations governing the transport of momentum, heat and mass for systems of interest. The need to measure and record residual acceleration data for use in post-flight assessment of low-gravity experiments has also been recognized. This has led to the development and recent use of several acceleration measurement systems. However, despite the subsequent abundance of acceleration data and the numerous g-jitter analyses, direct correlation of recorded experimental events with g-jitter events is lacking in all but a few cases.

The USMP-1 MEPHISTO experiment involved the directional solidification of a tin-bismuth alloy. During the experiments, acceleration disturbances due to firing of the Orbiter Maneuvering System (OMS) thrusters were recorded which resulted in measurable changes in composition at the growing melt-liquid interface. These changes in composition were measured in real time using Seebeck measurements [10]. In this paper results of simulations of the effect of such thruster firings on transport in Sn-Bi alloys are described. The objective of the work is to better understand the measured response of the interfacial concentration and to plan for future experiments on USMP-3 which will focus on quantifying residual acceleration (g-jitter) effects.

2. Formulation

A sketch of the experiment set-up is shown in Fig.1. There are two furnaces, one is fixed the other is translated through a temperature gradient. The applied temperature profile shown in Fig. 1 leads to a central cylindrical melt volume bounded by a moving and a stationary (or reference) solid-liquid interface. The melt composition at the moving and the stationary reference interfaces is not the same. For Sn-Bi there is a dependence of melting temperature on concentration. Thus, it follows that the melting temperature at the two interfaces will not be the same. The Seebeck effect [10, 11, 12] gives rise to a small but measurable voltage difference between these two interfaces. Measurement of this voltage difference allows the determination of the average temperature and, thus, the average composition of at the growing interface. The MEPHISTO set-up and the Seebeck measurements are discussed in more detail in [11, 12]. Figure 2 shows the response of the raw Seebeck signal to a $2.7 \cdot 10^{-2}$ g acceleration due to an OMS burn during the USMP-1 MEPHISTO experiment. The basic model system has been described elsewhere [7, 8]. The essential features are outlined below. Figure 1 shows the basic model geometry. Solidification takes place as the ampoule is translated along a temperature gradient. For this model system, translation of the ampoule is simulated by supplying a doped melt of bulk composition c_∞ at a constant velocity V_g at the top of the computational space (inlet), and withdrawing a solid of composition $c_s = c_s(\mathbf{x}, t)$ from the bottom. The crystal-melt

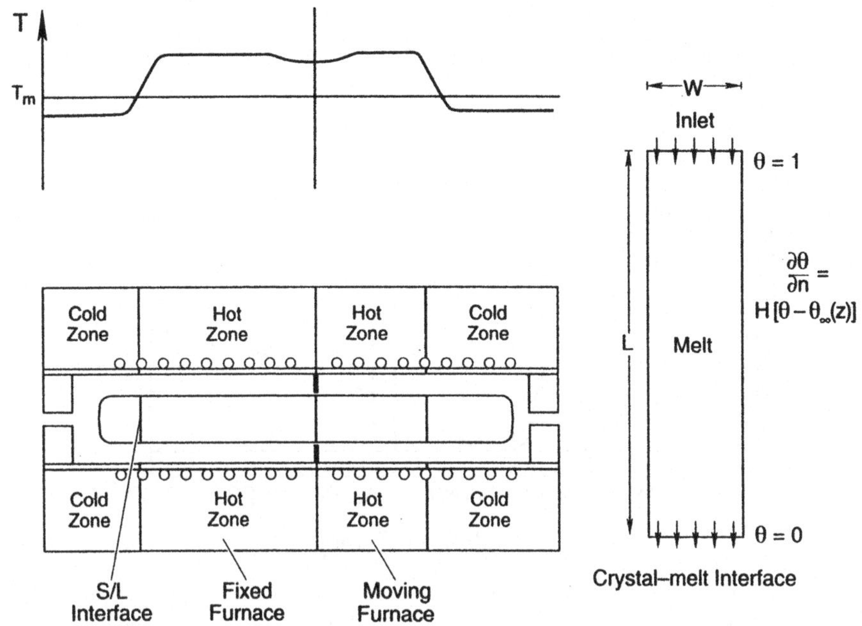

Fig. 1. The MEPHISTO set-up (left, with temperature profile on top) and model geometry (right).

interface is located at a distance L from the inlet; the width of the ampoule is W. The temperature at the interface is taken to be T_{m}, the melting temperature of the crystal, while the upper boundary is held at a higher temperature T_{h}. The ampoule wall temperatures are prescribed according to the particular situation to be modelled. Since we wish to confine our attention to compositional nonuniformities caused by buoyancy-driven convection, rather than variations resulting from non-planar crystal-melt interfaces, the interface is held flat. The slight change in melting temperature (and, thus, the interface shape) due to composition shifts at the interface following the onset of the convective response to impulsive acceleration are not accounted for in this model.

The governing equations are cast in dimensionless form using L, κ/L (κ is the melt's thermal diffusivity), $\rho_{\mathrm{m}}\kappa^2/L^2$, $T_{\mathrm{h}} - T_{\mathrm{m}}$, and c_∞ to scale the length, velocity, pressure, temperature, and solute concentration. The dimensionless equations that govern momentum, heat and solute transfer in the melt are then

$$\frac{\partial \mathbf{u}}{\partial t} + (\mathbf{u}\cdot\nabla)\mathbf{u} = -\nabla p + Pr\nabla^2\mathbf{u} + (RaPr\Theta + Ra_{\mathrm{s}}PrC)\mathbf{g} \qquad (1)$$

$$\nabla\cdot\mathbf{u} = 0 \qquad (2)$$

$$\frac{\partial \Theta}{\partial t} + \mathbf{u}\cdot\nabla\Theta = \nabla^2\Theta \qquad (3)$$

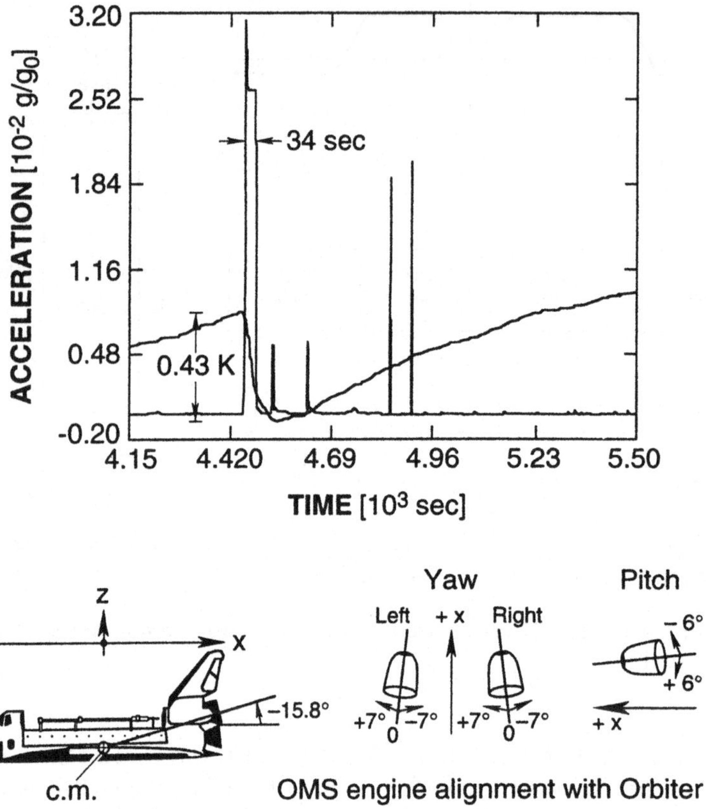

Fig. 2. Accelerations associated with the OMS firing (horizontal line with spikes, left scale) and the response of the raw Seebeck signal (curved line with drop by 0.43 K). The sketches indicate the orientation of the acceleration relative to the Orbiter x-axis which is parallel to the axis of the solidifying sample.

$$\frac{Sc}{Pr}\left(\frac{\partial C}{\partial t}+\mathbf{u}\cdot\nabla C\right)=\nabla^2 C \tag{4}$$

where $\mathbf{u}(\mathbf{x},\mathbf{t}), \Theta = (T(\mathbf{x},t)-T_{\mathrm{m}})/(T_{\mathrm{h}}-T_{\mathrm{m}})$ and $C = c_{\mathrm{m}}(\mathbf{x},t)/c_{\infty}$ respectively represent the velocity, temperature and solute concentration. The Prandtl, Rayleigh, Solutal Rayleigh and Schmidt numbers are $Pr = \nu/\kappa$, $Ra = \beta(T_{\mathrm{h}} - T_{\mathrm{m}})L^3 g/\nu\kappa$, $Ra_{\mathrm{s}} = \beta_{\mathrm{c}} c_{\mathrm{m}}^{\infty} L^3 g/\nu\kappa$, and $Sc = \nu/D$, respectively, and ν is the kinematic viscosity, β and β_{c} are the thermal and solute expansion coefficients and D is the solute diffusivity. Table 1 gives the values of the physical properties used in these calculations. The term $\mathbf{g}$ in (1) specifies the magnitude and orientation of the gravity vector which may be time dependent. The Rayleigh number is taken to be the value of Ra at the Earths surface. Thus the magnitude of $\mathbf{g}$ is the actual acceleration magnitude relative to 9.8 ms^{-2}.

Property	Symbol	Value
Kinematic Viscosity	ν	$2.6 \cdot 10^{-3}$ [cm^2s^{-1}]
Thermal Diffusivity	κ	0.17 [cm^2s^{-1}]
Solute Diffusivity	D	$1.4 \cdot 10^{-5}$ [cm^2s^{-1}]
Distribution Coefficient	k	0.26
Growth Velocity	V_g	2 [μms^{-1}]

Table 1. Thermophysical properties used in the calculations.

The following boundary conditions apply at the crystal-melt interface.

$$\Theta = 0, \quad \mathbf{u} \cdot \mathbf{N} = \frac{Pe_g Pr}{Sc}, \quad \mathbf{u} \times \mathbf{N} = 0, \quad \frac{\partial C}{\partial z} = Pe_g(1-k)C \tag{5}$$

where $Pe_g = V_g L/D$ and $\mathbf{N}$ is the unit vector normal to the planar crystal melt interface. The measure of compositional nonuniformity in the crystal at the interface to be the lateral range in concentration given by

$$\xi(t) = \frac{c_{smax} - c_{smin}}{c_{sav}} \% \tag{6}$$

where c_s is the (dimensional) solute concentration in the crystal, and c_{av} is the instantaneous average interface concentration. The following boundary conditions are applied at the "inlet"

$$\Theta = 1, \quad \mathbf{u} \cdot \mathbf{N} = \frac{Pe_g Pr}{Sc}, \quad \mathbf{u} \times \mathbf{N} = 0, \quad \frac{\partial C}{\partial z} = Pe_g(C-1) \tag{7}$$

At the side walls the conditions are (with $\mathbf{e}_w$ normal to the walls)

$$\mathbf{u} \cdot \mathbf{N} = \frac{Pe_g Pr}{Sc}, \quad \mathbf{u} \cdot \mathbf{e}_w = 0, \quad \nabla C \cdot \mathbf{e}_w = 0 \tag{8}$$

together with prescribed wall temperature and flux conditions [9, 10].

In an actual experiment, owing to the finite length of the ampoule, there is a gradual decrease in length of the melt zone during growth. In this model, transient effects related to the change in melt length are ignored. This assumption is referred to as the quasi-steady assumption and is frequently used in melt-growth modelling. The thruster firings are simulated using impulsive accelerations which are introduced through the time-dependent body-force term $\mathbf{g}$. For the OMS burn, the acceleration is suddenly raised up from a nominal level of $10^{-6}g$ to $2.7 \cdot 10^{-2}g$, and held constant for 35 seconds and then set back to the nominal level. Other magnitudes and orientations of impulsive accelerations are also examined.

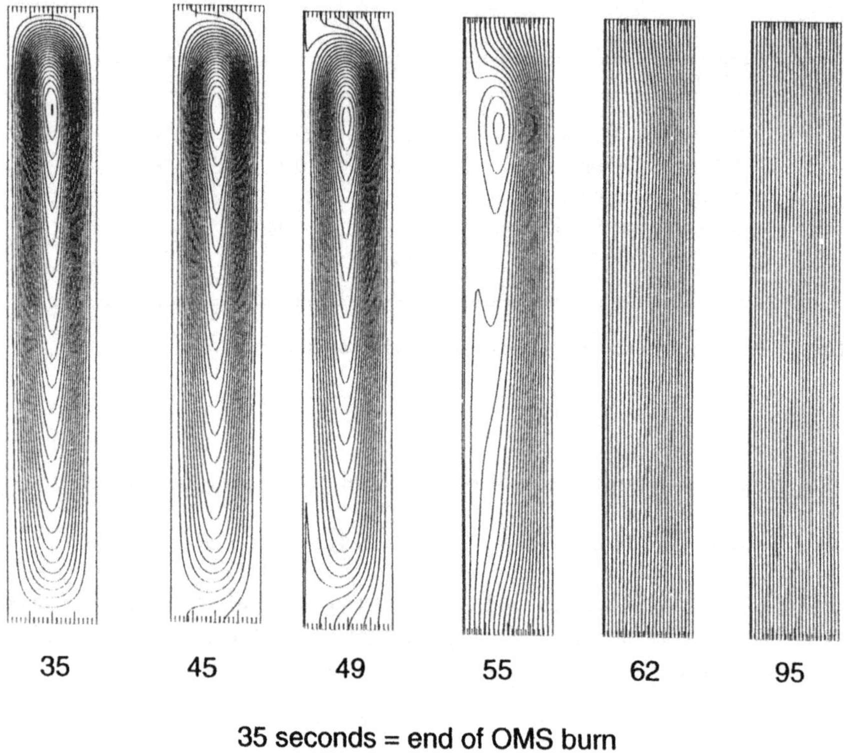

Fig. 3. The stream function at different times associated with a 35 second $2.7 \cdot 10^{-2}g$ disturbance oriented at 15° to the furnace axis.

3. Results

The stream function at different times associated with a 35 second $2.7 \cdot 10^{-2}g$ impulse acceleration oriented at 15° to the furnace axis is shown in Fig. 3. The initial condition was diffusion dominated. The $10^{-6}g$ background acceleration results in negligible flow velocities in comparison to the translation velocity of $2\mu ms^{-1}$. Under these diffusive conditions, the stream function is steady and is represented by streamlines parallel to the ampoule walls. The concentration profile is laterally uniform and follows the classical exponential profile as it decays into the melt. After the impulse begins, the stream function changes to a single cell pattern as shown in Fig. 3. At 35 seconds (the end of the impulse) the fluid velocities reach a maximum value and then the flow rapidly decays in strength until, at 62 seconds no noticeable deviation from purely diffusive conditions is apparent. The response of the concentration field is shown in Fig. 4 at different times. Note the increase in lateral composition nonuniformity even after the termination of the disturbance and after the decay of the velocity fields. Furthermore, it is evident that the system has not returned to its initial condition even after 1280 seconds. Indeed, Fig. 5 which depicts the average interface concentration (initial value 3.846) and lateral segregation as a function of time, shows that it will be well in excess of 3000 seconds before the initial conditions are retrieved.

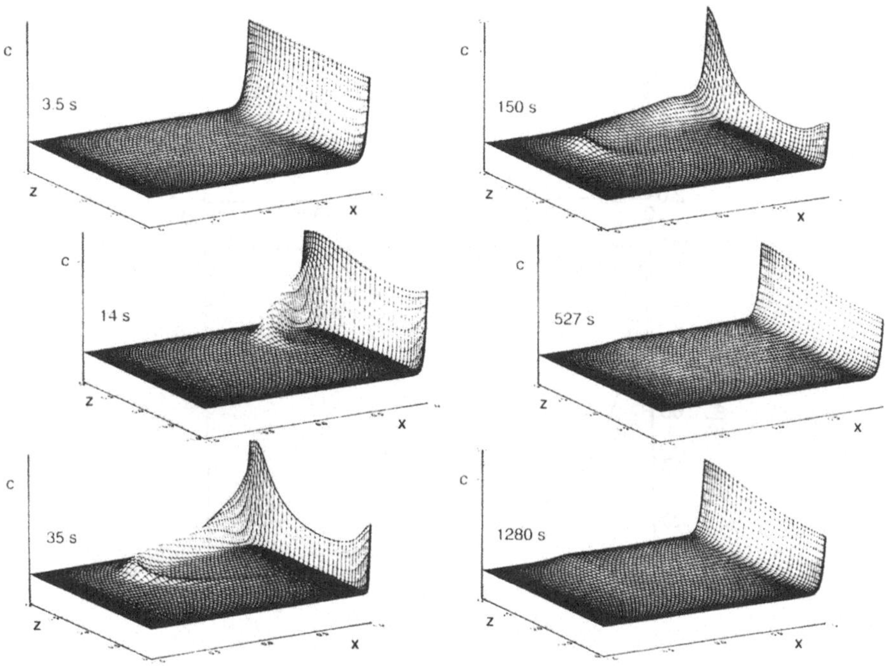

Fig. 4. The concentration field at different times associated with a 35 second $2.7 \cdot 10^{-2} g$ disturbance oriented at 15° to the furnace axis. The right hand side of each diagram reflects the uniformity at the solidification interface.

Figure 6 summarizes results for a growth rate of $2\mu ms^{-1}$ for different magnitude and duration impulses oriented parallel to the interface. Here the time has been made dimensionless using $D/k^2V_g^2$. This time scale is obtained as follows. The length that must be solidified before the interfacial concentration reaches its steady state value when starting from a melt of constant composition is given by [13] $x^* = D/kV_g^2$. Taking this to represent approximately the length that must be solidified before steady state is reached following the impulse disturbance we find that the characteristic diffusion time is x^*2/D or $D/k^2V_g^2$. The curves show that the time taken for the system to completely recover from the impulse acceleration is, at least for the smaller magnitude responses, on the order of $D/k^2V_g^2$. For the larger disturbances to the concentration a longer recovery time is necessary. This is probably related to the extent to which the convective flow has carried solute away from the crystal melt interface and into the melt.

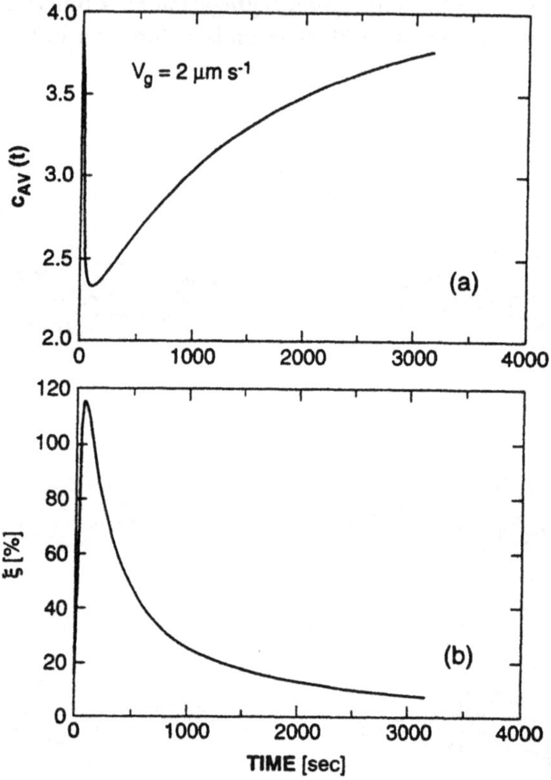

Fig. 5. Average interface concentration (a) and lateral non-uniformity (b) as a function associated with a 35 second $2.7 \cdot 10^{-2}g$ disturbance oriented at 15° to the furnace axis.

Two other features of the curves in Fig. 6 should be mentioned. The first is that, for a given growth velocity, the time taken to reach a minimum value of c_{av} is the same for all cases. The second feature is that for the same value of the product of the acceleration magnitude and duration, the history of c_{av} is the same. However, comparison of the actual concentration fields for any given time shows that they are different.

4. Discussion

Comparison of the simulated response of the average interfacial concentration to the average concentration obtained from the Seebeck measurement shows reasonable agreement. The time taken to reach a minimum average concentration (on the order of 100 seconds in the simulation) and the change in average concentration (38%) agree quite well with the measured response. Evaluation of the predicted time taken for the system to return to its initial state is more

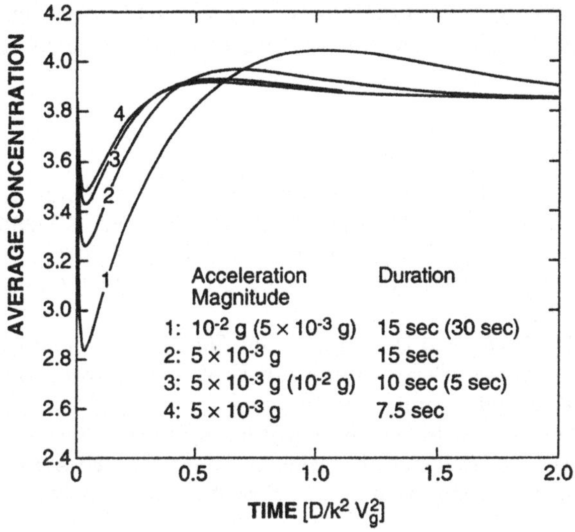

Fig. 6. Average interface concentration as a function of dimensionless time associated with various impulse disturbances.

difficult due to the fact that when the OMS burn ocurred the system was in a transient. The simulation predicts that even though the acceleration disturbance lasted only 35 seconds, the concentration disturbance lasted for several thousand seconds. This is consistent with our earlier results [7] for the impact of thruster firing events and is comparable to the characteristic diffusion time for the system. The behavior of the average concentration in response to impulse accelerations is analogous to the response first described by Tiller et al. [13] to a increase in solidification rate from V_1 to V_2 and back to V_1. An important feature of the simulated results which cannot be deduced from the Seebeck measurement is that the lateral interfacial composition nonuniformity, which for a flat interface would normally only arise from convective contributions, continues to increase even after the convective disturbance has died out. This behavior is explained in terms of the influence of diffusion on the coupling between rejection of solute at the (still planar) interface, and the local variation in concentration gradient initially produced by the convective disturbance which sweeps solute out into the melt and decreases the average concentration gradient. The nominal acceleration of $10^{-6}g$ results in insignificant convective velocities in comparison to the growth velocity and cannot account for this behavior.

Acknowledgments

The author would like to acknowledge support from the National Aeronautics and Space Agency under grant NAG3 1740 and from the state of Alabama through the Alabama Supercomputing network and the Center for Microgravity and Materials Research. I would also like to thank J.J. Favier and J.P. Garandet for fruitful discussions and for making their experimental data accessible. Thanks are due to Ahmed Elshabka who undertook some of the calculations.

References

1. H. Hamacher, R. Jilg and U. Merbold, in: *Proc. 6th European Symposium on Materials Sciences under Microgravity Conditions*, **ESA SP-256** (1987) 413.
2. J. I. D. Alexander and C.A. Lundquist *AIAA J.* (1988) 193.
3. M.J.B. Rogers and J.I.D. Alexander and J. Schoess, *AIAA J. Spacecraft and Rockets* **28** (1992) 52.
4. R. Monti, J.J. Favier and D. Langbein, in: *Fluid Sciences and Materials Sciences in Space, A European Perspective*, Ed. H.U. Walter Springer, Berlin 1987, 237.
5. J.I.D. Alexander, *Microgravity Science and Technology* **3** (1990) 52.
6. V. I. Polezhaev, A.P. Lebedev, and S.A. Nikitin, in: *Proc. 5th European Symposium on Materials Sciences under Microgravity Conditions*, **ESA SP-222** (1984) 237.
7. J.I.D. Alexander, J. Ouazzani and F. Rosenberger, *J. Crystal Growth* **97** (1989) 285.
8. J.I.D. Alexander, S. Amiroudine, J. Ouazzani and F. Rosenberger, *J. Crystal Growth* **113** (1991) 21.
9. D. Thevenard, and J.J. Favier,in: *Proc. 7th European Symposium on Materials Sciences under Microgravity Conditions*, **ESA SP-295** (1990) 243.
10. J.J. Favier, J.P. Garandet, A. Rouzaud and D. Camel, *J. Crystal Growth* **140** (1994) 237.
11. J.J. Favier and A. Rouzaud, *Revue Phys. Appl.* **22** (1987) 713.
12. J.J. Favier, P.Lehmann, B. Drevet, J.P. Garandet, D.Camel, and S.R. Coriell, this volume.
13. W. A. Tiller, K.A. Jackson, J.W. Rutter and B. Chalmers, *Acta Metall.* **1** (1953) 428.

Growth of 20 mm Diameter GaAs Crystals by the Floating Zone Technique During the D-2 Spacelab Mission

G. Müller and F.M. Herrmann

Kristallabor, Institut für Werkstoffwissenschaften,
Universität Erlangen-Nürnberg, Martensstr.7, D-91058 Erlangen, Germany

Abstract

Five GaAs single crystals Si-doped and undoped with diameters of 20 mm were grown by the floating-zone-technique (FZ) under microgravity during the second German Spacelab Mission D2. The GaAs rods were sealed in silica ampoules which contained an integrated As-source to provide controlled stoichiometry conditions. The heating system consisted of a specially designed mirror furnace. The results show that stoichiometric material was achieved within the limits of error of a coulometric analysis. The occurrence of dislocation networks could be strongly reduced and nearly avoided. Both oxygen and boron contents of the space-grown crystals are very low. Results of numerical calculations of the curvature of the interface, the distribution of temperature and stress in the crystal are shown to be in good agreement with the experiments.

1. Introduction

GaAs is the most important semiconducting material for the production of optoelectronic devices and high-speed electronic circuits. Further progress in these fields of applications, however, is strongly depending on improvements in the quality of the crystals with respect to density and distribution of dislocations, control of residual impurities and uniformity of the relevant properties. To achieve these goals, various problems in the production of low defect GaAs single crystals have to be solved.

The floating-zone-method (FZ) is a well suited growth technique in order to study some of the effects occuring during growth which are causing the above mentioned deficiencies of the GaAs single crystals. FZ has several advantages compared to the established crystal growth production techniques (LEC, HB and VGF [1]) such as:

[1] LEC - acronym of the Liquid Encapsulated Czochralski growth technique, HB stands for Horizontal Bridgman growth and VGF for Vertical Gradient Freeze method. (Eds. note)

- no direct crucible-crystal or crucible-melt interaction,
- advantageous segregation profiles of zone refining,
- excellent control of the melt composition by an As-vapour pressure source due to the large surface to melt volume ratio,
- no process induced B-contamination (B_2O_3-free technique).

Although the FZ-technique seems to be superior to the other growth techniques with respect to these advantages, it has not gained any importance for the growth of GaAs. The reason for that is the limitation by the hydrostatic pressure [1], permitting only the growth of crystals with a diameter smaller than 7-8 mm on earth. This drawback, however, can be totally overcome under microgravity conditions, where crystals can be grown with much larger dimensions, allowing to study phenomena and mechanisms which presently limit the terrestrial growth technology. In this paper we report for the first time about the growth of GaAs crystals of 20 mm diameter by the FZ-technique with a controlled As-vapour source. These crystals were grown during the German Spacelab mission D2 in 1993 and intensively characterized subsequently.

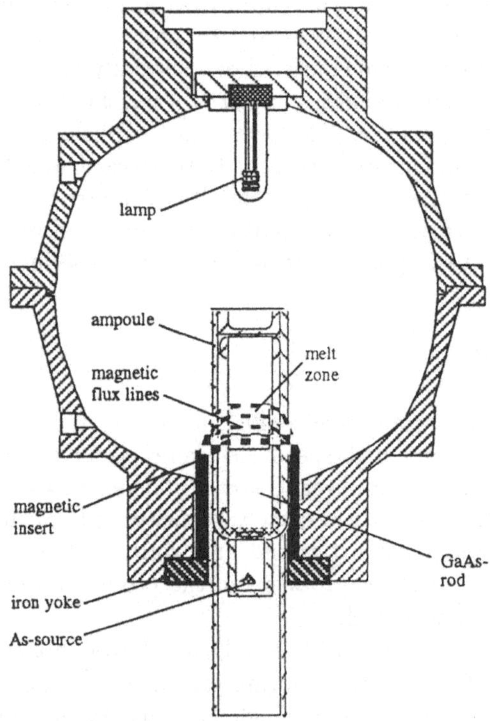

Fig. 1. Sketch of the mirror furnace PARELLI containing the growth ampoule. The magnetic insert (black and dark shaded) was adapted in the lower opening of the furnace.

2. Crystal growth experiments

The experimental set-up of the FZ-crystal growth configuration is illustrated by Fig.1 and described in more detail elsewere [2]. Cylindrical single crystalline (LEC) GaAs-rods doped with $[Si]=1\cdot 10^{18}cm^{-3}$ or undoped semi-insulating, with a length of 86 mm and a diameter $\phi = 20$ mm are fixed at both ends by graphite rings and centred inside a silica ampoule with a distance of 4 mm to the wall. A separately heated As-source was used to achieve a constant $T_{As} = 617°C$ which should correspond to an As-vapour pressure providing stoichiometric GaAs. The growth runs were carried out in a specially designed mirror furnace with a halogen lamp producing a ring-shaped focus on the rim of the GaAs zone (Fig.1). Growth is performed by pulling the ampoule out of the furnace via a translation mechanism with a constant rate of 6.0 or 8.4 mm/h, respectively. Figure 1 also shows the magnetic insert close to the ampoule wall, which was optionally used in order to damp the Marangoni-convection in the melt. The growth runs were controlled from the ground by the authors at the German Space Observation Center (GSOC) in Oberpfaffenhofen via telecommand and real time TV.

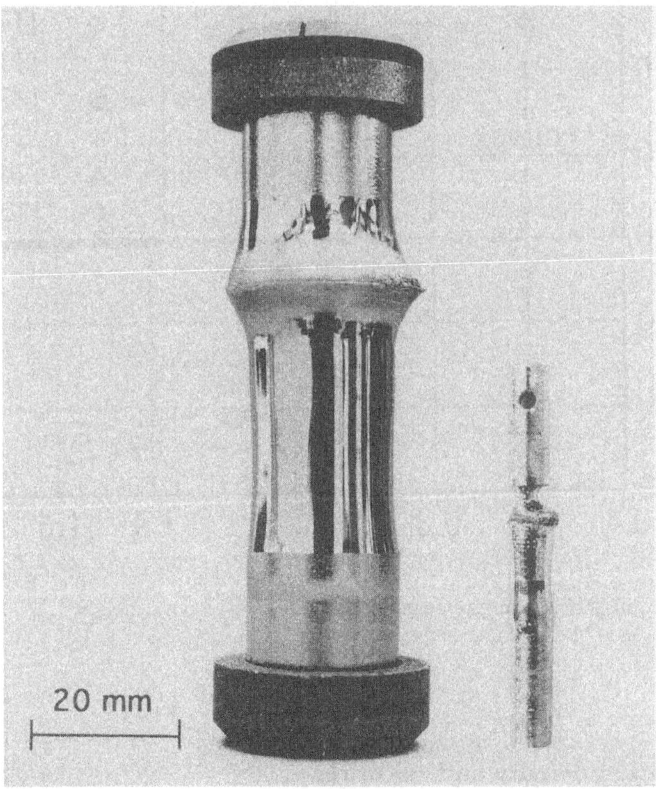

Fig. 2. Photograph of a GaAs crystal grown during the D2 mission by the floating zone technique (left) in comparison with an terrestrial specimen of maximum size (right).

Two undoped < 100 >-oriented and three Si-doped < 111 >-oriented GaAs single crystals were grown successfully. A specimen is shown in Fig.2 in comparison with a crystal with the largest possible diameter to be grown on earth by FZ. With the exception of one run (112F) all experiments were carried out using the magnetic field.

3. Experimental results

3.1 Shape of the phase boundary

All five as-grown crystals showed a glossy surface (compare Fig.2) without any traces of surface depositions. The shape of the initial phase boundary was studied by DSL-etching [3] of longitudinal sections. The correlation between the curvature of the interface in terms of the deflection DB (defined as maximum deviation from the flat interface) and the aspect ratio height over diameter h/d of the melt zone, is given in Fig.3.

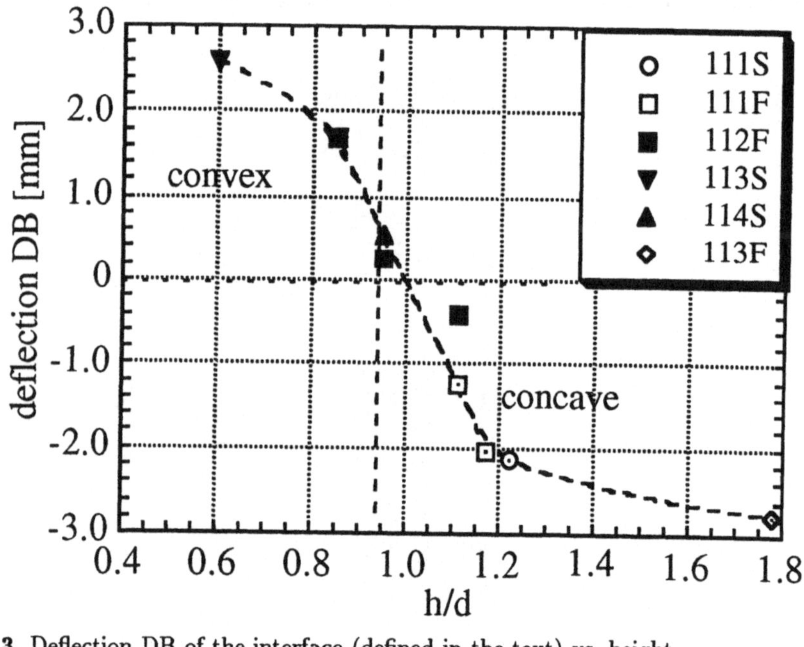

Fig. 3. Deflection DB of the interface (defined in the text) vs. height to diameter ratio h/d.

3.2 Dislocation density and networks

A reduction of the dislocation density (EPD) was not a major goal of these FZ experiments, because mirror furnaces typically create relatively high thermal gradients and stresses, respectively. Nevertheless, the best achieved value of EPD=

$5 \cdot 10^3 cm^{-2}$ is notable if one considers the high cooling rate ($\approx$ 40 K/min), which had to be used because of the tight timeline of the spacelab mission. Axisymmetric calculations of the temperature and stress distribution in the growing crystals were carried out using the commercial finite element code ABAQUS [4]. The model considers heat transport by conduction and radiation. Results of the calculations of temperature and stress in the crystal are given in detail elsewhere [2]. The calculated curvatures of the interfaces agree qualitatively well with the experimental results of Fig.3. The maximum of the von Mises stress for the convex interface σ_{max}= 13 MPa is at the rim of the crystal, whereas σ_{max} = 11 MPa was obtained in the center of the crystal for the concave interface. A comparison of the calculated radial distributions of the von Mises stresses and the dislocation densities (EPD) of crystals grown with concave and convex interfaces gives a qualitative correlation. PL-mapping at 77K were carried out by Baeumler et al. [5]. In Figs.4 and 5 the results of line-scans of wafers taken from the undoped LEC seed material (113S) (Fig.4) and a space-grown wafer of the same crystal are shown. The comparison clearly indicates an improvement of the lateral uniformity of the space grown material due to an increase in the average cell size of the dislocation network. An increase in the average size of the cell pattern in the PL-topogram was also reported by Cröll et. al. [6], who carried out a FZ-GaAs experiment without an As-source during the same mission. It could only be found in samples with a low EPD ($< 1 \cdot 10^4 cm^{-2}$). This observation supports the idea that the effect is correlated with low stress conditions, which has already been described in GaAs grown by VGF [7].

3.3 Doping striations

Pronounced dopant striations were revealed on the longitudinal sections by means of DSL-etching [3] in both kind of crystals grown with and without a magnetic field. The occurrence of doping striations in the space-grown crystals mainly originate from unsteady Marangoni-convection. The results of numerical simulations of the Marangoni-convection for the D2 conditions, published already earlier by our group predicted that a magnetic field of 130 mT would be necessary to suppress unsteady thermocapillary flows [8, 9]. Unfortunately it was not possible under the technical constraints of the D2 mission to build a magnet providing this desired field strength. Only a non-uniform field from 20 to 80 mT was available. Nevertheless, we expected a considerable damping of the flow fluctuations which obviously was not the case. Only one type of striations, correlated to a low fluctuation frequency of 0.008 Hz were suppressed. This result needs a more detailed analysis especially by numerical modelling, which will be presented in a forthcoming paper.

3.4 Residual impurities, stoichiometry and EL2

The results of spark source mass spectroscopy (SSMS) measurements by Wiedemann [10] and of local vibration mode (LVM) infrared absorption spectrometry by Alt [11] from crystal 113S are given in Tab.1. The detected concentrations

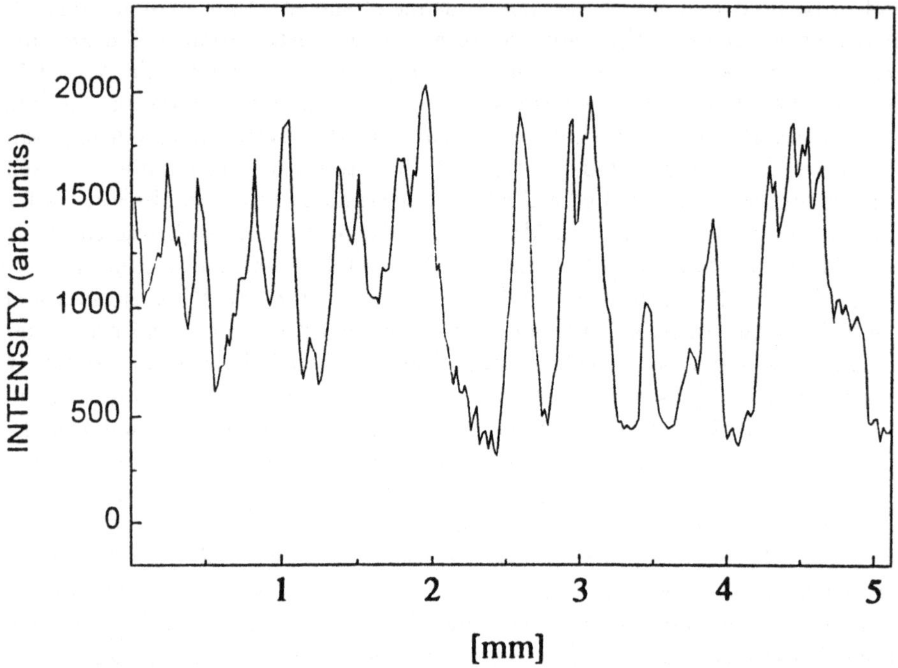

Fig. 4. Line-scan in the photoluminescence topogram measured at a sample taken from the seed part (LEC-GaAs) of the FZ crystal 113S.

for boron and oxygen are extremely low, in the case of boron even 3 orders of magnitude lower than in the starting material. In all nominally undoped samples a silicon content was found in the range of $[\mathrm{Si}] \approx 2 \cdot 10^{16}\mathrm{cm}^{-3}$ leading to the n-type behaviour mentioned above. The semi-insulating nominally undoped starting material was changed into semiconducting by the FZ growth conditions. The resistivity ρ of the space-grown crystals is obviously determined by the Si-contamination ($[\mathrm{Si}] = 10^{16} - 10^{17}\mathrm{cm}^{-3}$), which presumeably originates from the wall of the silica ampoule. Although such an effect is known from terrestrial horizontal Bridgman growth in resistance furnaces [12], it was not expected in the case of the mirror furnace, as this is a cold wall heating facility. The carbon concentration in our space-grown material was found to be slightly increased in comparison with the starting material, while the EL2 concentration measured by means of DLTS was found to remain unchanged ($[\mathrm{EL2}] \approx 1 \cdot 10^{16}\mathrm{cm}^{-3}$). Thus the ratio of the concentrations of C and EL2 would have been sufficient to grow semiinsulating FZ material if the Si-contamination could have been kept below $10^{16}\mathrm{cm}^{-3}$. The oxygen concentration found by SSMS [10] is very low when compared to GaAs grown by low pressure LEC. It is very likely to be interstitial oxygen and thus electrically inactive, because it could not be detected by means of LVM ($[\mathrm{O}] < 10^{14}\mathrm{cm}^{-3}$) [11]. The achieved B-concentrations, being below the detection limit of $2 \cdot 10^{14}\mathrm{cm}^{-3}$, shows that B can be removed to a high degree

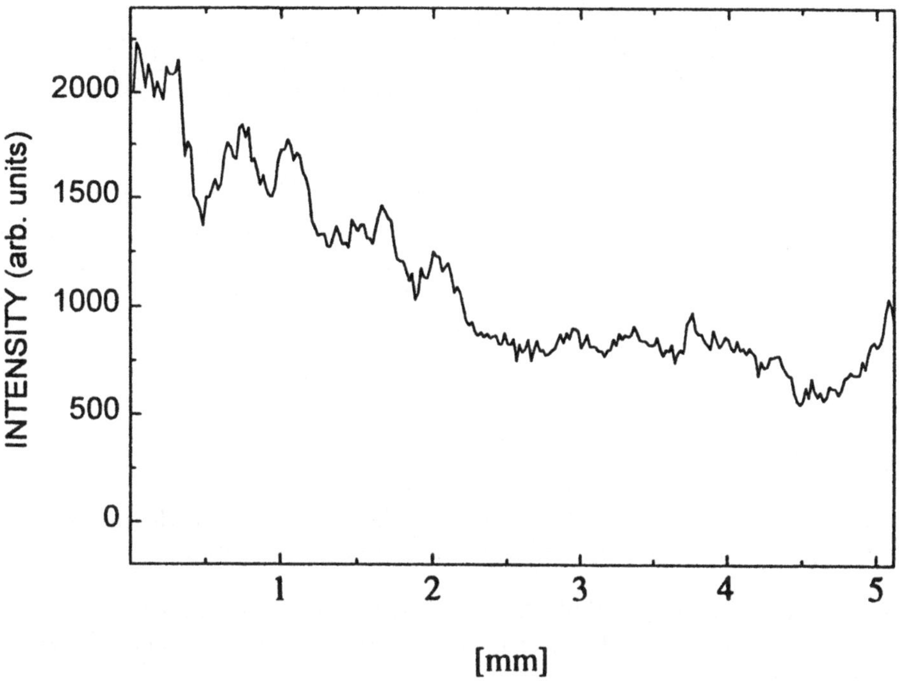

Fig. 5. Line-scan in the photoluminescence topogram measured at a space-grown sample of the FZ crystal 113S. The measurements were carried out by Baeumler et al. [6].

by the FZ technique. The space-grown samples have the lowest B-concentration ever obtained in GaAs to the knowledge of the authors. This is of technical interest, as B was identified to be detrimental to the thermal stability of semi-insulating GaAs during annealing steps [13] and is lowering the reproducibility of FET-performance [14]. Coulometric analysis by Oda and Kainosho [15] show that the use of the As-source leads to stoichiometric material within the limits of error of 10^{-5}.

4. Conclusions

The results of this paper demonstrate that crystals such as GaAs can be grown with larger diameters by the FZ-technique under microgravity with terrestrial control by TV transmission and telecommand power control without astronaut involvement. The properties of the space-grown crystals are remarkable with respect to the strong reduction of the formation of dislocation cell structure and the extremely low boron content. Thus the spacelab D2 experiments on the FZ growth of GaAs with controlled As-source contribute to the study and solution of crystal growth problems which are limiting the terrestrial production of high quality GaAs substrates.

sample	Si [cm^{-3}]	O [cm^{-3}]	C [cm^{-3}]	B [cm^{-3}]
113S/seed	$< 1.7 \cdot 10^{14}$*	- -	$0.9 \cdot 10^{15}$**	$4.8 \cdot 10^{17}$*
113S/A1 (a)	$2.5 \cdot 10^{16}$	$4.0 \cdot 10^{15}$	$3.1 \cdot 10^{15}$***	$< 2.2 \cdot 10^{14}$
113S/A1 (b)	- -	$< 3.7 \cdot 10^{15}$	$2.5 \cdot 10^{15}$	- -
113S/A2 (a)	$2.1 \cdot 10^{16}$	$3.7 \cdot 10^{15}$	$2.5 \cdot 10^{15}$***	$< 2.2 \cdot 10^{14}$
113S/A2 (b)	- -	$< 3.7 \cdot 10^{15}$	$2.5 \cdot 10^{15}$	- -

Table 1. Results of SSMS- (a) and LVM (b) measurements of samples of the nominally undoped crystal 113S (A1: axial position x = +5.5 mm, A2: axial position x = +17.8 mm) (*) SIMS, (**) LVM. (***) the values of C are given under the assumption of a relative sensitivity coefficient RSC of 3 [12].

Acknowledgement

The authors are very grateful to various people who supported the preparation of the space experiments and the characterization of the crystals:

- Mr. Zöbelein and Mr. Heil (Siemens AG) for the ampoule preparation.
- Mrs. G. Neuner and Mr. P. Olbricht (this institute) for assistance in sample preparation.
- the astronauts and the teams of DLR, Dornier, MBB ERNO and KT for their support before and during the mission.
- Mr. Hammer, Mrs. Kumann, Dr. Rotsch and Dr. Weinert (FEW, Freiberger Elektronikwerkstoffe) for sample preparation.
- Prof. Alt (FH Munich), Dr. Baeumler, Dr. Forker and Dr. Jantz (IAF Freiburg), Prof. Tanner (Uni Durham), Dr. Weyher (KFA Jlich) and Dr. Wiedemann (Uni Frankfurt) for characterization of the samples.
- Dr. Oda and K. Kainosho (Japan Energies) for coulometric stoichiometry analysis.

Financial support was granted by the federal minister of research and technology (BMFT) under the project management of the German space agency (DARA) , contract no. 50 QV 90580.

References

1. W. Heywang, *Z. Naturforschung* **11** (1956) 238 - 243.
2. F. M. Herrmann and G. Müller, *J. Cryst. Growth* (accepted for publication)
3. J. Weyher and J. van de Ven, *J. Cryst. Growth* **63** (1983) 285 - 291.
4. ABAQUS Manual, Version 4.8., Hibbitt, Karlson & Sorensen, Inc.
5. M. Baeumler, J. Focker and W. Jantz (private communication).
6. [6] A. Cröll, A. Tegetmeier, G. Nagel, K. W. Benz, *Cryst. Res. Technol.* **29** (1994) 335 - 342.
7. U. Voland, C. Frank, G. Gärtner, V. Klemm, J. Klöber, G. Kühnel, unpublished.

8. J. Baumgartl, M. Gewald, R. Rupp, J. Stierlen and G. Müller, Proc. 5th Europ. Symp. on Mat. and Fluid Sci. in Microgravity (ESA SP-295) (1989) 47- 58.
9. F. M. Herrmann, J. Baumgartl, T. Feulner and G. Müller, Proc. VIIIth Europ.Symp. on Mat. and Fluid Sci. in Microgravity (ESA SP-333) (1992) 57 - 60.
10. B. Wiedemann (private communication).
11. H. Ch. Alt (private communication).
12. C. N. Cochran and L. M. Foster, *J. Electrochem. Soc.* (1962) 149 - 154.
13. M. Baumgartner, K. Löhnert, G. Nagel, H. Rüfer and E. Tomzig, *Inst. Phys. Conf. Ser.*, **91** (1987) 97.
14. R. Anholt and T. W. Sigmon, *J. Electronic Materials* **17** (1988) 5.
15. O. Oda and K. Kainosho (private communication)

Microstructure Evolution in Immiscible AlSiBi Alloys Under Reduced Gravity Conditions

L. Ratke[1], *S. Drees*[1], *S. Diefenbach*[1], *B. Prinz*[2] *and H. Ahlborn*[3]

[1] Institut für Raumsimulation, DLR, 51140 Köln, Germany
[2] Metall-Laboratorium, Metallgesellschaft, Reuterweg 14, Frankfurt, Germany
[3] TU Hamburg Harburg, Germany

Abstract

Strip cast Al-Si-Bi alloys were directionally melted and solidified in the Isothermal Heating Facility (IHF) of the Werkstofflabor on board the Spacelab mission D2. The main objective of the experiment was to investigate the microstructure evolution during melting and solidification in alloys exhibiting a miscibility gap in the liquid state. The microstructure rearranges from an evenly distributed dispersion of soild Bismuth particles in a solid AlSi matrix due to thermocapillary motion of Bi-rich droplets during melting and solidification in microgravity. The strip casting process provides a material with a well-defined particle dispersion. Two experiment runs were performed successfully. Samples with a Bi-content of 7 wt.% and two different melting rates were melted and solidified. Their as-solidified microstructures clearly show that the Bi droplets moved by thermocapillary motion as anticipated. The mean size of the droplets increased by a factor of two and the droplet size distribution exhibits a maximum with very big droplets, revealing that the coalescence of drops contributed to the coarsening of the dispersion. The results are analysed with a numerical modell for such alloys: the Discrete Multi-Particle Approach (DMPA) is presented in some detail and applied to the results.

1. Introduction

Alloys being immiscible as melts like aluminium with additions of lead or bismuth are currently investigated as potential candidates for advanced bearings in car engines. These bearing materials are composed of a tough, hard matrix, which sustains the high dynamically varying combustion pressure and soft particles evenly distributed, which have to embed dirt and dust particles stemming from wear inside the motor. The alloys used today in all car engines are based on bronze with lead distributed irregularly in the interdenrtic regions. Future car engines operating at higher temperatures and pressure would require a bearing material with a lower coefficient of friction and wear and also one that can sustain higher dynamical pressures as compared to bronze-lead alloys. These engines could run with a higher efficiency, consume less gasoline and emit less pollutants. These new dispersion alloys are just on the verge of being developed industrially, originating from research results with immiscible melts obtained during the last twenty years under microgravity conditions [1, 2, 19, 20].

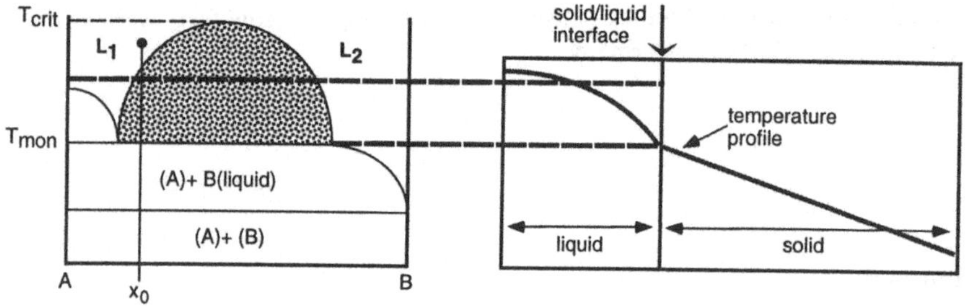

Fig. 1. Scheme of the relation between a typical phase diagram with a miscibility gap of a monotectic alloy and the solidification of the latter. Ahead of the solid/liquid interface is a reasonable large region where two melts of different composition are in equilibrium.

Casting of liquid immiscible alloys by classical casting has often been tried, but was never successful. The process of decomposition of the melt being homogeneous at high temperatures into two melts being immiscible at lower temperatures and which are generally of different density is very rapid on earth (within a few seconds). Gravity arranges the melts in two layers. The absence of gravity offers the unique possibility to disperse the soft minority phase evenly in the matrix melt. Therefore these alloys were investigated from the very beginning of research in space. The results of space research were quite surprising. Even in space a bulk separation of the immiscible alloy melts was observed. The reasons for this became clear through intensive experimental and theoretical research, revealing especially that Marangoni motion is of tremendous importance during decomposition and casting processes even on earth [1]. This led recently to the invention of a new casting process by the Metallgesellschaft, Frankfurt, especially designed such that the droplet motion due to gravity is counteracted by Marangoni motion. This so-called strip casting process is for the first time able to produce endless strips of aluminium-lead or aluminium-bismuth alloys [2, 3].

The full potential of these materials is still not exploited to its full extent, since this would require the casting of strips with a higher content of soft phases than possible today, i.e. the alloys should be hypermonotectic to a large extent (around 5 vol% soft phase at the monotectic temperature). In order to cast such alloys in a well controlled manner the coarsening of the dispersion by collision and coagulation events between droplets has to be controlled. These events, however, are determined by the drop motion relative to the matrix (Stokes and Marangoni motion). Concurrently with nucleation and growth coagulation processes occur in a zone ahead of the solidification front depicted schematically in Fig.1. In most technically important Al-based bearing alloys a volume fraction of the second phase liquid around 5 % leads to a temperature difference of the order of 200 to 300 °C between the entry into the miscibility gap during cooling and the monotectic temperature at which the alloy solidifies via the monotectic reaction. Even with steep temperature gradients of 100 to 200 K/cm ahead of the solidification front the two phase region has a size of about a centimeter.

Multi-droplet Marangoni motion (and coagulation), however, is rarely understood theoretically, although a theory developed by Anderson [16] seems to treat this problem at least for low volume fraction of dispersed phase. The theoretical approach of Anderson is nowadays questioned theoretically [21]. Especially there are no experimental investigations dealing with multi-droplet Marangoni motion. Only two-droplet interactions were treated theoretically by especially the group of Davis [22].

Therefore we designed an experiment for the German Spacelab mission D-2 flown in April 1993 as one step in a sequence of experiments to investigate the effect of Marangoni motion, drop dissolution, growth, coarsening and coagulation in technically important bearing alloys on the final solidified microstructure.

2. The experimental set up and performance

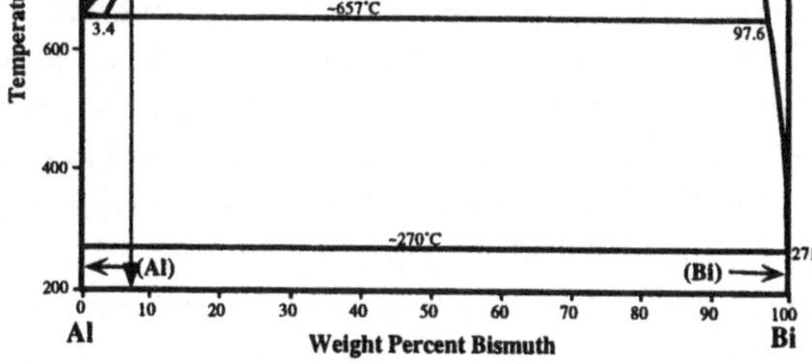

Fig. 2. Phase diagram of the binary alloy AlBi.

The investigations were performed with Al-Si-Bi alloys. Figure 2 shows the binary phase diagram Al-Bi. Above 1037 °C aluminium and bismuth completely dissolve in each other forming a homogeneous melt. Below, there exists a miscibility gap in the liquid state. An Al-rich melt coexists with a Bi-rich melt, which has the higher density. At 657 °C the Al-rich melt decomposes into solid Al and a Bi-rich melt (monotectic reaction). At 237 °C the Bi-rich melt itself solidifies.

In view of possible technical applications the base metal used was not pure Al but Al+5wt.% Si. Two samples were processed in the so-called Isothermal Heating Facility (IHF) of Spacelab. Within the samples (diameter 6.8 mm, length approx. 140 mm) there were two zones of different compositions (Fig. 3). One of them had an excess of Bi compared to the monotectic composition (7 wt.% Bi) and the other was of exact (ternary) monotectic composition (2.4 wt.%).

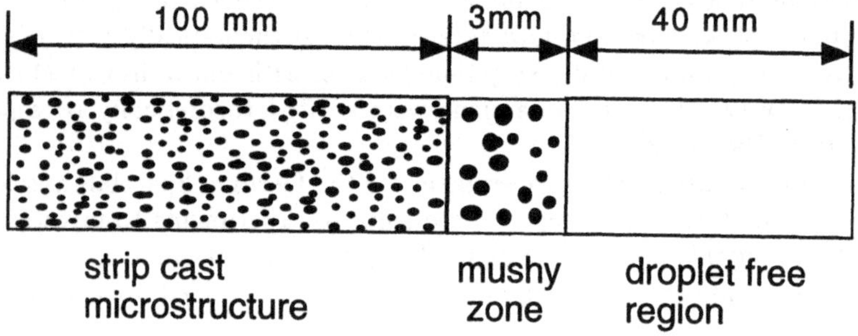

Fig. 3. Sample set-up for the experiment on D-2.

The samples were strip cast at the Metallgesellschaft as described in [2, 3] ensuring that the Bi-excess was distributed evenly in the material as droplets of well defined size (3 μm mean radius) as by a pre-flight numerical analysis of the experiment [4]. The samples were molten in a controlled way with an axial temperature gradient.

One of the measured temperature/time profiles is shown in Fig. 4. The temperatures were measured by four thermocouples fixed in small channels at the sample surfaces. The furnace was heated to the set temperature and then the cooling device (CD) was pulled 30 mm along the sample, thus exposing an equivalent length of the sample to the hot furnace. The temperature profile within the sample was then allowed to become stationary while holding the cooling device at this position for 15 min. Nominally the cooling device was at this time 10 mm ahead of the bismuth-rich zone. Then the cooling device was pulled further with the predefined speed (0.5 mm/min. and 1 mm/min.) for either 1 hour or 30 minutes such that in both cases a length of 20 mm of the bismuth-rich part of the samples was molten. Then the heater was switched off and after 10 minutes the direction of the cooling device was reversed.

In order to model the microstructure we have to know the temperature field as a function of position and time in the sample as accurate as possible. Therefore we set up the following thermal model to fit the measured temperature profiles.

During melting of a sample the temperature at the origin $z_0 = 0$ mm (hot end) increases continuously while the solid-liquid interface z_m moves with a predefined velocity $z_m(t) \approx \alpha_0 + \alpha_1 \cdot v_{CD}$. During solidification the temperature profile is inverted in its direction. A solution of the stationary heat diffusion

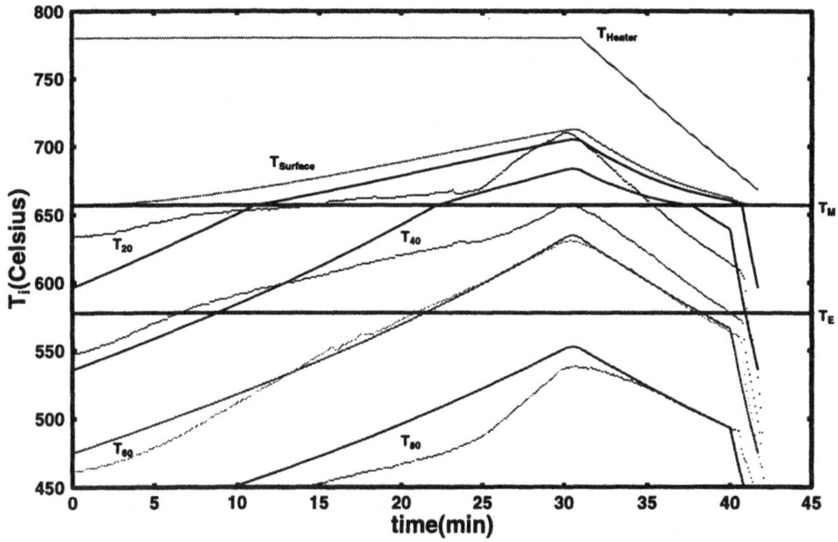

Fig. 4. Temperature profile as measured in the run with a translation speed of the cooling device of 1 mm/min. The experimental curves are indicated by points plotted every fifth second. The sampling rate was 1 Hz. The temperatures were measured at the surface of the sample located at positions given by the index numbers in mm (measured from the hot tip). The solid lines are the analytical fit based on to the heat diffusion equation with radiation boundary conditions.

equation in the molten part of the sample

$$\frac{\partial^2 T}{\partial z^2} = \frac{\varepsilon \sigma_B T_0^3 \cdot 2\pi R_s}{\lambda \cdot \pi R_s^2} \cdot (T - T_0) \ , \tag{1}$$

is fitted to the measured temperature profile. This equation is applicable if the rodlike sample is radially heated by insolation from the furnace heaters (kept at 780 °C) and the temperature difference between the sample and the heater $T - T_0$ is small. We assume as a boundary condition that the solid/liquid interface is kept at the monotectic temperature. The sample length exposed to the heater (and therefore the domain in which we solve (1)) changes as the cooling device moves. R_s in (1) denotes the sample radius, λ the heat conductivity, σ_B is the Stefan-Boltzmann constant and $\varepsilon = 0.3$ the emissivity. Within the solid part of the sample the temperature profile is assumed to be linear, decreasing from the monotectic temperature to approx. 50 °C at the cold junction being ($z_1 = 210$ mm away from the hot tip).

Carefully inspecting the measured temperature profiles shown in Fig. 4 we see that the thermocouples (TC) at 20 mm and 40 mm exhibit some bends after 5, 10 and 23 minutes. The thermocouple at 40 mm never records a temperature higher than the monotectic one. This would mean the sample should only be molten till that position. Inspection of the microstructure (see below) definitively shows that the sample was completely liquid over 55 mm measured from the hot tip. Thus the thermocouple denoted as T_{40} made false readings. Similar arguments hold for the thermocouple closest to the tip, T_{20}. The origin of such misreadings can be understood from the manner the TCs are fixed in the channels. If they become hot and the sample material melts, they can bend either towards the melt or the ceramic container which covers the sample. We observed in another sample a TC being 3 mm inside the sample. If a TC has contact to the ceramic tube it will record higher temperatures otherwise smaller ones. Therefore we took the thermocouple at 60 mm which has no serious breaks or bends in the recording as being most reliable and fitted the solution of (1) to it. The fits are shown in Fig. 4 as solid lines.

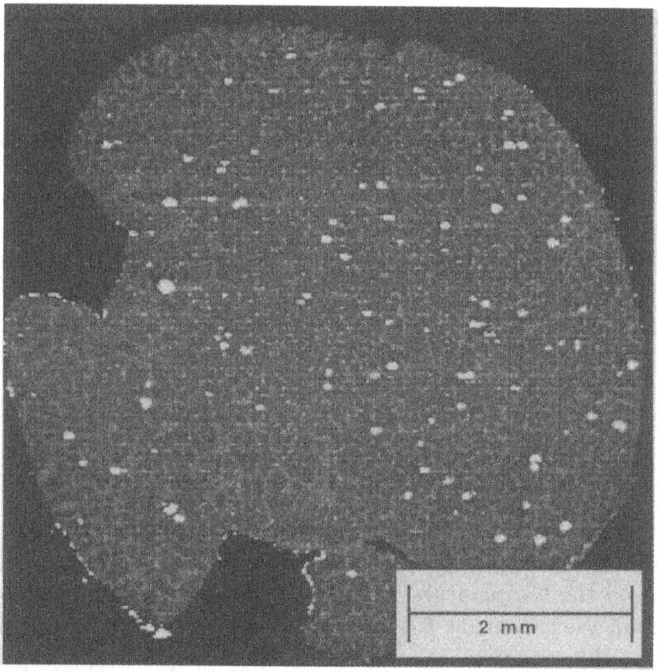

Fig. 5. SEM picture of the sample processed with a translation speed of 1 mm/min. The picture is of a cross section 50 mm away from the (hot) sample tip. The Bi-droplets are white. The dark grey features are the aluminium dendrites, the light grey features in between correspond to the ternary eutectic Al-Si-Bi. The two slots visible on the left and bottom side of the photo were milled into the sample material to hold the supplies of the thermocouples. One can clearly see, that during cooling they acted as heat sinks, because they are decorated by radial dendrites reaching far into the sample.

3. Experimental results

The microstructure in the samples was evaluated by Scanning Electron Microscopy (SEM). The rod-shaped samples were cut into 70 slices starting from the sample tip perpendicular to the samples long axis, each exactly being at a distance of 1 mm from each other. The slices were polished by a special micro-miller to a surface roughness better than 1 μm . Figure 5 shows a cross section taken 50 mm away from the tip. The bismuth droplets are light in a darker matrix. The features being darker grey are the aluminium dendrites. The interdendritic region consists of the Al-Si eutectic with very small Bi droplets stemming from the monotectic reaction. Thus the interdendritic regions are in a light grey. Near the channels one can observe that aluminium dendrites emerge from the channel surface radially into the sample, indicating a radial solidification there. This feature can be observed in any cross section having channels. The area of radial solidification with Al dendrites increases stepwise from the tip to the bottom of the sample in as much as the number of thermocouples increases.

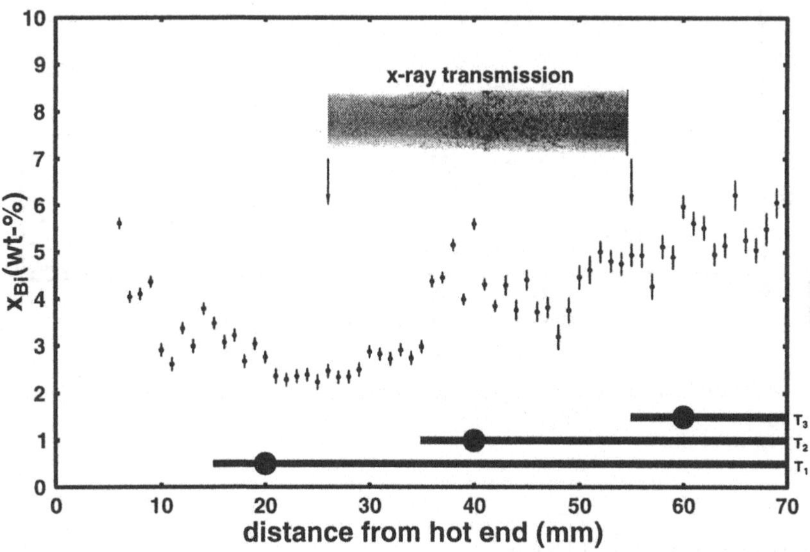

Fig. 6. Measured bismuth content in sample 3042 as a function of position. The insert shows an x-ray transmission bitmap of the corresponding part of the sample. The error bars denote the statistical error (2σ). The positions of the thermocouples are also plotted, labeled as T_i.

In each section the sizes of all droplets were measured, their mean radius and the size distribution determined. We also measured in each sample the Bi-content within the SEM by EDX. Figure 6 shows the measured concentration variation of bismuth for the sample processed faster. The Bi-content decreases from a value of approx. 7 wt.% at (z = 70 mm) to the monotectic composition at (z = 25 mm) . Near the hot end (z = 0 mm) the bismuth concentration increases again. There seem to be local enrichments of Bi at z = 40 mm. The composition shows a strong scatter from slice to slice which is far beyond the measurement error. This scatter is to be expected since the volume (area) fraction of bismuth is around 1-2 %. If by accident a cut did just not hit a big drop the bismuth content appears to be lower and a slice neighbouring will eventually show a larger bismuth concentration. This is also the reason that the maximum observable droplet size 7 shows a strong scatter.

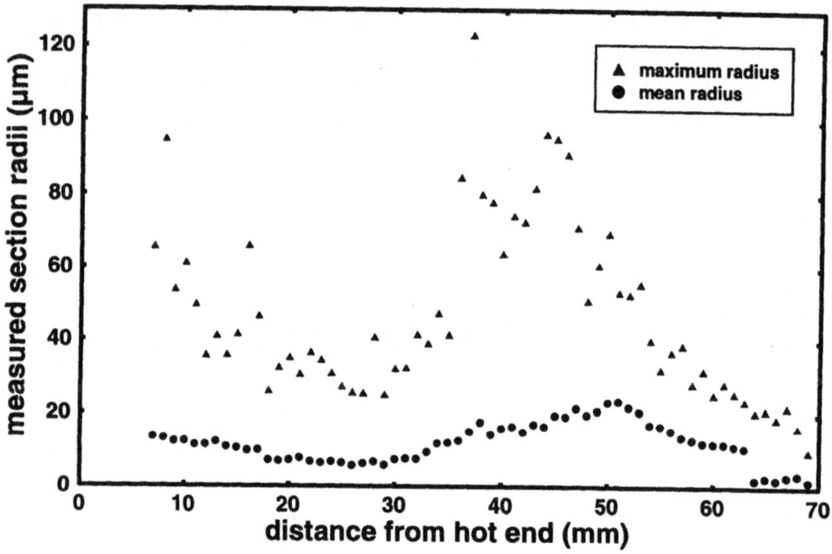

Fig. 7. Measured mean and maximum section radii as a function of position in the sample 3042.

4. Numerical modelling

The Discrete Multi-Particle Approach (DMPA) [12, 13], enables the numerical modelling of phase separation and solidification of immiscible alloys. The DMPA evolved from the desire to understand the complex interplay of the processes, mentioned in the introduction on the solid microstructure of cast immiscible alloys with a gross composition well above the monotectic one.

4.1 Modelling of the physical processes

Given a one dimensional temperature field $T(z,t)$ the unknown fundamental physical quantity in the sample is the matrix concentration field $c(\mathbf{r},t)$, if we neglect any concentration variation inside the drops. We artificially set the drop concentrations $c_i(\mathbf{r}), (1 \leq i \leq N(t))$ always to the value determined by the local temperature of the drop $T(z,t)$ and the phase diagram. Then the matrix concentration evolves in space and time according to:

$$\frac{\partial c}{\partial t} - \frac{\partial}{\partial z}\left(D(T)\frac{\partial c}{\partial z}\right) = Q\left(c, T, f\left(R, \mathbf{r}, t\right)\right) \tag{2}$$

Here $D(T)$ denotes the thermal diffusivity and Q describes the action of the moving drops as sources or sinks for solute atoms. Q itself is determined by c, T and the local number density of drops $f(R, \mathbf{r}, t)$, which also depends on radius R and time. The time dependence of the drop locations is ruled by thermocapillary motion (in the experiments considered here) and their hydrodynamic interaction; their radii change via growth or dissolution in the matrix concentration field. The solution of (2) is approximated through solving the homogenous part iteratively by a classical grid method while treating the source term Q with a discrete microscopic model.

4.2 Concentration field

The homogeneous equation is solved by a Gauss–Seidel iteration scheme of the discretized version of (2):

$$\frac{\partial \mathbf{c}}{\partial t} - \frac{\partial}{\partial z}\left(D(T)\frac{\partial \mathbf{c}}{\partial z}\right) = 0\,, \tag{3}$$

with $\mathbf{c} \equiv (c_0, c_1, \ldots, c_M)$, $M \approx 800$ and the differential operators discretized in the usual way, namely:

$$c_{i,k+1} = c_{i,k} + \frac{D_{i,k}\cdot\tau}{\zeta^2}\cdot(c_{i+1,k} + c_{i-1,k} - 2\cdot c_{i,k}) + \frac{F_{i,k}\cdot\tau}{2\zeta}\cdot(c_{i+1,k} - c_{i-1,k})\,. \tag{4}$$

Here $c_{i,k}$ denotes the matrix concentration in slice i at $z\,(i)$. The second index k stands for the time t. The $c_{i,k+1}$ is the concentration vector at $t = t + \tau$. $D_{i,k}$ corresponds to $D(T(z,t))$, and $F_{i,k} = \partial D(T(z,t))/\partial z$. The grid constant is ζ.

The source term is handled differently. the drops are treated as isolated, microscopic quantities compared to field variables like the concentration and temperature. In as much as they change their location they change their concentration and the local matrix concentration. They can either grow or shrink depending on the local matrix concentration. Growth or shrinkage of a drop is treated in the single particle approximation using the classical growth law of a drop in a supersaturated (or depleted) matrix:

$$\frac{dR}{dt} = \frac{D\,(T)}{R}\cdot\frac{\bar{c}\,(z,t) - c_{\mathrm{eq}}\,(\mathrm{L}_1, R)}{c_{\mathrm{eq}}\,(\mathrm{L}_1) - c_{\mathrm{eq}}\,(\mathrm{L}_2)}\,. \tag{5}$$

With $c_{\mathrm{eq}}\,(\mathrm{L}_1, R) = c_{\mathrm{eq}}\,(\mathrm{L}_1)\cdot\exp\,(\alpha/R)$, and $\alpha = 2\cdot\sigma\cdot\Omega/\,(k_{\mathrm{B}}\cdot T)$. Thus we take into account the effect of curvature on the solubility (Gibbs-Thomson effect). Here σ = interface tension, Ω = atomic volume, k_{B} and T have their usual meanings and "eq" denotes the equilibrium values. The equilibrium concentration depends on the local temperature and the shape of the binodal line. The introduction of the average matrix concentration $\bar{c}(z,t)$ is an approximation to the real multi-particle diffusion problem.

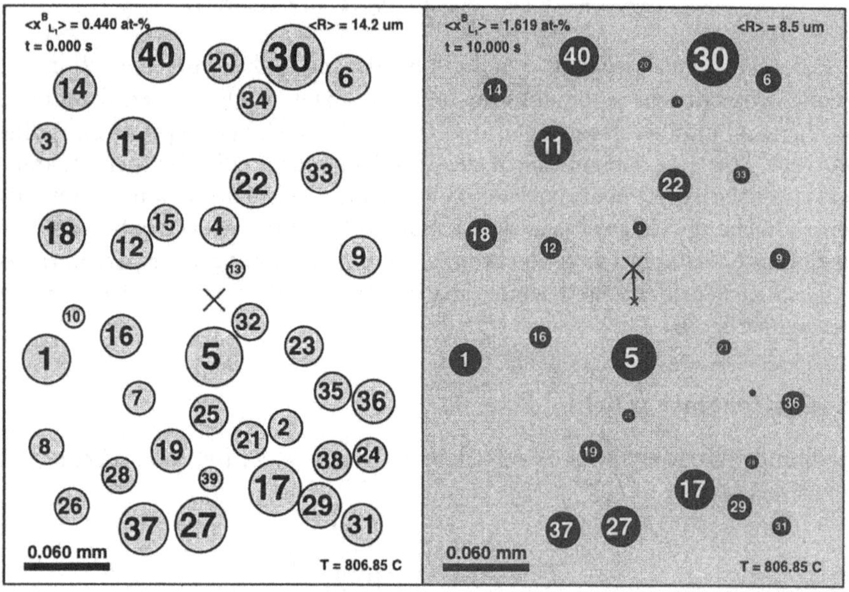

Fig. 8. Microstructure evolution due to radius dependent dissolution of a small model system (Al-Bi). The initial system symmetry plane and central sections of the drops together with some characteristic data are plotted on the left. In 10 s, the system evolves to the one, depicted on the right. Further details see text.

For illustration purposes (Fig. 8) we set up a small, simple model containing 40 drops of L_2-phase drawn from a Gaussian distribution for the radii with mean $\bar{R} = 15\ \mu$m and standard deviation $\sigma = 0.3\ \mu$m . This ensemble is contained in a rectangular box of width $x_{\mathrm{box}} = 300\ \mu$m , height $y_{\mathrm{box}} = 400\ \mu$m and depth $z_{\mathrm{box}} = 60\ \mu$m . The centers of the drops are located in the symmetry plane with the coordinates $\mathbf{r} = (x, y, z_{\mathrm{box}}/2)$. At $t = 0$s, the volume fraction of the L_2-phase is $\phi_{L_2}(t = 0) \approx 3.85$ vol-%. The initial matrix concentration on the left of Fig. 8 is the monotectic one of Al-Bi. This could be a representation of a small Al-Bi sample which was rapidly heated 200 K above the monotectic temperature with no resulting temperature gradients in the sample. The realized mean radius of the L_2-ensemble is 14.2 μm and its center of mass is indicated by the cross. Depicted are of course the central cuts of the drops. The system now tries to reach its equilibrium state via dissolution of the drops. The right part of Fig. 8 shows the state of the model 10 s later. The mean radius has been reduced roughly to the half of the initial one. The matrix phase now holds about four times of the initial Bi concentration. The center of mass changed significantly due to the radius dependence in (5) and its volume fraction was reduced to $\phi_{\mathrm{L}_2}(t = 10\mathrm{s}) \approx 0.85$ vol-%.

4.3 Temperature induced motion

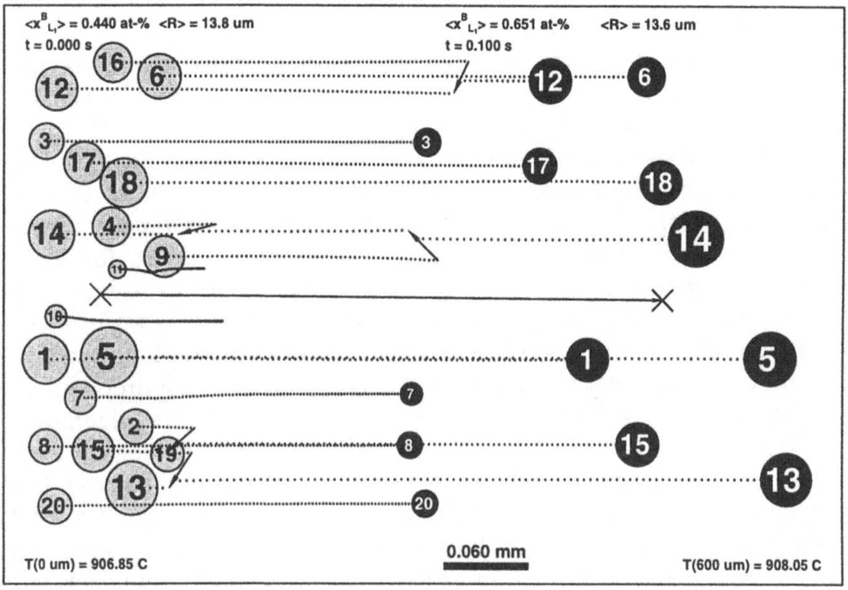

Fig. 9. Planar configuration of 20 drops with Gauss distributed radii at two different times. The drops move due to Marangoni motion induced by a typical temperature gradient of about 2 K/mm in horizontal direction. The large supersaturation (equilibrium matrix concentration of T_{m} at $T_{\mathrm{m}} + 250$ K) leads to a fast dissolution of the drops.

The right-hand side of (2) also depends on the movement of the drops present in the sample. Here we assume that the drops move solely by Marangoni motion. Their terminal velocities are approximated with the two droplet formulas from Anderson [16]. The influence of all the two-particle velocities $\mathbf{v}_{ij}^{2\mathrm{p}}$ on the time development of the i-th particle is then calculated using a pairwise additive method [18]. Convergence is assured by summing up the $\mathbf{v}_{ij}^{2\mathrm{p}}$ in the reference frame of the i-th particle, moving with its unperturbed one-particle velocity $\mathbf{v}_i^{1\mathrm{p}}$. After performing the sum, the $\mathbf{v}_i^{1\mathrm{p}}$ is added again. This leads to the following expression for the approximated N-particle velocity of the i-th particle in an infinite volume:

$$\mathbf{v}_i^{\mathrm{np},V=\infty} = \mathbf{v}_i^{1\mathrm{p}} + \sum_{\substack{1 \le j \le N(t) \\ j \ne i}} \left(\mathbf{v}_{ij}^{2\mathrm{p}} - \mathbf{v}_i^{1\mathrm{p}} \right). \tag{6}$$

Similar to the model in Fig. 8, a planar configuration of 20 drops from the same radius distribution is placed inside a rectangular box, its dimension were $x_{box} = 600\,\mu m$ wide. The drops are initially confined to the left part with $x < 150\,\mu m$, and their volume fraction is $\phi_{L_2}(t = 0) \approx 1.77$ vol-%. Again the initial matrix concentration (see Fig. 9) is the monotectic one of Al-Bi, so the physical equivalent should be a small Al-Bi sample rapidly heated 250 °C above the monotectic temperature. A constant temperature gradient $\nabla T = (2, 0, 0)$ K/mm is assumed. The mean radius of the L_2-ensemble is 13.8 μm and its center of mass is again indicated by the cross.

Firstly, since the drops experience a temperature gradient, they move due to Marangoni motion. The distance of the dots marking the trajectories is fixed in time ($\delta t = 1$ ms) to achieve a stroboscope like effect.

Secondly, they dissolve as before into the matrix to approach the equilibrium state corresponding to approx. 250 K above the monotectic temperature. The data on the top of the figure describe the ensemble and the matrix at $t = 0$ resp. $t = 100$ ms. As in the example discussed above, the center of masses for the final states are connected with the mean displacement vector. In this regime a drop, e.g. number 11, not coagulating with another one, but following a curved path is dissolved after $t < 100$ ms, whilst e.g. drop number 14 grows. Its history is as follows: move, coalesce with drop number 4, move, coalesce with drop number 9 and move. During these 100 ms, the mean radius has not changed significantly, but roughly one third of the drops has disappeared. Drops number 10 and 11 were dissolved, Drops number 2, 4, 9, 12, 13, 14, 15, 16 and 19 have coagulated. The collisions reduced these last nine drops after 100 ms to the number of four. The matrix phase now has about 3/2 of the initial Bi concentration. The ensemble travels with about 4 mm/s in direction of the temperature gradient, while spreading significantly in the same direction. Its volume fraction is reduced to $\phi_{L_2}(t = 100\,\text{ms}) \approx 1.16$ vol-%.

4.4 Finite volume and dendritic corrections on the drop velocity

The effect of the finite volume fraction on the drop velocity is taken into account via the empirical Richardson-Zaki formula [17]:

$$\xi_{\text{Zaki}} = (1 - \phi)^5 \tag{7}$$

In areas of the sample where a mushy zone exists the drop velocity is reduced treating the mushy zone like a porous medium with continuously varying pore size. This is modelled with parabolically varying mobilities on parabolically varying dendrite shapes (Kárman–Kozeny) [17].

$$\xi_{\text{Karm}} \propto 1 - \left(\frac{|\mathbf{r}(T_{\text{mon}}) - \mathbf{r}_i|}{|\mathbf{r}(T_{\text{mon}}) - \mathbf{r}(T_{\text{eut}})|} \right)^4 \tag{8}$$

Using (7) and (8) one arrives at the final formula for the N-particle velocity for a finite sample volume:

$$\mathbf{v}_i^{\text{np}} = \xi_{\text{Zaki}} \cdot \xi_{\text{Karm}} \cdot \mathbf{v}_i^{\text{np},V=\infty}, \tag{9}$$

where $\mathbf{v}_i^{\text{np},V=\infty}$ is the “unbounded” velocity.

4.5 Approximating the solution of the equations

Let the sample be in a state, where the two liquid phases L_1 and L_2 coexist. Three characteristic time scales are then to be treated seperately. Since not only continous field variables like the concentration, but also local variables induced by the particles dispersed in the matrix, and their couplings are to be considered, the maximal time step is determined by the fastest varying process. The typical values for systems like Al-Bi are: Δt (thermal) $\approx$ 1 ms, Δt (diffusive) $\approx$ 100 ms and Δt (external) $\approx$ 1 s for the external temperature field.

The DMPA-model couples all these different processes and interactions at any time with each other. To approximate the solution of this set of equations via iteration, a timestep function $\delta t\,(t)$ has to be chosen. Due to the strong couplings between the different interactions, one global timescale has to be established. The largest $\delta t\,(t)$ depends on the actual dimensions and velocities of the drops present. We use the (kinematic) relation: $\delta t\,(t) \leq \delta R / |\mathbf{v}_{\max}|$ with the length δR (radius of smallest drop) and the maximal present drop velocity $\mathbf{v}_{\max}$.

If using a dispersion as the initial microstructure, a Poisson process is used to distribute the center of masses of the particles over the sample in a homogeneous, uniform random way. Since the particles have a finite volume, the overlap of particles is avoided by pairwise checking and redistributing one of the two, if an overlap has been found. The initial radius distribution may be given in an analytic form, or as an explicit, "measured" one.

In the D-2 experiment we used a sample of hypermonotectic composition with $T\,(z,t) < T_c$ under reduced gravity (μ g-level). The forced external temperature field $T\,(z,t)$ leads to a time and space dependency of the concentration field $c\,(z,T;t)$ for the matrix phase and the $N(t)$ drop concentration fields $c_n\,(T(z,t))$. The drops move by thermal and solutal Marangoni motion. Given the initial values for all fields and variables, the iteration takes place in the following order:

For all t with $t_0 \leq t \leq t_1$ we evaluate with a maximal $\delta(t) \approx$ 1ms the temperature field $T(\mathbf{r},t)$ and its gradient $\nabla T(\mathbf{r},t)$. Then the $N\,(t)$ particle concentrations $c_n\,(T(z,t))$ and their radius distribution $f(R_{ij})$, are evaluated. Since drops may be dissolved, the number of drops $N(t)$ is re-calculated. Subsequently the concentration field $c(z,t)$ of the matrix and its gradient $\nabla c(z,t)$ and the M average matrix concentrations, $\bar{c}_m\,(z,t)$ with the number of (theoretical) subregions M of the sample [1] are calculated. The drop velocities $\mathbf{v}_{ij}^{\mathrm{NP}}$ are calculated and collision events are handled as instantaneous unions of drops at the center of mass, to a new one replacing them with their combined volume. Consequently the new locations $N(\mathbf{r}_i)$ of the drops are found via a linear extrapolation using their velocities.

Following this procedure the balances for mass δM, and momentum $\delta \mathbf{P}$ are checked.

---

[1] Modelling the D2-experiments we used cylindrical slices of the cylindrical samples. These had a length of about 100 μm . So $M \approx 800$.

4.6 Solidification

The solidification process is treated in two different ways. The first way is a stepwise one: If the monotectic temperature has just passed a slice (see footnote 1) it is allowed to solidify and during that process all excess of bismuth is diffusively redistributed to the droplets being present in that slice. The solidification proceeds axially starting from the cold end till it reaches the hot tip within approx. 10 minutes. The second method simulates a radial solidification: Each slice is cooled down independently from each other with a fit of the measured temperature profile (see above).

4.7 Ternary systems

Up to now we did not take explicitly into account that we investigated a ternary alloy. This is matched into the process in the following way. In the ternary Al-Si-Bi alloy the material solidifies according to this sequence: first a homogeneous liquid of Al, Si and Bi decomposes into two liquids with different Bi-contents (i.e. similar to a binary system, here both liquids contain Si). Then the aluminium-rich liquid precipitates pure aluminium in form of dendrites and the liquid enriches its Bi-content (still the other Bi-rich liquid is present in form of droplets). This is a monotectic reaction taking place until the ternary eutectic temperature of about 575 °C is reached. The tip of the Al-dendrites occurs at the monotectic temperature of the binary Al-Bi system. Thus solidification as well as melting does not proceed with a planar solid-liquid interface but is dendritic in nature. At the monotectic temperature the sample is fully liquid, at the eutectic temperature it is fully solid (besides the Bi-droplets which are still liquid). This transition zone is treated as a mushy zone of a length dictated by the linear temperature profile between the monotectic temperature and the temperature of the cold junction. Within this mushy zone the Marangoni motion is described by (9) together with (8). The droplets being in the interdendritic liquid are allowed to grow or to shrink by diffusion.

5. Results of the numerical simulation and discussion

The locally varying bismuth concentration after a simulated D-2 experiment run is shown in Fig. 10. The bismuth concentration shown is the sum of the Bi dissolved in the matrix and contained in the mobile droplets. In the case of axial solidification the drops were mobile until the solidification process freezed the drop distribution in a slice (see footnote 1), whereas in radial solidification any slice solidified independently. The mushy zone extends from $z = 75$ mm to approximately $z = 56$ mm. In this region the bismuth content decreases almost linearly from 7 wt% to 2 wt% (axial solidification) or 3.4 wt% (radial solidification. From $z = 55$ mm the bismuth content increases continuously till the hot end, where with radial solidification the Bi content would slightly decrase. In the axial solidification mode the Bi content is approximately constant in the region from $z = 40$ to 10 mm and then increases to large values (layer build up). This is due to the fact that during axial solidification from $z = 75$ mm all drops being mobile will still move in the molten part into the direction of the hot tip and the hot tip is solidified last.

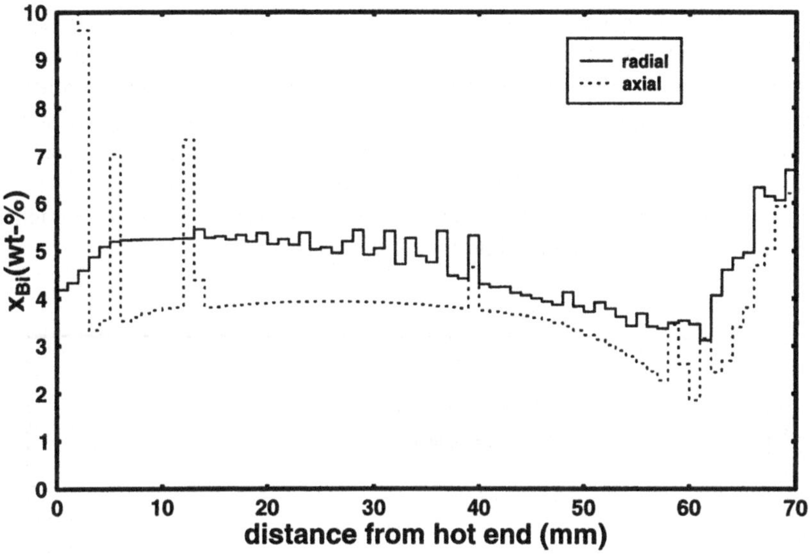

Fig. 10. Bismuth content as simulated using the following set of parameters: interfacial tension at T_m is set to 30 mJ /m^2 (see [6]), mean initial radius is 5 μm , distributed Gaussian, spread = 1.73 · mean. Melting speed = 1 mm/min (= sample 3042) and diffusion coefficient $D(T) = 1.13 \cdot 10^{-14} T^2$.

As mentioned in the section 3. the real sample exhibited a combination of axial and radial solidification. Near the thermocouples the solidification was radial, which also means that drops were shifted by the advancing dendrites into the center of the sample. The amount of radial solidification is larger farther away from the tip (compare the number and locations of the thermocouples in Fig. 6). Comparing the simulated bismuth concentration in the samples with the real experiment (see Fig. 6) we see that in the real experiment the Bi content decrease only slightly in the region from $z = 70$ to $z = 40$ mm (from 6.5 to 3.5 wt%) and has a marked sudden increase around $z = 40$ mm. In the region from $z = 35$ to 20 mm the Bi content is approximately only the monotectic one. Near the tip the bismuth concentration increases again. Thus we may conclude that a suitable combination of the bismuth profile obtained in the simulation with the radial and axial solidification modes may reproduce the experimental result. Although the general features of the simulation agree with the experiment, there are two points which definitively disagree: First, in the real experiment the profile of the Bi concentration has a linear shape in the range from $z = 70$ to $z = 40$ mm, whereas in any simulation run we performed with different parameters (like initial drop size distribution, interfacial tension etc.) the profile is of concave nature. Second, the hump around 40 mm was never seen in any simulation. We attribute both effects to the action of the thermocouples. The microstructure clearly reveals that they acted as heat sinks inducing radial solification and therefore a local temperature gradient altering the Marangoni motion of the drops. A part of the drops was shifted back into the direction of the mushy zone, another part accelerated to the hot tip. Therefore the region after this hump ist depleted of drops, in contrast to the simulation. There is, however, an additional artefact in the samples, that induces the disagreement. The SEM pictures, x-ray transmission and x-ray computer tomography showed that at approximately $z = 35\,\mathrm{mm}$ an oxide skin folded into the sample till approx. $z = 20\,\mathrm{mm}$. This skin is enriched with Bi. Therefore we made a second simulation run introducing at $z = 35\,\mathrm{mm}$ a sink for droplets. The result is shown in Fig. 11.

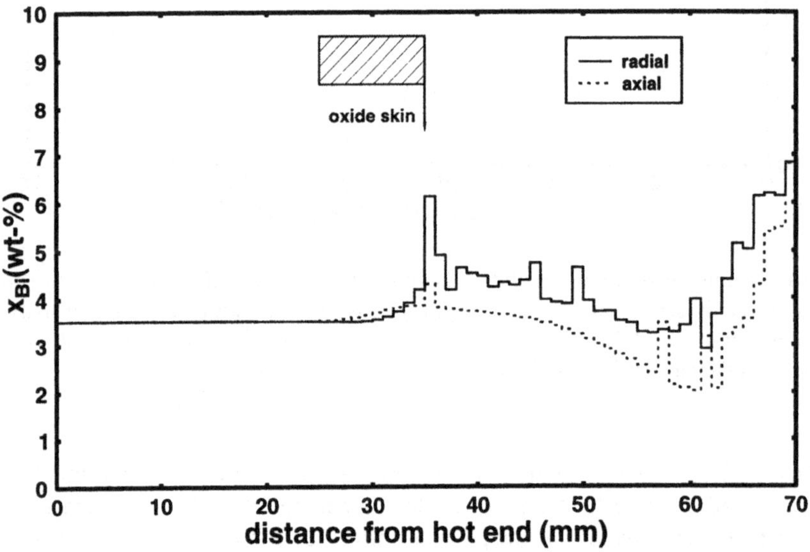

Fig. 11. Bismuth content as simulated using the following set of parameters: interfacial tension at T_m is set to 30 mJ /m^2 (see [6]), mean initial radius is 5 μm , distributed Gaussian, spread = 1.73 · mean. Melting speed = 1 mm/min (= sample 3042) and diffusion coefficient $D(T) = 1.13 \cdot 10^{-14} T^2$.

The Bi enrichment at that skin was modelled by taking out all drops which moved in the upper half of the sample (this definitely is an overestimation of the effect, but shall only serve to illustrate the action of the interior sink for bismuth). The result of the simulation shows that especially with radial solidification switched on the profile of the bismuth concentration is less concave and reflects better the experimentally observed one.

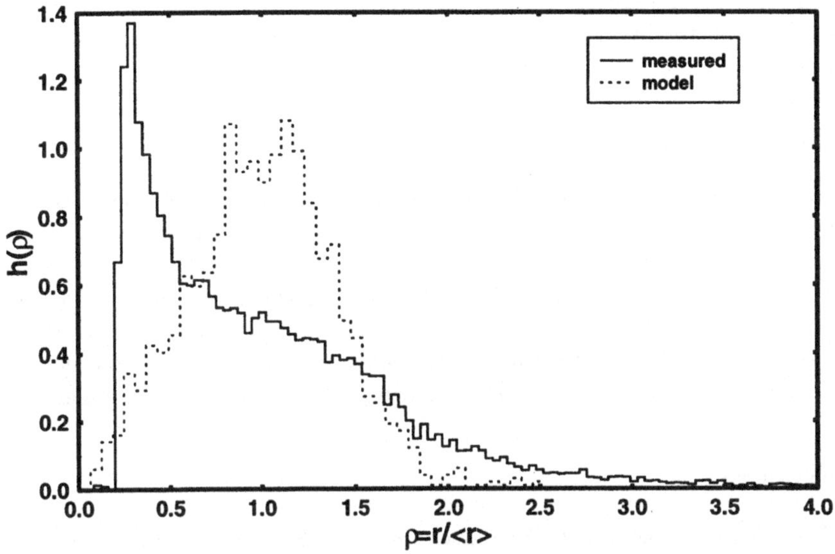

Fig. 12. Normalized histograms of section radii. The experimental data consists of two distributions: monotectic and hypermonotectic particles. About 15000 section radii have been measured, ten times more than calculated. The maximal section radii, in both cases, are ten times the mean.

Another comparison between simulation and experiment is shown in Fig. 12. The solid line is the histogram of all bismuth drops in the interval from $z = 0$ to 65 mm of the real D-2 experiment and the dashed line represents the simulation run (see Fig. 10). In the simulation we counted only the drops being present in the two phase regime, whereas in the experiment the monotectic reaction also produces drops, whose size distribution normally is log-normal with diameters smaller than the hypermonotectic one (the size distribution depends on the cooling conditions (see [23])). Thus the experimental histogram is the sum of two: the drops that emerged from the monotectic reaction with a maximum around $\rho = 0.3$ and the hypermonotectic ones yielding the shoulder at approx. $\rho = 1.1$. The simulation gives a size distribution of hypermonotectic drops with a maximum in the range of $0.9 \leq \rho \leq 1.2$. Comparing both size distribution one has to keep in mind that both are normalized. Deconvolution of the monotectic drops yields a close agreement between experiment and simulation.

Therefore we conclude from the comparison of the local bismuth concentration and the size distributions that the principle features of the numerical model are correct. The concave shape of the bismuth concentration with a minimum at the end of the mushy zone indicates that we either do not treat the multi-droplet Marangoni motion correctly or our treatment of the mushy zone as simply reducing locally the drop velocity, see (8) and (9), is incorrect. Finally we may remark that a simulation of the experimental results was made rather difficult due to the action of the thermocouples and the presence of the oxide skin. In future experiments under reduced gravity conditions we try to avoid these complications and hope to reveal the discrepancies between experiment and numerical simulation.

Acknowledgements

We gratefully acknowledge the financial support of the German Space Agency DARA within the joint project "Monotectic Alloys" and the financial support some of us obtained by DLR (L.R., S.Dr., S.Di.). We also wish to express our gratitude to the team at MUSC performing the Bodenbegleitprogramm (Dr. F. Gillesen) and the D2-management team responsible for the Werkstofflabor (Dr. Kramp, Dr. Meusemann) who always reacted promptly to every wish we had even just prior to the mission regarding some changes to the experiment.

References

1. B. Predel, L. Ratke and H. Fredriksson, in: *Fluid Sciences and Materials Science in Space*, Ed. H.U. Walter, Springer-Verlag, Berlin (1987) 517–565
2. B. Prinz and A. Romero, in: *Immiscible Liquid Metals and Organics*, Ed. L. Ratke, DGM-Informationsgesellschaft, Oberursel (1993) 281–289
3. B. Prinz, A. Romero and L. Ratke, *J.Mat.Sci.* (in press)
4. S. Diefenbach, L. Ratke, B. Prinz and H. Ahlborn, in: *Immiscible Liquid Metals and Organics*, Ed. L. Ratke, DGM-Informationsgesellschaft, Oberursel (1993) 291–298
5. L. Landau and I. Lifshitz, in: *Hydrodynamik*, Akademie Verlag, Berlin (1981)
6. S. Diefenbach, PhD-thesis, Ruhr-Universität Bochum (1993) (*For a copy, contact the authors*).
7. F. Sommer, in: *Immiscible Liquid Metals and Organics*, Ed. L. Ratke, DGM-Informationsgesellschaft, Oberursel (1993) 79–90
8. F. Falk, in: *Immiscible Liquid Metals and Organics*, Ed. L. Ratke, DGM-Informationsgesellschaft, Oberursel (1993) 93–100
9. J. Alkemper, S. Diefenbach and L. Ratke, *Scripta Metall. et Mater.* **29** 11 (1993) 1495–1500
10. J. Legros, A. Sanfeld and M. Velarde, in: *Fluid Sciences and Materials Science in Space*, Ed. H.U. Walter, Springer-Verlag, Berlin, (1987) 83–
11. N.O. Young, J.S. Goldstein and M.J. Block, *J. Fluid Mech.* **6** (1959) 350–356
12. S. Drees and L. Ratke, *Mat. Sci. Forum*, Trans. Tech. Publ., Zürich (in press)
13. L. Ratke and S. Diefenbach, *Reports Mat. Sci. Eng.* (in press)
14. L. Gránásy and L. Ratke, *Scripta Metall. et Mater.* **28** 11 (1993) 1329–1334
15. J. Happel and H. Brenner, *Low Reynolds Number Hydrodynamics*, First Paperback Edition, Ed. R.J. Moreau, Martinus Nijhoff Publishers, The Hague (1983)
16. J.L. Anderson, *Int.J.Mult.Flow* **11** 6 (1985) 813–824
17. R.F. Probstein, in: *Physiochemical Hydrodynamics*, Butterworths, Boston (1989) 98–100
18. G.K. Batchelor, *J. Fluid Mech.* **119** (1982) 379–408

19. H.U. Walter, in: *Proc. of an RIT/ESA/SSC Workshop, Sweden* (1984), ESA SP-219 47–64
20. H.U. Walter, in: *Material Sciences in Space* Ed. B. Feuerbacher, H. Hamacher, R. Naumann, Springer-Verlag, Berlin (1986) 343–378
21. B. U. Felderhoff, private communication, 1995
22. R. H. Davis, X. Zhang, H. Wang, in: *Immiscible Liquid Metals and Organics*, Ed. L. Ratke, DGM-Informationsgesellschaft, Oberursel (1993) 163–180
23. W. K. Thieringer, L. Ratke, Acta metall. **35** (1993) 1237–1244

Part III
Crystallization

Crystallization in Solutions: Effects of Microgravity Conditions

A.A. Chernov

Institute of Crystallography, Russian Academy of Science Moscow 117333 Leninsky Prosp. 59, Russia

1. Introduction

Absence or drastic reduction of flows in a liquor in which crystallization occurs results in at least two well known consequences: 1. Virtually purely diffusion-controlled mass and heat transfer, 2. elimination of the flow-induced hydrodynamic forces which may cover weak mechanical forces acting on and between the growing crystallites. Experimental parameters determining the diffusion-limited growth are overviewed in Sec.2., the diffusion-controlled mass crystallization of calcium phosphates is discussed in Sec.3., the one of proteins in Sec.4., morphological stability of polygonal crystals and their interfaces in Sec.5.. The mechanical interactions are briefly discussed in Sec.6..

2. Numerical data for diffusion and interface kinetic growth modes

If a crystal growing in solution or from the melt has a typical size L, then the typical diffusion rate of material or heat transport to or from the crystal is D/L where D is the mass or heat diffusivity. The rough interface is characterized by linear dependence of the growth rate and thus the diffusion flux on supersaturation or supercooling of the interface. For solution growth

$$D\frac{\partial n}{\partial z} = \beta(n - n_e) \equiv \beta n_e \sigma, \qquad \sigma = (n - n_e)/n_e \tag{1}$$

Here n and n_e (cm^{-3}) are the actual and equilibrium number densities of the solution from which a crystal grows, σ is the relative supersaturation, the z coordinate is normal to the interface and the kinetic coefficient β is associated with desolvation of the incorporated species and its steric (entropic) restrictions on the way to the lattice. A rough interface, which prevails in the melt growth, is rare in solution growth. The NH_4Cl is such an example. However, no precise measurements of the kinetic coefficient β for rough interfaces in solutions (e.g. NH_4Cl) are known. On the contrary, the layerwise growth from solutions was extensively studied during the last years [1, 2]. The average kinetic coefficient for steps, β_{st} is defined similarly to (1) as the proportionality factor between the step velocity, v, the diffusion flux and the supersaturation at the step:

$$v = \beta_{st}\omega(n - n_e) = \omega D\frac{\partial n}{\partial r} \tag{2}$$

Here $\partial/\partial r$ is the derivative with respect to the radial distance r from the straight step considered as a linear sink for crystallizing species for growth or source for dissolution. A step experiences strong thermal fluctuations and is analogous to the rough interface. This analogy makes β_{st} a reasonable approximation to β. The kinetic coefficient β_{st} may be presented as:

$$\beta_{st} = (a^2/\lambda_o)\nu\theta e^{(-E/kT)} \tag{3}$$

Here a is the typical size per particle in the lattice,$\lambda_o \geq a$ is an average interkink distance,ν is the attachment attempt frequency, θ is the steric factor and E is the potential barrier on the way of the species from the solution to the kink. For inorganic crystals in aqueous and hydrothermal solutions, typically $E \simeq 10 - 20$ kcal/mol [3].

The steric restrictions are probably not very important for relatively small inorganic species but might reduce kinetic coefficients by several orders of magnitude for large molecules, e.g. proteins. For instance, the value $\theta \simeq 10^{-3}$ was estimated for the lysozyme ellipsoidal molecules; these molecules may be approximated by a sphere $\sim 35\AA$ in diameter [4]. Notwithstanding that it resembles a sphere, such a molecule should enter the lysozyme crystal lattice in a precise orientation determined by the most effective specific hydrogen and salt bridge bonds [5, 6], along with less specific hydrophobic and van der Waals attractions.

The frequency factor, ν, in (3) may be approximated by either the reaction frequency $kT/h \simeq 10^{13}\mathrm{s}^{-1}$ at T=300K, $h = 6.62 \cdot 10^{-27}\mathrm{erg} \cdot \mathrm{s}$ or the thermal rate over the typical lenght $l \simeq 0.1 - 0.3$nm characterizing the potential barrier width

$$\nu = \sqrt{kT/2\pi m}/l \tag{4}$$

An even more elemental approximation for ν is the average vibration frequency which for small inorganic species is usually taken to be $10^{12} - 10^{13}\mathrm{s}^{-1}$. Some experimental values of β_{st} are listed in the Tab. 1.

A rough interface is similar to the rough step risers since both possess high density of kinks generated by thermal fluctuations, i.e. the kink densities on both are comparable to the density of lattice sites. Therefore, for both the steps and rough interfaces the main resistance for incorporation is associated with the kinks. Therefore one might estimate

$$\beta \simeq \beta_{st}(a/\lambda_o) \simeq (1 - 0.3)\beta_{st}$$

For a crystalline face growing layerwise

$$\beta \simeq \beta_{st} p \tag{5}$$

substance, face	β_{st}, cm/s	p	$p\beta_s t$, cm/s
ADP, KDP, DKDP(100)	$(5-12)\cdot 10^{-2}$	$3\cdot 10^{-4} - 8\cdot 10^{-3}$	$10^{-4} - 10^{-3}$
ADP(101)	0.4 - 1	$10^{-4} - 5\cdot 10^{-3}$	$4\cdot 10^{-5} - 5\cdot 10^{-3}$
$BaNO_3(111)$	$1.3\cdot 10^{-2}$	$(3-15)\cdot 10^{-4}$	$4\cdot 10^{-6} - 2\cdot 10^{-5}$
$KAl(SO_4)_2\cdot 12H_2O(111)$	$8\cdot 10^{-2}$	$(0.4-3.5)\cdot 10^{-3}$	$3\cdot 10^{-5} - 3\cdot 10^{-4}$
$Y_3Fe_5O_12(110), (211)$		$(0.3-3)\cdot 10^{-2}$	$(0.4-1)\cdot 10^{-3}$
$(YSm)_3(FeGa)_5O_12(111)$	$1.3\cdot 10^{-2}$		10^{-2}
$(EuYb)_3Fe_5O_12(111)$			$(0.1-3)\cdot 10^{-3}$
lysozyme(101)	$4.6\cdot 10^{-5}$	$(1.1-1.5)\cdot 10^{-2}$	$6\cdot 10^{-7}$

Table 1. The data are taken from [7, 8] and references therein

where p is the vicinal slope, i.e. the average step density on the face, i.e. the ratio of the step height to the average interstep distance. For many inorganic crystals at low supersaturations $10^{-3} \leq \sigma \leq 10^{-1}$ the steps are generated by dislocations, their density increases with the supersaturation and the slope p was measured to be roughly $10^{-3} \leq p \leq 10^{-2}$ (see Table 1). At $\sigma > 15\%$, one should expect the two dimensional nucleation to substantially contribute to the growth rate [9, 10].

For the growth of proteins such as lysozyme the only interferometric measurements performed so far [11] demonstrated that the vicinal hillock slope $p \simeq (1.1-1.5)\cdot 10^{-2}$ was independent on supersaturation at $\sigma \leq 400\%$, i.e.$(n/n_e \leq 5)$. The $v(\sigma)$ dependence is much steeper than the linear one at $15\% \leq \sigma \leq 330\%$ with the short eventually linearized $v(\sigma)$ dependence at $0.33 \leq \sigma \leq 0.5$ for which range $\beta_{st} = 2.8\cdot 10^{-4}$cm/s. The latter figure listed in the Table 1 ($4.6\cdot 10^{-5}$cm/s) is obtained from this author's value defined by (2) with the driving force $\sigma = \ln n/n_e$ rather by replacing it by $\sigma = (n-n_e)/n_e$. The relevance of this replacement is discussed in [4, 12].

Thus the data presented in the Tab. 1 suggest that the interface kinetic coefficients are of the order of $\beta \simeq 1 - 10^{-2}$cm/s for a rough interface and $\beta \simeq 10^{-3} - 10^{-4}$cm/s for smooth crystal faces for some inorganic crystals and $\beta \simeq 5\cdot 10^{-5}$cm/s for lysozyme.

The growth kinetics are limited by the interface incorporation if the crystal size

$$L \ll L_c = D/\beta \tag{6}$$

and the kinetics are diffusion-limited if

$$L \gg L_c \tag{7}$$

With $D \simeq 10^{-7}\text{cm}^2/\text{s}$ (inorganic salts) one gets for polyhedral crystals $L \simeq 10^{-2} - 10^{-1}$cm and larger $L_c \simeq 10^{-1} - 1$cm for lysozyme with $D \simeq 10^{-7}\text{cm}^2$.

The diffusion-limited growth mode at $L \gg D/\beta$ is well known to occur at low supersaturations always decreasing with the crystal size. The consequences are discussed below.

3. Mass crystallization (calcium phosphates)

3.1 The problem and the experiment

Low supersaturations are of special interest to analyze the growth kinetics of slightly soluble salts like calcium phosphates, carbonates, oxalates, urates etc. These biomineralogical materials prevail in teeth and bones, they are formed in the human body and other living organisms (see, e.g. [13]-[18]).

Calcium phosphates [19]-[27] are widely used in tooth and bone repair and their efficiency as such depends on the crystalline modifications. The calcium phosphates are crystallized in at least 10 modifications with the Ca/P ratio varying between 0.5 for $Ca(H_2PO_4)_2 \cdot H_2O$ to 2 for $Ca_4(PO_4)_2O$. The most typical modifications in the biologically relevant neutral solutions (pH7) are octa-calciumphosphate (OCP) $Ca_8H_2(PO_4)_6 \cdot 5H_2O$, amorphous calcium phosphate (ACP), $Ca_3(PO_4)_2 \cdot H_2O$ and hydroxyapatite (HAP), $Ca_5OH(PO_4)_3$ [28]. The latter seems to be the thermodynamically most stable phase. The solubility of all these compounds is very low, Therefore, even a delicate mixing of the calcium and phosphorus containing solutions results in supersaturations $\Delta\mu/kT$ as high as 10^9. Here $\Delta\mu$ is the gain in chemical potential provided by the reaction. The slowest mixing might be achieved under purely diffusion conditions, e.g. in a gel. To exclude any chemical influence of the latter, microgravity conditions have been employed in the Solution Growth Facility (SGF) [1] flown on EURECA. A schematic view of the SGF growth reactor is shown in Fig.1. The charge reservoirs have been filled with $5.4 \cdot 10^{-2} M CaCl_2 + 5.8 \cdot 10^{-2} MKCl$ and with $6.7 \cdot 10^{-3} M KH_2PO_4 + 4.7 \cdot 10^{-2} MK_2HPO$, 2 liters each. The middle reaction chamber contained 0.5 l of $0.16MKCl$, all aqueous solutions. The systems with NH_4 along with K were also used in terrestrial experiments.

The experiment was started by slowly opening the valves. During the EURECA flight in 1993 the valves have been open for 5 months. The crystalline powders grown during the flight have been analyzed by optical microscopy, X-ray and electron diffraction, electron microscopy in conventional imaging, diffraction and high resolution modes by fluorescent X-ray analysis and computer simulation [29, 30].

3.2 The terrestrial vs. space-grown crystallites

The terrestrial powder consists of polycrystalline agglomerates (Fig.2), and occasionally single crystalline platelets (Fig.3). The techniques listed at the end of Sec.3.1 shows a clear preference of the HAP with traces of OCP and amorphous phase.

The space sample is also a polycrystalline powder. It consists, however, of even optically observable crystalline "stars", i.e. incompletely filled spherolites (Fig.4 and Fig.5) built of HAP needles up to 10 μm long. Even more interesting is that the OCP rather than the HAP phase dominates in the space sample with crystalline needles and "stars" up to 3mm long, typically $\simeq 0.5mm$. The

[1] Design and construction for ESA by Oerlikon Contraves AG, Zurich

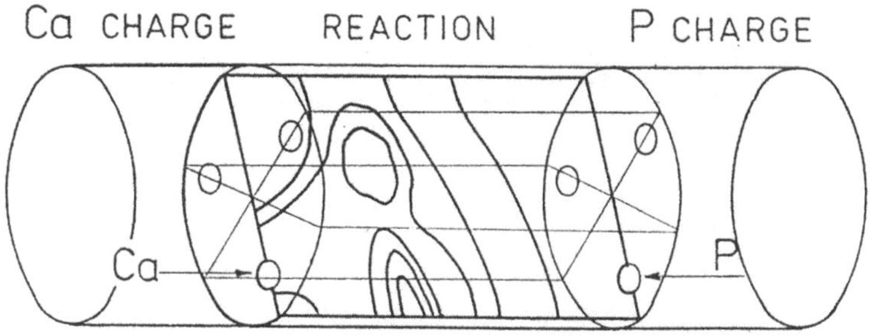

Fig. 1. The ESA Solution Growth Facility (SGF) built by Oerlikon Contraves. Isothermal Chamber Nutrient Ca and P aqueous salt solutions are kept in the left and the right hand side charge chambers. Between there is a reaction chamber. The two membranes separating the reaction chamber from the Ca and P reservoirs contain three holes each. Before the experiment starts the holes are closed by valves.One of three identical planes crossing the chamber axis and the Ca and P holes indicated by arrows is drawn. The lines within this plane symbolize iso-concentration lines.

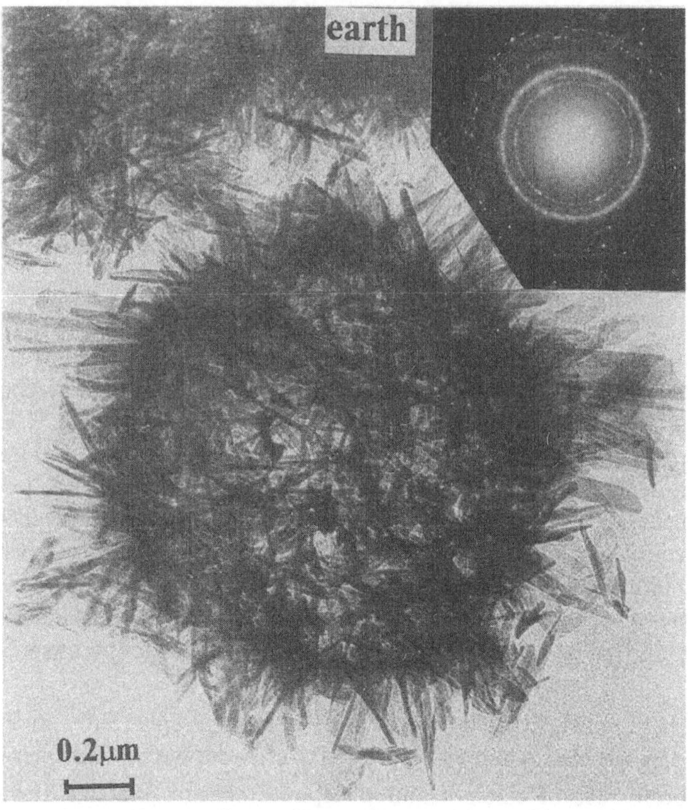

Fig. 2. Transmission electron microscopic image of a terrestrial HAP aggregate and electron microdiffraction pattern (the insert) taken from the same aggregate.

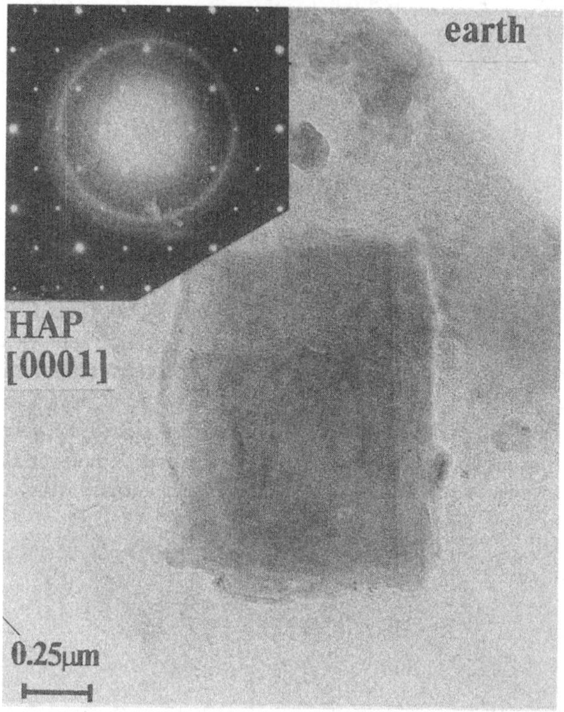

Fig. 3. Electron microscopic image of a single terrestrial crystal and the electron microdiffraction with the incident beam along [00.1] (the insert).

identification of the OCP phase was made by the techniques listed at the ending Sec.3.1. The HAP crystals become optically invisible in an immersion liquid with refractive index 1.58, while the standard indices for the OCP are $n_\alpha = 1.576, n_\beta = 1.583$ and $n_\gamma = 1.585$. The HAP index was found to be close to 1.65. The OCP spherolitic stars are typically 0.5-1.5 orders of magnitude larger than the HAP ones. Fig.6 presents a high resolution electron micrograph showing the complex structure of an OCP platelet. The upper and lower parts present the OCP lattice divided by a grain boundary and, probably an amorphous layer. In the lower part one can see a dislocation. In other samples dislocations possessing both screw and edge components including dislocation loops have been also found.

The average number density of the HAP and OCP spherolites in the space-grown samples was estimated to be $3.6 \cdot 10^4 cm^{-3}$ and 80cm^{-3}, respectively.

Only about 15% of calcium and phosphate charges reacted during the 5 month space experiment while the calcium and phosphate charges reacted completely in less than 6 weeks of terrestrial experiment. We believe that the dramatic difference between the products grown in space and on earth is caused by the difference in supersaturation.

Fig. 4. Scanning electron micrograph of the space grown HAP spherolite.

3.3 Computer simulation

The aim of this simulation was to quantitatively probe numerous factors which should influence the masscrystallization: diffusion field, supersaturation, heterogeneous vs homogeneous nucleation parameters, kinetic coefficients of incorporation of the species on the growing interface.

The diffusion field was calculated taking into account two effective diffusivities for the calcium and phosphate containing clouds:

$$D[\mathrm{CaCl_2}] = 1.88 \cdot 10^{-5}\mathrm{cm}^2/\mathrm{s}$$

and

$$D[(\mathrm{NH_4})_{1.66}(\mathrm{H}_{1.34}\mathrm{PO_4})] = 1.7 \cdot 10^{-5}\mathrm{cm}^2/\mathrm{s} \quad (8)$$

respectively. These effective diffusivities have been found making use of the known relationship for the cation-anion coupled diffusion for the fully dissociated electrolyte and introducing the effective diffusing phosphate ion indicated in (8). Its composition and charge 3-x=1.66 were found making use of the equilibrium constants regulating the concentrations of the $H^+, PO_4^{3-}, HPO_4^{2-}, H_2PO_4^-$ and H_3PO_4 species. The diffusivities in solutions consisting of potassium rather than NH_4^+ ions do not differ more than couple of percent.
The supersaturation was defined as

$$\Delta\mu/kT = \ln \Pi/\Pi_e \Pi \equiv C_1^{\nu_1} C_2^{\nu_2} \Pi_e = C_{1e}^{\nu_1} C_{2e}^{\nu_2} \quad (9)$$

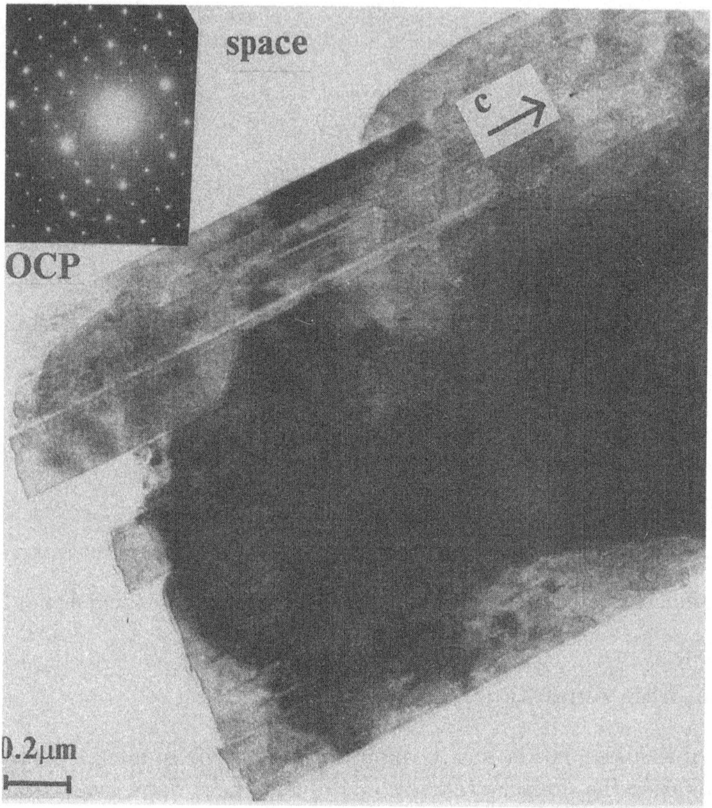

Fig. 5. Electronmicroscopical picture of a section of a space-grown OCP crystal with its electron diffraction shown in the insert.

where C_1 and C_2 are the concentrations of the $CaCl_2$ and the effective molecules $(NH_4)_{3-x}H_xPO_4$ (x=1.66) and Π_e is their equilibrium product. For the OCP and HAP precipitation, it was taken $\nu_1 = 4, n_2 = 3$ and $n_1 = 5, n_2 = 3$, respectively. For the OCP and HAP

$$\begin{aligned} C_{1e}^4 C_{2e}^3 &= 0.9 \cdot 10^{-29} (\mathrm{mol/l})^7 \\ C_{1e}^5 C_{2e}^3 &= 3 \cdot 10^{-40} (\mathrm{mol/l})^8 \end{aligned} \tag{10}$$

respectively. Spherical shape of the crystals obeying linear interface kinetics with the kinetic coefficients β_i, i=1,2 was assumed:

$$D_i \partial C_i / \partial r = \beta_i (C_i - C_{ie}) \qquad i = 1, 2 \tag{11}$$

where $\partial/\partial r$ is the derivative with respect to the distance r from the crystal center, and C_i are taken at the growing interface $r = R$.

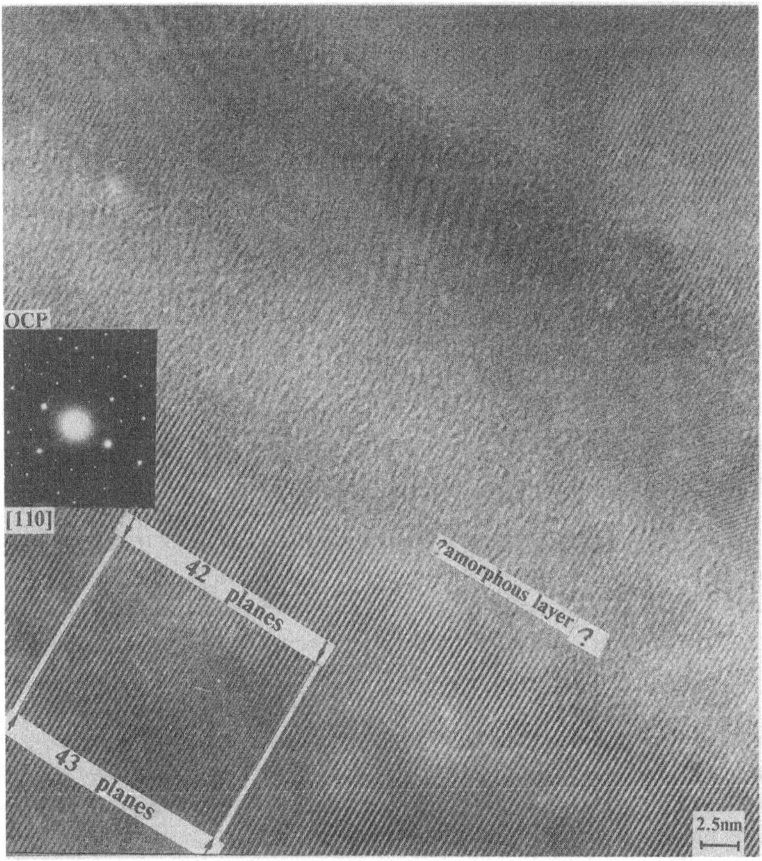

Fig. 6. High resolution electron micrograph of the OCP platelet. The contour in the lower left corner reveals a dislocation with the edge component. Moire patters is seen in the upper part of the grain boundary between the uper and lower grains. The layer adjacent to the boundary might be an amorphous material.

The crystal growth rate

$$dR/dt = \omega D_1\beta_1(C_1 - C_{1e})/\nu_1(1+\beta_1 R/D_1) = \omega D_2\beta_2(C_2 - C_{2e})/\nu_2(1+\beta_2 R/D_2) \quad (12)$$

Here the factors $(1+\beta_i R/D_i)$ follow from (11) in the quasi steady state Laplace approximation to the diffusion equation, while ν_1 and ν_2 enter to express stoichiometry of fluxes into the OCP or the HAP phases.

The concentration distribution in the diffusion clouds spreading one toward the other from the holes were calculated (see Fig.1). As soon as the clouds overlap the supersaturation arises - first around the middle of the reaction chamber (Fig.7, 3 left patterns). Later on, when each of the counterdiffusing species reaches the opposite walls, the supersaturation maxima appear at the holes (Fig.8).

We do not know whether the crystallization starts hetero- or homogeneously, and we have therefore considered the both processes. For heterogeneous nucleation, 10, 30 and 100 centers/cm^3 was assumed on which the growth is supposed to begin as soon at the supersaturation is achieved.

The solution depletion caused by consumption of the ions by the growing crystals in the cases of homogeneous and heterogeneous nucleation follows from comparison of the left and right hand patterns in Fig.7 for the OCP crystallization. The dashed lines are the lines $\Pi = \text{const} < \Pi_e$, the solid lines correspond to $\Pi = \text{const} > \Pi_e$. The full lines depicting the regions where the crystals grow are evidently spreading with time. The regions where the supersaturation and its spacial gradients are relatively low occupy a substantial part of the growth chamber. Both features would hardly be observed if convection would be involved since in the latter case the typical inhomogeneity scale is reduced from the chamber size to the much smaller flow pattern size and the effective diffusivity rises.

With an interfacial energy of approx. 80 erg/cm^2 the homogeneous nucleation rate becomes competitive with the growth on heterocenters if concentration of the latter is below several tens of centers/cm^3. The homogeneous nucleation rate of the OCP reaches maximum of $\sim 10^{-29}\text{cm}^{-3}\text{s}^{-1}$ if the density of heterocenters is 10^{-2}cm^{-3}. If the heterocenter density is 10^2cm^{-3} , the homogeneous nucleation rate of the OCP reaches its maximal value of $\simeq 10^{-29} 1/\text{cm}^3\text{s}$ after ~ 5 days of diffusion and decreases later since the material supplied by diffusion is consumed by the already born and growing crystals. Under these conditions, the overall number of homogeneously created crystals even during the 5 month experiment is much less than 10^2 cm^{-3} of heterocrystals. On the contrary, homogeneous nucleation should prevail if the heterocenter density is below several tens per cm^3. The much higher supersaturations reached if there is convection, is an essential result in the more important role of homogeneous nucleation. Thus, much lower levels of active heterocenters have to be achieved to realize homonucleation in space and we may conclude that heterogeneous nucleation prevailed during the EURECA SGF experiment in which no special precautions have been taken to deactivate the heterocenters, such as precrystallization heat treatment.

The OCP is thermodynamically less stable than the HAP (cf.(10)) therefore the nucleation of the HAP seems to occur at much higher rates. Indeed, the number density of the HAP crystallites (ca $\sim 3.6 \cdot 10^4\text{cm}^{-3}$) greatly exceeds the one of OCP (ca 80 cm^{-3}). However, the gross amount of the OCP grown in space is much larger than that of the HAP. To explain this fact we have to assume faster growth of the OCP. The fitting suggest $\beta_{\text{HAP}} \simeq 10^{-3}\text{cm/s}$, while the infinite incorporation rate, i.e. purely diffusion growth mode for all crystal sizes, should be taken for the OCP. For comparison, at low supersaturations the true kinetic coefficients (ignoring the impurity effects) for growth steps on the $NH_4H_2PO_4$ and KH_2PO_4 single crystals are $10^{-1} - 1\text{cm/s}$ [1]. The calcium phosphates grow at very high supersaturations and thus the step and kink density is expected to be high on the growing faces and should not be orders of magnitude lower than

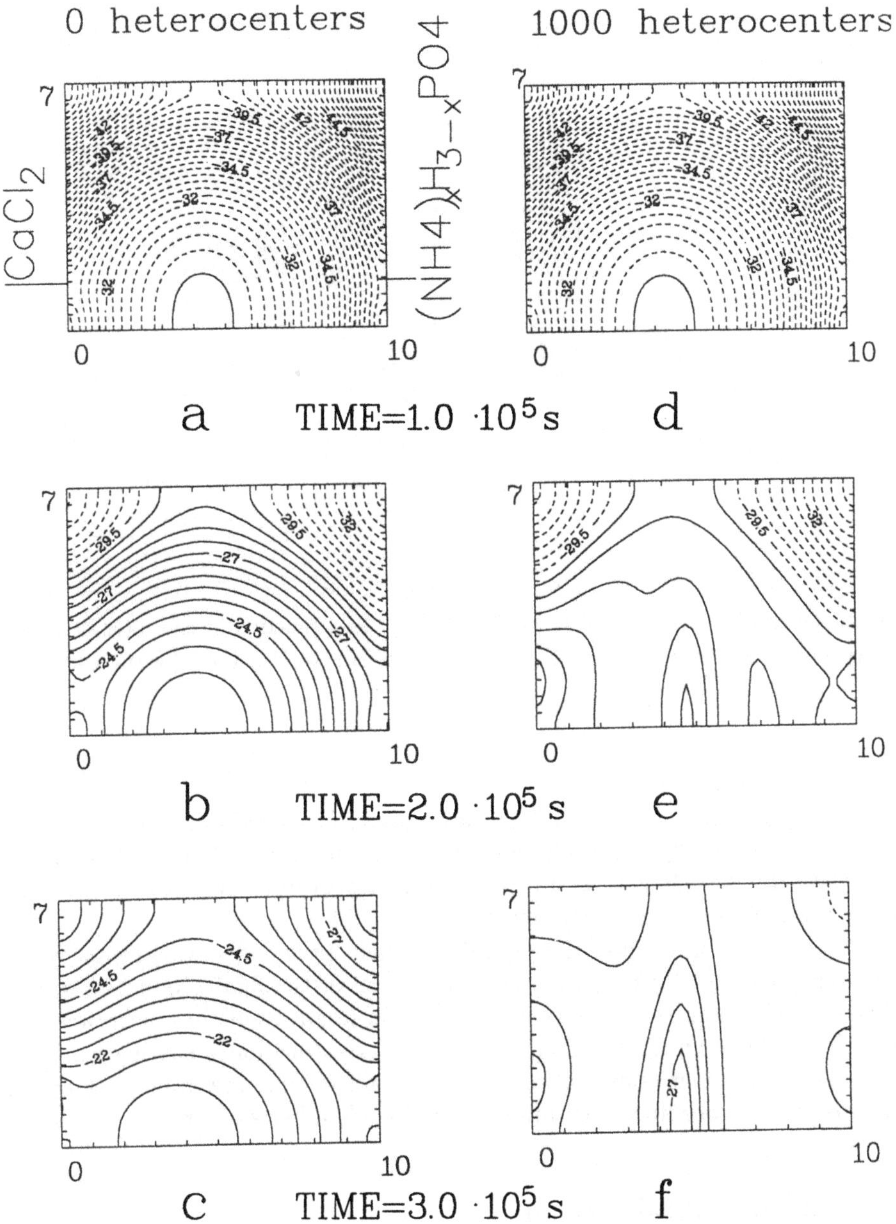

Fig. 7. Line of constant concentration product at different time elapsed after the onset of diffusion. Growth of the OCP. Dashed and solid lines correspond to the under - and supersaturated solutions. *a* to *c*: no crystals are allowed in solution, *d* to *f*:: crystals start growing on hetero centers homogeneously distributed in the chamber with the number density $10^3 \mathrm{cm}^{-3}$. The growth starts as soon as the solution becomes supersaturated.

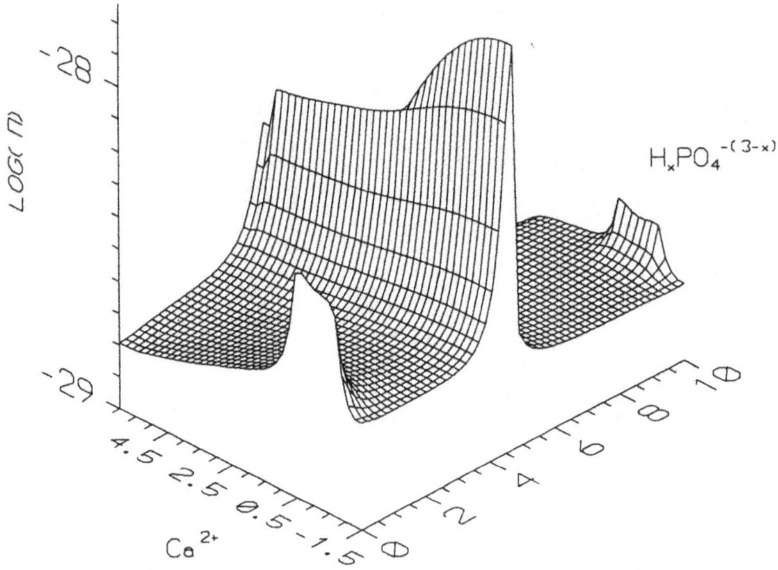

Fig. 8. Concentration product $\Pi = C^4_{\mathrm{Ca}} C^3_{\mathrm{H_x(PO_4)}}$ as a function of coordinates within the cross-section plane shown in Fig.1. The arrows indicate the holes for the Ca^{+2} and the $H_x(PO_4)^{-(3-x)}$ supply.

the ones measured for steps.

Thus under terrestrial conditions one may assume that about all the dissolved ions are consumed by the easily homogeneously nucleated HAP while in space the relatively rare OCP crystallites are able to prevail in overall mass, though being relatively low in number density.

The complex three dimensional diffusion field provided a wide range of supersaturations and thus a wide size distribution of the final product in space. Therefore, the simpler onedimensional diffusion might result in a narrower distribution. A better understanding of the mass crystallization of sparingly soluble salts needs also in situ observation of the nucleation and growth kinetics - both with respect to coordinates within the reaction chamber and time.

4. Protein crystallization

The experiments on protein crystal growth in space [31]-[35] suggest that, sometimes, the best of the space grown crystals are larger and provide better resolution for X-ray structure analyses than the best of the ground-grown samples. More precisely [36], about 20% of the protein crystals grown in space allow the 0.2-0.3Åbetter resolution and 10-15% higher number of reflections than their earth-grown counterparts. About 20% of the space crystals were worse and ca

60% showed the same diffraction quality. This ambiguity might be associated with the wide range of nucleation and growth conditions employed (see Sec.3.).

Therefore, the issue needs more experimental work supported by attempts to understand possible reasons for the quality improvement. In analogy with conventional inorganic crystal growth one should expect that the most important reason for the perfection improvement is low supersaturation achieved rather soon after nucleation in microgravity.

The X-ray diffraction resolution is controlled by the long range order in the lattice, translational and rotational. The former is controlled by homogeneity of molecular sizes, along with traditional mosaicity, dislocations etc. The orientational order is violated if molecules join the lattice in wrong orientations. The latter might even be accompanied by chemical (hydrogen or salt bridge) binding, though weaker than the ones for the properly oriented molecules. These misoriented molecules in the crystal may be considered as an impurity the attachment energy of which in the kink position is lower than the one for the molecule properly oriented with respect to its neighbors in this kink. To the accuracy of this analogy, the concentration n_f of the misoriented molecules is expected to be proportional to [37]

$$n_{\mathrm{f}} \propto \frac{1+\sigma}{[1+(w_{-\mathrm{AA}}/w_{-\mathrm{AB}})(1+\sigma)]\sigma+(w_{-\mathrm{BA}}/w_{+\mathrm{BA}})} \; . \tag{13}$$

Here A and B symbolize the properly and improperly oriented molecules, respectively: $w_{\pm\alpha\beta}$ denote the attachment (+) and detachment (-) frequency of the species β=A,B to the kink, i.e. the end of the atomic row along a step ended by the species α=A,B. In these notation the supersaturation

$$\Delta\mu/kT \simeq \sigma = (w_{+\mathrm{AA}} - w_{-\mathrm{AA}})/w_{+\mathrm{AA}} \; . \tag{14}$$

For weakly adsorbed misoriented molecules one has $w_{-\mathrm{BA}}/w_{+\mathrm{BA}} > 1$ (at equilibrium temperature) and $w_{-\mathrm{AA}}/w_{-\mathrm{AB}} < 1$. Therefore one may expect that n_f increases with supersaturation σ at least at low σ.

The physical reason for such dependence is that the misoriented impurity particle is "buried" by the next properly or improperly oriented particles before it desorbes from the kink. The higher the supersaturation the higher is the probability of the improper burying and consequently the lower is the crystal perfection. The larger the concentration of the faulted molecules in the crystal the more probable is the presence of such molecules in the kink position. Therefore the probability for the next molecule to be attached permanently in the faulted orientation is even larger. Therefore one should expect very steep concentration increase of the faulted species at supersaturations above a certain critical level - the kinetic phase transition occurs [37]. At even higher supersaturations the crystal may be completely orientationally disordered, though translationally ordered. We might suspect that this kind of disordering happens e.g. in thermolysin, which exhibits polygonal shape, optical birefrigence, but no X-ray diffraction [38]. The phenomenon may be also due to building of the thermolysin in from aggregates. The authors support this model by electron microscopic photographs [39].

Alternatively, the orientational disordering may be described treating a misoriented molecule as a colloidal inclusion [37, 3]. Namely, if a step propagates at a rate v it needs an average time $\sim a/v$ to cover a lattice space. assuming that the misoriented particle has the residence time τ_k in a kink site on the step, then the probability that this "wrong" particle is buried by the step should be

$$e^{(-a/\tau_k v)} \equiv e^{(-v_c/v)} \tag{15}$$

with $v_c = a/\tau_k$. Equation (15) means a steep increase in the relative amount of misoriented particles when the step rate v rises together with supersaturation whether non-linearly or linearly. The absolute concentration of misoriented species seems to be proportional, besides to the factor (15), to the effective frequency of various faulted attachments to the kinks. Much more experimental data on the nature of defects responsible for the diffraction deterioration, step propagation and impurities are needed for deeper insight into the influence of supersaturation and discrimination between possible models, including the ones discussed above.

5. Stability of crystal shape and interface

5.1 The crystal shape

Let a polygonal crystal grow layerwise in about purely diffusion condition. Then the surfaces of constant concentration are about spherical and therefore the supersaturation at the crystal corners and edges are higher than the one near the face centers. At supersaturations allowing only the dislocation but not the 2D nucleation growth mode, each face remains practically flat. The reason is that the steps are generated by dislocations somewhere in the middle part of the face at minimal supersaturations. The steps forming a growth hillock accelerates when moving to the face periphery and thus compensate the supersaturation inhomogeneity, however, by decreasing their density. If supersaturation at the crystal corners and edges exceeds the one needed for the 2D nucleation, then these are the vicinities of the corners and edges leading the growth. In such a case, the steps are spreading towards the face center rather than to the face periphery and depressions arises around the face centers. Such initial stages of skeleton growth are known to exist under diffusion condition. These depressions might be expected on proteins growing typically at high supersaturation are well known on the terrestrial crystals and are even more probable in microgravity, provided, however, sufficiently high supersaturation is sustained. Impurities "scraped" by the growth steps may also be responsible for the skeleton shapes.

5.2 Step bunching

A growing vicinal face formed by elementary steps may be morphologically unstable, i.e. the steps may be piled up into bunches. If solution flows in the same direction in which the steps move this flow dramatically enhances this instability. If the solution flows up of the step stream the bunches disappear and the vicinal remains stable [2, 39]. This phenomenon arises since the diffusion fields around the steps are drifted by solution flow within the boundary layer, in the immediate vicinity of the growing interface [7].

Let us now assume that the solution is stagnant. Then the step flow is equivalent to the (ideal) liquid flow in the direction opposite to the one of the steps and, as such, it should stabilize the vicinal face against the step bunching [40]. This effect was called self stabilization. It exists not only in solution but also in growth from the impure [41] and pure melt [42]. The higher the step rate, the stronger is the self stabilization. The estimates shows that this kinetic stabilization and self-stabilization most effective at long perturbation wavelengths are much stronger than capillarity stabilizing only the short-wave perturbations [41, 42]. Thus, together, these two factors at high rates should provide absolute stability of a layerwise growing vicinal face. The higher is the unperturbed step density, the weaker is the self-stabilization. It does not naturally operate on the slightly anisotropic rough faces.

To observe the self-stabilization one needs the solution in which the flows have rates much smaller than the ones of the growth steps, i.e.,$\ll 10^{-5} - 10^{-3}$cm/s for solution growth at conventional supersaturations up to several percents. Such stagnation may be achieved in microgravity even in large volumes but needs probably very special arrangements on earth even in microvolumes, if at all. The step rates in supercooled melts may reach 1-50 cm/s so that the effect might be tested in convecting melt.

Succinonitrile dendritic velocities measured recently in microgravity have been found to be several times lower than on earth demonstrating the importance of convective flows [43]. This achievement obtained in the understanding of the morphology of dendritic growth of substances with non-singular rough interfaces encourages more systematic analyses of flows and of purely diffusion growth conditions on morphological stability of faceted crystals and their singular faces growing layerwise.

6. Mechanical forces

In the growth from pure vapor, without inert gas, each of the molecule incorporated into the crystal brings the momentum ($\simeq \sqrt{3mkT}$), one half of the momentum the reflected or reevaporated molecule brings to an inert or at least not growing surface. This effect has been used to measure the growth rate from the vapor phase using a torsion balance with a growing crystal platelet on one side and the opposite side mate inert [44]. Similar to that, a free crystal should experience a net force if its opposite halves grow at unequal net rates. Real growth rate of different, even crystallographically equivalent faces, are not equal. First of all, because of different defect structure and impurities and, on top of that fluctuations in time. Therefore one may expect enhanced chaotic motion of growing microcrystals - as compared to the conventional Brownian motion [45, 46].

In a solution or in a gas in which a noticeable amount of inert species is present the difference in partial pressure of the crystallizing species is, at least partly, compensated by the inert species. Nevertheless, diffusiophoresis is well known to exist in both electrolyte and non-electrolyte solutions [47, 48]. The diffusiophoresis of the crystallites growing in a mass-crystallization chamber in microgravity may be expected to occur under the influence of microscopic concentration gradient in the chamber and, also, because of local gradients inevitably arising around crystals separated by distances comparable to the crystal size. The sign of the diffusiophoresis depends on the molecular field and subsurface ordering in solution. For instance, balls of japanese wax are moving towards the lower concentration of glucose in a water-methyl-alcohol mixture [47].

There might be also a contribution from the unbalanced momentum transfer to the crystal side growing or dissolving at different rates - similar to the vapor growth, though strongly reduced by the solvent pressure compensation. This effect might be analogous to the osmotic pressure since the non-equilibrium crystal-solution interface is "transparent" for the crystallizing solute in the super- or undersaturated solutions but is not "transparent" for the solvent. One might estimate the momentum transfer to be similar to the one resulting from the osmotic pressure. Let m_0 and m_1 and n_0, n_1 are the amsses and number densities of the solvent and solute, respectively. Then the growth induced pressure might be expected to be

$$\propto \sqrt{2\pi kT}\left[\sqrt{m_0}D_0\frac{\partial n_0}{\partial z}+\sqrt{m_1}D_1\frac{\partial n_1}{\partial z}\right]$$

where D_0 and D_1 are the diffusivities and the gradients $\partial n_0/\partial z$ and $\partial n_1/\partial z$ have opposite signs. It is also known that the crystals growing in solutions may be electrically charged. This charge depends on the growth rate and may also cause crystal motion in the process of mass crystallization in microgravity.

7. Conclusions

Growth of inorganic salt from aqueous solutions in microgravity is expected to be purely diffusion controlled beginning with a crystal size of $10^{-2} - 10^{-1}$cm, while for the protein crystals the contribution of interface incorporation kinetics may be important up to the cm size scale.

Low supersaturations seems to be responsible for the growth of the octacalciumphosphate (OCP) and hydroxyapatite (HAP) phases in microgravity, opposite to practically pure HAP obtained in terrestrial experiments. The space-grown OCP and HAP crystals are at least 1-2 orders of magnitude larger than their terrestrial counterparts. Low supersaturation in microgravity seems to be responsible for these differences.

Low supersaturation might be responsible for the sometimes observed higher perfection of protein crystals due to the trial and error mechanism known to be the more effective the lower the supersaturation in trapping of impurities.

In a completely stagnant solution, the motion of growth steps may cause morphological self-stabilization hardly observable in the presence of convection, because of low ($< 10^{-4}$cm/s) step velocities.

Diffusiophoretic forces may redistribute crystals suspended in solution. Momentum transfer accompanying the growth and dissolution processes may also contribute to the mechanical interactions.

Acknowledgements

The calcium phosphate samples experimental conditions, diffusivities of specific ions and optical and X-ray difraction powder analyses have been provided by Prof. H.E. Lundager Madsen. The characterization and analyses of terrestrial and space-grown phosphate crystals have been also performed in collaboration with E.I.Suvorova, M.O.Kliya, V.V.Klechkovskaya, A.A.Timofeev and O.Lebedev. The computer simulation have been done in cooperation with by L.E.Polyak. Cooperation with all these colleagues and discussion with Mr. H.Schneiter and K.Bentz as well as financial support from Oerlikon Contraves AG, Zurich is highly appreciated. I am very thankful also to Ms L.A.Solomentseva and L.S.Spiridonova for technical assistance.

References

1. A.A.Chernov, Progress in Crystal Growth and Characterization of Materials, vol **26** (1993) 121.
2. A.A.Chrenov, Contemp.Phys. **30** (1989) 251.
3. A.A.Chernov, Modern Crystallography III. Crystal Growth. Spr.Ser.Sol.St. vol **36**, Berlin, 1984).
4. A.A.Chernov and H.Komatsu, in: Science and Technology of Crystal Growth. Eds J.P.van der Eerden and O.S.L.Bruinsma (Kluwer Acad.Pub.Dordrecht, 1995, ch.6.4).
5. L.A.Monako and F.Rosenberger, J.Cr.Gr.**129** (1993) 465.
6. Y.Matsuura, H.Komatsu and A.A.Chernov, Abstracts, ICCG11, The Hague, June 18-23, 1995.
7. A.A.Chernov, J.Cr.Gr. **118** (1992) 333.
8. P.G.Vekilov, Yu.G.Kuznetsov, A.A.Chernov, J.Cr.Gr. **121** (1992) 643.
9. A.I.Malkin, A.A.Chernov and I.V.Alexeev, J.Cr.Gr. **97** (1989) 765.
10. J.J.de Yoreo, T.A.Land and B.Dair. Phys.Rev.Lett. **73** (1994) 838.
11. P.G.Vekilov, M.Ataka and T.Katsura, J.Cr.Gr. **130** (1993) 317.
12. A.A.Chernov and H.Komatsu, in: Science and Technlolgy of Crystal Growth. Eds J.P.van der Eerden and O.S.Bruinsma (Kluwer Acad.Pub.Dordrecht, 1995, ch.2.2).
13. H.A.Lowenstam and S.Weiner, On Biomineralization (Oxford Univ.Press. Oxford 1991),
14. Biological Mineralization and Demineralization, Ed. G.H.Nancollas, Springer, Berlin, 1982,
15. The Chemistry and Biology of Mineralized Tissues, Ed.W.T.Butler (Ebsco Media, Birmingham, AL, 1984,
16. Biological Mineralization. Ed.G.H.Nancollas, J.Cr.Gr. **53** (1981), special issue,
17. R.S.Bockman, A.L.Boskey, N.C.Blumental, N.W.Alcock and R.P.Warrell Ir., Calcif. Tissue Intern. **39** (1986) 376,
18. M.A.Repo, R.S.Bockman, E.Betts, A.L.Boskey, N.W.Alcock and R.P.Warrell Jr., Calcif. Tissue Intern. **43** (1988) 300.

19. H.E.Lundager Madsen, Acta Chemica Scandinavica **24** (1970) 1677.
20. H.E.Lundager Madsen, I.Lopez-Valero, V.Lopez-Acevedo and R.Boistelle, J.Cr.Gr. **75** (1986) 429.
21. G.Vereecke and J.Lemaitre, J.Cr.Gr. **104** (1990) 820.
22. J.A.Budz and G.H.Nancollas, J.Cr.Gr. **91** (1988) 490.
23. J.Zhand and G.H.Nancollas, J.Cr.Gr. **123** (1992) 59.
24. J.Zhang and G.H.Nancollas, J.Cr.Gr. **125** (1992) 251.
25. F.J.G.Cuisinier, J.-C.Voegel, F.Apfelbaum and L.Mayer, J.Cr.Gr. **125** (1992) 1.
26. R.M.H.Vereecke and A.H.Devenys, J.Cr.Gr. **121** (1992) 335.
27. M.R.Cristoffersen and J.Cristoffersen, J.Cr.Gr. **121** (1992) 617.
28. H.E.Lundager Madsen and F.Cristensson, J.Cr.Gr. **114** (1991) 613.
29. H.E.Lundager Madsen, F.Christensson, L.E.Polyak, E.I.Suvorova, M.O.Kliya and A.A.Chernov, Accepted to J.Cr.Gr.
30. H.E.Lundager Madsen, F.Christensson, A.A.Chernov, L.E.Polyak, E.I.Suvorova, M.O.Kliya. Adv.Space Res. Vol.**16** (1995) 8(65)-8(68).
31. W.Littke and C.John, Science **225** (1984) 203.
32. L.J.DeLucas, F.L.Suddath, R.Shyder, R.Naumann, M.B.Broom, M.Pusey, V.Yost, Blair Herren, D.Carter, B.Nelson, E.J.Meehan, A.McPherson, C.E.Bugg. J.Cr.Gr. **76** (1986) 681.
33. S.D.Trakhanov, A.I.Grebenko, V.A.Shirokov, A.V.Gudkov, A.V.Egorov, I.N.Barmin, B.K.Vainshtein and A.S.Spirin, J.Cr.Gr. **110** (1991) 317.
34. J.Cr.Gr. **122** (1992) 310-340, J.Cr.Gr. **110** (1991) 302-333.
35. A.McPherson, A perspective on macromolecular crystal growth in microgravity. The abstract and report presented at the Colloque Annuel Group Francais de Croissance Cristalline, Gif-sur-Yvette, 23-25 March 1994.
36. B.L.Stoddard, R.K.Strong, J.Cr.Gr. **360** (1992) 293.
37. A.A.Chernov, Sov.Phys.Uspekhi **13** (1970) 101.
38. G.Sazaki, H.Ooshima, J.Kato, Y.Harano and N.Hirokawa, J.Cr.Gr. **130** (1993) 357.
39. A.A.Chernov, Yu.G.Kuznetsov, I.L.Smol'sky and V.N.Rozhansky, Sov.Phys.Cryst. **31** (1986) 14.
40. A.A.Chernov, S.R.Coriell and B.T.Murray, J.Cr.Gr. **132** (1993) 405.
41. S.R.Coriell, B.T.Murray and A.A.Chernov, J.Cr.Gr. **141** (1994) 219.
42. A.A.Chernov, S.R.Coriell and B.T.Murray, J.Cr.Gr. **149** (1995) 120.
43. M.E.Glickman, M.B.Kossand, E.A.Winsa, Phys.Rev.Lett. **73** (1994) 573.
44. R.F.Strickland - Constable Kinetics and Mechanism of Crystallization (Academic Press. London and New York, 1968), see p.255.
45. A.A.Vedernikov and I.V.Melikhov, Proc.Int.Workshop on Short Term Experiments inder Strongly Reduced Gravity Conditions (Bremen, 1994), p.263.
46. I.V.Melikhov and A.A. Vedernikov, Vestnik MGU, Sec.2, Khimiya **36** (1995) N 1 (Trans. of the Moscow State University, Sec.2, Chemistry) - in Russian.
47. B.V.Deryaguin, N.V.Churaev and V.M.Muller, Surface Forces (Moscow, Nauka, 1985) in Russian.
48. S.S.Dukhin and B.V.Deryaguin, Electrophoresis (Moscow, Nauka, 1976) in Russian.

An Investigation of the Perfection of Lysozyme Protein Crystals Grown in Microgravity and on Earth

J.R. Helliwell, E. Snell and S. Weisgerber

Chemistry Department, University of Manchester, Oxford Road, Manchester, M13 9PL, UK

Abstract

Lysozyme has been used to investigate the effect of microgravity crystallisation on protein crystal perfection. Crystals were grown in the European Space Agency's Advanced Protein Crystallisation Facility onboard the NASA Space Shuttle. Two missions of differing duration took place, Spacehab-1 and IML-2. The microgravity crystallisation time in each was 7 days and 10 hours, and 12 days and 11 hours respectively. The IML-2 crystals had grown much larger than the Spacehab-1 crystals (2.5 mm versus 0.8 mm at maximum). The earth grown control crystals, in each case, reached a size of 0.8 mm at maximum. The perfection of the crystals was evaluated with collimated, intense, synchrotron radiation. This was done using the Laue method, via the spot size, and by monochromatic rocking widths directly. For the Spacehab-1 crystals spot size measurements were carried out on station 9.5 of the SRS, along with an analysis of intensity to sigma ratio, immediately after the mission. Five months later rocking widths were measured at LURE. The IML-2 crystals were evaluated at the ESRF, on BL3 three months after their return to earth and also a further three months later on the joint Swiss-Norwegian beamline. Both the Spacehab-1 and IML-2 crystals were of exceptional perfection with the crystal mosaicity reaching values as small as 0.0010° and 0.0017° respectively. Earth-grown control crystals had values as small as 0.0032° and 0.007° respectively. There is no evidence of 'shelf-life' ageing of the crystals, at least over a period of 6 months, since there is close agreement of the mosaicity values from the Spacehab-1 crystals tested within weeks of that mission and the IML-2 crystals tested 6 months after that mission. The perfect mosaic block size has evidently increased over that realised in the earth-grown controls.

1. Introduction

The growth of a protein crystal is arguably the most critical step on the path to the determination of a protein structure at atomic resolution by X-ray crystallography. The determination of such structures has a tremendous impact on biological research by revealing the functions of macromolecules. These methods can be applied also to nucleic acids and their complexes with proteins including viruses. As a result a detailed understanding has been gained of various living processes such as oxygen storage and transport, enzyme catalysis, the immune response, the encoding of hereditary information, viral infection and photosynthesis. It should be emphasised that the field of macromolecular crystal structures and the need for the growth of crystals is distinct from the field of crystal growth of semiconductors etc. where the crystals themselves perform some function. By contrast, in protein crystallography, the crystal is the vehicle to allow determination of the molecular structure.

The presence of gravity in the ground-growth of crystals is manifest in several ways [11]. These are, firstly, convective flow and mixing of the liquid due to density gradients at the growing crystal surface, secondly, sedimentation due to density differences between the crystal and its 'mother' liquor and, thirdly, the need for a container, which adds a further surface(s) impacting on crystal nucleation and/or growth. These effects can, respectively, cause some degree of turbulence in the growth of a crystal, damage to crystals on sedimentation and extra complications. Microgravity therefore offers a medium where convection driven turbulence is absent, the growth process is not impeded by sedimentation and, in principle, 'containerless' crystal growth is possible. There is scope then for investigating microgravity as an environment to improve the size, quality and resolution of crystals for X-ray analysis as well as producing much larger crystals suitable for neutron protein crystallography.

2. Crystallization Apparatus and Method Used

The European Space Agency (ESA) and, on contract, Dornier GmbH have developed the Advanced Protein Crystallization Facility (APCF) as a standard tool for microgravity crystallization experiments aboard the NASA Space Shuttle [16, 2]. Each APCF facility supports 48 crystallization reactors at a constant process chamber temperature with video CCD monitoring of 12 of the reactors at variable depths of focus. The crystallization reactors support three types of crystallization, vapour diffusion, free interface liquid diffusion and dialysis crystallization - the latter is the method we have utilised. The dialysis method has the advantage over the vapour diffusion method of avoiding Marangoni convection [13]. The dialysis reactor consists of two quartz glass blocks separated by a dialysis membrane. The upper block contains the protein solution (188μl in our case), the lower block the salt solution. The salt and buffer are separated by a cylindrical quartz glass plug which also contains salt solution. To activate (or deactivate) the reactor this plug is rotated by 90° so that all chambers become contiguous (or blocked). In preparation for each mission, laboratory trials were used to determine the optimum crystallization conditions. These conditions are summarised in table 1. Because of the longer duration of the IML-2 mission, compared with Spacehab-1, a slightly reduced salt concentration was used so as to prolong the crystal growth process.

Reactor Part	Filling Solution
Protein Chamber (188μl)	15.79mg lysozyme (3× crystallized, dialysed and lyophilized powder of chicken egg white lysozyme) dissolved in 188μl 0.04M acetate buffer (pH 4.7)
Buffer Chamber (59μl)	0.04M acetate buffer (pH 4.7)
Salt plug and reservoir (541 μl)	7.89g NaCl in 100ml H_2O (Spacehab-1) 7.34g NaCl in 100ml H_2O (IML-2)

Table 1. Filling solutions for dialysis reactor

3. Evaluation of Perfection

The quality of crystals for structural analysis is characterized by a number of factors namely size, mosaicity and resolution limit. An objective of this program of work is to establish if, for tetragonal lysozyme crystals (a=b=79.1 Å , c=37.9 Å, $P4_32_12$) at least, a reduction of the number of mosaic blocks (ideally to one) can occur with perfect growing conditions, and to see if this then improves the diffraction quality and resolution.

Crystal mosaicity can be evaluated by measurement of the rocking width [6, 4]. Rocking width measurements need to be done at a synchrotron to ensure that the angular width of the diffraction profile is dominated by the mosaicity of the crystal rather than the X-ray beam divergence or spectral spread. One possible method is to use a diffractometer with a small angular step size combined with a monochromatic synchrotron beam of very low divergence and small $\delta\lambda/\lambda$. Then a representative sample of reflections at various Bragg angles are scanned and the full width at half maxima (FWHM) determined. The crystal sample mosaicity, η, can be obtained by deconvoluting the instrument resolution function, IRF, out of the overall, as measured, reflection width, φ_R, using equation (1), (IRF' in [4]).

$$\eta = \sqrt{\varphi_R{}^2 - IRF^2} \tag{1}$$

An alternative method to evaluate the mosaicity of the crystals is to make use of Laue diffraction. The mosaicity of the sample can be derived from the size and shape of the Laue diffraction spots. Their radial extension is given by equation (2) [1] for the case of an incident X-ray beam of zero size and divergence;

$$\Delta_{radial} = 2\eta \frac{D}{cos^2 2\theta} \tag{2}$$

where D is the crystal to detector distance, and θ the Bragg angle. In practice for η to be measurable by the Laue method, as low a divergence as practicably possible is established by slitting down the beam, and by using a small collimator (0.2 mm) the beam cross section is also kept small. In addition a long crystal to detector distance is used (a few metres) and a detector such as film with a fine grain size is employed. The size of the direct, non-diffracted, beam measured at the film position allows for proper account of the practical size and divergence of the beam to be made. This is then subtracted from the measured diffraction spot size to finally yield Δ_{radial} as given by equation 2 and from which η is extracted, initially at full width (ie. 6σ) then converted to a FWHM (ie. 2.3σ).

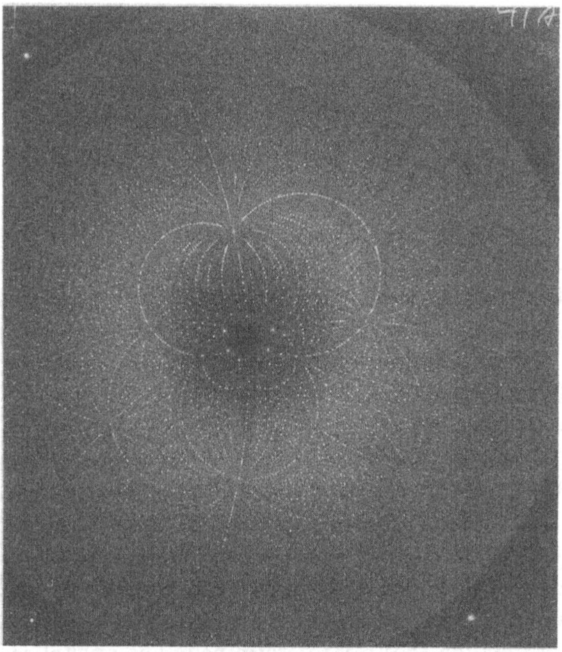

Fig. 1. a/1. Spacehab-1 evaluation via the synchrotron Laue method at SRS 9.5. Laue patterns recorded at 90 mm from the crystal samples. Earth grown sample.

4. Spacehab-1 Investigation

Laboratory trials prior to the mission had yielded crystals growing on the walls of the protein chamber which led to difficulties in harvesting useable crystals. To overcome this problem, in preparation for the mission itself, the protein chambers were siliconized, 10 minutes siliconization treatment for the flight reactors, 5 minutes for the ground control reactors - too much siliconization on the ground

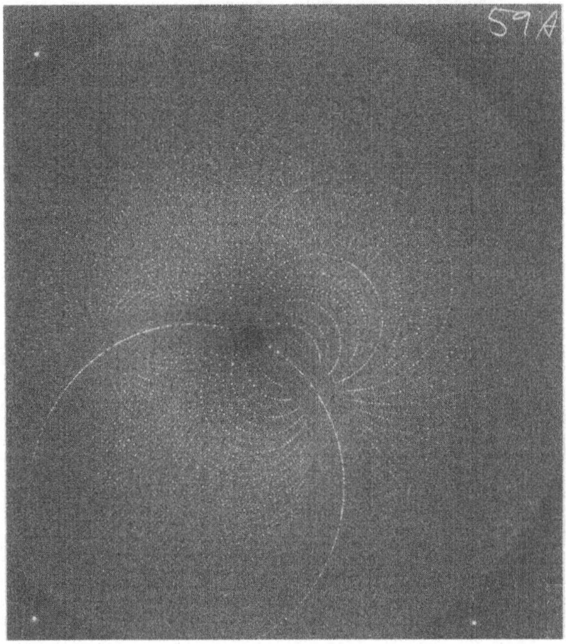

Fig. 1. a/2. Spacehab-1 evaluation via the synchrotron Laue method at SRS 9.5. Laue patterns recorded at 90 mm from the crystal samples. Microgravity grown sample. At this distance from the crystal the size and shape of the spots in the pattern look essentially identical to the earth grown sample.

control reactors had led to sedimentation during further laboratory trials. This was not expected to be a problem for the microgravity reactors. The reactor was filled according to the conditions in table 1.

Due to a launch delay of the STS-57 Space Shuttle carrying the Spacehab-1 mission the reactors stayed in a preprocess state (protein and salt solution separated) for 28 days. During this time precipitate was observed in some of the reactors within the APCF facility. This was more noticable in the flight reactors rather than in the ground control reactors. The reactors for the mission were therefore swapped over. The mission allowed 7 days and 10 hours of microgravity growth. The number and sizes of crystals in each reactor were assessed under an optical microscope. In each of them, ca. 100-150 crystals up to a size of 0.8 mm were observed; most were 0.6-0.7 mm long. In the flight reactors ca. 80% of the crystals were growing on the walls, the rest on the top and bottom membranes. The ground control experiment had yielded the same number of crystals and of the same size, but the vast majority had sedimented so that only the few crystals that were growing on the walls (ca. 10-20) could be harvested.

The crystal mosaicity values ([18, 17] were derived from the spot size using the Laue method on station 9.5 of the SRS Daresbury. The beam was trimmed

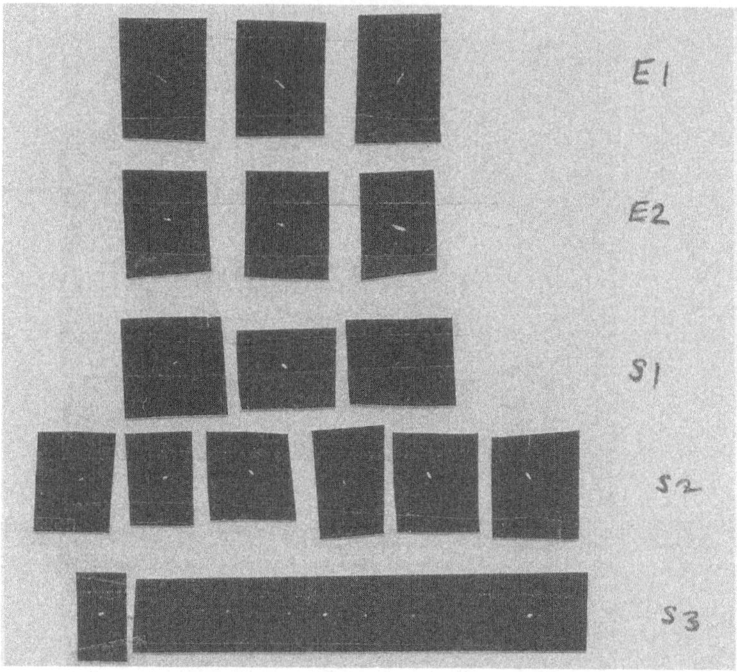

Fig. 1. b. Spacehab-1 evaluation via the synchrotron Laue method at SRS 9.5. Laue spots recorded on film 2.4 m, from the crystal samples. Now the effect of the different mosaicities of each crystal are revealed. From top to bottom the rows are; Earth 1, Earth 2, Space 1, Space 2 and finally, bottom row, Space 3 (this labelling corresponds to that given in Tab. 2).

down by a factor of ≈50 in the horizontal and ≈10 in the vertical direction to obtain a very low divergence beam (0.03 mrad and 0.012 mrad respectively). The crystal to film distance was 2433 mm. The direct beam size itself was recorded on self-adhesive paper coated with a radiation sensitive layer at this position of 2.4 m from the crystal. Five crystals were evaluated altogether, namely three microgravity grown and two earth grown. Three other crystals (two earth grown and one microgravity grown) were examined but were badly damaged on harvesting from the reactors and subsequent crystal mounting. In Weisgerber [18] this was referred to as 'twinning'. Only medium intensity diffraction spots were selected in all cases, to avoid any flaring of spots, and their full size and shapes measured under an optical microscope. Fig. 1 compares the Laue spots obtained for the earth grown and the microgravity grown crystals. The size of the direct beam was subtracted from the Laue spot sizes , and η then calculated from equation (2) and converted to a FWHM value. Table 2. shows a summary of the results.

Full Laue intensity data sets were then also recorded on a MAR image plate scanner situated on station 9.5 of the SRS from one earth-grown and one mi-

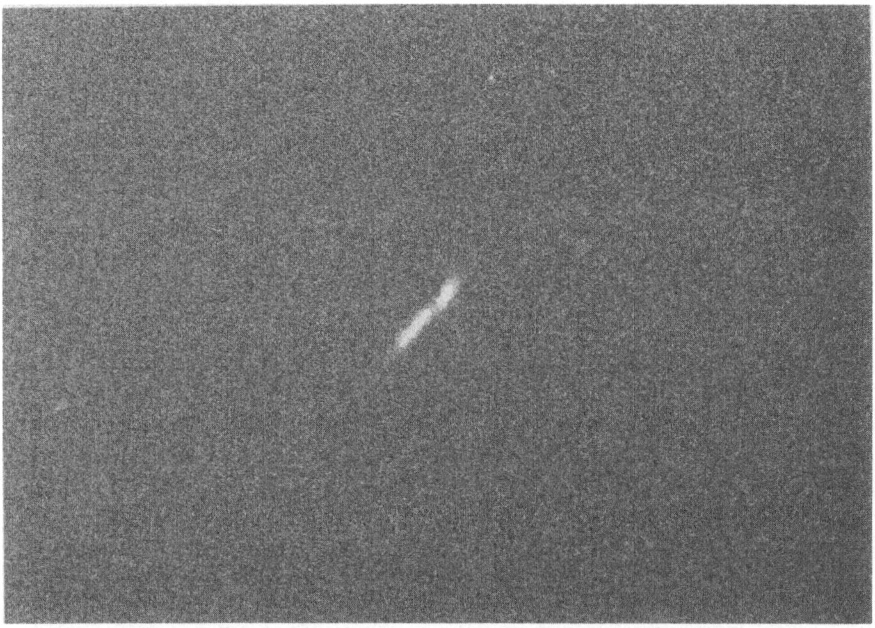

Fig. 1. c/1. Spacehab-1 evaluation via the synchrotron Laue method at SRS 9.5. An enlarged Laue spot for sample Earth 1 (this figure) and sample Space 3 (Fig. 1c/2) representing the extreme cases in detail (see Tab.2).

crystal	Earth 1	Earth 2	Space 1	Space 2	Space 3
$<\eta>^{\circ}$	0.0062	0.0032	0.0012	0.0022	0.0010
$\sigma(<\eta>)^{\circ}$	0.0006	0.0001	0.0002	0.0001	0.0001
Spots	7	7	3	14	7

Table 2. Crystal mosaicities estimated from Laue spot sizes at the SRS. For examples of the actual spots see figures 1b and 1c.

crogravity grown crystal. Eleven images were collected at 6° intervals for both. A 0.4 mm thick Al foil was used to partially attenuate the longer wavelengths in the beam. Hence λ_{max} was 1.4 Åand λ_{min} was 0.5 Å, the latter set by the cutoff of the focusing mirror. The circulating current in the ring was approximately 130 mA and exposure times of 500 msec were used throughout. The data were processed using the Daresbury Laue suite of software ([7] and further reduced using CCP4 software ([3]. With Laue singlet reflections only the earth-grown crystals produced a data set with completeness d_{min}-∞ 76.5%, d_{min}-2 d_{min} 79.7% and 2d_{min}-∞ 47.6%. The corresponding microgravity-grown crystal produced a data set of completeness d_{min}-∞ 77.7%, d_{min}-2 d_{min} 80.9% and

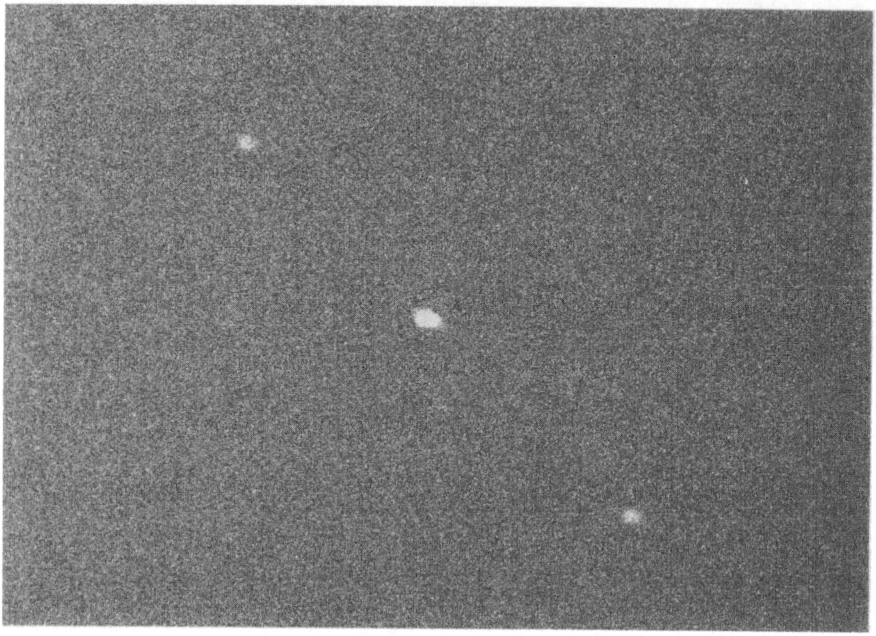

Fig. 1. c/2. Spacehab-1 evaluation via the synchrotron Laue method at SRS 9.5. An enlarged Laue spot for sample Space 3 (see Tab.2).

$2d_{min}$-∞ 48.5%. These values are essentially identical; the small % differences being what one would expect due to different crystal orientations. An analysis of d_{min} vs $I/\sigma(I)$ is given in table 3. These are very similar up to 2.70 Å resolution. The microgravity data is better for the 2.70 to 2.50 Å resolution range. Any difference looks marginal however. The effects are probably being masked by the use of the image plate whose detector quantum efficiency (DQE) performance with weak signals and small spots (0.2 mm) is adversly effected by its large point spread factor (0.4 mm), see [10].

Five months later rocking widths were measured monochromatically using a diffractometer at LURE, Paris. These results are summarised in table 4. The average values of φ_R and their recorded distributions are not significantly different from each other. These values, however, are a combined value of sample mosaicity along with instrument smearing and diffractometer step size.

5. IML-2 Investigation

The crystallization recipe for the IML-2 mission (table 1) was identical to that for the Spacehab-1 mission with the exception of a slightly reduced salt concentration (7.34g in 100ml) to allow for the longer projected period for microgravity

Earth Grown controls			Microgravity (Spacehab-1)		
d	$I/\sigma(I)$	N_{ref}	d	$I/\sigma(I)$	N_{ref}
8.60	4.1	43	8.90	5.3	37
6.86	7.5	146	7.01	5.8	96
5.88	7.9	214	5.96	7.2	196
5.22	11.5	262	5.28	6.5	258
4.75	4.5	561	4.79	3.4	509
4.38	5.2	1009	4.41	2.8	992
4.09	6.3	1258	4.11	5.4	1231
3.85	7.1	1419	3.86	5.8	1392
3.65	7.2	1635	3.66	8.0	1648
3.47	6.2	1766	3.48	5.7	1712
3.32	9.2	1980	3.33	6.5	1848
3.19	8.1	1781	3.19	6.6	1805
3.07	7.7	1959	3.07	7.2	1879
2.96	7.7	1791	2.97	6.8	1821
2.87	7.9	1880	2.87	6.5	1811
2.78	2.3	1927	2.78	4.9	1854
2.70	3.7	1754	2.70	4.1	1805
2.63	3.1	1922	2.63	5.5	1890
2.56	2.2	1775	2.56	4.7	1747
2.50	0.9	1770	2.50	3.3	1709

Table 3. Analysis of $I/\sigma(I)$ vs resolution, d.

Crystal	Earth grown control	Microgravity (Spacehab-1)
1	0.0165°	0.0075°
2	0.0115°	0.0090°
3	0.0230°	0.0080°
4	0.0082°	0.0080°
5	0.0085°	0.0160°
6	0.0073°	0.0075°
7	0.0160°	0.0113°
8	0.0155°	0.0073°
9	0.0105°	0.0160°
10	–	0.0155°
11	–	0.0105°
	$<\varphi_R> = 0.0091°$ $\sigma(<\varphi_R>) = 0.0025°$	$<\varphi_R> = 0.0130°$ $\sigma(<\varphi_R>) = 0.0049°$

Table 4. Rocking widths, φ_R, measured at LURE

growth. The mission lasted nearly 15 days allowing 12 days and 11 hours of microgravity growth. The mission produced good crystals of average size 1.8 mm along the longest dimension realising a maximum value of 2.4 mm. For this mission the ground control reactors stayed on the ground and the flight ones flew, consequently there were few sedimentation problems. Of note with the flight reactors was the presence of a minority of orthorhombic crystals. Two distinct groups were in evidence, clumps of orthorhombic crystals (maximum length approx 1.0 mm) and single 'needle' shaped crystals (up to a length of 2.1 mm) extruding from the surfaces of a minority of the tetragonal crystals. The ground control crystals, grown for an identical period with identical materials, were smaller, 0.6 mm on average, 0.8 mm at maximum. No orthorhombic crystals were present in the ground control reactors; the reason for this difference is not known.

The tetragonal crystals, harvested from both sets of reactors, were initially evaluated on station ID9 (BL3) of the ESRF, three months after their return to earth. Laue exposures were made onto film situated 2120 mm from the crystal. Again the beam was trimmed down to obtain a very low divergence, but sharper values were obtained on the ESRF (0.01 mrad horizontally and 0.01 mrad vertically) than the SRS.

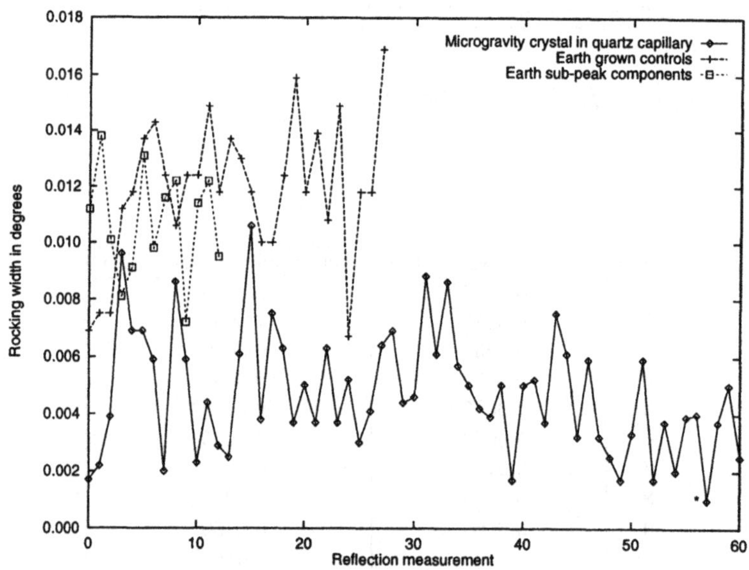

Fig. 2. Mosaicity of a microgravity crystal in quartz capillary and an earth grown crystal mounted in glass capillary, secondary peaks also plotted extracted from a sequence of reflections through the earth run. Microgravity crystal approximately double the volume of the earth crystal. The reflection indicated by '*' (reflection number 57 on the abscissa) has a measured φ_R of 0.0020°, deconvoluting the IRF of 0.00195° from this gives the indicated point, but this is clearly at the limit of the instrument measuring ability. Further details to be found in Snell *et al.* (1995).

There was some problem in the mounting of such large crystals into glass capillaries. Capillary sizes of 2.5-3.5 mm diameter were needed. These have a smaller wall thickness to diameter ratio than the more typically used capillary sizes (eg. 1 mm diameter or less). The large, standard, glass capillaries proved very fragile and suffered a high failure rate. Furthermore, microcracks in the glass capillaries, that initially appeared undamaged, unfortunately caused the protein crystals that could be mounted to dry out over a period of time within the warm conditions of the BL3 hutch.

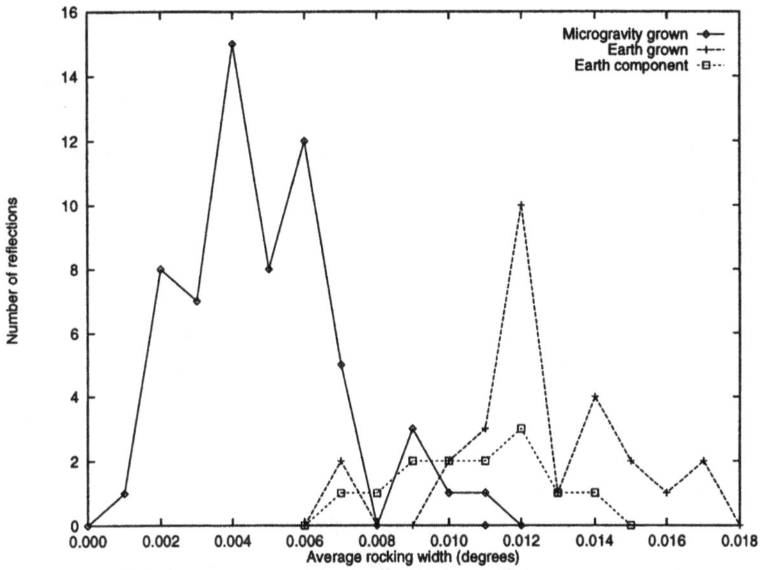

Fig. 3. Distribution of mosaicity values for a microgravity and an earth grown crystal. In the earth grown case the rocking curve consisted of resolved components. The mosaicity values of these components are also plotted for the earth crystal reflections.

Three months later, direct measurements of the rocking width were carried out on station A of the joint Swiss-Norwegian beam line at ESRF using a Huber ψ circle diffractometer from the University of Karlsruhe. The advantage of this diffractometer was that there were 5 axes to control the orientation of the crystal. In particular rocking widths for one reflection could be recorded at different azimuthal settings around a given diffraction vector. Also the diffractometer step size used was very small at 0.0002°. The beamline optics, unfocused, consisted of a double crystal Si(111) monochromator set at a Bragg angle of $\theta = 9.18°$, (ie. for a wavelength of 1 Å), with $\delta\lambda/\lambda = 2 \times 10^{-4}$. The sample slits, at 45 m from the source, were set at 0.6 mm. The ESRF vertical source size (2.3σ) is 0.2 mm. The instrument resolution function, IRF, is 0.00195° (equation (1)).

A ground control crystal exhibited mosaicity values (ie. after deconvolution of the IRF of 0.00195°) ranging from 0.0067° to 0.0169° degrees, calculated at FWHM. In the case of some reflections a composite structure of the peak was resolved. The mosaicity values of these sub-peaks or components were also evaluated at FWHM as shown in Fig. 2. There was no evidence of drying out of the crystal, of approximately 0.8 mm in the largest dimension, in the glass capillary (diameter 1 mm).

A microgravity crystal was mounted in a glass capillary for data collection but dried out quickly during data collection. A second microgravity crystal, again in a glass capillary, was measured and exhibited mosaicities ranging from 0.005° to 0.0213° (noticably increasing toward the end of that data collection run due to the crystal drying out).

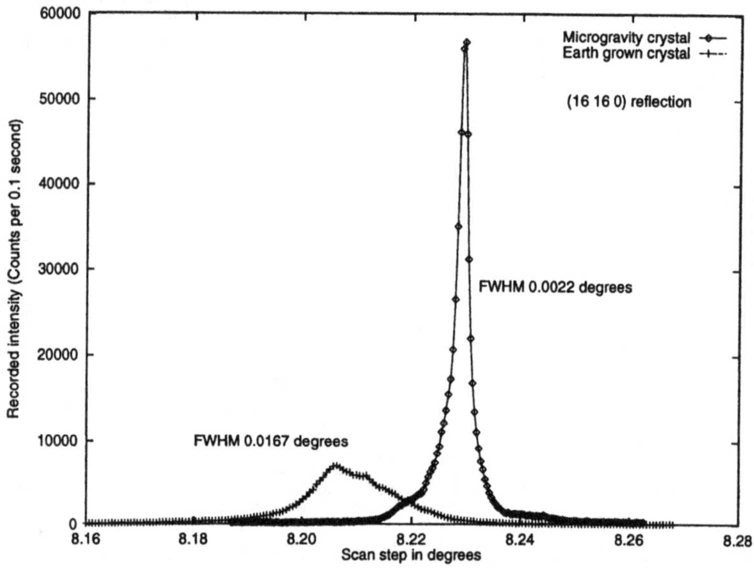

Fig. 4. Scans of an identical reflection (16 16 0) from a microgravity and an earth crystal. The labelled FWHM values have the instrument resolution function value, IRF, (0.00195°) deconvoluted out. The slight difference in Bragg angles for the two crystals is significant; the least squares refined cell parameters (from a group of reflections) are, earth-grown a=79.149, b=79.154, c=37.992 Å and microgravity-grown a=79.073, b=79.067 and c=38.051 Å . The a and b lattice parameters are unconstrained and so the difference of a and b in each gives a measure of the accuracies of the determined values.

A third microgravity crystal was then mounted in a quartz capillary. These capillaries did not exhibit the property of microcracks. Mosaicity values were again measured ranging from 0.0017° to 0.010° and with no evidence of any drying out (Fig. 2).These were clearly improved over the earth control. By plotting the distribution of the mosaicity values in Fig. 3 it is seen that the average value for the earth grown crystals is approximately three times that of the microgravity grown crystal. This factor of three difference in the mosaicity value is essentially the same as for the Spacehab-1 tests (Table 2).

It was noticed that the space crystals exhibited far higher peak intensities for reflections than the corresponding earth grown reflections. A typical reflection (16 16 0) is shown in Fig. 4, the integrated intensity of the space grown crystal reflection is almost exactly double that of the earth grown (corresponding to their relative volumes) but the peak intensity of the reflection from the space grown is a factor of 8 greater than the earth grown crystal. This is the effect partially of an increased crystal size but largely of a narrower rocking width. The diffraction quality of the microgravity grown crystal was further explored by a limited peak search in reciprocal space. It was readily possible to find evidence of diffraction to 1.2 Å, Fig. 5. A more detailed report of the ESRF Swiss-Norwegian beamline evaluation will be published elsewhere [14, 15].

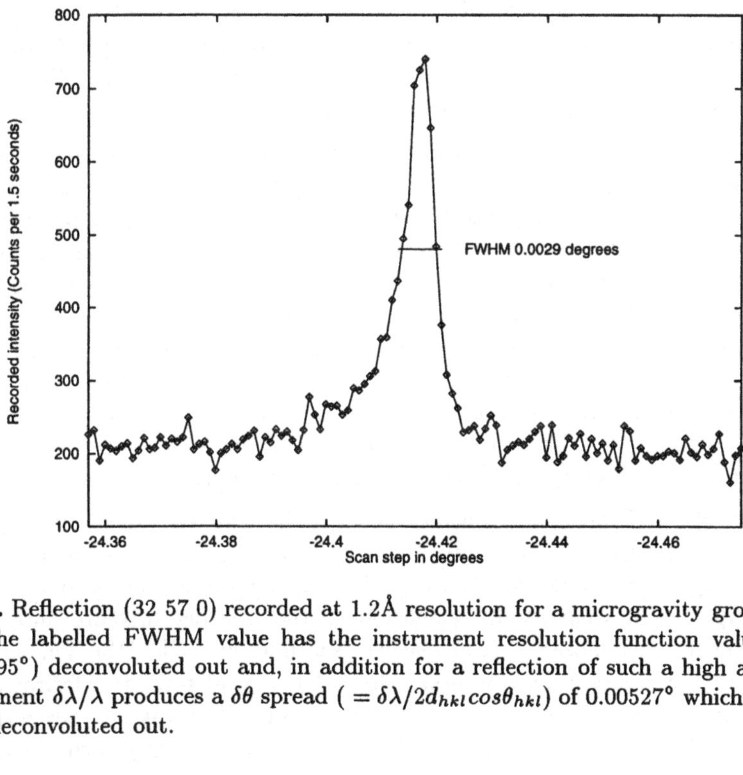

Fig. 5. Reflection (32 57 0) recorded at 1.2Å resolution for a microgravity grown crystal. The labelled FWHM value has the instrument resolution function value, IRF, (0.00195°) deconvoluted out and, in addition for a reflection of such a high angle the instrument $\delta\lambda/\lambda$ produces a $\delta\theta$ spread ($= \delta\lambda/2d_{hkl}cos\theta_{hkl}$) of 0.00527° which has also been deconvoluted out.

6. Discussion and Concluding Remarks

We believe that we have shown that the mosaicity values of microgravity grown tetragonal crystals of lysozyme are improved by a factor of three over the earth grown control values (Table 2 and Figs. 1 and 3). This is evidently accompanied by a reduction in the number of mosaic blocks in a crystal, as we have seen directly in the texture of the Laue spots (Figs. 1b and c) and via rocking curves

(Fig. 3 and [14, 15]), ie. by an increase in size of a mosaic block to essentially reach the whole crystal size. These results have been reproduced on two separate missions, Spacehab-1 and IML-2, and involve a total of 4 microgravity grown and 3 earth grown crystals. The increased peak count (Fig. 4), far greater than that expected from crystal volume considerations, is most startling. In data collection, with an appropriate detector and a synchrotron monochromatic, very fine step scan (0.0002°) method, weak reflections could be recorded with a much greater signal to noise ratio with such perfect crystals. Hence the methods of collecting intensity data for X-ray structural analysis still have considerable scope for further improvement.

There are a number of reports of enhanced $I/\sigma(I)$ for microgravity-grown over earth-grown crystals for some proteins [5, 11]. There are also opposite conclusions drawn for cases of other protein crystals [8]. Our results point to the physical basis of why $I/\sigma(I)$ might be better as well as crystals being larger. Namely this can be attributed to fewer, more perfectly arranged mosaic blocks. We can also see that such effects may not be critical in some cases.

If diagnostic feedback control of the crystal growth process were available, the termination of growth at a crystal size equal to exactly one mosaic block might be made. Feedback control could be based on laser light scattering to monitor the nucleation stage and then interferometry (eg. [15]) for the ensuing stages. Such diagnostics should clearly allow more insight into both microgravity and earth based crystal growth. It is not inconceivable that procedures might be routinely developed, on earth, where more perfect crystals could be grown so as to match the standard set by the microgravity grown crystals. It could also be the case that essentially perfect crystals do grow for some or even a large part of the time on earth. The precise measurement of perfection advocated here is relatively new [6, 4] and should be more routinely applied.

A limitation of mosaicity considerations is that, at least in X-ray data collection, it is now routine to rapidly freeze crystals (eg. see [9]) so as to greatly reduce X-ray radiation damage but whereby mosaicity is considerably increased. For example, even with careful attention to the freezing mixture and conditions the minimum mosaic spread reached is still about 0.25° [12]. This is much larger than the values under discussion here. Clearly, at present, lifetime in the X-ray beam is a more important parameter than initial angular perfection of the crystal. It is interesting to wonder, however, if crystals grown containing just one mosaic block would show a more limited increase in mosaicity of a sample on freezing; optimisation of the freezing of crystals would also need to be tested as a function of crystal size, as well as mosaic block composition. Also if, by present crystallization knowledge and methods, only small crystals could be grown (eg. $<20\mu$m) attention to mosaic block composition/angular rocking widths as a function of growth conditions might yield larger crystals in the end. This would also be useful in neutron crystallography if crystals were initially small (eg. $<$ 1 mm) but could likewise be grown larger by optimisation of the growth process via such diagnostic methods. These techniques would then become extremely valuable in structural research.

Acknowledgements

We are grateful to the Daresbury SRS for the provision of synchrotron facilities for the Spacehab-1 crystal evaluation and the help of Dr. S. McSweeney. We are grateful to Dr's M. Wulff and T. Ursby for their participation in the ESRF BL3 perfection tests. We would like to express our thanks to the Swiss-Norwegian CRG at the ESRF, Grenoble for providing access to their beamline facilities and in particular to Dr. P. Pattinson and his colleagues as well as Dr. E. Weckert and his colleagues for their participation with the Karlsruhe diffractometer work (details of which will be published seperately). We are especially grateful to Dr's H.U. Walter, K. Fuhrmann and O. Minster of ESA for their constant help and support with this research. Dr's R. Bosch, L. Potthaust and P. Lautenschlager of Dornier are also thanked for discussions on methods to improve diagnostic facilities for monitoring the crystal growth process.

References

1. S.J. Andrews, J.E. Hails, M.M. Harding and D.W.J.Cruickshank, *Acta. Cryst.* **A43** (1987) 70-73.
2. R. Bosch, P. Lautenschlager, L. Potthast and J. Stapelman *Journal of Crystal Growth* **122** (1992) 310-316.
3. CCP4 *Acta. Cryst.* **D50** (1994) 70-73.
4. G. Colapietro, G. Cappuccio, C. Marciante, A. Pifferi, R. Spagna and J.R. Helliwell *J. Appl. Cryst.* **25**(1992) 192-194.
5. L.J. DeLucas, C.D. Smith, W. Smith, S. Vijay-Kumar, S.E. Senadhi, S.E. Ealick, D.C. Carter, R.S. Snyder P.C. Weber, F.R. Salemme, D.H. Ohlendorf, H.M. Einspahr, L.L. Clancy, M.A. Navia, B.M. McKeever, T.L. Nagabhushan, G. Nelson, A. McPherson, S. Koszelak, G. Taylor, D. Stammers, K. Powell, G. Darby, and C.E. Bugg, *Science* **246**(1989) 651-654.
6. J.R. Helliwell *Journal of Crystal Growth* **90** (1988) 259-272.
7. J.R. Helliwell, J. Habash, D.W.J. Cruickshank, M.M. Harding, T.J. Greehough, J.W. Campbell, I.J. Clifton, M. Elder, P.A. Machin, M.Z. Papiz, and S. Zurek, *J. Appl. Cryst.* **22** (1989) 483-497.
8. R. Hilgenfeld, A: Liesum, R. Storm,and A. Plaas-Link, *Journal of Crystal Growth* **122** (1992) 330-336.
9. H. Hope, F. Frolow, K. von Böhlen, I. Makowski, C. Kratky, Y. Halfon, H. Danz, P. Webster, K.S. Bartels, H.G. Wittmann, and A. Yonath *Acta Cryst.* **B45** (1989) 190-199.
10. R. Lewis, *J. Synchrotron Rad.* **1** (1994) 43-53.
11. A. McPherson, *J. Phys. D: Appl. Phys.* **26** (1993) B104-112.
12. E.P. Mitchell, and E.F. Garman, *J. Appl. Cryst.* **27**(1994) 1070-1074.
13. T. Molenkamp, L.P.B.M. Janssen, and J. Drenth, *ESA SP-1132* **Vol. 4** (1994) 22-43.

14. E. Snell, J.R. Helliwell, and P. Lautenschlager, in preparation for *Acta Cryst D* (1995).
15. E. Snell, J.R. Helliwell, E. Weckert, K. Hölzer, M. Masson, M. Masson, K. Schroer, and J. Zellner, in preparation for *Acta Cryst D* (1995).
16. R. Snyder, K. Fuhrmann and H.U. Walter *Journal of Crystal Growth* **110** (1991) 333-338.
17. S. Weisgerber, and J.R. Helliwell, *Joint CCP4 & ESF-EACBM Newsletter on Protein Crystallography*, November issue, Vol **29** (1993) 10-13.
18. S. Weisgerber, Ph.D. Thesis, University of Manchester, 1993.

Fluid-Dynamic Modelling of Protein Crystallyzers

R. Monti and R. Savino

Università degli Studi di Napoli "Federico II" - Dipartimento di Scienza e Ingegneria dello Spazio "Luigi G. Napolitano" - P.le V.Tecchio 80, 80125 Napoli (Italy)

Abstract

A fluid-dynamic model of the hanging (or sitting) drop is proposed to study the time evolution of the thermo-solutal flow fields in protein crystallyzers. An Order of Magnitude Analysis of the vapour phase surrounding the drop shows that buoyancy effects are negligible in the vaporization chamber and that the evaporation is a very fast process, so that the rate of evaporation is controlled essentially by water diffusion through the air space. The liquid drop is modelled as a mixture of water, precipitating agent and protein bounded by an undeformable interface with a surface tension depending on the concentrations. The cases considered refer to the crystallization of lysozyme in a solution of NaCl in water. Preliminary numerical results are shown corresponding to zero-g and to one-g (with and without Marangoni effect). The computations seem to indicate that in the pre-nucleation phase the Marangoni effects may be relevant at 0-g for the full drop. For the half-drop geometry Marangoni flows may plays only a role on ground. In the post-nucleation phase the comparison between 0-g and 1-g shows that in ground conditions the mass transfer is enhanced by convection and therefore the crystal growth rate increases, but the nonuniformity in the interface concentration gradients around the growing crystal may have a detrimental effect on the growth kinetics. A preliminary validation of the code is accomplished comparing the numerical results with experimental ones obtained in a facility at MARS center.

1. Introduction

Protein crystallization in microgravity (μg) environment is receiving increasing attention substantiated by the success of "better" protein crystals obtained in space. A number of experiments performed aboard shuttle flights (see e.g. [4, 5] and MIR missions [1] (see e.g. [15]) showed that microgravity-grown crystals, compared to the best crystals obtained by standard ground-based laboratory

[1] MIR - acronym for the Russian space station. (Eds. note)

techniques, are larger and display more uniform morphologies and diffraction qualities. More recently crystallization experiments using the so called hanging (or sitting) vaporizing drop have been performed with the Advanced Protein Crystallization Facility (APCF, see [1] during the Spacehab 1 mission (1993) and the IML-2 mission 1994 [2]). Microgravity experimentation growth techniques aim at improving the size and the quality of protein crystals required for X-ray crystallographic analyses or for other fundamental studies of biological molecules [9]. One of the main objectives of this paper is the study of the fluid-dynamic aspects of these crystallization processes with the aim of identifying the experimental parameters that would optimize the microgravity results.

Apart from the choice of the protein and of the composition of the mixture (vaporizing liquid, precipitating agent), the parameters to be decided by the Principal Investigator and that play a role in the growth process are the following ones:

1. initial protein concentration;
2. initial volume of the drop;
3. initial shape of the drop;
4. ambient boundary conditions (geometry, temperatures, vapour pressure);
5. temperature and material of the drop support.

For ground crystallization all these parameters are typically being defined by a brute empirical methodology that consists in accumulating statistical data on the most likely experimental conditions leading to crystals of sufficient size. However the large amount of the parameters involved and the obvious interplay among them very often imply time consuming experimentation and discourages optimization processes. Thermofluidynamic (TFD) analysis of the process may provide extremely useful guidelines for the identification of these parameters especially for microgravity experimentation that offers only few opportunities to a relatively small number of samples.

The numerical modelling must clarify the differences between ground and μg environment processes and assess the benefits of the mg with respect to ground activities (i.e. the microgravity relevance). The difficulties of the TFD computation for protein crystal growth are related to:

1. long duration of the process;
2. simultaneous solution of the TFD fields inside and outside the hanging (or sitting) drop;
3. uncertainties on the physical properties of the mixture, especially related to the surface tension and to its variation with temperature and concentration;
4. partial knowledge of the diffusion coefficient of the protein molecules and of its aggregates;
5. necessity of two different models for the pre-nucleation and for the post-nucleation (crystal growing) phase;

[2] IML - the acronym for the International Microgravity Laboratory; Spacelab missions performed by the National Aeronautics and Space Administration NASA. (Eds. note)

6. difficulty in finding the initial location of the growing crystal inside the drop and the subsequent crystal positions.

Here a first numerical model of the hanging-drop fluid-dynamics is being proposed to study the time evolution of the thermo-solutal flow fields during the process. The liquid drop is modelled as a mixture of water, precipitating agent and protein, bounded by an undeformable spherical surface of radius R, with density and surface tension exhibiting a linear dependence on the temperature and on the concentrations. Due to the small temperature differences the liquid is supposed to exhibit constant transport properties and the Boussinesq approximation for the density is considered; the effects of moving surface due to evaporation are considered negligible (quasi-stationary evaporation). Axial symmetry is assumed with respect to the drop axis. Two geometrical configurations are considered: the half drop and the full drop (Fig.1). The differences between the two geometries are that the contact boundary with the drop support is a circular disk of radius R, for the half sphere, and is limited to a small spherical cap, for the full size drop ($0 \leq \varphi \leq \varphi^*$).

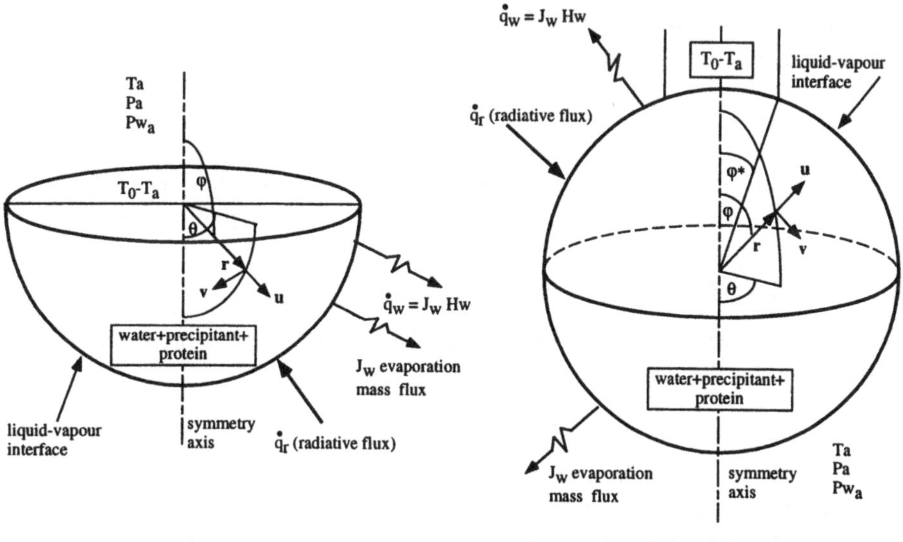

Fig. 1. Geometry of the problem and coordinate system. a) half drop; b) full drop.

The continuity, Navier-Stokes, energy and mass species equations in the liquid phase together with the appropriate boundary and symmetry conditions are numerically integrated in a staggered grid in polar coordinates system by a numerical algorithm based on finite-difference schemes. The radius of the drop is computed, at each time, by the mass flow rate of evaporation at the drop surface.

Two different codes have been implemented for the pre-nucleation (PREN) and the post-nucleation (POSTN) phases. The first phase is characterized by water evaporation from the surface, causing the concentrations of protein and precipitant to increase in the drop until appropriate values are reached and protein molecules begin to nucleate. The most likely location for nucleation is the zone of maximum protein and precipitant concentration, so that the study of the pre-nucleation phase allows to predict the location of the crystal nucleation. The growing crystal phase is modelled assuming a given location of the crystal and appropriate boundary conditions for the concentrations of protein and precipitating agent in the neighbourhood of the crystal.

To validate the model the numerical results have been compared to experimental results obtained on ground during crystallization of Lysozyme in a solution of NaCl in water. This process has been selected as test-case because the lysozyme crystallization is a relatively easy one and because this system is often employed as reference model in crystallization studies, so that the thermodynamics of nucleation is available. Once the model is validated it will be extended to the crystal growth of the SsADH enzyme to be crystallised in the APCF. This experiment, recently selected for flight on STS-72 (1995) [3] has been proposed by the University of Naples [4].

2. Order of Magnitude Analysis (OMA) for the ambient phase

Mass transfer in the vaporization chamber is characterized by three different processes (Fig.2): 1) water evaporation from the drop surface; 2) water vapour diffusion from the neighbourhood of the drop to the layer adjacent to the reservoir liquid; 3) water condensation at the interface of the liquid reservoir.

2.1 Kinetics of the evaporation processes

For the evaporation and condensation at the liquid-vapour interfaces the mass flux is given by the well known equation:

$$J_w = K_w(p_v^* - p_v) \tag{1}$$

where K_w is the water evaporation constant, p_v^* and p_v are, respectively, the equilibrium saturation pressure and the vapour partial pressure at the interface surrounding the drop and/or the reservoir. If the process is steady, for the overall mass fluxes at the two surfaces we have, assuming the same conditions along the entire interface:

$$K_w(p_v^* - p_v)_d S_d = K_w(p_v^* - p_v)_r S_r = \int_{S_i} \rho D_v \frac{\partial c_v}{\partial n} dS \tag{2}$$

[3] STS - acronym for Space Transportation System, meaning here the Space Shuttle flight of the NASA. The flights are simply numbered in sequence of occurence. (Eds. note)

[4] Principal Investigator: A. Zagari, Chemistry Dept., with the support of the authors at the Dept. of Space Science and Engineering and of Mars Center in Naples, Dr. L. Carotenuto

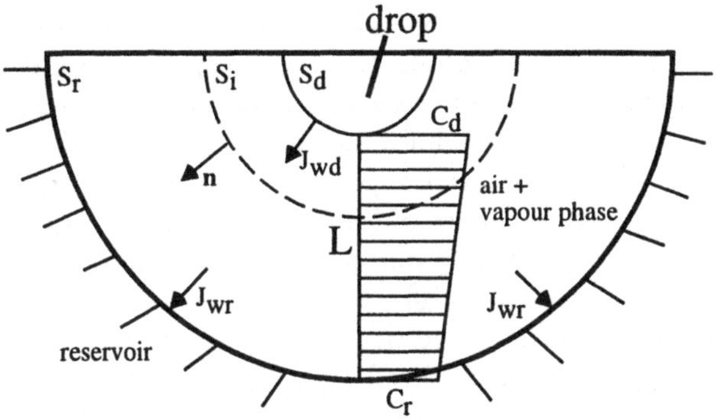

Fig. 2. Schematics of the one-dimensional diffusion process in a hanging drop crystallyzer.

where the subscripts (d) and (r) denote the drop and reservoir conditions, S_d and S_r are, respectively, the drop surface and the surface area of the external liquid and S_i is any surface surrounding the vaporizing drop (Fig.2). In fact for the geometry of Fig.2 assuming steady state and negligible effects of curvature

$$\frac{\partial c_v}{\partial n} \cong (c_{\rm vd} - c_{\rm vr})/L$$

the diffusion flux in the vapour phase is simply $J \propto \rho D_v$, where D_v is the diffusion coefficient of water vapour in air. For $\Delta p = (p_{\rm vd} - p_{\rm vr}) \cong 10^{-2} p_0^*$, where $p_0^* \cong 0.02$ [Atm] is the vapour pressure of water (see e.g. [10]), $(p_{\rm vd} - p_{\rm vr})/p \cong 2 \cdot 10^{-4}$. For $\rho \cong 10^{-3}$g/cm^3, $D_v \cong 0.22$cm^2/s, L=1 cm, the diffusion flux is about $J_w \cong 2 \cdot 10^{-8}$ g/(cm^{2}s). Assuming for the water evaporation constant a mean value $K_w \cong 1.08 \cdot 10^{-4}$ s/cm, we find from (2) a difference $(p_v^* - p_v) \cong 2.16 \cdot 10^{-4}$ Pa, that is negligible (of order 10^{-5} when compared to the difference $(p_{\rm vd} - p_{\rm vr}) \cong 20$ Pa. We may therefore conclude that the evaporation is a very fast process compared with diffusion, and that the rate of evaporation is controlled by diffusion through the air space between the drop and the reservoir. Consequently one may assume:

$$p_{\rm vd}^* \equiv p_{\rm vd} \quad \text{and} \quad p_{\rm vr}^* \equiv p_{\rm vr}$$

3. Mass transport in the vapour phase

The thermal expansion of the fluid is negligible due to the small temperature differences involved (estimated temperature differences are less than 1 K). Buoyancy effects in the vapour phase result from concentration differences between the drop and the reservoir. Assuming ideal gas behaviour, the density of the mixture of air (A) and aqueous vapour (B) is:

$$\rho = \rho_{\rm A}\left[1 + \frac{M_{\rm B} - M_{\rm A}}{M_{\rm A}}\right] = \rho_{\rm A}\left[1 + \frac{1}{\rho_{\rm A}}\frac{\partial \rho}{\partial x}x\right] \tag{3}$$

where ρ_A is the air density at the same temperature and pressure, M_A and M_B are the molecular weights of air and water vapour and x is the mole fraction of water vapour. Thus $\Delta\rho = \rho_A \beta' \Delta x$, where $\beta' = M_B - M_A/M_A$ and the appropriate formulation of the solutal buoyancy velocity is $V_g = (g\beta' \Delta x L^2)/\nu$.

For a mixture of air and water vapour at the ambient conditions, $\beta' \approx 0.4$, $\nu \approx 0.125 \mathrm{cm}^2/\mathrm{s}$; in correspondence of the typical partial pressure differences between drop and reservoir occurring during protein crystallization experiments by vapour diffusion techniques (see [10]), we have that the maximum values of the molar fraction differences, in the vertical direction, are $\Delta x = O(10^{-4})$. For L=1 cm and considering the diffusion speed $V_D = D_v/L$ (the diffusion coefficient of water vapour in air is $D_v \approx 0.22 \mathrm{cm}^2/\mathrm{s}$) we find that the solutal Peclet number $Pe_c = V_g/V_D$ is order unity. These values of the Peclet number are much below the critical values for the onset of Rayleigh instability. On the other hand, if horizontal gradients are considered, the characteristic concentration differences are smaller, so that the order of magnitude of the solutal Peclet number, that gives the measure of the relative importance of convection and diffusion effects, is less than one. In any case, from this Order of Magnitude Analysis, we can conclude that in the vapour phase surrounding the droplet diffusion effects prevail and no relevant differences have to be expected between ground and microgravity conditions.

4. Fluid-dynamic model of the evaporating drop

The liquid drop has been modelled as a mixture of water, precipitant and protein bounded by an undeformable interface with a surface tension exhibiting a linear dependence on the temperature and on the concentrations.

$$\sigma = \sigma_0 \left[1 - \sigma_{rmT}(T - T_0) - \sigma_{c1}(c_1 - c_{10}) - \sigma_{c1}(c_2 - c_{20})\right] \tag{4}$$

where c_{10} and c_{20} are the initial values of the precipitant and protein mass concentrations. Due to the small temperature differences the liquid is supposed to exhibit constant transport properties (viscosity μ, thermal diffusivity λ, diffusion coefficients, D_i); the effects of moving interface due to evaporation are considered negligible (quasi-stationary evaporation). For the density a linear dependence is also assumed with the temperature and the concentrations of protein and precipitant:

$$\rho = \rho_0 \left[1 - \beta_{rmT}(T - T_0) - \beta_{c1}(c_1 - c_{10}) - \beta_{c1}(c_2 - c_{20})\right] \tag{5}$$

The flow in the liquid phase is governed by the Boussinesq form of the continuity, Navier-Stokes, energy and mass species equations that are written below in the polar coordinates system shown in Fig. 1 (the geometry considered and the axial symmetry suggest to use the polar (r, φ) coordinates in order to easily impose the symmetry and boundary conditions):

$$\frac{1}{r^2}\frac{\partial(r^2 u)}{\partial r} + \frac{1}{r \sin\varphi}\frac{\partial(v \sin\varphi)}{\partial\varphi} = 0 \tag{6}$$

$$\frac{\partial u}{\partial t}+u\frac{\partial u}{\partial r}+v\left(\frac{1}{r}\frac{\partial u}{\partial \varphi}-\frac{v}{r}\right)+\frac{1}{\rho_r}\frac{\partial p}{\partial r}=$$
$$=\nu\left(\nabla^2 u-2\frac{u}{r^2}-\frac{2}{r^2\sin\varphi}\frac{\partial(v\sin\varphi)}{\partial\varphi}\right)+\frac{\Delta\rho}{\rho_r}\mathbf{g}\cdot\mathbf{j}_r \quad (7)$$

$$\frac{\partial v}{\partial t}+u\frac{\partial v}{\partial r}+v\left(\frac{1}{r}\frac{\partial v}{\partial \varphi}+\frac{u}{r}\right)+\frac{1}{\rho_r}\frac{\partial p}{\partial \varphi}=$$
$$=\nu\left(\nabla^2 v-2\frac{v}{r^2\sin\varphi}+\frac{2}{r^2}\frac{\partial u}{\partial\varphi}\right)+\frac{\Delta\rho}{\rho_r}\mathbf{g}\cdot\mathbf{j}_\varphi \quad (8)$$

$$\frac{\partial T}{\partial t}+u\frac{\partial T}{\partial r}+\frac{v}{r}\frac{\partial T}{\partial \varphi}=\alpha\left[\frac{1}{r^2}\frac{\partial}{\partial r}\left(r^2\frac{\partial T}{\partial r}\right)+\frac{1}{r^2\sin\varphi}\frac{\partial}{\partial\varphi}\left(\sin\varphi\frac{\partial T}{\partial\varphi}\right)\right] \quad (9)$$

$$\frac{\partial c_1}{\partial t}+u\frac{\partial c_1}{\partial r}+\frac{v}{r}\frac{\partial c_1}{\partial \varphi}=D_1\left[\frac{1}{r^2}\frac{\partial}{\partial r}\left(r^2\frac{\partial c_1}{\partial r}\right)+\frac{1}{r^2\sin\varphi}\frac{\partial}{\partial\varphi}\left(\sin\varphi\frac{\partial c_1}{\partial\varphi}\right)\right] \quad (10)$$

$$\frac{\partial c_2}{\partial t}+u\frac{\partial c_2}{\partial r}+\frac{v}{r}\frac{\partial c_2}{\partial \varphi}=D_2\left[\frac{1}{r^2}\frac{\partial}{\partial r}\left(r^2\frac{\partial c_2}{\partial r}\right)+\frac{1}{r^2\sin\varphi}\frac{\partial}{\partial\varphi}\left(\sin\varphi\frac{\partial c_2}{\partial\varphi}\right)\right] \quad (11)$$

where

$$\nabla^2=\frac{1}{r^2}\frac{\partial}{\partial r}\left(r^2\frac{\partial}{\partial r}\right)+\frac{1}{r^2\sin\varphi}\frac{\partial}{\partial\varphi}\left(\sin\varphi\frac{\partial}{\partial\varphi}\right) \;;$$

the subscripts (1) and (2) denote the precipitating agent and the protein, u and v are the radial and azimuthal velocity components, ν is the kinematic viscosity, α the thermal diffusivity, D_i the diffusion coefficient of the component i.

Equations (6) - (11) must be solved together with the following initial and boundary conditions:

Initial conditions (t=0)

$$u=v=0 \qquad T=T_0 \qquad c_1=c_{10} \qquad c_2=c_{20} \quad (12)$$

Boundary conditions ($t>0$)

on the support

$$0\le\varphi\le\varphi^*,\ r=R \quad \text{for the full drop: } u=v=0;\ T=T_0;\ \frac{\partial c_1}{\partial r}=0\,;\ \frac{\partial c_2}{\partial r}=0 \quad (13)$$

$$\varphi=0,\ r\le R \quad \text{for the half drop: } u=v=0;\ T=T_0;\ \frac{\partial c_1}{\partial \varphi}=0;\ \frac{\partial c_2}{\partial \varphi}=0 \quad (14)$$

on the liquid interface ($r=R$; $\varphi^*<\varphi<\pi$):

$$u=0;\quad \mu\left[\left(\frac{\partial v}{\partial r}\right)_R-\frac{v}{R}\right])\frac{1}{R}\frac{\partial\sigma}{\partial\varphi} \quad (15)$$

$$-\lambda\frac{\partial T}{\partial r}=h_r(T-T_a+J_w H_w \quad (16)$$

$$-\rho D_1\frac{\partial c_1}{\partial r}=\rho c_1\frac{\partial R}{\partial t}=-J_w\frac{c_1}{c_w} \quad (17)$$

$$-\rho D_2 \frac{\partial c_2}{\partial r} = \rho c_2 \frac{\partial R}{\partial t} = -J_w \frac{c_2}{c_w} \tag{18}$$

on the symmetry axis ($\varphi = 0$; $\varphi = \pi$)

$$v = 0; \ \frac{\partial u}{\partial \varphi} = 0; \ \frac{\partial T}{\partial \varphi} = 0; \ \frac{\partial c_1}{\partial \varphi} = 0; \ \frac{\partial c_2}{\partial \varphi} = 0 \tag{19}$$

where H_w is the latent heat of vaporization; the water evaporation flux is proportional to the difference between the partial pressures of water on the drop and in the reservoir $J_w = K_w(p^*_{vd} - p^*_{vr}) = K_w(p_{vd} - p_{vr})$, as shown at Sect.3.. Equations (17) and (18) indicate that, as the evaporation proceeds and the drop volume is reduced, the interface concentrations of protein and precipitating agent increase; assuming quasi-steady conditions, the corresponding diffusive mass fluxes are proportional to the time rate of change of the drop radius. According to the Raoult's law the vapour partial pressure in the adjacent phase surrounding a liquid interface is proportional to the equilibrium saturation pressure at the temperature of the liquid and to the concentration of the volatile species in the liquid ($p^*_v/c_w = p^*_{v0}/c_{w0}$). Thus the evaporation flux can be written also as

$$J_w = J_{w0} \frac{c_{wd} - c_{wr}}{c_{w0} - c_{wr}}$$

where c_{wd}, c_{wr} are the water mass concentrations at the drop surface and in the liquid reservoir and the subscript (0) denotes initial conditions.

5. Numerical model

The numerical solution of the field equations along with the appropriate boundary and symmetry conditions has been obtained using an explicit time-dependent method in primitive variables, splitting at each time the computation of the velocity field in two steps. First, an approximate vector field ($\mathbf{V}^*$) corresponding to the correct vorticity of the field (but with $\nabla \cdot \mathbf{V}^*$) is obtained from the momentum equation for $\nabla p = 0$. This vector is then corrected with the pressure field obtained solving with a SOR (Successive Over Relaxation) iterative method the Poisson equation deriving from the divergence of the momentum, accounting for the continuity equation. The field equations have been discretized using a staggered grid and finite-difference methods. Central differencing scheme was used for the diffusion and convective terms. A uniform mesh was employed with (20x20) grid points for the numerical solutions of the pre-nucleation phase, whereas a finer grid (40x40) was used in the post-nucleation phase to accurately solve the field near the crystal. The radius of the drop is computed, at each time, by the integration of the rate of evaporation of the drop.

6. Results and discussion

Numerical experimentations have been carried out for the Lysozyme-NaCl system, which is the most commonly employed as reference model in protein crystallization studies. For this liquid mixture, the relevant physical properties are available in literature, e.g. densities, kinematic viscosity, expansion coefficients (see e.g. [7]), mass diffusivities, solubility [16], [6], surface tension dependence on the concentrations [11], critical supersaturation conditions for nucleation [12]. In the pre-nucleation phase we have analyzed the half and the full drop geometries (Figs. 1) and compared the "ideal" case of zero-g conditions with the ground (1-g) situation. For the half drop both the cases of hanging and sitting drops have been considered. The initial radius of the drops is R=2 mm, corresponding to an initial volume of about 16 μm^3 for the half drop and 32 μm^3 for the full drop. The initial salt concentration is c_{10}=0.029 and the initial lysozyme concentration is c_{20}=0.025. Since the numerically computed temperature differences are very small (less than 1 K) only concentration differences are considered as possible sources for buoyancy-driven and surface tension-driven convection (typically the dimensionless parameter $\beta_T \Delta T / \beta_c \Delta c$ giving the relative importance of thermal and solutal buoyancy effect is of order of magnitude 10^{-2}). In the post-nucleation phase we have considered the simplest model consisting in a macroscopic single crystal located near the surface of the hanging drop in correspondence of the symmetry axis (region of maximum probability for nucleation). Mass transport and interface kinetics are related using the boundary conditions for the changing concentrations on the crystal surface proposed by [7] on the basis of experimental results on the growth and kinetics of lysozyme crystals [12].

6.1 Prenucleation phase in a half drop

In Fig. 3 the concentration contours of NaCl (Fig.3-left) and Lysozyme (Fig.3-right) are shown after 5 minutes of evaporation for the half drop at 0-g (top), the hanging (middle) and the sitting drop (bottom) at 1-g. In the purely diffusive situation (zero-g, absence of any buoyancy convection), water evaporates from the drop surface, the surface concentrations of salt and protein become higher and a diffusion front moves from the surface towards the center of the drop. However the mass transfer rate for the protein is lower than for the precipitating agent, being the diffusivity of protein one order of magnitude smaller than that of the NaCl ($D_1 \approx 10^{-5}$cm^2/s; $D_2 \approx 10^{-6}$cm^2/s). For this geometrical configuration (angle φ^* between the symmetry axis of the drop and the liquid surface contacting the solid wall of the tip equal to $\pi/2$) if diffusion conditions prevail, the concentrations are homogeneous along the surfaces parallel to the drop surface and there are only radial variations (central symmetry).

Under normal gravity conditions, both in the hanging drop and in the sitting drop, the density gradients resulting from the concentration differences are sources of natural convection, with typical buoyant plumes rising along the symmetry axis and flowing down along the drop surfaces (Fig.4). Comparing the

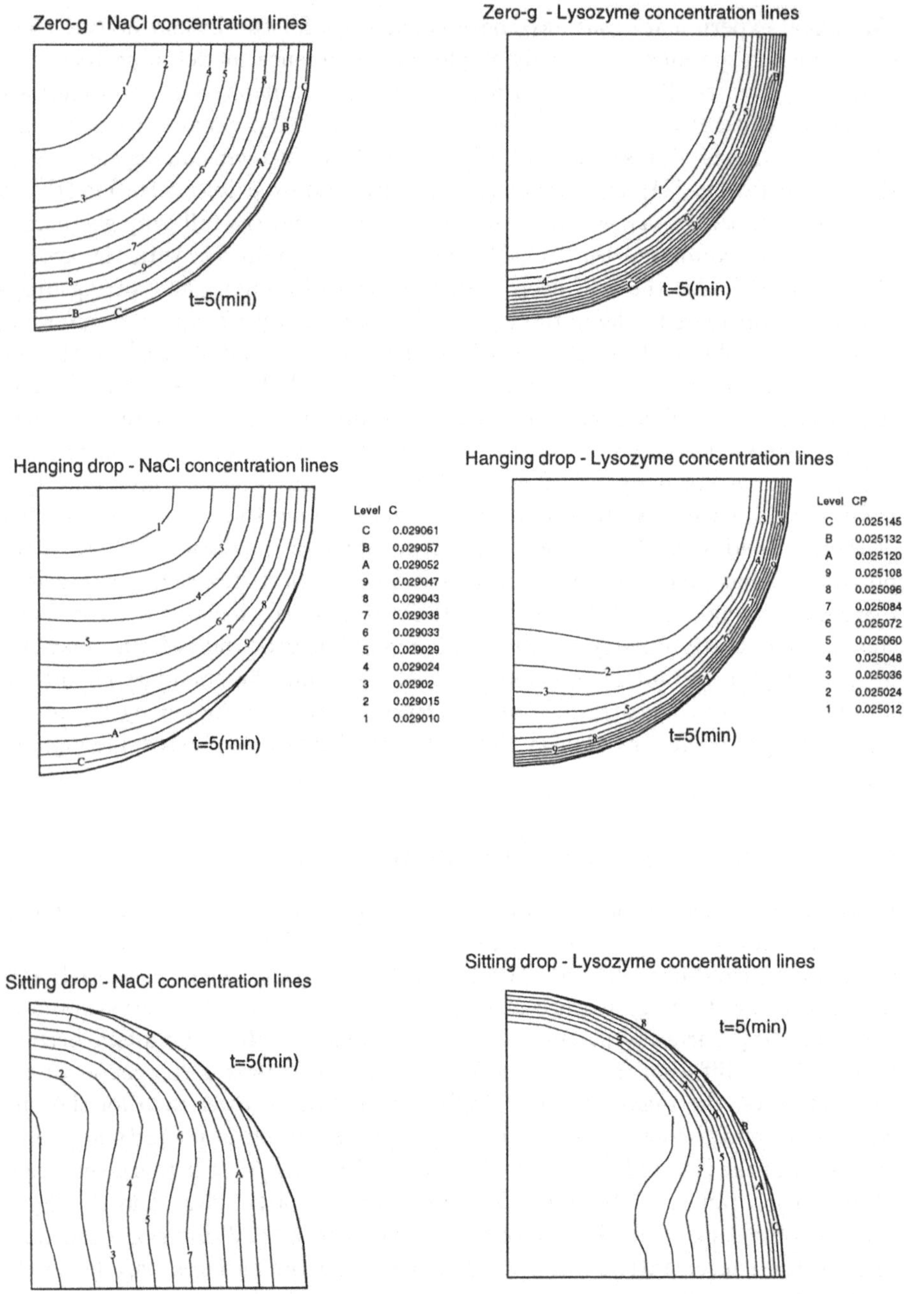

Fig. 3. Left: Iso-concentration contours of NaCl at 0-g for the half drop (top), and at 1-g for the hanging drop (middle) and sitting drop (bottom), after 3 min. Right: Iso-concentration contours of Lysozyme at 0-g for the half drop (top), and at 1-g for the hanging drop (middle) and sitting drop (bottom), after 3 min.

stream-lines of Fig.4(left, hanging drop) with those of Fig.4(right, sitting drop) we recognize that the stream function has negative sign in the first case and positive sign in the second one, as consequence of the fact that for the hanging drop the flow along the symmetry axis is from the surface to the solid support, whereas it is opposite for the sitting drop. The maximum calculated flow velocities are about V= 5 μm/s for the hanging drop and V= 15 μm/s for the sitting drop. A comparison of the concentration fields at 0-g and 1-g (Figs. 3) shows that these relatively small velocities are responsible for large concentration distortions and that these distortions are more relevant for the concentration of Lysozyme. In fact, the relative importances of the convective and diffusive terms in the solute transport equations are measured by the solutal Peclet numbers $Pe_{ci} = VR/D_i$; for V=10 μ m/s and R=2 mm, this numbers are $Pe_{c1} = 20$ for the NaCl and $Pe_{c2} = 200$ for the Lysozyme.

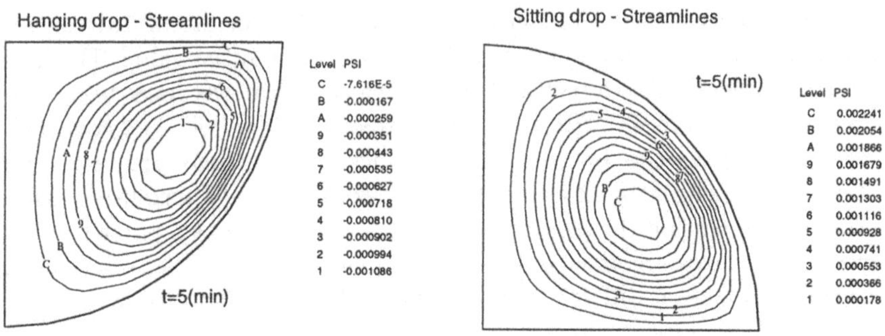

Fig. 4. Stream-lines at 1-g for the hanging (left) and sitting drop (right), after 3 min.

In particular, due to the convective circulation cell, fluid with lower concentrations flows from the bottom to the top, and fluid with higher concentrations flows down along the drop surface, so that the surface concentrations become higher on the bottom and lower near the top. This non uniformity in the surface compositions is illustrated in Fig.5, showing the surface concentration profiles of NaCl (left) and Lysozyme (right); $\varphi = \pi/2$ denotes conditions at the solid support and $\varphi = \pi$ at the symmetry axis (see Fig.1). In the purely diffusive case (zero-g) the concentrations are uniform along the surface, so that Marangoni effects should be completely absent. In the sitting drop the concentrations are higher for $\varphi = \pi/2$, i.e. near the solid wall; the opposite takes place for the hanging drop. These distributions may cause Marangoni flows in 1-g of opposite direction.

In Figs.6the concentration contours of NaCl and Lysozyme are shown, for the same three cases (0-g (top), hanging drop (middle), sitting drop (bottom)]

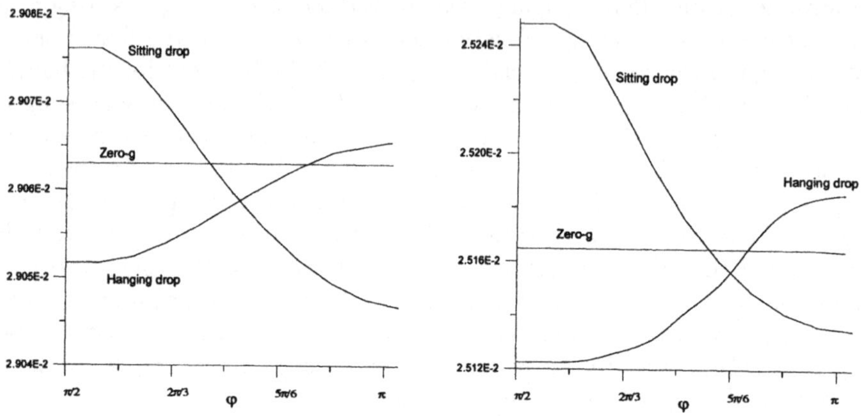

Fig. 5. Profiles of surface concentration of NaCl (left) and Lysozyme (right) at 0-g for the half drop and at 1-g for the hanging and sitting drop, after 3 min.

after 48h). The volume of the drop is about one half of the initial volume and the drop radius has been reduced to about 80% of its initial value. The diffusion front is completely developed and quasi-steady conditions have been reached. In the 1-g cases the velocities are still responsible for relevant concentration distortions and, as can be seen in Figs.6, these distortions are larger for the protein concentration. If Marangoni effects are taken into account, considering for the surface tension the dependence on the concentrations reported by [11], no relevant differences are observed. This can be explained by the relatively small surface concentration differences in this geometrical configuration.

6.2 Termination of the prenucleation phase

The first phase of the process terminates when conditions for nucleation prevail at some locations inside the drop. The contours of "isoprobability" for the crystal nucleation have been computed on the basis of the couples of values for the concentrations of precipitating agent and protein. Generally the nucleation occurs when the solubility limit is trespassed and the concentrations of protein and salt are close enough to the precipitation conditions, that could be represented by suitable supersaturation conditions (see Fig. 7).

According to [12] the critical supersaturation for protein crystallization is about two orders of magnitude higher than for crystal growth of inorganic systems, so we assumed as most likely precipitation conditions those corresponding to the curve, in the plane (c_{Lys}, c_{NaCl}), defined by the equation $\sigma = (c_{Lys} - c_{sat})/c_{sat}=10$ (c_{sat} is the protein saturation concentration at the same temperature and NaCl concentration), that has been found to be a likely supersaturation value for the nucleation of only a few crystals at T=18 °C. With reference to Fig.7, the probability for nucleation in the generic point P is defined

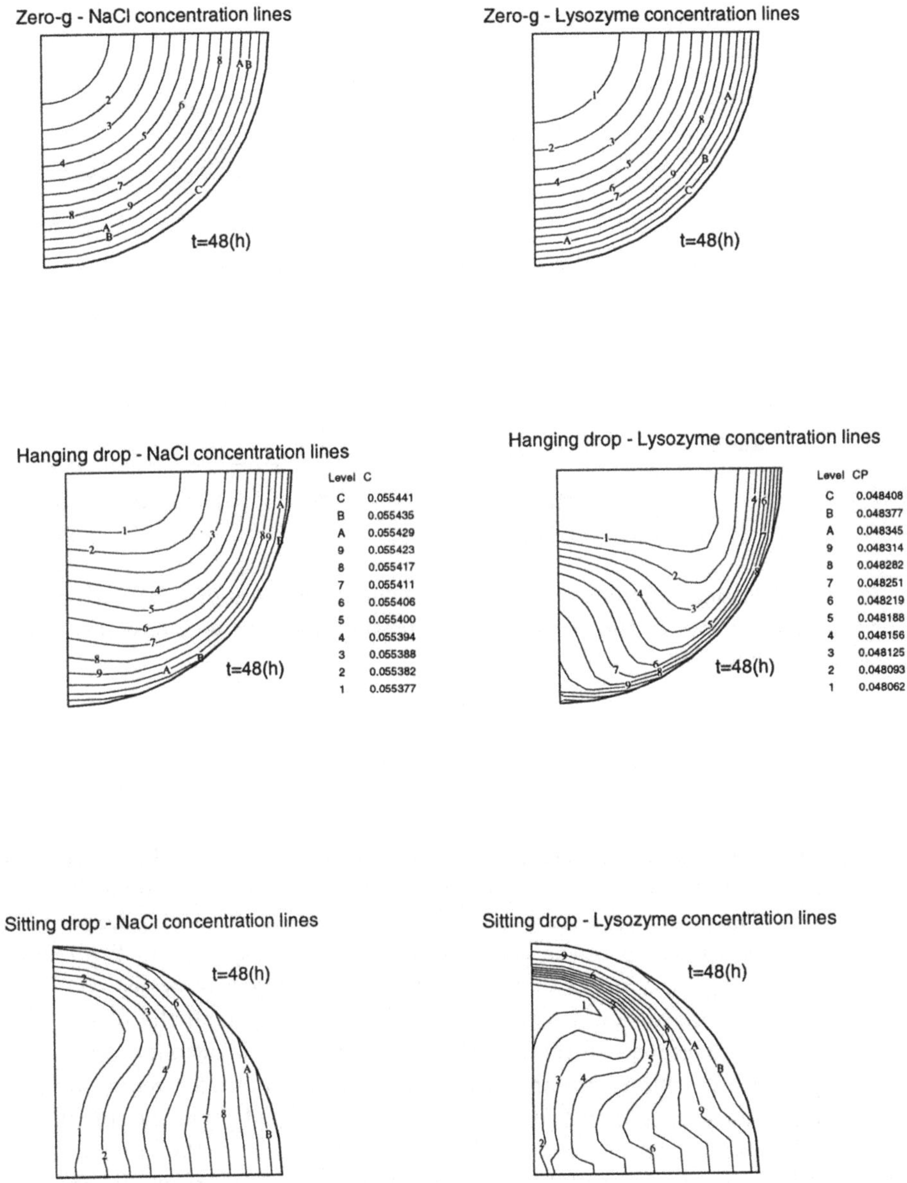

Fig. 6. Left: Iso-concentration contours of NaCl at 0-g for the half drop (top), and at 1-g for the hanging drop (middle) and sitting drop (bottom), after 48 h. Right: Iso-concentration contours of Lysozyme at 0-g for the half drop (top), and at 1-g for the hanging drop (middle) and sitting drop (bottom), after 48.

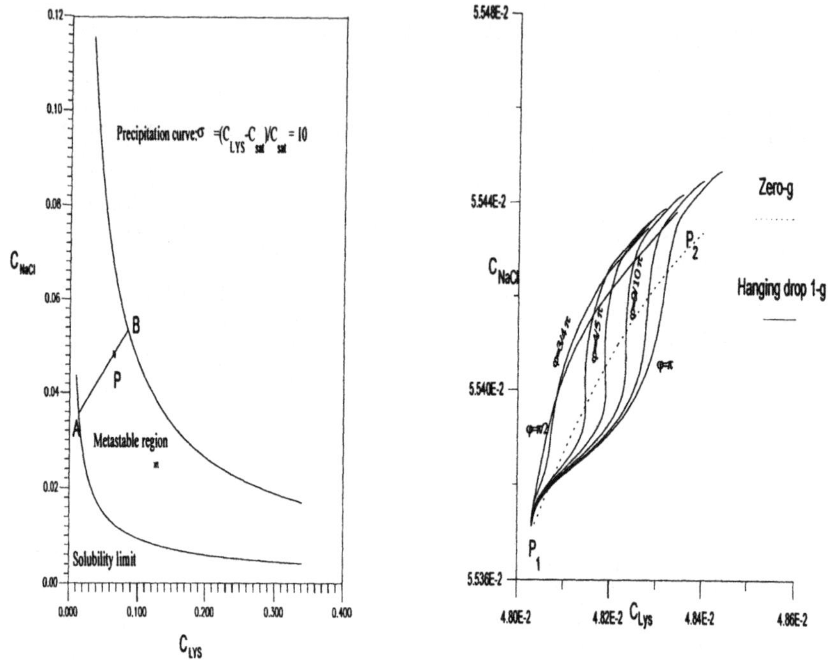

Fig. 7. Nucleation thermodynamics for Lysozyme-NaCl in water. (left): solubility and precipitation curves; (right): envelope of concentrations in the drop after 48 h, for 0-g and for hanging drop at 1-g.

as

$$\mathcal{P} = \frac{\overline{\mathcal{PA}}}{\overline{\mathcal{PA}} + \overline{\mathcal{PB}}} \tag{20}$$

where $\overline{PA}$ and $\overline{PB}$ are the distances of the point P from the solubility curve and from the precipitation curve, respectively. For $P \equiv A$, $\mathcal{P} = /$ whereas for $P \equiv B$, $\mathcal{P} = \infty$. The concentrations conditions of the points inside the drops, in the diffusive case (0-g) and in the hanging drop (1-g) are reported in Fig.7, in the plane (c_{Lys}, c_{NaCl}). We see that when central symmetry is preserved, all the points of the drop are on a segment line ($P_1 P_2$) and that all the points of the drop surface correspond to the point of maximum concentrations (P_2).

In Fig. 8 the contours of same probability for nucleation are reported for the three different situations (0-g; hanging and sitting drop at 1-g). The conclusion is that in the 0-g case all the points of the exposed surface have the same maximum probability, whereas in ground conditions we may identify more restricted regions of maximum probability (e.g. in the hanging drop this probability increases from the support to the symmetry axis) where we expect nucleations will take place.

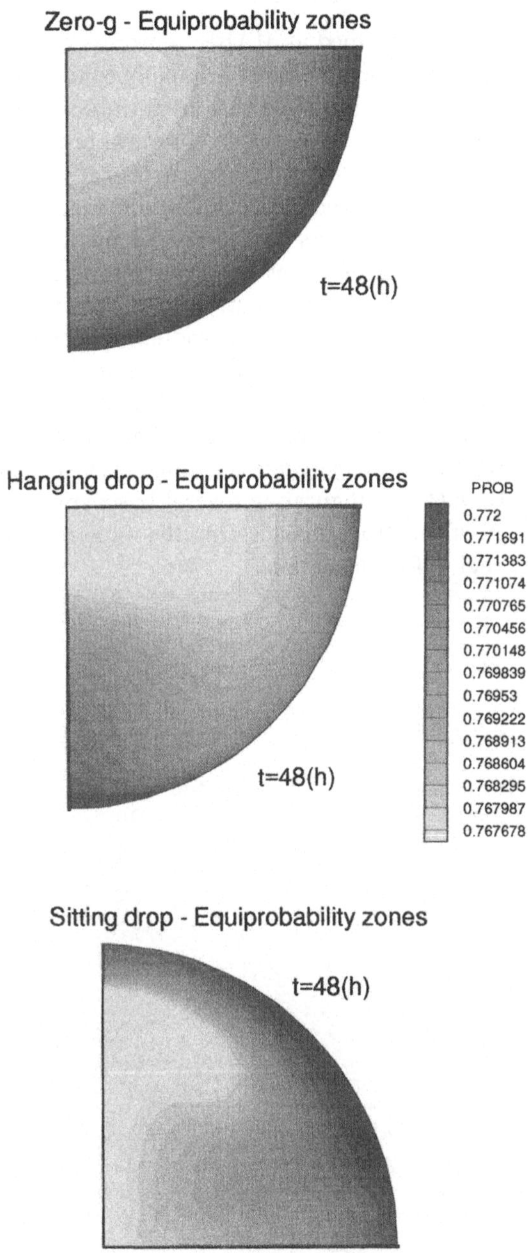

Fig. 8. Equiprobability contours at 0-g for the half drop (top) and at 1 g for the hanging drop (middle) and the sitting drop (bottom), after 48 h.

6.3 Comparison with experimental results

Laboratory tests have been performed at MARS Center to validate the numerical results [2]. For this purpose the apparatus originally conceived to investigate thermocapillary convection in small samples has been improved to obtain qualitative and quantitative estimation of the time-evolution of protein crystallization in the hanging drop configuration. The experiment operative sequence follows the typical procedure usually employed in the vapour diffusion techniques of protein crystallization. Lysozyme solutions were prepared by dissolving 50 mg/ml of the protein in sodium acetate buffer at PH=4.5. The final pending drop is achieved by mixing a droplet of protein solution with an equal volume of the reservoir solution at the proper concentration (sodium chloride in water 1 M). The initial concentrations of the numerical calculations are identical to those of the experimental hanging drop (c_{10}=0.029 and c_{20}=0.025). In Figs.9 and 10 two photographs show the hanging drop in its initial configuration (Fig.9) and after 48 h, when proteins have begun to aggregate forming nuclei that are the precursors to crystals (Fig.10); in this configuration several Lysozyme nuclei are visible near the bottom portion of the droplet confirming the numerical prediction that this is the most likely region for nucleation.

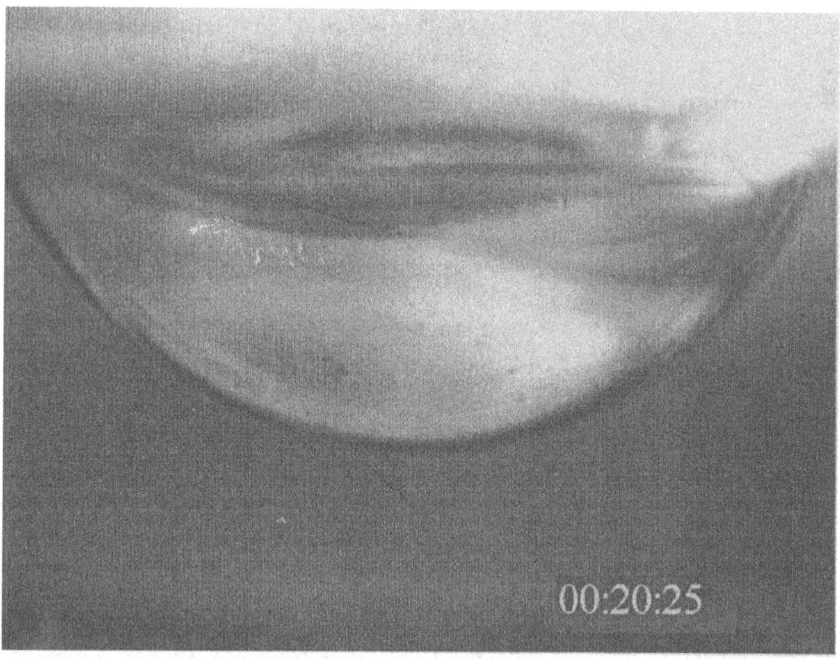

Fig. 9. Photograph showing the hanging drop in its initial configuration.

Fig. 11 shows a sequence of images taken at different times after the first appearence of the nucleation. The first nuclei aggregate close to the surface of the

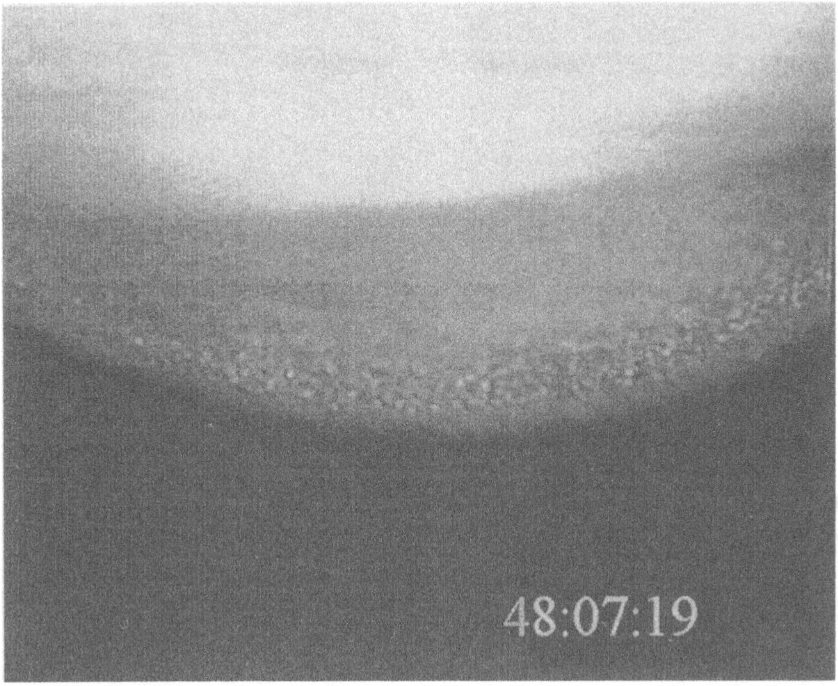

Fig. 10. Photograph showing the crystal nucleation in the hanging drop after about 48 h (initial state see 9).

drop and then their population increases with time, forming a typical "sickle" arrangement. The focused plane is not the meridian one, where crystals were not observed, but a section closer to the drop surface (at about one half of radius from the drop axis). These experimental results seem to be in accordance with the numerical computations, confirming that in the hanging drop configuration the probability for crystal nucleation is higher in the region close to the interface and increases in the bottom region (see Fig. 8). This is also confirmed by Fig.12, where the space distribution of the protein crystal nuclei in the drop is shown in a sequence of images corresponding to different transversal sections. The first focused plane (Fig.12a) is the meridian section; the following ones are at increasing distances from the symmetry plane. We see that in any case the crystal population is always higher near the surface of the drop and that the crystals distribution inside the drop has the same qualitative behaviour of the numerically computed iso-probability curves of Fig.8, with increasing population at the bottom portion of the drop. Therefore we conclude that the agreement between numerical and experimental results is reasonable.

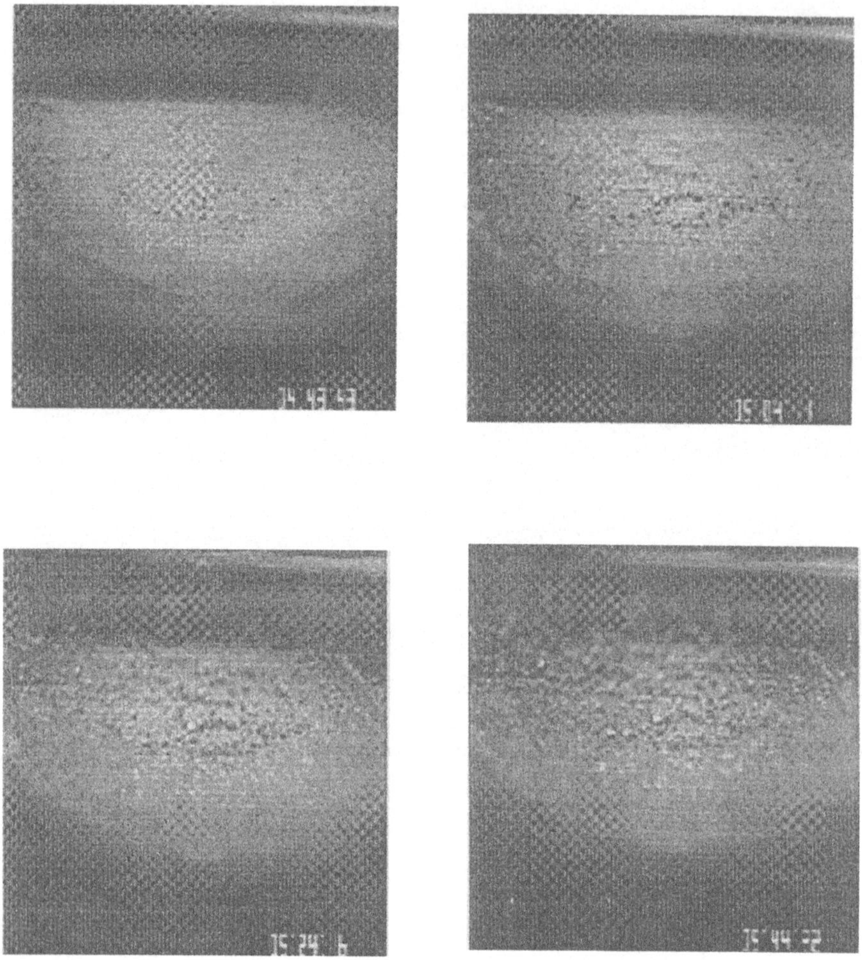

Fig. 11. Sequence of images at different times after the first appearence of the crystals nucleation.

6.4 Prenucleation phase in a full drop

In zero-g it would be possible to chose any shape of the drop without the risk of the drop "falling". The case of "almost" full drop, anchored to a support of size small with respect to the drop radius (see Fig. 1), has been considered to compare the results with the half drop case and to make assessment on the microgravity relevance of the hanging drop crystallization. Figures 13 - 15 show the results of numerical computations in the case of a full drop, after 48 h of evaporation. The angle between the symmetry axis and the liquid surface contacting the solid support is $\varphi^* \approx 30°$. In this case, even in the ideal diffusive situation (Figs.13, 14) central symmetry no longer holds and the concentration contours show a distortion with respect to the sphero-symmetrical configuration

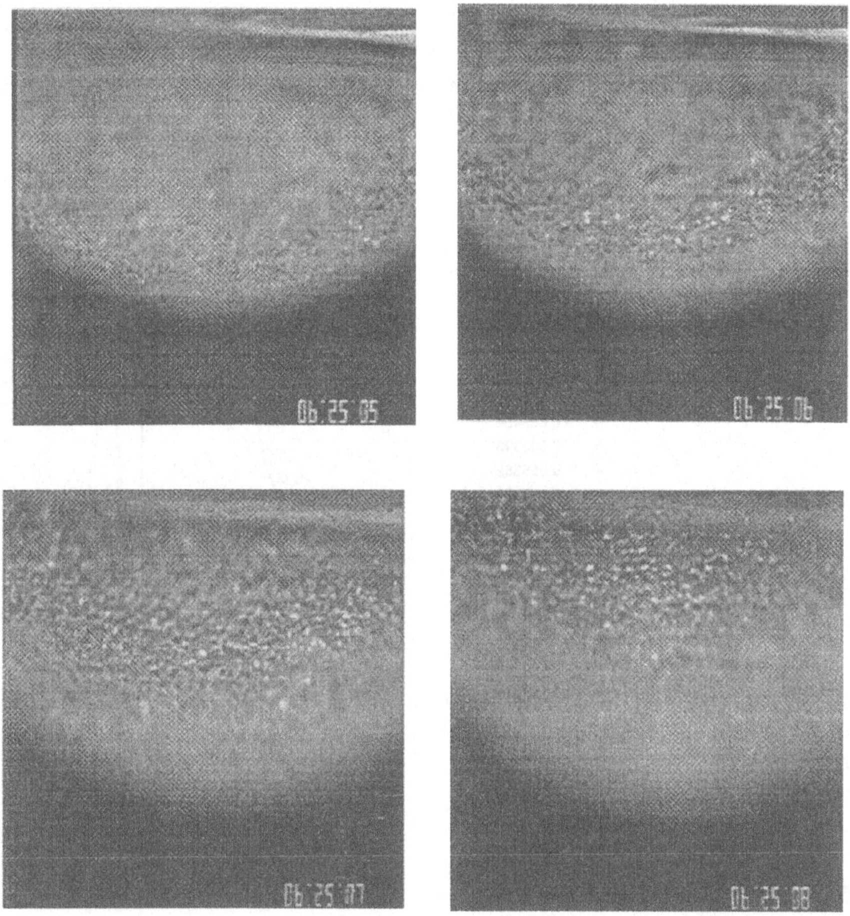

Fig. 12. Sequence of images showing the space distribution of the crystals at about 6h, 25' after the first appearence of the crystal nucleation, in different sections:(top left) meridian plane; (top right to bottom right) section at increasing distance from the axis of the drop.

("onion" configuration). The effect of the geometry on the diffusive field causes the surface concentration to be lower in the region close to the top wall than the surface concentrations at the bottom of the drop. As the surface tension decreases with the concentration [11] if Marangoni effects are taken into account (Figs.13, 14) surface tension-driven flows arise (Fig.15). The fluid in the drop is driven by the surface tension gradient, flowing from the bottom of the surface (where the concentrations are higher and the surface tension lower) towards the syringe tip (where the concentrations are lower and the surface tension higher). The velocities induced by capillary effects are responsible for relevant concentration distortions, mainly in the Lysozyme field. These distortions from Marangoni effects in 0-g, seem to be larger that those corresponding only to buoyancy effects at 1-g (even though a full drop geometry would be difficult to establish on ground); we might therefore conclude that the microgravity relevance (for the range of the assumed values of σ_c) is less evident for this configuration than for the half drop.

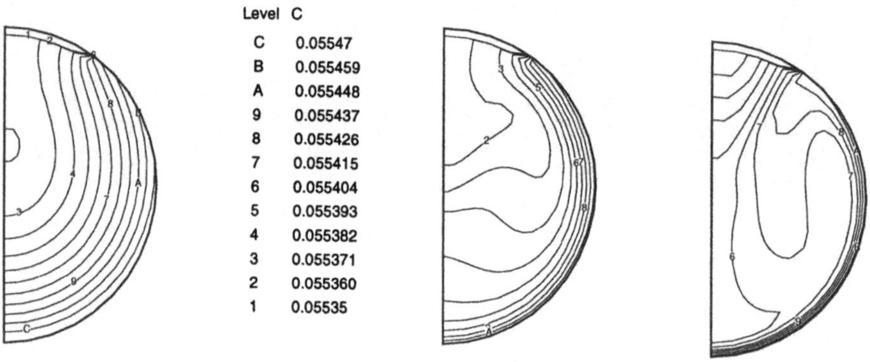

Fig. 13. Iso-concentration contours of NaCl for the full drop at zero-g (left) and 1-g (middle) without Marangoni effect, and at zero-g with Marangoni effect (right), after 48h.

6.5 Post-nucleation phase

In order to analyze the buoyancy effects in the crystallization phase numerical experimentations have been performed for the model consisting of a hanging drop with a single crystal located on the bottom, in the region of maximum probability of nucleation. A fine mesh with 40x40 grid points was used in order to solve with a good accuracy the fluid-dynamic field in the neighbourhood of the crystal. As initial conditions we have assumed quiescent conditions for the velocity and uniform values for the concentrations (c_1=0.055, c_2=0.048). As boundary conditions at the crystal interface we have assumed no-slip conditions for the velocity and the relations proposed by [7] for the changing interfacial lyzozyme and sodium chloride mass concentrations. For the lysozyme concentration, using mass balance and Fick's law, and assuming a linear dependence of the growth rate by the interface supersaturation, one obtains:

$$\beta \left[\frac{c_2 - c_2^{eq}}{c_2^{eq}} - \sigma_0 \right] = - \frac{D_2}{\rho_2^c - \rho^c c_2} \frac{\partial c_2}{\partial n}$$

where the kinetic coefficient $\beta = 8.25 \cdot 10^{-8}$ cm/s, σ_0=2.9 is the width of the zone in which no growth occurs, the superscript "eq" denotes the equilibrium

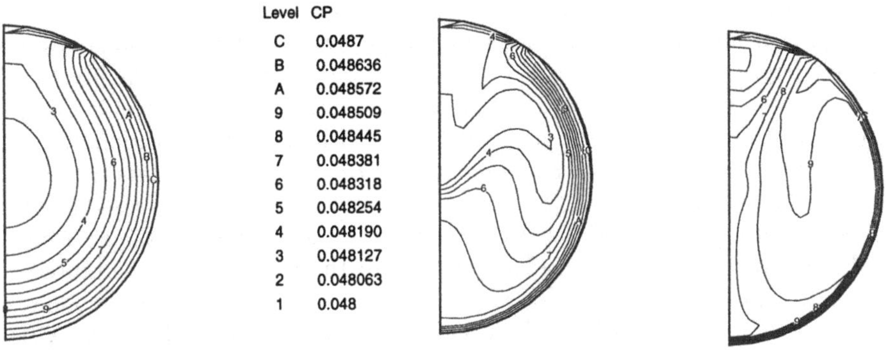

Fig. 14. Iso-concentration contours of Lysozyme for the full drop at zero-g (left) and 1-g (middle) without Marangoni effect, and at zero-g with Marangoni effect (right), after 48h.

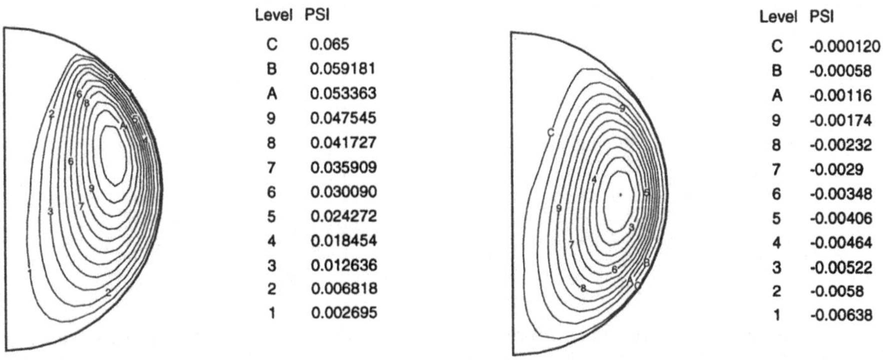

Fig. 15. Stream-lines for the full drop at zero-g with Marangoni effect (left) and 1-g without Marangoni effect (right) after 48h.

protein solubility at the temperature considered, ρ_2^c and ρ^c the protein mass density and the total mass density in the crystal. According to [7], we assumed ρ_2^c=820 mg/ml and ρ^c=1.233 g/ml. For the interface precipitant concentration we assumed:

$$\rho c_1 V_f \left[\frac{\rho^c}{\rho} - \frac{k \rho_2^c}{\rho c_2} \right] = -D_1 \frac{\partial c_2}{\partial n}$$

where V_f=80 A/s is the face growth (salt rejection) rate and k=0.01 is the segregation coefficient. In Fig.16 the concentration contours of Lysozyme are illustrated some minutes after the nucleation, in 0-g and 1-g. In 0-g the protein concentration at the crystal interface decreases with time, causing a depletion zone with high concentration gradients where protein diffusion mass transfer takes place. In normal gravity conditions this depletion region is distorted by buoyancy-driven convection (Fig.17): the lighter fluid surrounding the crystal, with lower protein concentration, rises from the upper face of the crystal along the symmetry axis, whereas the heavier fluid, with higher concentration, flows down along the drop surface towards the lateral face of the crystal. Mass transfer is enhanced by convection and therefore the crystal growth rate increases, but the non uniformity in the interface concentration gradients around the growing crystal might have a detrimental effect on the growth kinetics. Since the protein concentration at the surface nearest to the crystal is lower than in the surrounding areas, Marangoni effects could play an important role in this phase. We will analyze these effects in future simulations.

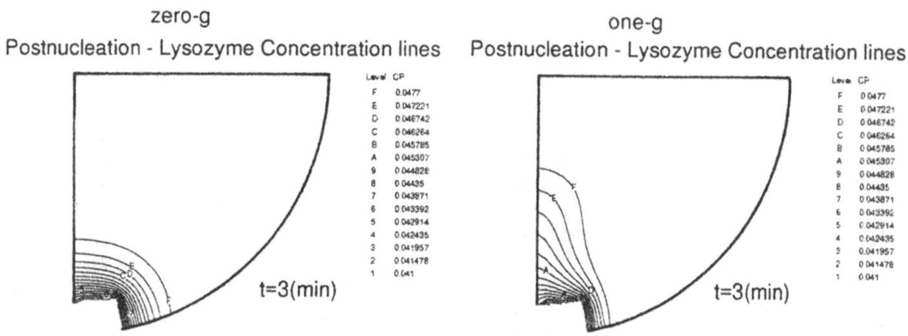

Fig. 16. Iso-concentration contours of Lysozyme 3 min after the crystal nucleation, at "0-g" and "1-g" (without Marangoni effect).

7. Conclusions

Numerical computations of the velocity and concentration fields in a hanging-drop vapour diffusion facility for protein crystal growth have been obtained for a half drop and for a full-size drop, at different conditions corresponding to 0-g and 1-g, with and without Marangoni effect. A preliminary validation of the code has been accomplished by comparing the numerical results with experimental ones obtained at MARS Center, using a facility for a qualitative and quantitative estimation of the protein crystallization in the hanging drop configuration. The agreement between numerical computations and experimental

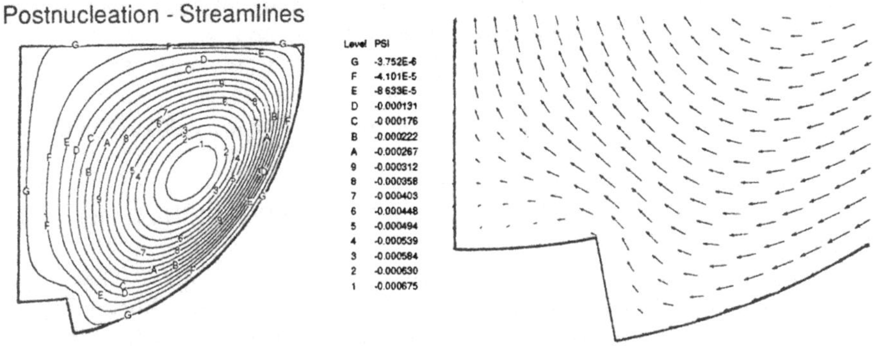

Fig. 17. Stream-lines and vector plots near the crystal 3 min after the crystal nucleation, at "1-g" (without Marangoni effect).

results is reasonable. The on-ground experiments with a pending half drop confirm that the bottom portion near the surface of the drop is the most likely region for crystal nucleation and the space crystals distribution inside the drop has the same qualitative behaviour of the numerically computed iso-probability curves. A more quantitative validation of the PREN (Prenucleation) code is being carried out by an appropriate experimental laboratory equipment. The numerical results concerning the full drop configuration indicate that, for the range of the assumed values of the surface tension derivative with concentration, capillary effects may be relevant at 0-g, whereas for the half drop geometry, Marangoni effects may play a relevant role on ground but less important in microgravity. The numerical computations obtained for the post-nucleation phase show that in ground conditions mass transfer is enhanced by convection and therefore affects the crystal growth and the non uniformity in the interface concentration gradients around the growing crystal may have a detrimental effect on the growth kinetics. Since the protein concentration at the surface nearest to the crystal is lower than in the surrounding areas, Marangoni effects could play an important role in this phase. We will analyze these effects in future simulations.

Acknowledgements

The authors wish to thank Dr. L. Carotenuto, Prof. A. Zagari and Dr. F. Peluso for the support in the obtainment of the experimental data.

References

1. R. Bosh, P. Lautenschlager, L. Potthast and J. Stapelmann, *J. Crystal Growth* **122** (1992) 310-316

2. L. Carotenuto et al. : "Model protein crystal growth experimental apparatus for numerical validation". MARS Center Report, (March 1995)
3. G.A. Casay, W.W. Wilson, *J. Crystal Growth* **122** (1992) 95-101.
4. L.J. De Lucas et al., *J.Crystal Growth* **110** (1991) 302-311
5. L.J. De Lucas, C.E. Bugg, *Adv. Space Biol. and Med.* **1** (1991) 249-278.
6. Y.C. Kim, A.S. Myerson, *J. Crystal Growth* **143** (1994) 79-95.
7. H. Lyn, F. Rosenberger, J.I.D. Alexander and A. Nadaraja, To be published (1995).
8. J.R. Luft et al., *J. Appl. Cryst.* **27** (1994) 443-452.
9. A. McPherson, A. Greenwood and J. Day, Adv. Space Res. **11**, No.7, (1991) 343-356
10. V. Mikol, J. Rodeau, R. Gieg, *Analytical Biochemistry* **86** (1990) 332-339.
11. T. Molekamp, L. Janssen, J. Drenth, Final Reports of Sounding Rocket Experiments in Fluid Science and Materials Sciences, ESA SP-1132 4, 22-43 (1994).
12. L.A. Monaco, F. Rosenberger, *J. Crystal Growth* **129** (1993) 465-484.
13. R. Monti, R. Savino, ELGRA Biennal meeting and General Assembly, Madrid (Spain), 11-14 December (1994), unpublished work.
14. F. Rosenberger, S.B. Howard, J.W. Sowers., T.A. Nyce, *J. Crystal Growth* **129** (1993) 1-12 .
15. R.K. Strong et al., *J. Crystal Growth* **119** (1992) 200-214.
16. H.J.V. Tyrrel , K.R. Harris in "Diffusion in liquids", Butterworths Monographs in Chemistry (1961)
17. A. Zagari., R. Savino, L. Carotenuto: "Protein crystallization in microgravity: Sulfolobus Solfataricus Alcohol Dehydrogenase". Experiment Proposal for APCF on Mission STS-72 (1994).

Plasma Crystals

H.M. Thomas and G.E. Morfill*

Max-Planck-Institut für extraterrestrische Physik, Giessenbachstrasse, 85740 Garching, Germany

Abstract

We present observations of a new type of model system of crystalline structures, the so called plasma crystal, and compare that system with other well studied model systems - electron crystals, ion crystals and colloidal crystals in aqueous solutions. Plasma crystals combine some unique properties, which the other systems do not have, such as global charge neutrality, very little damping, optical imaging of individual particles ("atoms"), stability, easy and flexible production in the laboratory. These properties make them exciting systems for studying basic procceses. The possibility of combining measurements of the dynamics of single "atoms" with different large scale structure analyses should provide the essential information necessary for a detailed investigation of phenomena associated with self-organisation, phase transitions, energetics etc. Plasma crystals, given their special properties, can trigger significant progress in this area.

1. Introduction

The search for models of any type of physical system is a classical and successful scientific approach aimed at achieving a better understanding of nature. Such models can be either numerical (computer models) or laboratory analogs. They are usually simpler than the original system and may allow measurements of processes, which in the original system may not be accessible to observations. Thus the study of model systems helps to increase our knowledge of physics in general and of the physics of the special system the model is developed for in particular.

Crystalline systems are of very high complexity and due to the small dimensions of atoms and the interatomic distances, especially dynamical observations of the atoms are impossible. Already in 1948, Bragg [1] developed a soap bubble model to simulate crystalline lattices and to study the processes responsible for

* Permanent address: DLR-Institut für Raumsimulation, Linder Höhe, 51140 Köln, Germany. Electronic address: Hubertus.Thomas@europa.rs.kp.dlr.de

crystallization and deformation of these lattices. The crystalline planes are represented by layers of soap bubbles of 0.1-0.2 mm diameter, which float densely packed in a regular structure on the surface of a suitable fluid. Every single bubble represents a lattice site. Certainly, this 2-dimensional system constituted only a rough approach to the 3-dimensional reality of a natural lattice, however, these soap bubble layers show for example many of the imperfections (dislocations, boundary layers, textures), which can be found in real crystals, too. Important processes can be observed and studied in detail with the soap bubble model, such as deformations, sliding processes caused by an external shear potential, crystallization, influences of impurities or annealing.

The model systems of crystalline structures, which we will discuss briefly in this paper are closer to real crystals than the soap bubble model. These systems are ion crystals [2, 3, 4] and electron crystals [5, 6] on the atomic scale and colloidal crystals in aqueous solutions [7, 8, 9, 10] and the recently discovered plasma crystals [11, 12, 13, 14, 15, 16] on the macroscopic scale.

1.1 One-Component Plasmas

In 1939 Wigner developed a theory for the electron gas in a metal [17]. In this work he discribed the condensation of a cooling electron gas and the possibility of forming regular structures due to the Coulomb interaction between the electrons. Since that time this "electron solid" is known as "Wigner crystal". In the classical limiting case it is expected that the Wigner crystal condense since its density is very low. At low temperatures (approach to 0 K) the energy of the system will be determined by long-range Coulomb interaction. The potential energy will be minimized when the electrons order themselves in a regular lattice. Although the Coulomb interaction between individual electron pairs is repulsive, a fixed crystalline structure will form. An experimental verification of the existence of the Wigner crystal was first obtained in 1990 by Jiang et al. [5] and Goldman et al [6] in a 2-dimensional system (thin cold electron layer on the surface of liquid Helium). In this experiment the density is so high that quantum effects are present. This is of great interest because it allows the study of the fractional Quantum-Hall-Effect (QHE) [5, 6].

The electron crystals belong to the field of one-component plasmas (OCP's) [18]. An OCP is a system consisting of a single species of charged particles embedded in an uniform background of neutralizing charges. The important quantity for its thermodynamic description is the so-called Coulomb coupling parameter Γ which is the ratio of the Coulomb energy to the kinetic energy. For charged particles obeying classical Boltzmann statistics, the kinetic energy per particle becomes $k_B T$ and

$$\Gamma = \frac{Q^2}{4\pi\epsilon_0 \Delta k_B T}. \tag{1}$$

For a degenerate electron system, like the Wigner crystal, one instead uses the Fermi energy. Here Q is the charge of a particle, T its temperature, Δ the interparticle distance, ϵ_0 the permitivity of free space and k_B the Boltzmann constant. Those plasmas with a coupling parameter less than unity are called weakly coupled, those with values greater than unity are called strongly coupled.

An OCP is a substantially idealized model for real plasmas, but some plasmas in nature do indeed satisfy the conditions for such idealization [18]. A most typical example for this is the plasma inside a highly evolved star. The interior of such a star is in a compressed, high-density state where all the atoms are in ionized states. The electron system constitutes a weakly coupled degenerate plasma with an immensely large Fermi energy (electron gas). It makes an ideal neutralizing background of negative charges to the ion system. Those atomic nuclei stripped of the electrons form an ion plasma obeying the classical statistics; their de Broglie wavelength are much smaller on the average than the interparticle distance. The Coulomb coupling constant Γ for such an ion plasma is usually greater than unity; in a white dwarf one estimates that $\Gamma = 10 - 200$ [18].

Numerical simulations have been performed for the classical ion OCP [18, 19, 20]. Especially, the transition from liquid to solid has been observed in these simulations and the calculations led to a critical value of the Coulomb coupling parameter $\Gamma = \Gamma_c = 172$, where Coulomb crystallization occurs [19]. This value could be experimentally veryfied with so-called ion crystal. Ion crystals are formed from singly ionized atoms, mostly Beryllium or Magnesium ions are used. The ions repel each other and can be trapped electromagnetically, e.g. in Paul traps [2] and quadrupole storage rings [4]. They have to be cooled to temperatures in the milikelvin range with laser excitation before the ions can arrange themselves into a regular structure, which, however, depends on the external fields and the number of captured ions.

1.2 Colloidal Crystals in Aqueous Solutions

Another crystalline model system was proposed by Deryagin and Landau in 1941 [21]. They presented the theory of the stability of strongly charged sols. The coalescence of strongly charged particles suspended in electrolytes - the interaction between plane surfaces and the interaction between curved surfaces - in combination with the stability criterion (minimum energy criterion) could lead to crystal formation. There has been a lot of experimental verifications of colloidal crystals in aqueous solutions since that time [10, 22, 23] etc. Colloidal crystals in aqueous solutions consist of monodispersive particles, usually manufactured from plastics, with maximum sizes in the range of a few hundred of nanometers. The particles are suspended in an electrolyte and are charged up through the interaction with the free charges. Around the negatively charged particles a cloud of opposite charge forms, screening the charge on the surface of the particle. Such a system is globally charge-neutral and does not need any external supporting or confining fields like the OCP's. Since the charge density in the fluid is very high, the screening length (and therefor the lattice distance in the colloidal crystals) is very low. This implies an optically thick system. Analysis of the lattice structure can then only be done with light scattering experiments [24, 25], comparable to the methods employed in solid bodies. Further, the motion of the particles in the fluid is heavily overdamped. Changes in the system structure as well as phase transitions take a very long time (weeks).

In 2-dimensional colloidal crystals, where the particles embedded in the fluid are placed between two glass plates, thus forming a monolayer, it is possible to work with particle sizes that can be observed with optical microscopy [26, 27]. This opens up the possibility to investigate the detailed dynamics of 2-dimensional crystals [28, 7]. Interesting phenomena have already been discovered, for example the phase transition from solid to liquid passes through an intermediate phase, the so called "hexatic phase", that retains remnants of the crystalline long-range orientational order but has liquidlike short-range translational order. This transition in a 2-dimensional colloidal crystal can be interpreted within the context of a 2-dimensional melting theory developed by Kosterlitz, Thouless, Halperin, Nelson, and Young, known as the KTHNY theory [29, 30, 31, 32, 33]. This theory predicts that 2-dimensional crystals undergo two continuous melting transitions, in contrast to the single first-order transition of 3-dimensional crystals. First, a 2-dimensional crystal melts continously into the hexatic phase. Second, the hexatic melts continuously into an isotropic liquid phase, whose orientational and translational order are both short range. The predictions of KTHNY for translational order apply specifically to a translational correlation function g_G defined in terms of reciprocal lattice vectors [29, 30, 31, 32, 33]. However, the pair-correlation function g is used more commonly than g_G by colloidal experimenters for comparison with theory [22, 23, 34, 35]. This pair correlation function represents the probability of finding two particles separated by a distance r, and it measures the translational order in the structure.

The other tool is the bond-orientational correlation function, $g_6(\mathrm{r})$ [9, 28]. This function is defined in terms of the nearest-neighbor bond angles of a lattice, and it measures hexagonal orientational order (sixfold symmetry) in the structure. Its calculation is based on the principle that all nearest-neighbor bonds in a perfect hexagonal lattice should have the same angle, modulo $\pi/3$, with respect to an arbitrary axis. The bond angles and locations are determined from the particle locations by Delaunay triangulation.

We calculate the bond correlation function by

$$g_6(r) = \frac{1}{N} \sum_{i,j=1;i\neq j}^{N} \delta(\mathbf{r} - \mathbf{r}_i - \mathbf{r}_j)\Theta_i\Theta_j^* \tag{2}$$

with

$$\Theta_j = \frac{1}{n} \sum_{m=1}^{n} exp(i6\Phi_{j,m}). \tag{3}$$

N is the number of particles in the field, n is the number of nearest neighbors of particle j and $\Phi_{j,m}$ is the angle between the bond joining the jth and the mth particle and some fixed axis.

Here we list commonly-accepted criteria [9] involving g and $g_6(r)$ for identifying the various phases. In the crystalline phase, $g(r) \propto r^{-\eta(T)}$ and $g_6(r) = const$, where $\eta(T) \leq 1/3$ and is weakly temperature dependent. In the hexatic phase,

$g(r) \propto exp(-r/\xi)$ and $g_6(r) \propto r^{-\eta}$ with $0 < \eta \leq 1/4$. In the liquid phase, $g(r) \propto exp(-r/\xi)$, $g_6(r) \propto exp(-r/\xi_6)$ and $\xi = \xi_6$. Here ξ and ξ_6 are scale lengths for translational and orientational order respectively. In the liquid phase ξ is smaller than in the hexatic. These criteria are based on the KTHNY theory; other empirical criteria, such as the numbers of nearest neighbors are sometimes used as well [35].

It would appear straightforward to simply determine the quantities η and ξ from the data and then use the KTHNY criteria to determine which state a given observed "crystal" occupies. Unfortunately, dislocations and other structure defects also affect η and ξ so that care must be taken not to confuse the analysis.

1.3 Plasma Crystals

The "plasma crystals" [11, 12, 13, 14, 15, 16] mentioned at the beginning differ from the colloidal crystals in aqueous solutions through the diffuse medium between the particles and their sizes, which are in the micron range. The medium is a low density plasma and the colloidal crystals formed do not exhibit the disadvantages of the fluid systems mentioned above, namely the optical thickness and the strong overdamping. For example, equilibration times are typically a million times faster.

The possible existence of plasma crystals was predicted by Ikezi [36] in 1986. He calculated the Coulomb coupling parameter Γ for different conditions in a colloidal laboratory plasma (by using the theory of OCP's) and found parameters which in principle suggested that Coulomb crystallization may occur. It took eight years more to verify this effect experimentally [11, 13].

Plasma crystals are a special form of "dusty plasmas" which have been observed in space as well as in the laboratory. In space, dusty plasmas are ubiquitous, including interstellar clouds, circumstellar and protoplanetary accretion disks, nova ejecta and planetary magnetospheres [37, 38, 39]. In the laboratory, dusty plasmas were first investigated by Langmuir at the end of the 1920s. Since then the study of dusty plasmas in the laboratory has become an important field of investigation because dust particles can condense and grow in plasmas used for microelectronics fabrication [40, 41]. These particles may contaminate the processed surfaces and decrease the yield of the manufacturing process. Thus it is not surprising that astrophysicists and industrial researchers have investigated many physical processes [40, 41, 42, 43, 44, 45, 46] including especially the problem of dust charging [47, 48, 49, 50].

Dust particles embedded in a plasma are continously charged and discharged due to the incidence of electrons and ions. In the absence of secondary electron emission and photo emission (which are both negligible due to the low energy of the plasma and the low intensity of the UV-radiation in our experiment) the particle charge reaches an equilibrium negative value due to the much larger thermal velocity of the electrons compared to that of the ions. The charge can reach several thousand elementary charges on a particle of a few micrometer diameter. Once equilibrium is reached, which takes only a fraction of a second, low energy electrons are repelled from the strong field at the surface of the dust particle resulting in a zone of reduced electron density. At the same time ions are attracted and their density is correspondingly enhanced. In this zone, called sheath, the free charge distribution in the plasma organises itself to neutralize the charge of the dust particle. In a way, such a system is comparable with a very heavy atom whose positive core is neutralized by an electron cloud. The effective 'atom radius' in the case of the dust particle is called 'Debye length' λ_{D}. Coulomb interaction between neighboring particles and therefore crystallization can occur only if their distance is in the range of the Debye length, or less.

Most dusty plasma charging theories are based on theories of electrostatic probes in plasmas [50]. These theories predict the electron and ion currents to the probe. The currents are termed "orbit-limited" when the condition $a \ll \lambda_{\mathrm{D}} \ll \lambda_{\mathrm{mfp}}$ applies, where a is the particle radius and λ_{mfp} is a collisional mean free-path between neutral gas atoms and either electrons or ions.

$$\lambda_{\mathrm{mfp},\alpha} = \frac{1}{\pi n_g (r_\alpha + r_g)^2 (1 + \bar{c_{\mathrm{g}}^2}/\bar{c_\alpha^2})^{1/2}}, \tag{4}$$

where $n_{\alpha,\mathrm{g}}, r_{\alpha,\mathrm{g}}$ and $\bar{c^2_{\alpha,\mathrm{g}}} = (3k_{\mathrm{B}}T_{\alpha,\mathrm{g}}/m_{\alpha,\mathrm{g}})$ are the density, hard sphere radius and mean thermal velocity of electrons and ions ($\alpha = e, i$), and neutrals (g), respectively.

The charge on a particle is shielded in a plasma through the free charges. The linearized sreening length (Debye length) $\frac{1}{\lambda_{\mathrm{D}}} = \sqrt{\frac{1}{\lambda^2_{\mathrm{D,e}}} + \frac{1}{2\lambda^2_{\mathrm{D,i}}}}$ is calculated from the Debye length for electrons $\lambda_{\mathrm{D,e}}$ and that for ions $\lambda_{\mathrm{D,i}}$.

$$\frac{1}{\lambda_{\mathrm{D}}} = \sqrt{4\pi e^2 n (\frac{1}{k_{\mathrm{B}}T_{\mathrm{e}}} + \frac{1}{2k_{\mathrm{B}}T_{\mathrm{i}}})}. \tag{5}$$

In Fig.1a-b are shown the mean free path (due to collision with neutrals) and the Debye length as a function of neutral gas pressure and plasma density, respectivcely, for typical parameters of the experiment described later. In order to validate the applicability of "orbit limited" theory, λ_{D} and λ_{mfp} must be compared. To do this it is important to know that the plasma density decreases with increasing neutral gas pressure in a typical low power rf-discharge at a neutral gas pressure between 0.1 and 0.5 mbar [51]. At a neutral gas pressure of about 0.1 mbar the electron mean free path is a few mm and the ion mean free path $\approx$ 0.5mm. For the appropriate plasma density of $\approx 3 \times 10^9$ cm^{-3} the Debye length is 0.03 mm. Here, the "orbit limited" condition is guaranteed for both plasma species and a typical particle radius of a few μm. At a pressure of 0.5 mbar, and the corresponding plasma density of $\approx 1 \times 10^9$ cm^{-3}, we calculate an electron/ion mean free path of 0.4/0.09 mm and a Debye length $\approx$ 0.07 mm. So, in this case the collisionless theory begins to fail because the ions might be scattered by neutrals in the Debye sphere of the particle. Nevertheless the calculation of the charge on the particle using this theory gives us a useful first approximation even at this high pressure end of our operating range.

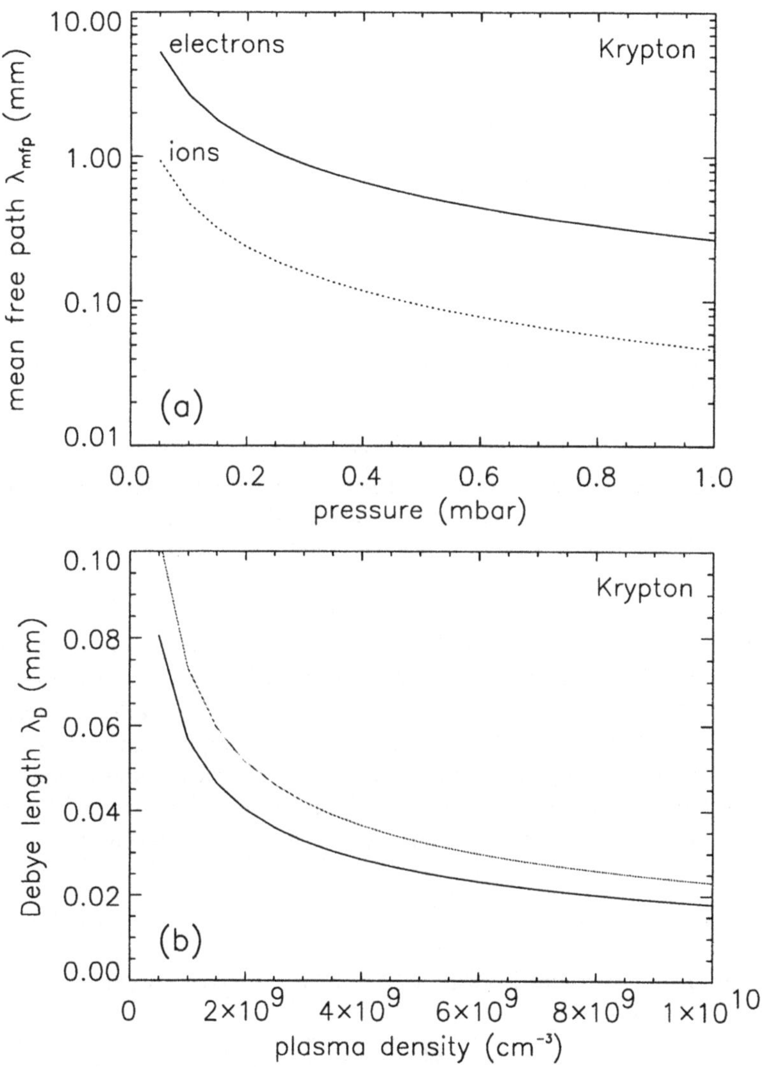

Fig. 1. a) mean free path of electrons and ions as a function of the neutral gas (Krypton) pressure for $k_BT_e = 3eV$ and $k_BT_i = 0.03eV$. b) linearized Debye lenght λ_D as a function of the plasma density for $k_BT_e = 3eV$ and $k_BT_i = 0.03eV$ (black line) and $k_BT_e = 0.5eV$ and $k_BT_i = 0.06eV$ (grey line)

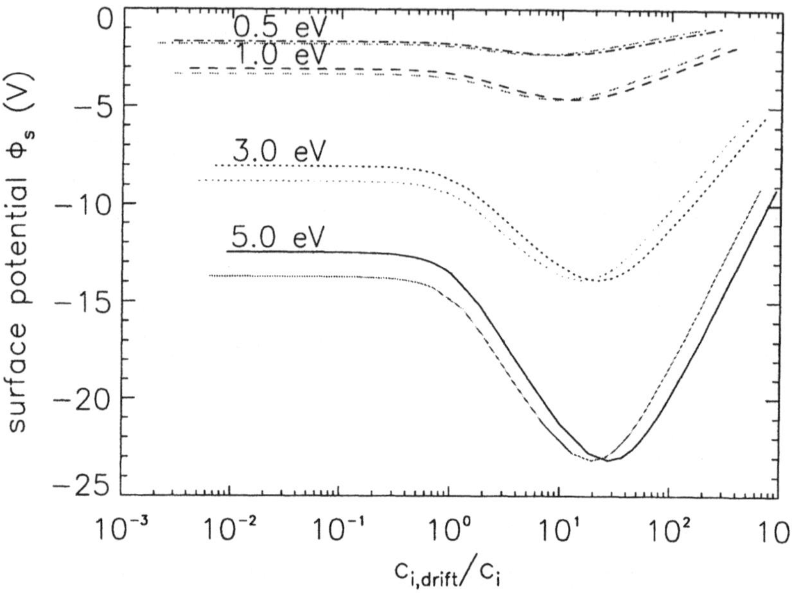

Fig. 2. Surface potential as a function of the ratio of ion drift velocity to ion thermal velocity for different electron temperatures (indicated in the figure) and different ion temperatures (black curves with $k_B T_i = 0.03$eV and grey curves with $k_B T_i = 0.06$eV)

In the "orbit-limited" theory the currents to the particle are calculated by assuming that the electrons and ions are collected if their collisionless orbits intersect the probe's surface. It is assumed that the currents are infinitely divisible; that is, the discrete nature of the electronic charge is ignored. The latter assumption must be dropped if one is interested in charge fluctuations on the particle. A particle with zero charge that is immersed in a plasma will gradually charge up, by collecting electron I_e and ion currents I_i, according to

$$\frac{dQ}{dt} = I_e + I_i. \tag{6}$$

The charge of the particles is correlated with the surface potential Φ_s through its capacity $C = 4\pi\epsilon_0 a$:

$$Q = C\Phi_s. \tag{7}$$

For electrons and ions with maxwellian distribution function the orbit-limited currents for a negative surface potential $\Phi_s < 0$ are

$$I_e = -e n_e \pi a^2 (\frac{8 k_B T_e}{\pi m_e})^{1/2} e^{u_e} \tag{8}$$

$$I_i = e n_i \pi a^2 [(1 + \frac{c_i^2}{2c_{i,drift}^2} + \frac{u_i^2}{c_{i,drift}^2}) erf(\frac{c_{i,drift}}{c_i}) + \frac{c_i}{\sqrt{\pi} c_{i,drift}} e^{-\frac{c_i}{c_{i,drift}}}], \tag{9}$$

where $c_{i,\text{drift}}$ is the drift velocity between the ions and the particles. $u_{e,i} = e\Phi_s/k_B T_{e,i}$ is the normalized surface potential and $c_i = (2k_B T_i/m_i)^{1/2}$ is the thermal velocity of the ions. The averaged surface potential Φ_s will be calculated numerically by setting

$$dQ/dt = 0. \tag{10}$$

The potential can be approximated by a constant value in the case of small drift velocities, i.e. $c_{i,\text{drift}}/c_i \ll 1$. Then the ion current to the particle can be simplified by

$$I_i = en_i\pi a^2 (\frac{8k_B T_i}{\pi m_i})^{1/2}(1 - u_i). \tag{11}$$

On the other hand for large drift velocities $c_{i,\text{drift}}/c_i \gg 1$ the potential strongly increases and I_i becomes

$$I_i = en_i\pi a^2 c_{i,\text{drift}}(1 - \frac{2e\Phi_s}{m_i c_{i,\text{drift}}^2}). \tag{12}$$

A calculation of the surface potential for different electron temperatures and ion temperatures is shown in Fig.2. It is obvious that a change of the ion temperature by a factor of 2 has only a small influence on the slope and amplitude of the surface potential while changes in the electron temperature have a strong effect on the surface potential. As electron temperature is decreased the amplitude of the surface potential also decreases. The influence of the ion drift velocity becomes less important for decreasing electron temperatures.

The transition from random distributions of dust particles in plasmas to a more ordered structure requires special conditions. Early investigations on plasma crystallization were done in the context of one-component plasmas (OCP) as mentioned above. Here, the thermodynamics is described by one parameter: the Coulomb coupling parameter Γ (Eq.1).

Unlike an OCP, a dusty plasma has a "neutralizing background" that is not fixed. The background plasma of ions and electrons adjusts self-consistently to provide Debye shielding. The thermodynamics is thus more complicated than for OCP's. Borrowing from colloidal suspension theory, Farouki et al. [52] proposed to describe the thermodynamics of the charged dust by two dimensionless parameters, Γ and the ratio $\kappa = \Delta/\lambda_D$. Their numerical simulations showed a transition between fluid and solid phases at a critical value Γ_c, depending on the interaction range κ. Although some of the assumptions in the simulation, such as the expression for the inter-grain potential energy and a small simulation box with cubic boundary conditions, may be unsuited to our experiment, the results are revealing. For a certain particle density, they indicate a value of $\Gamma_c \geq 100$ for $\kappa = 0.5$ for the liquid-solid transition.

2. Experimental Setup

The plasma crystal experiment is performed in a radio frequency discharge chamber typical for etching processes in microelectronics fabrication (GEC RF Reference Cell). The GEC RF Reference Cell is discribed elsewhere in detail [53].

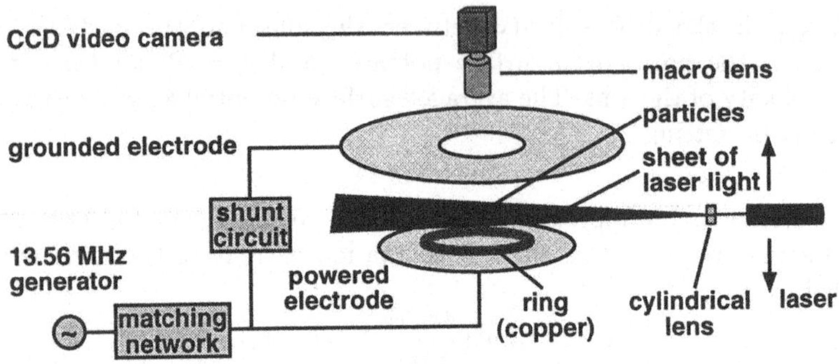

Fig. 3. Schematic of apparatus. A discharge is formed by capacitively-coupled rf power applied to the lower electrode. A vacuum vessel, not shown, encloses the electrode assembly. A cylindrical lens produces a laser sheet in a horizontal plane, with an adjustable height. The dust cloud is viewed through the upper ring electrode.

We modified the original Cell for the plasma crystal experiment as follows: (1) we removed the upper electrode system and installed a ring electrode without insulator, (2) we installed a window on the top chamber flange, and (3) we built a dust dispenser (sieve, mounted movable over the hole in the upper electrode) for particle injection. These changes allow a vertical view into the plasma chamber. In Fig.3 a sketch of the experimental setup is shown. The rf-field between the two electrodes (distance of 2 cm) partially ionizes the neutral gas (Krypton) at a pressure between 0.1-0.5 mbar (ionization fraction of $10^{-7} - 10^{-6}$). Monodispersive melamine/formaldehyde spheres of 6.9 μm diameter are ejected into the plasma, where they become charged and are levitated by a constant, selfsustained electrostatic field in the sheath of the lower electrode. For suitable plasma parameters the particles order themselves in a regular lattice and form a disk shaped cloud of more than hundred of lattice distances in the horizontal direction and a few lattice planes in the vertical direction. Gravity affects the structure in the vertical direction, the lattice planes are compressed and the number of planes decreases from the center to the edge. The 'edge on' view is that of a lense shaped or pyramid shaped structure. Observations were made by illuminating a plane with a sheet of laser light. The 2-dimensional structure can be observed with a CCD camera with a macro lens positioned over the upper ring electrode and can be stored on a VCR. A typical digitized image of a plasma crystal is shown in Fig.4. The CCD camera as well as the illumination system is adjustable in the vertical direction so that one can assemble a 3-dimensional picture or one can follow particles if they move about. The latter is of interest when the plasma parameters are changed.

3. Plasma Crystal Experiments

In the experiment described here, we study the phase transition of a plasma crystal to its liquid and gas phase. The particles which constitute the crystal are

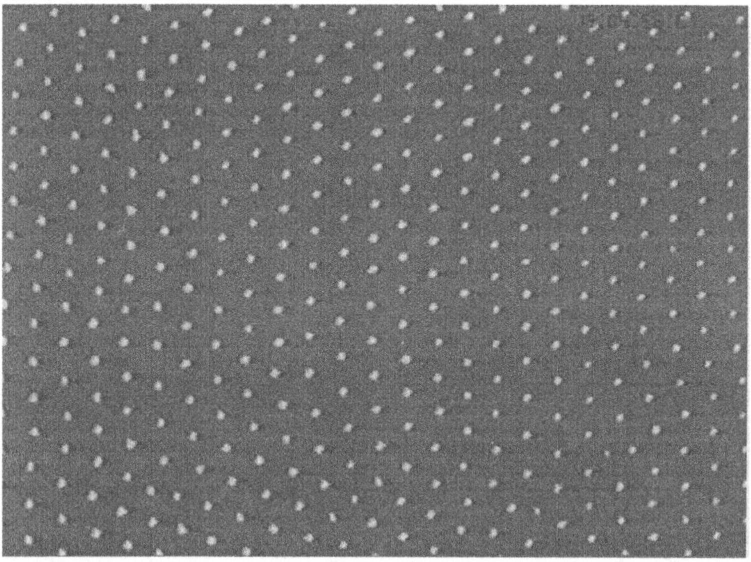

Fig. 4. Image of the particle cloud in a plane above the lower electrode. The area shown is $3.2x4.9mm^2$ and contains 392 particles of $6.9\mu m$ diameter.

observed individually and their motion is followed in two dimensions. The phase transition is initiated by a decrease of the neutral gas pressure (see Fig.5,top). If we decrease the neutral gas pressure, the combination of the associated changes in the plasma parameters [51], the charge reduction on the particles, as well as the decrease of the friction with the neutrals eventual lead to the "melting" of the plasma crystal. Figure 5(bottom) shows the slope of the magnitude of the bond orientational correlation function $g_6(0)$. The decrease in this slope clearly indicates that the system melts to the liquid and the gas phase during the time (of about 300 sec) the neutral gas pressure is decreased.

The changes in the plasma parameters are expressed by the lowering of the electron temperature from approximately 3 eV down to 0.5 eV and the increase of the plasma density from approximately 1×10^9 to $3 \times 10^9 \mathrm{cm}^{-3}$. These values are measured in a comparable experimental setup in a low power rf-discharge [51]. These changes are due to the transition of the electron heating process from ohmic to stochastic heating, since the electron mean free path (see Fig.1a) becomes comparable to the vertical scale of the bulk plasma, which is less than 1 cm. The decrease of the particles' surface potential (proportional to the charge on the particles, Eq.7) due to these changes is shown in Fig.2. If we take the low ion drift velocity situation ($c_{\mathrm{i,drift}}/c_{\mathrm{i}} \ll 1$) we see a change of the surface potential from -8 V to -1.8 V caused by the decrease of the electron temperature from 3.0 eV to 0.5 eV. Thus, the Coulomb coupling parameter decreases by a factor of 16 as a consequence of this charge reduction. This trend is anhanced further due to the increase of the kinetic temperature of the particles - there is less damping of the particles by collisions with the neutrals (the neutrals are

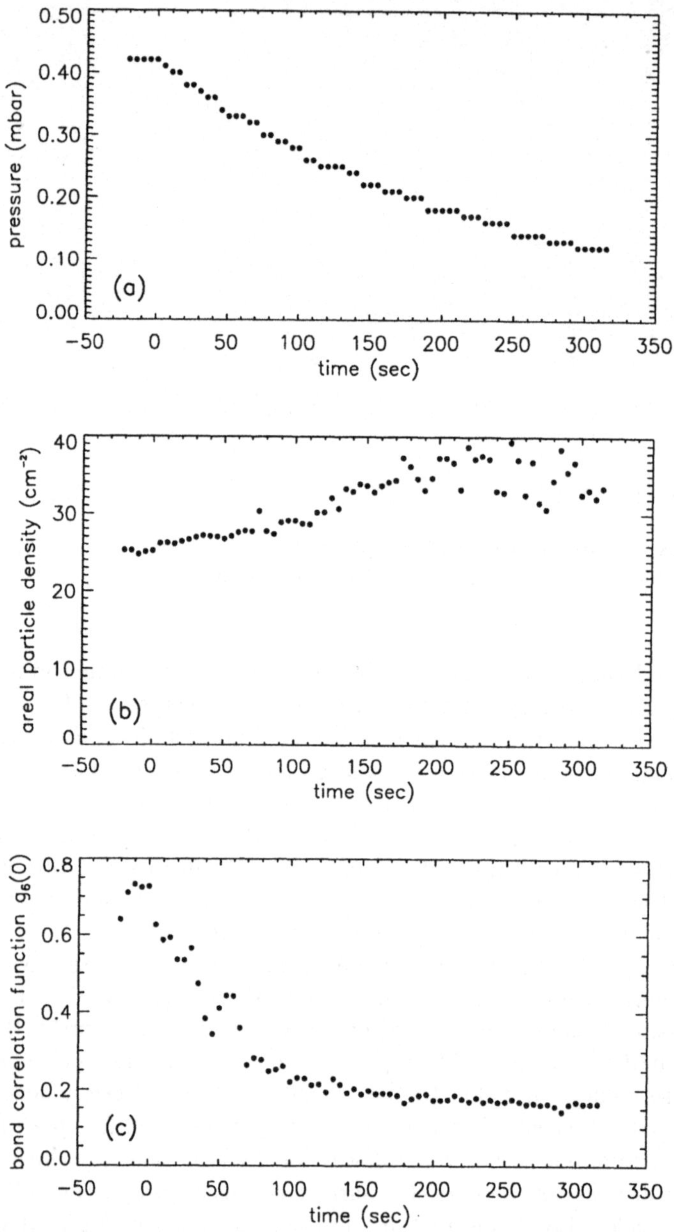

Fig. 5. Time dependent decrease of the neutral gas pressure (top) the surface density of the particles (middle) and the magnitude of the bond-orientational order parameter averaged over the field of view $g_6(0)$ (bottom). The start time (0 sec) corresponds to the beginning of the depressurization and the initiation of the melting transition.

at room temperature). Raising the plasma density decreases the Debye length in the plasma (see Fig.1b). Of course, the particles become more tightly packed when λ_D is decreased. Comparing the measured surface density (see Fig.5) and the calculated change in λ_D shows, that $\kappa = \Delta/\lambda_D$ is increased slightly. Both changes, the decrease of the Coulomb coupling parameter Γ and the increase of the interaction parameter κ lead to the melting of the crystalline structure.

Figure 11 shows the trajectories of the particles in the crystalline phase at a neutral gas pressure of 0.42 mbar. A static analysis of the particle positions according to the KTHNY 2-D melting theory clearly indicates the crystalline structure. The fits to the bond-orientational correlation function leads to a scale lenght of over 100 lattice distances and $\eta \approx 0$ (see Fig.12). A further empirical test is to count the fraction of particles having 6 bonds (6-fold coordination). This fraction is above 90 % for these data, also indicating the crystalline phase. Instead of analysing the data statically only, in the plasma crystal experiment we have the opportunity to follow the dynamics of single particles and the whole cloud of particles, respectively. It is evident that the hexagonal structure for nearly all of the 392 particles (mean particle number, averaged over all frames) in the frame is stationary. A few single particles near dislocations in the lattice exhibit minor changes in their positions during the observation time of 1 sec. In the marked window a larger motion can be seen. This motion is initiated by the disappearance of a single particle. This particle moves vertically in and out of the field of view until it disappears. Then, reordering of the neighboring particles take place. At the edges of the frame out of plane motion ocurrs due to the motion of the particles out and into the field of view ($\times$'s and $\Box$'s).

In Fig.7 the trajectories of the particles are shown for a lower pressure (0.38 mbar). It is obvious that the hexagonal structure isn't as well established as in the previous figure. The bond-orientational correlation function gives us a scale length of $\xi_6 = 14$ lattice constants and $\eta = 0.17$ that is comparable with a so called hexatic phase (intermediate phase, which has quasi-long-range orientational order, but short-range translational order). The acount of particles with sixfold symmetry is 83 %. The mobility of the particles has increased and some local changes in the structure appeared. In the marked window a position change coupled with an out-of-plane motion can be observed. A particle moves towards a neighboring particle which disappears in the vertical direction (marked with a $\times$). At the same time a particle appears ($\Box$) close to the starting position of the moving one. Then reordering takes place.

The changes in the structure due to a further decrease of the pressure are shown in the following two Figs. 89, both at a pressure of 0.36 mbar. Figure 9 shows the sequence following directly that shown in Fig.8. Looking at Fig.8 it is obvious that the local motion of the particles has increased although the the fit to the bond-orientational correlation function (scale length of 14 and 15 lattice constants, respectively, and $\eta = 0.15$) and the sixfold symmetry (over 80 %, for both) indicate the hexatic phase for both, also. More and more out of plane motion occured, leading to a reordering and flow of the particles in the observed plane. In the marked window (dashed lines) a particle appears in the observed plane leading to the displacement of the particles around it. In the

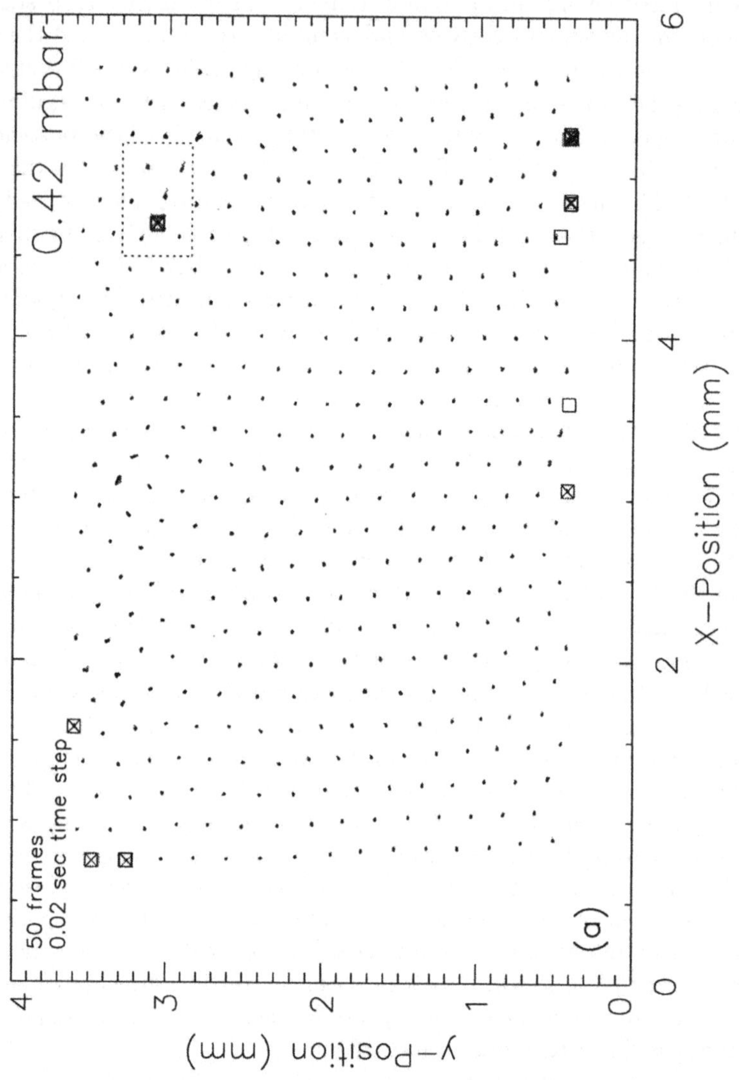

Fig. 6. Trajectories (marked from bright grey at its starting point to black at its end) of located particles calculated over succesive video frames (the number of frames and the time step between following frames are indicated in the upper left corner) at different neutral gas pressures (indicated in the upper right corner) corresponding to different phases of the plasma crystal. ***This figure displays the crystalline phase.*** The ×'s/□'s mark particles which disappear/appear from/in the field of view; in the center of the observed plane this is associated with the vertical (out of plane motion) of the particles.

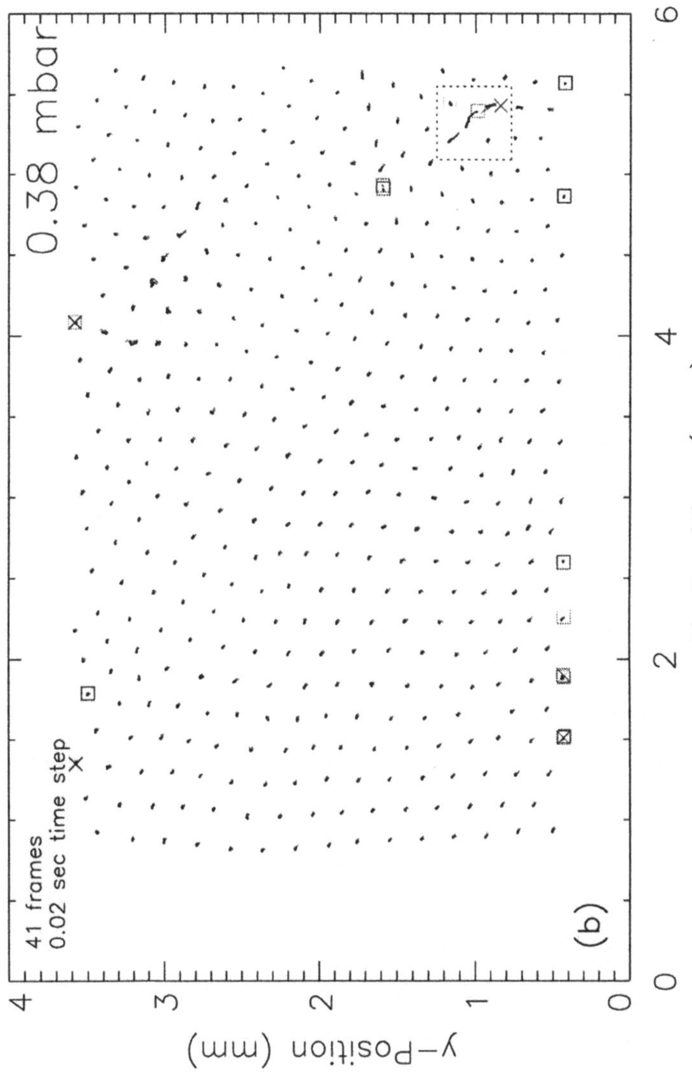

Fig. 7. Trajectories (marked from bright grey at its starting point to black at its end) of located particles calculated over succesive video frames (the number of frames and the time step between following frames are indicated in the upper left corner) at different neutral gas pressures (indicated in the upper right corner) corresponding to different phases of the plasma crystal. ***This figure displays the hexatic phase.*** The ×'s/□'s mark particles which disappear/appear from/in the field of view; in the center of the observed plane this is associated with the vertical (out of plane motion) of the particles.

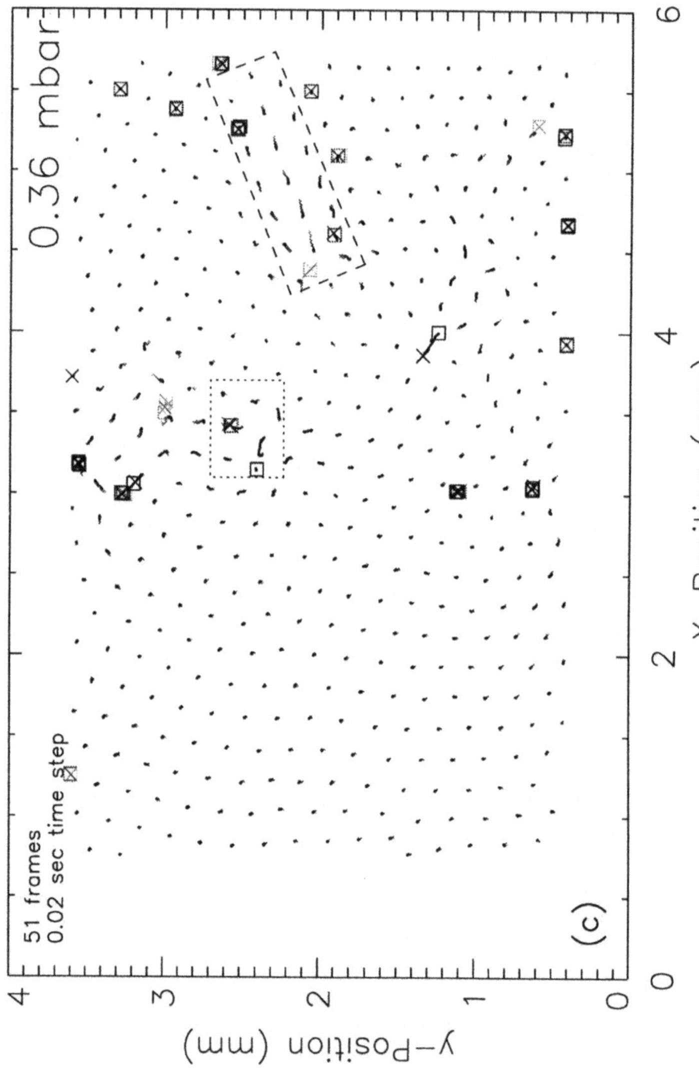

Fig. 8. Trajectories (marked from bright grey at its starting point to black at its end) of located particles calculated over succesive video frames (the number of frames and the time step between following frames are indicated in the upper left corner) at different neutral gas pressures (indicated in the upper right corner) corresponding to different phases of the plasma crystal. *This figure displays the hexatic phase.* The ×'s/□'s mark particles which disappear/appear from/in the field of view; in the center of the observed plane this is associated with the vertical (out of plane motion) of the particles.

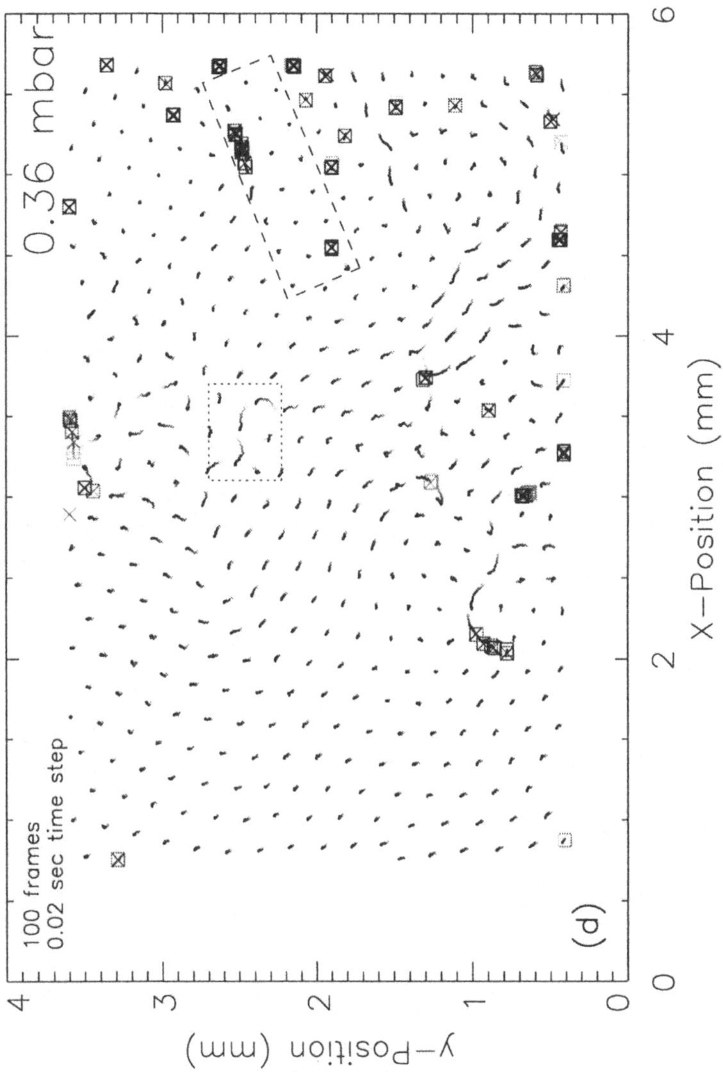

Fig. 9. Trajectories (marked from bright grey at its starting point to black at its end) of located particles calculated over succesive video frames (the number of frames and the time step between following frames are indicated in the upper left corner) at different neutral gas pressures (indicated in the upper right corner) corresponding to different phases of the plasma crystal. ***This figure displays the hexatic phase.*** The ×'s/□'s mark particles which disappear/appear from/in the field of view; in the center of the observed plane this is associated with the vertical (out of plane motion) of the particles.

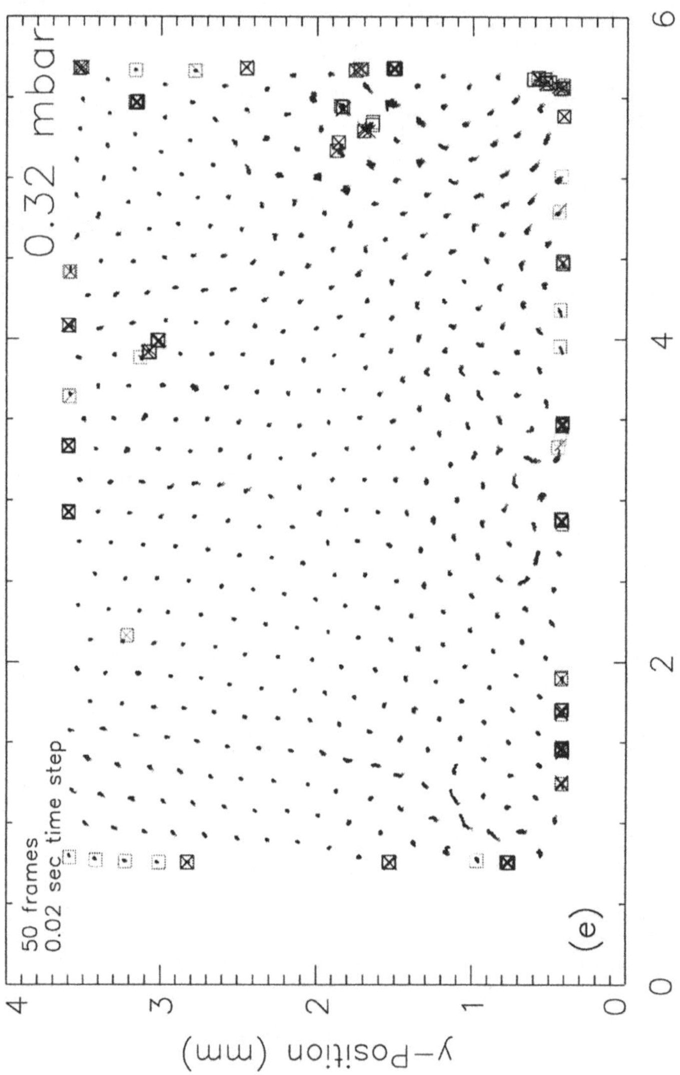

Fig. 10. Trajectories (marked from bright grey at its starting point to black at its end) of located particles calculated over succesive video frames (the number of frames and the time step between following frames are indicated in the upper left corner) at different neutral gas pressures (indicated in the upper right corner) corresponding to different phases of the plasma crystal. ***This figure displays the fluid phase.*** The ×'s/□'s mark particles which disappear/appear from/in the field of view; in the center of the observed plane this is associated with the vertical (out of plane motion) of the particles.

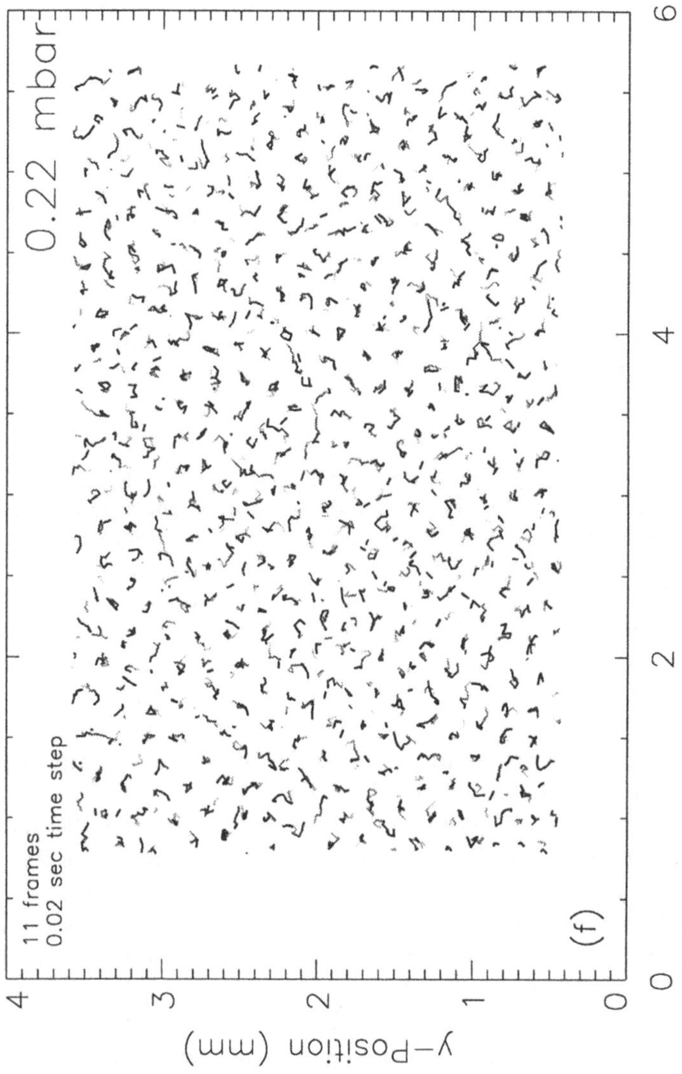

Fig. 11. Trajectories (marked from bright grey at its starting point to black at its end) of located particles calculated over succesive video frames (the number of frames and the time step between following frames are indicated in the upper left corner) at different neutral gas pressures (indicated in the upper right corner) corresponding to different phases of the plasma crystal. *This figure displays the gas phase.* The ×'s/□'s mark particles which disappear/appear from/in the field of view; in the center of the observed plane this is associated with the vertical (out of plane motion) of the particles.

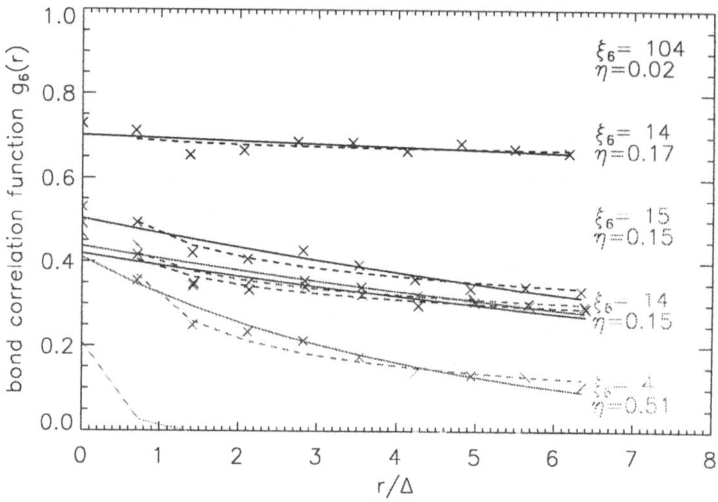

Fig. 12. Bond-orientational correlation function as a function of the normalized radius r/Δ for the different phases. The fit parameters to the power law (dashed lines, η) and exponential fits (solid lines, ξ_6) for the calculated bond-orientational correlation functions are shown in the right of the figure (corresponding to: 0.42 mbar, 0.38 mbar, 0.36 mbar, 0.36 mbar and 0.32 mbar from top to bottom). At a pressure of 0.22 mbar there is no longer bond orientational order.

second window (dotted lines) a particle moves at the beginning (grey colors) vertically in and out of the field of view until it disappears totally. The particles to the right move towards the new formed dislocation to restore order tend to fill up the vacant spot. This motion influences particles at a distance as far as about 6 lattice distances. A reverse flow occurred in the lattice line above the latter.

In Fig.9 we recorded 2 seconds to show not only local motion but directed motion of larger parts of the structure, also. The particles from the dashed window in the last figure exhibit an 'eddy-like' reordering, now. Many particles participate in ordered flow while many others seem to stay still and are unaffected by the flow. This transition stage between solid and liquid has not been observed before. It is reminiscent of an ocean with ice floes - and may be desribed as the "flow and floe" stage.

Figure 10 clearly depicts a liquid phase. At this neutral gas pressure of 0.32 mbar we found a scale length of 4 lattice constants, $\eta = 0.51$ and about 75 % particles with six neighbors. At some locations more random-like motion can be observed, now.

The last phase that we observed is a gas phase of the plasma crystal Fig.11. At a pressure of 0.22 mbar the particles move randomly through the field of view and out of it. Here, we did not show the out of plane motion because most of the particles went in and out during the observation time and so the whole diagram would be filled with crosses and squares. The Coulomb interaction between the particles is so weak at this stage, that particles only interact when they come close together. Their kinetic temperature, calculated from the distances travelled in a given time, has increased. The temperature would increase further if the neutral gas pressure were increased even more, but the particles cannot be followed then because their motion is too rapid for our CCD-camera. The particles show large trajectories of a few lattice constants during the exposure time of 0.02 sec.

4. Conclusion

We presented a new type of model systems of crystalline structures, the so called plasma crystal, and compared this system with the well studied model systems of electron crystals, ion crystals and colloidal crystals in aqueous solutions. We showed the advantages of observations performed on this optically thin and only weakly damped system, in two dimensions. The possibility of combining measurements of the dynamics of single "atoms" with different large scale structure analyses is a powerful tool to extract the essential information necessary for a detailed investigation of phenomena associated with self-organisation, phase transitions, energetics etc. The plasma crystal, given its special properties of global charge neutrality and very little damping (through particle-gas frictional coupling) of the particle motion, can trigger significant progress in this area.

Other interesting effects have already been tested experimentally, e.g. the excitation of oscillations in the plasma crystal. From such experiments we hope to obtain exact information on the pair interaction potential and to calculate the dispersion relation.

In the laboratory we are restricted to basically two dimensional plasma crystals with perhaps a few lattice planes in the vertical. This restriction is due to gravity. To overcome this, a space shuttle experiment under microgravity conditions will be performed. Under those conditions we hope to be able to study a 3-dimensional optically thin plasma crystal where edge effects and their influence on the dynamics of the particles can be neglected. Thus fundamental phenomena of the Coulomb crystallization might be visualized which could lead to a better understanding of solid state physics.

Acknowledgements

This work is supported by the German Space Agency DARA.

References

1. W. Bragg and W. Lomer. it Proc. Roy. Soc., **A196** (1948) 171.
2. F. Diedrich, E. Peik, J. Chen, W. Quint, and H. Walther. it PRL, **59** (1987) 2931–2934.

3. S. Gilbert, J. Bollinger, and D. Wineland. it Phys. Rev. Lett., **60** (1988) 2022–2025.
4. I. Waki, S. Kassner, G. Birkl, and H. Walther. it PRL, **68** (1992) 2007–2010.
5. H. Jiang, R. Willet, H. Stormer, D. Tsui, L. Pfeiffer, and K. West. it Phys. Rev. Lett., **65** (1990) 633–636.
6. V. Goldman, M. Santos, M. Shayegan, and J. Cunningham. it Phys. Rev. Lett., **65** (1990) 2189–2192.
7. D. Grier and C. Murray. it J. Chem. Phys., **100** (1994) 9088–9095.
8. H. Löwen. it J. Phys.: Condens. Matter, **4** (1992) 10105–10116.
9. C. Murray. In K. Strandburg, editor, *Bond orientational order in condensed matter systems'.* Springer Verlag, 1992.
10. A. Kose, M. Ozaki, K. Takano, Y. Kobayashi, and S. Hachisu. it J. Coll. Int Sci., **44** (1973) 330–338.
11. H. Thomas, G.E. Morfill, V. Demmel, J. Goree, B. Feuerbacher, and D. Möhlmann. it PRL, **73** (1995) 652–655.
12. J. Chu, Ji-Bin Du, and Lin I. it J. Phys. D: Appl. Phys., **27** (1994) 296–300.
13. J. Chu and Lin I. it PRL, **72** (1994) 4009–4012.
14. A. Melzer, T. Trottenberg, and A. Piel. it Phys. Lett. A, **191** (1994) 301.
15. Y. Hayashi and K. Tachibana. it Jpn. J. Appl. Phys., **33** (1994) L476–L478.
16. R. Quinn, J. Goree, C. Cui, H. Thomas, and G.E. Morfill. it PRL, subm.
17. E. Wigner. it Trans. Faraday. Soc., **34** (1939) 678–685.
18. S. Ichimaru. it Rev. of Mod. Phys., **54** (1982) 1017–1059.
19. D. Dubin and T. O'Neil. it Phys. Fluids, **B2** (1992) 460–464.
20. H. DeWitt, W. Slattery, and J. Yang. In H. Van Horn and S. Ichimaru, editors, *Strongly coupled plasma physics*, pages 425–434. University of Rochester Press, 1993.
21. B. Deryagin and L. Landau. it Acta Phys.-Chem. USSR, **14** (1941) 331–354.
22. C. Murray and R. Wenk. it PRL, **58** (1989) 1643–1646.
23. Y. Tang, A. Armstrong, R. Mockler, and W. O'Sullivan. it PRL, **62** (1989) 2401–2404.
24. W. Luck, M. Klier, and H. Wesslau. it Ber. der Bunsengesellschaft, **67** (1963) 75–83.
25. W. Luck, M. Klier, and H. Wesslau. it Ber. der Bunsengesellschaft, **67** (1963) 84–85.
26. A. Skjeltorp. it Physica, **127B** (1984) 411–416.
27. I. Dyradl. it J. Appl. Phys., **60** (1986) 1913–1915.
28. H. Löwen. it J. Phys.: Condens. Matter, **4** (1992) 10105–10116.
29. J. Kosterlitz and D. Thouless. it J. Phys. C, **6** (1973) 1181–1203.
30. B. Halperin and D. Nelson. it PRL, **41** (1978) 121–124.
31. D. Nelson and B. Halperin. it Phys. Rev. B, **19** (1979) 2457–2484.
32. A. Young. it Phys. Rev. B, **19** (1979) 1855.
33. D. Nelson. In C. Domb and J. Leibowitz, editors, *Phase transitions and criticalPhenomena, Vol. 7.* Academic Press, London, 1983.
34. C. Murray, W. Sprenger, and R. Wenk. it J. Condensed Matter, **2** (1990) SA385–SA388.
35. C. Murray, W. Sprenger, and R. Wenk. it Phys. Rev. B, **42** (1990) 688–703.
36. H. Ikezi. it Phys. Fluids, **29** (1986) 1764–1766.
37. E. Grün, G. Morfill, and A. Mendis. In A. Brahic and R. Greenberg, editors, *Planetary rings*, page 275. Univ. Arizona Press, Tucson, 1984.
38. C. K. Goertz. it Rev. of Geoph., **27** (1989) 271 –292.

39. T. Hartquist, O. Havnes, and G. Morfill. it Fundamentals Cosmic Phys., **15** (1992) 107.
40. G. Selwyn, J. Heidenreich, and K. Haller. it Appl. Phys. Lett., **57** (1990) 1876–1878.
41. L. Boufendi, A. Bouchoule, R. Porteous, J. Blondeau, A. Plain, and C. Laure. it J. Appl. Phys., **73** (1993) 2160 –2162.
42. G. Morfill, E. Grün, and T. Johnson. it Planet. Space Sci., **28** (1980) 1087.
43. C. K. Goertz and G. Morfill. it Icarus, **53** (1983) 219.
44. W. Pilipp et al. it Astrophys. J., **314** (1987) 341.
45. J. Burns, M. Showalter, and G. Morfill. In A. Brahic and R. Greenberg, editors, *Planetary rings*, page 200. Univ. Arizona Press, Tucson, 1984.
46. G. Morfill, C. Goertz, and O. Havnes. it JGR, **98** (1993) 1435.
47. E. Whipple. it Rep. Prog. Phys., **44** (1981) 1197–1250.
48. O. Havnes, T. Aanesen, and F. Melandsø. it JGR, **95** (1990) 6581 - 6585.
49. B. Draine and E. Salpeter. it APJ, **231** (1979) 77 - 94.
50. J. Goree. it Plasma Sources Sci. Technol., **3** (1994) 400.
51. V. Godyak and R. Piejak. it Phys. Rev. Lett., **65** (1990) 996–999.
52. R. Farouki and S. Hamaguchi. it APL, **61** (1992)2973 - 2975.
53. P. Hargis et al. it Rev. Sci. Instrum., 1993.

Part IV

Fluid Statics and Thermophysical Properties

Are Liquids Molten Solids or Condensed Gases?

F.S. Gaeta[1], *F. Peluso*[1], *C. Albanese*[1], *and D.G. Mita*[2]

[1] MARS Center, v. Comunale Tavernola, 80144 Napoli, Italy
[2] Department of Human Physiology, II University of Naples, via Costantinopoli 16, I-80138 Naples, Italy

Abstract

The kinship between liquids and solids, advocated by the similarity of the respective densities, did not lead, until now, to a theory of transport processes in liquids in analogy to the phonon theory of solids. The ubiquitous presence of scattering centers due to the disordered structure of fluids makes a quantitative appraisal of energy and momentum exchange in the course of phonon-particle interactions pivotal for such an approach. The new physical concepts of radiant vector and of thermal radiation force provide the appropriate fundamentals for a quantitative treatment. The experimental foundations of the proposed theory reside in a clear-cut assessment and in the appropriate measurement of thermal radiation forces in the appropriate material system. For reasons discussed here such an unambiguous result may be only obtained in conditions of reduced gravity.

1. Introduction

The differences of transport phenomena in gases and solids correspond to different underlying microscopic mechanisms of molecular interaction.

In gases the constituent particles move at random with high speeds. Diffusion of the molecules or of their kinetic energy and momentum is slow owing to intermolecular collisions which cause complicated zig-zag motions. The applied generalized gradients ensure only a slightly higher probability of transport of the quantity not at equilibrium in the direction of the gradient.

In solids, the constituent particles are not dislodged from lattice sites by gradients of concentration or by moderate gradients of electric potential. Accordingly one expects - and indeed finds - exceedingly small rates of diffusion and electrical conduction, unless there are free carriers, as in metals.

What does not fit in this picture is the high rate of heat transport in nonmetallic solids. A moderate temperature gradient certainly does not dislodge atoms from their lattice sites in diamond. Still diamond, particularly in isotopically pure crystals, exhibits elevated thermal conductivities, higher than those of metals. This can be explained within the frame of the Debye model, thermal energy in solids being distributed amongst the normal modes of vibration of the crystal. Each normal mode can be analysed into elastic waves travelling in opposite directions; these wave-packets of acoustic energy are called phonons. Thermal excitations of a solid are considered as a gas of phonons [1].

If the potential energy of lattice particles is quadratic in the distance from equilibrium positions, we get harmonic normal modes, incapable of interacting; phonons, travelling at the speed of sound, would diffuse heat through solids nearly instantaneously. If not, it is because potential energy of lattice particles is not exactly quadratic in displacement and thus anharmonic terms arise. The Lennard-Jones potential is an example of a skew potential well, wherein a molecule would oscillate anharmonically, when displaced from its equilibrium position (Fig.1).

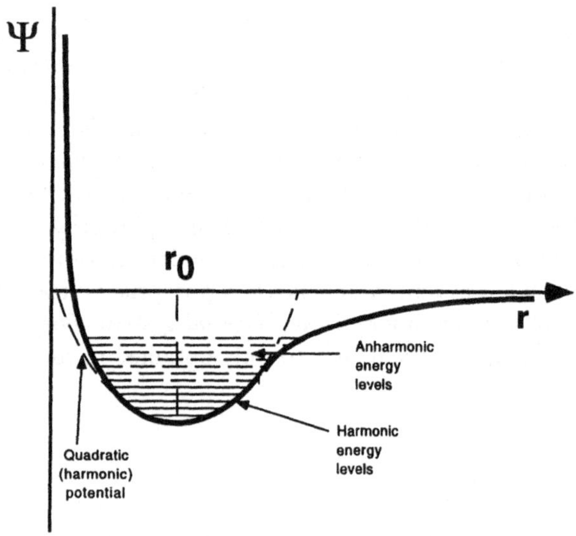

Fig. 1. Lennard-Jones potential (full line) compared with a quadratic distribution (dashed)

Similar and even larger anharmonicities are to be expected where the lattice is distorted as, for instance, at vacancies, interstitial atoms, grain boundaries or impurities. As well known, anharmonicity leads to phonon-phonon and to phonon-lattice interactions, allowing exchange of energy and momentum. On this basis, real-life values can be calculated for the heat conductivity (phonon diffusivity) of solids. The modern physics of transport in solids is built on these ideas [2, 3, 4, 5, 6].

Now, since the densities of liquids are very similar to those of their solid counterparts, the forces acting on particles displaced from their instantaneous equilibrium positions have intensities close to those occurring in solids. Potential energy however shall be non-quadratic practically everywhere, owing to the prevailing molecular disorder; anharmonicity will therefore be ubiquitous. Anyone aware of the difficulties of dealing with phonon-phonon-lattice (Umklapp) interactions in solids [7] will consider exceedingly complex the intricacies of a phonon theory in systems lacking long-range order and assume doomed from the start the attempt to construct a theory of liquids on such basis.

The new unifying concept of elastic and thermal radiation forces circumvents the main difficulties, thus consenting the extension of phonon theory to transport processes in liquids [8, 9, 10, 11].

2. The radiant vector R

The energy associated with a mechanical wave is $\langle E\rangle = \langle E_{\text{kin}}\rangle + \langle E_{\text{pot}}\rangle = \rho\langle\dot{\xi}\rangle$, where $\dot{\xi}$ is average particle velocity. The flux of elastic energy $\mathbf{J}_{\text{el}}$ due to wave propagation is $\mathbf{J}_{\text{el}} = (\langle E\rangle u_s)\mathbf{r}$, where u_s is group propagation velocity and $\mathbf{r}$ the unit vector along the direction of wave propagation. On the other hand

$$\mathbf{J}_{\text{el}} = \Delta p \cdot \dot{\boldsymbol{\xi}} = \left(\Delta p \dot{\xi}\right)\mathbf{r} = \mathbf{R} \tag{1}$$

where the radiant vector $\mathbf{R}$ is never negative, because both Δp and $\dot{\xi}$ change sign simultaneously at every half period. Dividing equation (1) by u_s, one has $(\mathbf{J}_{\text{el}}/u_s)\mathbf{r} = (\Delta p \cdot \dot{\xi}/u_s)\mathbf{r} = (R/u_s)\mathbf{r} = \boldsymbol{\Pi}_{\text{el}}$, $\boldsymbol{\Pi}_{\text{el}}$ being elastic radiation force per unit of surface normal to wave propagation. Between positions (1) and (2) this force will be

$$\boldsymbol{\Pi}_{\text{el}}^{1,2} = \left[\left(\frac{J_{\text{el}}}{u_g}\right)_1 - \left(\frac{J_{\text{el}}}{u_g}\right)_2\right]\mathbf{r} = \left[\left(\Delta p\frac{\dot{\xi}}{u_g}\right)_1 - \left(\Delta p\frac{\dot{\xi}}{u_g}\right)_2\right]\mathbf{r} = [\langle E_1\rangle - \langle E_2\rangle]\mathbf{r} \tag{2}$$

It is obvious that radiant pressure is expected to appear whatever the cause of the change of acoustic energy density. If this is sound absorption, it will always be $\langle E_1\rangle < \langle E_2\rangle$; if it is the temperature dependence of the properties of a nonisothermal medium, then $\boldsymbol{\Pi}_{\text{el}}$ will be a rather small quantity. Finally, if in between positions (1) and (2) there is a surface of discontinuity of two different, adjoining media, then $\langle E_1\rangle - \langle E_2\rangle$ depends on the constitutive properties of the media in contact.

3. Thermal radiation forces $\boldsymbol{f}^{\text{th}}$

We make now a strong assumption on the mechanism of heat propagation in liquids as due to high frequency elastic waves. The presence of local imperfections, causing non-harmonic interactions between microscopic heat currents and fluid particles, will give rise to transfer of momentum and to radiation pressure. In a straightforward manner, starting from the concepts of radiant vector and acoustical radiation pressure, it is easy to define, by analogy, a thermal radiation pressure.

The total heat content per unit of volume of the liquid is given by the quantity $q = \int_0^1 \rho C_V dT$, where T is absolute temperature and C_V specific heat at constant volume. A fraction m^* of this quantity pertains to the phonon gas, whose energy density is: $q^{ph} = m^* \rho C_V T = N^{ph}\langle\epsilon^{ph}\rangle = N^{ph} h \langle\nu^{ph}\rangle$, where h is Plancks constant, $\langle\nu^{ph}\rangle$ average frequency of thermal excitations, N^{ph} and ϵ^{ph} the phonon's number per unit of volume and their average energy, respectively.

In nonisothermal liquids there will be a gradient of density of phonon energy, which results in diffusion of wave packets along the temperature gradient. Where the propagating phonons sweep through regions of anharmonic intermolecular potential, they give rise to local interactions, with exchange of energy $\Delta\langle h\nu\rangle$ and of momentum $(\Delta\langle h\nu\rangle)/U_g = f^{th} \cdot \Delta t$ with particles, f^{th} being thermal radiation force exerted by the wave-packets on molecules in the course of interactions, lasting a time interval Δt, with the latter. Figure 2 represents the process in which momentum lost by a phonon is transformed into translational momentum $m\nu$ of the molecule (2a), and the inverse event, in which the phonon acquires momentum after a collision with a molecule (2b).

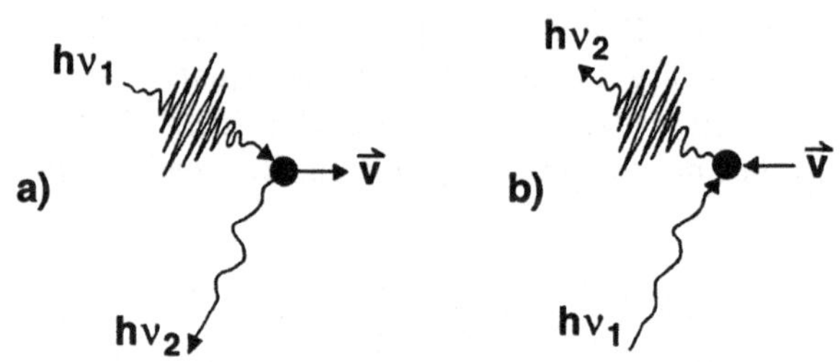

Fig. 2. In a) an energetic phonon transfers energy $h(\nu_1 - \nu_2)$ and momentum $h(\nu_1 - \nu_2)/u$ to a particle in the course of an interaction lasting Δt; in b) the event is reversed. Upon inversion of the time axis, each of these events is commuted into the other (microscopic time reversibility).

The heat flux $\mathbf{J}_q^{ph}$ due to a gradient $(\partial T/\partial z)\mathbf{r}$, i.e. the phonon contribution to heat flowing per unit of surface along z, is:

$$\mathbf{J}_q^{ph} = -\frac{\partial}{\partial T}(m^* \rho C_V T) u_\phi \langle\Lambda^{ph}\rangle \frac{\partial T}{\partial z}\mathbf{r} = -\frac{\partial}{\partial T}(N^{ph}\langle\epsilon^{ph}\rangle) u_\phi \langle\Lambda^{ph}\rangle \frac{\partial T}{\partial z}\mathbf{r} , \quad (3)$$

$\langle\Lambda^{ph}\rangle$ being phonon mean free path. Substituting for m^* its expression in terms of Einstein distribution, $m^* = m(\int_0^T \rho C_V dT)/(\rho C_V T)$ one obtains

$$\frac{\partial}{\partial T}(m^* \rho C_V T) = \frac{\partial}{\partial T}(N^{ph}\langle\epsilon^{ph}\rangle) = m\rho C_V \left(\frac{m^*}{m^2}\frac{dm}{dT}T + 1\right) ,$$

where m is the ratio of translational to total degrees of freedom of the molecules of the medium. (3) may now be written in the form:

$$\mathbf{J}_{\mathrm{q}}^{\mathrm{ph}} = -K^{\mathrm{ph}} \frac{\partial T}{\partial z} \mathbf{r} \tag{4}$$

where, from the above:

$$K^{\mathrm{ph}} = m\rho C_{\mathrm{V}} u_{\phi} \langle \Lambda^{\mathrm{ph}} \rangle \left(\frac{m^*}{m^2} \frac{dm}{dT} T + 1 \right) \tag{5}$$

represents a microscopic expression of thermal conductivity. Equation 5, derived on the basis of phonon-particle interactions, can be shown to correspond to the microscopic evolution mechanism to which Onsager refers in his work [12]. The time-reversal invariance of the phonon-particle scattering was shown to be the right candidate for a truly microscopic reformulation of nonequilibrium thermodynamics [10, 11].

The flux of thermal energy $\mathbf{J}_{\mathrm{q}}^{\mathrm{ph}}$ consists of elastic energy propagating down a temperature gradient. The considerations for the radiation pressure of elastic waves thus apply also here. A "thermal radiation pressure" is $\Delta\Pi_{\mathrm{th}}^{1,2}$ expected to appear at the macroscopic level between two successive positions along the propagation direction of heat flux:

$$\Delta\Pi_{\mathrm{th}}^{1,2} = \left[\left(\frac{J_{\mathrm{q}}^{\mathrm{ph}}}{u_{\mathrm{g}}} \right)_1 - \left(\frac{J_{\mathrm{q}}^{\mathrm{ph}}}{u_{\mathrm{g}}} \right)_2 \right] \mathbf{r} = \left[\left(\frac{K^{\mathrm{ph}}}{u_{\mathrm{g}}} \frac{\partial T}{\partial z} \right)_2 - \left(\frac{K^{\mathrm{ph}}}{u_{\mathrm{g}}} \frac{\partial T}{\partial z} \right)_1 \right] \mathbf{r} \tag{6}$$

$\Delta\Pi_{\mathrm{th}}^{1,2}$ is the resultant along z of all $\mathbf{f}^{\mathrm{th}}$ the due to the elementary phonon-particle interactions occurring per unit of time within the unit of volume of the medium. At the macroscopic level, forces due to heat propagation are thus expected to appear. Expressions analogous to (6) have been previously derived from a macroscopic analysis [9] based on the Boltzmann-Ehrenfest theorem. A solid, flat plate suspended in a nonisothermal liquid for instance was shown to be subject to a thermal radiation force given by:

$$F_{\mathrm{l,p}}^{\mathrm{th}} = 2\sigma_{\mathrm{p}} H^* \left[\left(\frac{K}{u} \right)_{\mathrm{l}} - \left(\frac{K}{u} \right)_{\mathrm{p}} \right] \frac{dT}{dz} \tag{6'}$$

In this expression K and u are the experimentally determined thermal conductivities and velocities of sound in (l) and (p); $(K/u)_{\mathrm{l}}$ and $(K/u)_{\mathrm{p}}$ are the momentum conductivities of the liquid and of the suspended object; σ_{p} is the objects cross section and $H^* = 2\pi R$ a numerical quantity proportional to the acoustic reflection coefficient R at the (l, p) boundary.

Equations (6) and (6′), obtained in completely independent ways, also have other non trivial differences, besides the substitution of phonon thermal conductivity with respect to the measured conductivity. The temperature gradient in the Boltzmann-Ehrenfest approach is the one externally applied to the system, irrespective of the local composition(s) of its components. The constant H^* accounts for the variation of acoustic impedance of the medium with temperature. It has been shown [9] that

$$R = \left[\frac{d(\rho u)}{2\rho u + d(\rho u)}\right]^2 ,$$

the change of acoustic impedance in this case being continuous. The concept of reflection of elastic waves is the macroscopic equivalent of the phonon-particle interactions in the approach which led to (6). Within this framework the comparison of (6) and (6′) relates and the experimentally determined thermal conductivity is $K^{\mathrm{ph}} = K[H^* u_{\mathrm{g}}/u]$.

4. Heat propagation in liquids

Our approach suggests that Fourier' law should be modified as:

$$J_{\mathrm{q}} + \langle\tau\rangle \frac{\partial J_{\mathrm{q}}}{\partial t} = -K \frac{\partial T}{\partial z} ,$$

where the (negative) term $\partial J_{\mathrm{q}}/\partial t$ represents the flux temporarily frozen and $\langle\tau\rangle$ accounts for the average time spent by heat as non-propagating molecular excitations. This expression, combined with the continuity condition

$$\rho C_{\mathrm{p}} \frac{\partial T}{\partial t} + \frac{\partial J_{\mathrm{q}}}{\partial z} = 0 ,$$

yields the hyperbolic propagation equation:

$$\frac{\partial^2 T}{\partial z^2} = \frac{\rho C_{\mathrm{p}}}{K} \frac{\partial T}{\partial t} + \tau \frac{\rho C_{\mathrm{p}}}{K} \frac{\partial^2 T}{\partial t^2} \tag{7}$$

in place of Fourier's classical parabolic form. Integration of (7) requires initial conditions for both $T(z,t)$ and $\partial T(z,t)/\partial t$. Its solution, dealt with elsewhere [13], shows that diffusive-type heat conduction is preceeded by a damped undulatory phase with velocity

$$\nu = \left(\frac{K}{\tau \rho C_{\mathrm{p}}}\right)^{1/2} . \tag{8}$$

In the trail of the advancing thermometric fronts, diffusive heat propagation occurs, with:

$$D^{\mathrm{ph}} = \frac{K}{\rho C_{\mathrm{p}}} = \frac{m}{\gamma} u_\phi \langle \Lambda^{\mathrm{ph}} \rangle \left(\frac{m^*}{m^2}\frac{dm}{dT} T + 1\right) ,$$

where $\gamma = C_{\mathrm{P}}/C_{\mathrm{V}}$. In terms of these quantities, derived from the phonon approach, and calling l the distance over which the directional input of momentum by phonons is randomized, in place of the velocity of (7) one gets:

$$\nu = \frac{l}{\tau} = \frac{D^{\mathrm{ph}}}{l/2} = 2\frac{m}{\gamma}\frac{\langle \Lambda^{\mathrm{ph}} \rangle}{l} \left(\frac{m^*}{m^2}\frac{dm}{dT} T + 1\right) . \tag{9}$$

Within the thickness of the advancing thermometric fronts l, the individual wavelets, which have not yet reconstituted envelop wavefronts, locally determine very steep temperature gradients, given by:

$$\frac{\Delta T}{l} = \frac{1}{6}\frac{N^{\rm ph}\langle\epsilon^{\rm ph}\rangle}{\rho C^* l}\left(\frac{u_\phi}{\nu^*} - 1\right) \tag{10}$$

The local rate of temperature change in the advancing fronts depends on constitutive properties of the medium and on the dynamics of phonon-particle interactions. Interestingly, it is this circumstance that makes physically understandable the surprising mathematical boundary condition on $\partial T(z,t)/\partial t$.

5. The experimental situation

In Table 1 we give the values of the ratios of measured thermal conductivities to the propagation velocity of elastic waves for some liquids and solids.

Thermal radiation forces (TRFs) produced on surfaces between two adjoining media are expected to be small, being proportional to the difference of the respective momentum conductivities K/u.

6. Measurements of TRFs and microgravity relevance of the problem

Various effects due to thermal radiation forces have been experimentally observed; we shall concisely describe two of them.

6.1 Thermal diffusion

Solute drift in nonisothermal solutions can be correctly explained by the TRF theory. Equation (6′) is valid in principle for any kind of object suspended in a nonisothermal liquid. Its application to disperse particles of molecular dimensions however raises some problems. One has to do with the definition of momentum conductivity K/u of a molecule or a solvated ion. Additionally there is a difficulty of describing quantitatively the interactions of phonons with obstacles having diametral dimensions comparable with the wavelength of the impinging thermal excitations. At a qualitative level, however, it can be said that the force $F^{\rm ph}_{\rm l,p}$ will cause solute drift with a velocity $\nu_{\rm d} = D^{\rm th} dT/dz$ where $D^{\rm th}$ is the thermal diffusion coefficient of the particle, defined as drift velocity in a unit temperature gradient. The value of $\nu_{\rm d}$ shall be determined by the balance of the power input $W^{\rm th}_{\rm p}$ due to the work performed by thermal radiation forces and the energy dissipated per second by the particle through hydrodynamic friction $W^{\eta}_{\rm p}$. Assuming the solvated solute particle to be spherical,

$$D^{\rm th}_{\rm p} = H^* \frac{r_{\rm p}}{3\eta_{\rm l}}\left[\left(\frac{K}{u}\right)_{\rm l} - \left(\frac{K}{u}\right)_{\rm p}\right] , \tag{11}$$

Substance	$K \cdot 10^{-4}$ erg.cm^{-1}s^{-1}K^{-1}	$u \cdot 10^{-5}$ cms^{-1}	K/u erg.cm^{-2}K^{-1}	T oC
water	5.791	1.448	0.4000	10
water	6.147	1.509	0.4072	30
water	6.424	1.543	0.4164	50
water	6.693	1.555	0.4305	80
ethylene glycol	2.540	1.717	0.1479	0
ethylene glycol	2.567	1.667	0.1540	20
ethylene glycol	2.594	1.617	0.1604	40
ethylene glycol	2.620	1.567	0.1672	60
ethanol	1.740	1.229	0.1416	0
ethanol	1.681	1.159	0.1450	20
ethanol	1.621	1.096	0.1479	40
ethanol	1.562	1.033	0.1512	60
benzene	1.483	1.372	0.1081	10
benzene	1.432	1.276	0.1122	30
benzene	1.356	1.140	0.1189	60
plexiglass	2.063	2.680	0.0780	17
glass	4.200	5.630	0.0750	17
n-heptane	1.327	1.237	0.1073	0
n-heptane	1.393	1.207	0.1168	7
n-heptane	1.407	1.125	0.1245	27
n-heptane	1.470	0.990	0.1426	57
polystyrene	1.50	2.39	0.063	5
polystyrene	1.52	2.37	0.064	15
polystyrene	1.52	2.35	0.065	20
polystyrene	1.55	2.32	0.067	30

Table 1. Values of thermal conductivities K, sound velocity u and of the ratios K/u for some substances (liquids and solids) at various temperatures.

η_l being the solution viscosity. From (11) it is immediate to deduce the expression for Soret coefficient $s = D_p^{th}/D_p$, where D_p is the diffusion coefficient. Making use of the Stokes-Einstein equation, we get

$$s = H^* \frac{2\pi r_p^2}{k_B T T_{av}} \left[\left(\frac{K}{u}\right)_l - \left(\frac{K}{u}\right)_p \right] , \tag{12}$$

where k_B is the Boltzmann constant and T_{av} solution average temperature. A qualitative prediction thus is that upon change of sign of the quantity $\left[(K/u)_l - (K/u)_p\right]$ an inversion occurs in the sense of solute drift relative to the temperature gradient. It is remarkable that studies of thermodiffusive behaviour of solutions of polyvinylpyrrolidone K90 of 360000 amu in various solvents evidenced an inversion in the sense of solute drift in different solvents (see Table 2) [14], precisely as expected from the comparison of the respective values.

Polyvinylpyrrolidone K90 in:	$(K/u)_l$ $erg/cm^{-2}(^\circ C)^{-1}$	$(K/u)_p$ $erg/cm^{-2}(^\circ C)^{-1}$	$(K/u)_l - (K/u)_p$	$10^3 s_p$ $(oC)^{-1}$
water	0.4031	0.137	0.266	19.82
methanol	0.1830	0.137	0.046	0.38
ethanol	0.1440	0.137	0.007	2.31
butanol	0.1205	0.137	-0.016	-5.78
propanol	0.1197	0.137	-0.017	-6.01

Table 2. Macromolecules of polyvinylpyrrolidone of 360000 a.m.u. drift in the direction of the heat flux in water, methanol and ethanol. Drift in the opposite direction occurs in butanol and propanol. The inversion is correlated with the change of sign of the difference $(K/u)_{\text{liq}} - (K/u)_{\text{poly}}$.

Another successful prediction of the TRF approach to thermal diffusion in liquids was that the Soret coefficient in macromolecular solutions should be proportional to the power 2/3 of molecular mass, as experimentally found [8, 9, 14]. Gas kinetic theories of thermal diffusion, applied to macromolecular solutes, on the contrary, lead to Soret coefficients inversely proportional to solute mass. Notwithstanding these significant successes, experiments of thermal diffusion cannot yield quantitative measurements of TRF's owing to the fundamental difficulties cited above.

6.2 Direct measurement of TRFs

Forces produced on solid non-metallic slabs suspended in nonisothermal liquids have been measured with a simple apparatus (Fig.3).

Qualitative agreement on the orientation of forces acting on the slab and on their dependence on T_{average} and gradT has been found [9]. Quantitative observations are plagued by buoyancy effects, unavoidable in normal gravity.

The experimental results obtained with many different solid-liquid systems provided full confirmation of the theoretical predictions [9], namely:

1. proportionality of measured forces to gradT;
2. inversion of the direction of $F^{\text{th}}_{\text{l,p}}$ when the term $\left[\left(\frac{K}{u}\right)_{\text{l}} - \left(\frac{K}{u}\right)_{\text{p}}\right]$ changes sign;
3. dependence of $F^{\text{th}}_{\text{l,p}}$ on average temperature coincident with the temperature dependence of $\left[\left(\frac{K}{u}\right)_{\text{l}} - \left(\frac{K}{u}\right)_{\text{p}}\right]$.

From the above it clearly appears that reliable, quantitative measurements of the intensity of thermal radiation forces have not yet been obtained. It will be exceedingly difficult to get such data in the presence of generally larger buoyancy effects.

An interesting corollary of the assumed existence of very steep temperature gradients $\Delta T/l$ within the advancing thermometric fronts is that they

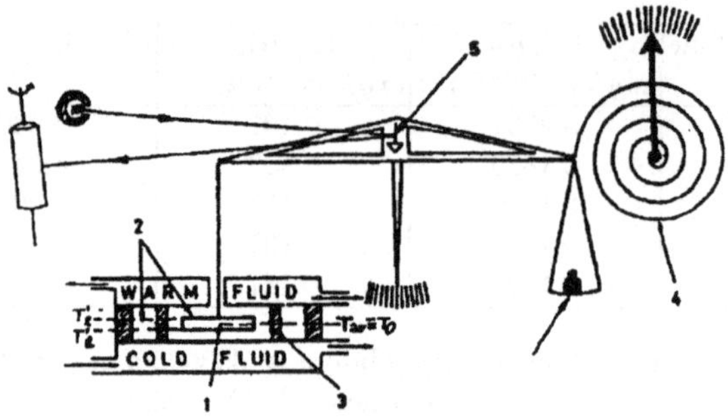

Fig. 3. Schematic representation of our apparatus. Plunger (1) is accurately counter-balanced at temperature T0: then the upper flow chamber is heated to T2 and the lower flow chamber cooled to T1 subject to the condition. Heat flows through the liquid (2) and the plunger. The guard ring (3) prevents convective disturbances. Thermal radiation force is compensated by means of dynamometric measuring device (4). Plunger movement can be detected by optical device (5) to better than +/- 0.005 cm.

could involve striking effects in biological systems. For instance, propagation of thermal energy in the non-periodic anharmonic structures of protein macromolecules should lead to strong thermal radiation forces. Interactions with prosthetic (heme) groups [15] or amino acids at enzymatic sites, can strongly affect protein-mediated catalysis [16]. A great enhancement of the catalytic activity rate of immobilised enzymes under nonisothermal conditions indeed has been recently demonstrated [17]. This finding, apart from its importance in biotechnology, hints to the possibility that temperature gradients might have fundamental functions in the living cell. Biological membrane transport of ions and molecules invariably occurs through biomembranes crossed by fluxes of thermal energy, owing to the different rates of metabolic heat production in the compartments separated by such membranes.

Clearly, it would be important to prove that thermal radiation forces do exist and to measure their values. Unfortunately, as we have seen, molecular probes cannot yield quantitative measurements of TRF's and, on the other hand, experiments with macroscopic objects will be invariably affected by unwanted buoyancy-induced perturbations. This justifies the effort - and the expense - of exploiting the unique possibility offered by microgravity to get a clear-cut, quantitative verification of the TRF theory.

References

1. J.M. Ziman, Electrons and phonons, Clarendon Press, IVed.,(1967)
2. G. Leibfried, E. Schlömann, *Nachr. Gtt. Akad.*, **2a** (1954) 71.
3. J. Bardeen, *Phys. Rev.*, **52** (1937) 688.
4. R.E.B. Makinson, *Proc. Camb. Phil. Soc.*, **34** (1938) 474.
5. C. Herring, *Phys. Rev.*, **96** (1954) 1163.

6. J.P. Jan, *Solid State Phys.*, **5** (1957) 1.
7. R. Peierls, *Ann. Phys.* (5), **3** (1929) 1055.
8. F.S. Gaeta, *Phys. Rev.* **182** (1969) 289.
9. F.S. Gaeta, E. Ascolese, B. Tomicki, *Phys. Rev. A*, **44** (1991) 5003.
10. F.S. Gaeta, F. Peluso, D.G. Mita, C. Albanese, D. Castagnolo, *Phys. Rev. E*, **47** (1993) 1066.
11. F.S. Gaeta, C. Albanese, D.G. Mita, F. Peluso, *Phys. Rev. E*, **49** (1994) 433.
12. L. Onsager, *Phys. Rev.*, **37** (1931) 405.
13. F.S. Gaeta, F. Peluso, C. Albanese, D.G. Mita, Thermal radiation forces and heat conduction, in preparation.
14. F.S. Gaeta, G.Scala, G. Brescia, A. Di Chiara, *J. Polym. Sci., Polym. Phys. Ed.*, **13** (1975) 177.
15. R.D.J. Miller, Ann. Rev. Phys. Chem., **42** (1991) 581.
16. M. Karplus, J.A. McCammon, Ann. Rev. Biochem., **53** (1983) 263.
17. D.G. Mita, M.A. Pecorella, P. Russo, S. Rossi, U. Bencivenga, P. Canciglia, F.S. Gaeta, J. Membrane Sci., **78** (1993) 69.

Containerless Processing in Space: Recent Results

Team TEMPUS *

Abstract

During the International Microgravity Laboratory Mission 2 (IML-2) on board the Space Shuttle Columbia, the containerless processing facility TEMPUS had its maiden flight. A team of 8 Principal Investigators from the U.S. and Germany performed a number of experiments on liquid metals and alloys. TEMPUS is an electromagnetic levitation facility designed to operate under microgravity conditions. It allows to melt and undercool metallic samples of 8 mm diameter, with no contact to a crucible. Thus, experiments on nucleation statistics, non-equilibrium solidification, and measurements of thermophysical properties are possible. During IML-2, 22 samples of different compositions have been processed, including pure metals such as Au, Ni, Zr, as well as eutectic alloys, for example NiNb and ZrNi. In addition, the solidification behaviour of quasicrystal forming alloys (AlCuFe and AlCuCo) was studied. The specific heat and surface tension of a number of samples could also be measured. This paper discusses the results of these experiments, and the difficulties in obtaining them.

1. Introduction

Containerless processing is an attractive way to provide a high purity environment to high-temperature, highly reactive materials. It is particularly suited to give access to the metastable state of an undercooled melt. In the absence of container walls, the nucleation rate is greatly reduced and deep undercoolings up to $(T_m - T_n)/T_m \approx 0.2$ can be obtained, where T_m and T_n are the melting and nucleation temperatures, respectively. Among the different containerless techniques, electromagnetic levitation is especially suitable for the study of metallic melts. It allows obtaining high temperatures, up to 2600 °C , to levitate bulk samples of a few grams, and to maintain the undercooled state for an extended period of time (up to hours). However, some limitations exist for electromagnetic levitation when used on ground: the required high electromagnetic fields

* For correspondence, write to: Ivan Egry, Institut für Raumsimulation, DLR, 51140 Köln, Germany, Fax: xx49-2203-61768

deform the shape of a molten sample, and induce turbulent currents inside the sample. In addition, the fields are so strong that the samples must be cooled convectively using a high-purity inert gas, like He or Ar . All these drawbacks can be avoided if electromagnetic levitation is performed under microgravity conditions. Microgravity offers the following advantages:

- There is practically no deformation of the sample, the spherical shape is maintained.
- There is less chance of turbulent flow in the melt.
- The heat generated in the sample by the positioning field is negligible and, therefore, no gas stream is required for cooling, and undercooling becomes possible in ultra high vacuum.
- Due to the small power dissipation induced by the positioning field, a wider temperature range becomes accessible, and temperature control is facilitated.
- Terrestrial levitation experiments are essentially restricted to refractory metals and good conductors. In space, processing and undercooling of metals with a low melting point becomes possible, giving access to alloys with a deep eutectic temperature and to glass-forming alloys.

The TEMPUS facility (Tiegelfreies ElektroMagnetisches Prozessieren Unter Schwerelosigkeit) was built to exploit this opportunity and was flown on Spacelab mission IML-2 in July 1994. The details of TEMPUS will be discussed in the next section.

The TEMPUS team consists of the Principal Investigators (PI's), their coworkers, and a team providing mission support. The members of the team, their affiliation, their experiment or function, respectively, are listed in Tab. 1.

It became clear soon that the experiments proposed could be grouped into 3 classes, according to their scientific objectives and their operational requirements. These classes are shown in Tab.2.

The characteristics of these different experiment classes will be discussed in chapter 3..

2. The TEMPUS Facility

The TEMPUS Spacelab module, which will be described in the following, fits into a single rack and consists of the following subsystems:

- Process control and Data acquisition Module (PDM)
- Experiment Unit (EU)
- High Power Supply (HPS)
- Cold plate/ Heat EXchanger (CHEX)

The EU consists of an Ultra High Vacuum chamber which houses the levitation coils. Attached to the chamber through different ports are the axial temperature measurement and video system, the radial recalescence detector and a second video camera, the evaporation shield magazine and the sample storage magazine. In addition, the chamber is connected to the vacuum and gas systems, and it

Team	Experiment/Task
R. Bayuzick T. Bassler, W. Hofmeister, C. Morton,J. Olive Vanderbilt University, Nashville	Effects of Nucleation by Containerless Processing
I. Egry B. Feuerbacher, G. Jacobs, G. Lohöfer, P. Neuhaus DLR Cologne	Viscosity and Surface Tension of Undercooled Metallic Melts
H. Fecht R. Wunderlich TU Berlin	Thermodynamics and Glass Formation of Undercooled Metallic Melts
M. Flemings J. Lum, D. Matson MIT Cambridge	Alloy Undercooling Experiments
D.M. Herlach M. Barth, K. Eckler, B. Feuerbacher DLR Cologne	Non-Equilibrium Solidification of Deeply Undercooled Melts
W. Johnson D. Lee CalTech, Pasadena	Thermophysical Properties of Metallic Glasses and Undercooled Alloys
J. Szekely E. Schwartz MIT Cambridge	Measurement of the Viscosity and Surface Tension of Undercooled Metallic Melts and Supporting MHD Calculations
K. Urban D. Holland-Moritz Forschungszentrum Jülich	Structure and Solidification of Deeply Undercooled Melts of Quasicrystal-Forming Alloys
R. Willnecker E. Bennet, A. Diefenbach, M. Kratz, W. Koerver, B. Pätz, D. Uffelmann, R. Fischer, R. Grümann, L. Mies, .K.D. Schmidt DLR Cologne	Experiment Support
R. Knauf H. Lenski, U. Lössner, Th. Nacke, P. Neuhaus, A. Seidel, M. Stauber, U. Zell Dornier-Deutsche Aerospace, Friedrichshafen	Engineering Support
W. Dreier, M. Turk, DARA Bonn M. Robinson, NASA Huntsville	Management

Table 1. TEMPUS team members and experiments on IML-2

Class	Name	PI's
A	Nucleation	Bayuzick, Flemings, Herlach, Urban
B	Specific Heat	Fecht, Johnson
C	Surface Tension	Egry, Szekely

Table 2. Experiment classes of TEMPUS IML-2

contains a light guide for illumination and a permanent magnet for oscillation damping. The schematics and interfaces of the EU are shown in Fig.1.

Some of the most important subsystems are discussed in the following.

1. The Coil System
   In microgravity, the fields need not produce a lifting force; restoring forces are only required to counteract the effect of the random g-jitter and residual accelerations. TEMPUS provides stable positioning against $10^{-2}g_0$, reducing the power absorbed in the sample by a factor of 100, as compared to the 1-g case.
   The TEMPUS facility [1, 2] uses a quadrupole field for positioning, which is established by two sets of parallel coils of identical dimension. The current through the coils has a phase shift of $\phi = 180°$, which produces an axially symmetric quadrupole field [3]. For such a field, the stabilizing field gradients are very strong, and the power absorption is very low. This implies that the positioning field is not sufficient to melt the sample. Therefore, an additional heating coil is used in the TEMPUS design. It consists of a single coil located around the equilibrium position of the sample producing a dipole field with high field strength and low gradient near the positioning point. This yields a high heating efficiency and only small forces. The photograph in Fig.2 shows heater and positioner coils, fitted to the RF-module, and a sample integrated into a sample holder.
2. Temperature Measurement
   Temperature is measured in TEMPUS by a number of pyrometers. Axially, i.e. from the top, two pyrometers measure the temperature at different wavelengths. In addition to a two-colour broadband pyrometer, equipped with an *InAs* detector, also a single colour, *Si*-pyrometer can be used. Radially, an *InAs*-detector is mounted to measure relative temperature changes; this is used especially for transient temperature measurements during solidification.
   One major difficulty in the temperature measurement of liquid high temperature melts is their high evaporation rate. If no precaution is taken, the evaporating atoms contaminate the optical path between pyrometer detector and target and lead to a decrease in the detected signal, which in turn would be erroneously interpreted as a decrease in temperature. To avoid this

Fig. 1. TEMPUS Experiment Unit interfaces and subsystems

TEMPUS is equipped with exchangeable evaporation shields, mounted on a revolving magazine between the sample and the pyrometer window. These evaporation shields consist either of windows, or of periscopically aligned mirrors. Each experiment has its assigned evaporation shield(s) which is brought into place automatically before or during processing.

3. Video System
For the observation of the sample behaviour, two video cameras are mounted to the EU, sharing the optical path with the pyrometers and providing top and side view of the sample. Both video cameras can operate in a high speed mode; the sampling rate can be set from 60 Hz to 480 Hz.

Fig. 2. TEMPUS coils and TEMPUS sample holder

3. The Experiments

For the IML-2 mission, the 8 science teams have prepared 26 experiments on 22 samples. Owing to the evaporation problem, which not only affects the optical systems (pyrometers and video), but also, more severely, the levitation coils, a prioritized sequence of the experiments had to be established by the PI's. This was based primarily on the evaporation rates of the different samples, and on the sticking properties of the layers deposited on the coils.

From the very beginning, the science teams combined their efforts and ressources to maximize the science output of the mission. In particular, they agreed on sharing their samples and performing different experiment classes on the same sample. During the mission, this spirit of cooperation became vital to overcome the difficulties encountered. The TEMPUS team had to cope with two major problems:

- *Sample contamination.* All samples were heavily contaminated. This was obvious from their visual appearance on the video pictures: oxide particles or even a closed oxide layer on the surface was clearly visible.
- *Sample stability.* The radial translational oscillations of the (liquid) samples had a larger amplitude than anticipated. In addition, all samples were positioned off-center with respect to the center of the sample cage, leaving only little space for sample excursions to one side of the cage. As the amplitude

of these oscillations increased, a sticking of the (liquid) samples to the cage resulted in most cases, and the sample was no longer available for further experiments.

Despite these problems, TEMPUS operated throughout the entire IML-2 mission for about 200 hours, and 26 experiments could be performed. Of course, the original experiment sequence could not be maintained, and massive replanning was necessary. The actual experiments that were performed during the mission, are shown in Tab.3 with a short summary of their result.

The experiments will be discussed in detail in the following according to experiment classes.

3.1 Class A: Undercooling Experiments

The experiments of this class were concerned with nucleation and growth phenomena from the undercooled melt.

The Vanderbilt group planned to perform a statistical study on the nucleation behaviour of Zr and a ZrNi-alloy. According to classical nucleation theory, the nucleation rate I is given by:

$$I = \frac{\Omega}{\eta} \exp(-\frac{16\pi}{3} \frac{\sigma^3}{\Delta G^2 kT}) \tag{1}$$

A statistical analysis according to Skripov's theory [4] not only yields the exponent, i.e. the interface tension σ and the free energy difference ΔG, but also the pre-exponential factor. By comparing the microgravity results to results obtained on ground, also some hints on the role of convection on nucleation can be expected. Unfortunately, it was only possible to perform one single undercooling cycle on Zr. The temperature vs time diagram of this cycle is shown in Fig.3. As can be seen, Zr was heated up to 2000 °C , and undercooled subsequently by $\Delta T = 160$ °C . Of course, no statistical analysis is possible, but some information can be extracted from Fig.3.

From the cooling curve, c_p/ϵ_{tot} can be determined. The cooling rate for a sphere in vacuo is approximately:

$$\frac{dT}{dt} = \frac{\epsilon_{tot}}{c_p} \frac{3\sigma_{S.B.}}{R\rho} T^4 \tag{2}$$

Here, ϵ_{tot} is the total hemispherical emissivity, $\sigma_{S.B.}$ is the Stefan-Boltzmann constant, R is the radius of the sample, ρ its density, and T its temperature. The latter quantities are all known, and hence we obtain:

$$\frac{c_p}{\epsilon_{tot}} = 1.58 \pm 0.02\,\mathrm{J/(gK)} \tag{3}$$

From the melting and recalescence plateaus, the heat of fusion can be determined, because the power input is known. The result is:

$$\Delta H_f = 14.8\,\mathrm{kJ/mol} \tag{4}$$

Sample composition (at%)	Experiment time	Experiment class	Remarks	Results
W	03 : 08	N/A	nominal calibration run	no deviations from premission data
Al60Cu34Fe6	01 : 57	A	Sample hit cage in 4th cycle, 1 melting cycle	microstructural analysis
Au	01 : 32	C	Sample hit cage in 9th cycle, 6 melting cycles	surface tension, viscosity
Au56Cu44	00 : 15	C	sample hit cage after 3 cycles	surface tension, viscosity
Zr76Fe24	01 : 53	B	sample hit cage after 13 cycles	specific heat
Ni79Si21	02 : 01	A	sample was not melted due to heavy rotations	none
Zr76Ni24	06 : 20	B	sample hit cage after 17 cycles	specific heat
Ni86Sn14	00 : 36	A	4 melting cycles	10° undercooling, microstructural analysis
Ni86Sn14	00 : 22	A	sample hits cage after 3 cycles	microstructural analysis
Cu	00 : 20	C	sample hits cage in 1st cycle	none
Zr78Co22	02 : 31	B	nominal experiment run	specific heat
Ni60Nb40	03 : 25	B	maximum experiment time exceeded	specific heat
Zr64Ni36	05 : 47	B	nominal experiment run	specific heat
Al67Cu21Co12	01 : 59	A	sample hit cage after 6 cycles, 1 melting cycle	microstructural analysis
Zr64Ni36	01 : 37	C	AuCu sample was lost and was replaced by ZrNi	surface tension, viscosity, dynamic nucleation
Ni79Si21	03 : 28	A	2 melting cycles	10° undercooling, microstructural analysis
Ni60Nb40	03 : 28	B	nominal experiment run	specific heat, 50° undercooling, metastable phase formation
Zr	00 : 09	A	sample hit cage in solid state	none
Ni79Si21	00 : 10	C	sample hit cage in 1st cycle	none
Ni81Sn19	00 : 29	A	2 melting cycles	10° undercooling, microstructural analysis
Al65Cu25Co10	00 : 42	A	run performed in UHV, sample hit cage after 1st cycle	slight undercooling, microstructural analysis
Zr	00 : 10	A	sample hit cage after 1st cycle	160° undercooling, heat of fusion
Ni99C1	00 : 52	A	telescience experiment, sample hit cage after 2 cycles	microstructural analysis, growth velocity
Fe75Ni25	00 : 16	A	sample hit cage in first melting cycle, severe sample contamination	none
Ni	04 : 33	A and C	sample could not be melted	none
Total	48 : 00			

Table 3. TEMPUS IML-2 experiment sequence - as flown

Closer inspection of Fig.3 reveals an oscillatory contribution to the signal. This is due to surface oscillations. We have therefore attempted to determine the frequency of these oscillations and applied Rayleigh's formula (9). to derive a value for the surface tension of liquid Zr. The frequency was found to be 27 Hz; the mass of the sample was 1.72 g. For a temperature of T=1960 °C , we obtain

$$\sigma_{\mathrm{Zr}}(1960°\mathrm{C}) = 1.47\,\mathrm{N/m} \tag{5}$$

There are only few measurements of the surface tension of Zr, and none above the melting point. The value quoted by Keene [5] for the surface tension at the melting point ($T_l = 1852°\mathrm{C}$) is $\sigma_{\mathrm{Zr}}(T_l) = 1.48\,\mathrm{N/m}$. This agrees very well with our result. The analysis of the experiment on Zr shows that, even if the major objective could not be reached, a great deal of interesting results can be obtained by careful inspection of the available data.

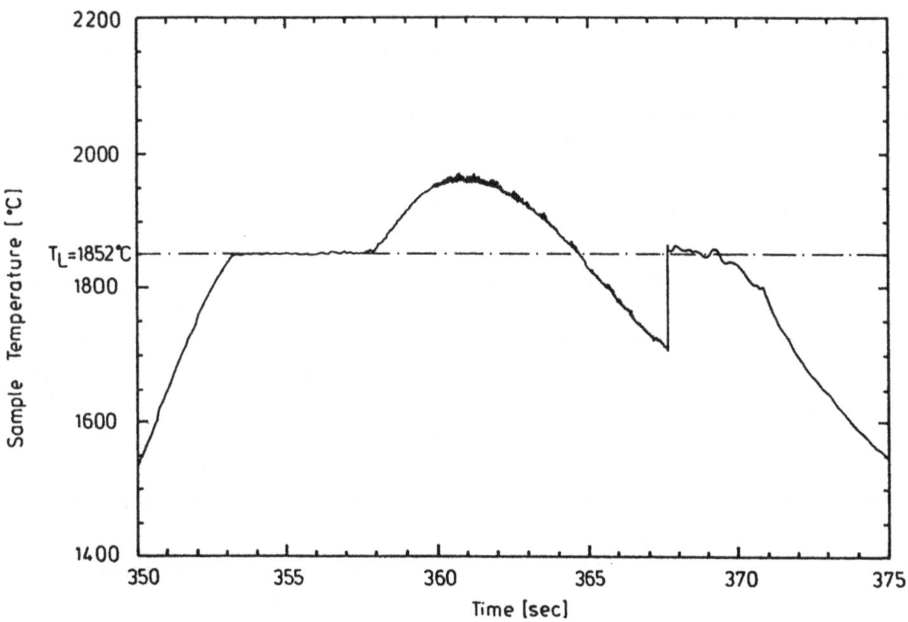

Fig. 3. Temperature vs time diagram for Zr

The experiments of Herlach and Flemings aimed to measure growth velocities as a function of undercooling and to produce metastable solid states by solidification of deeply undercooled metallic melts. NiSn-alloys of two different compositions, Fe-Ni, Ni-Si and Ni-C were chosen as samples.

In the system FeNi a metastable bcc phase is predicted to be nucleated if a critical undercooling is exceeded and may be conserved upon cooling to room temperature if the hypercooling limit can be reached.

On the other hand, metastable solid phases can be formed also by rapid crystal growth. This was the aim in the experiments using NiSi and NiC. In the first case a metastable anomalous eutectic microstructure is predicted by the theory of dendritic growth, [6], provided the crystal growth velocities (or undercoolings) are high enough. The system NiC offers the additional opportunity to test an interesting phenomenon which is predicted by modern theories of crystal growth [7], but has never been observed in metallic systems, probably owing to the strong convection and electromagnetically driven fluid flow in terrestrial levitation experiments. This concerns the enhancement of the crystal growth velocity by the addition of a small amount of a strongly partitioning element. Measurements of the crystal growth velocity as a function of undercooling both on $Ni_{99.4}$ $C_{0.6}$ and, for comparison, on pure Ni were scheduled for TEMPUS IML-2. None of these samples could be undercooled, and, hence, no results are available.

Finally, the experiments of Urban were intended to study the undercoolability and solidification behaviour of melts of quasicrystal-forming alloys. Molecular dynamic calculations predict that an icosahedral short range order prevails in the undercooled melt [8]. Such an icosahedral short range order in the undercooled melt is similar to the short range order of quasicrystalline phases. Therefore the interfacial energy between melt and quasicrystalline nucleus is expected to be low, leading to a small activation barrier for the nucleation of quasicrystalline phases and by means of this to a low undercoolability of the melt [9]. The TEMPUS experiments were performed on melts of Al-Cu-Fe and Al-Cu-Co. Al-Cu-Fe forms a stable icosahedral quasicrystalline phase (I-phase), Al-Cu-Co a stable decagonal quasicrystalline phase (D-phase). During the accompanying ground based research, specimen of these alloys could be successfully undercooled in an electromagnetic levitation facility [10]. The space experiments aimed to investigate the maximum undercoolability of these melts under improved purity conditions (in UHV). Moreover a possible impact of reduced convection and fluid flow on the microstructure of the as solidified samples should be studied.

The microstructural analysis of the $Al_{60}Cu_{34}Fe_6$-specimen, which did not significantly undercool in TEMPUS, showed two different phase selection sequences. On the one hand we observed the primary solidification of $\lambda - Al_{13}Fe_4$, which is followed by the formation of the I-phase and several low-temperature phases. The same phase selection sequence was also observed at specimen solidified on terrestrial conditions from the not significantly undercooled melt. On the other hand the flight sample exhibited the primary formation of the CsCl-type β-phase again followed by the solidification of the I-phase and several low-temperature phases. Under terrestrial conditions this phase selection sequence was only observed in Al-Cu-Fe-samples with a higher Fe-content. This result might be interpreted by a local enrichment of the melt with Fe, which is facilitated by the reduced convection and fluid flow during the μg-experiment.

During the process of the $Al_{65}Cu_{25}Co_{10}$-sample in TEMPUS, a small undercooling of about 40 K was achieved. This undercooling was sufficient to obtain the primary formation of D-phase. The formation of the CsCl-type β-phase, which solidifies under equilibrium conditions at the liquidus temperature of the alloy, was bypassed. The primarily solidified D-phase formed large rod-like structures. These rods were surrounded by several low-temperature phases, which were formed later during cooling. Between the rods large pores were observed. The extremely high porosity resulted in an increase of the sample diameter from originally 8 mm to 9.3 mm after the process in TEMPUS. Such a porous morphology was up to now not observed in reference samples that were processed in the earthbound levitator under similar conditions as in TEMPUS. Futher investigations are necessary to understand the reason for this effect.

In summary, the class A experiments suffered most severely from the contamination of the samples. The impurities served as heterogeneous nucleation sites, and only very small undercoolings could be obtained with the exception of Zr.

3.2 Class B: Specific Heat Measurement

The experiments of this class were concerned with the measurement of the specific heat of a number of glass-forming alloys. The method used was developed by Johnson and Fecht and is described in detail in [11]. It is a variant of non-contact a.c. calorimetry, normally used in low temperature physics. The heater power is modulated according to $P_\omega \sim \cos^2(\omega t/2)$ resulting in a modulated temperature response ΔT_ω of the sample. Temperature gradients inside the sample relax quickly, due to the high thermal conductivity of metals. This relaxation can be described by a relaxation time τ_{int}. On the other hand, relaxation to the equilibrium temperature is governed by radiation under UHV conditions, and is therefore slow. It can be described by a relaxation time τ_{ext} . If the modulation frequency ω is chosen such that $1/\tau_{\mathrm{ext}} \ll \omega \ll 1/\tau_{\mathrm{int}}$, a simple relation for the temperature variation can be derived:

$$\Delta T_\omega = \frac{P_\omega}{2\omega c_{\mathrm{p}}} \tag{6}$$

from which c_{p} can be determined.

This non-contact method is applicable in the undercooled state, too. *Zr* -alloys of eutectic composition have been selected by Fecht for investigation in TEMPUS, namely ZrCo, Zr36Ni, ZrFe. These alloys are good glass-formers, and the study of the anomaly of the specific heat near the glass transition was a major objective of these experiments. Owing to the before mentioned contamination problems, however, it was not possible to obtain enough undercooling to approach the glass transition temperature. Nevertheless, the a.c. calorimetry method worked perfectly in the melt, and data of the specific heat of these alloys near their melting point have been obtained. A plot of the heater modulation and the temperature response during the experiment on ZrNi is shown in Fig. 4. The melting temperature of this alloy is $T_{\mathrm{l}} = 1010°\mathrm{C}$. From the data, the specific heat as function of temperature can be determined. The values given in Tab.4 are averages of several measurements and have been derived from (2) and from the increase in average sample temperature. The quantitative evaluation is based on

- the specific heat at one calibration point in the solid ($c_p(960°C)$ = 32J/(Kmol)). This determines the heating efficiency of the TEMPUS generators (i.e. P_ω).
- the change of the heating efficiency through the solid-liquid phase transition. This is based on the change of the sample resistivity upon melting.

The data given in Tab.4 can be described by the following expression for the undercooled liquid:

$$c_{\mathrm{ZrNi}}(T) = 57.8 - 1.002 \cdot 10{-}2T \quad \mathrm{J/(Kmol)} \tag{7}$$

The values of the total hemispherical emissivity were obtained from measurements of the external relaxation time, according to (2).

T (K)	ϵ_{tot}	c_p(J/(Kmol))
1485	0.368	43.4 ± 2.0
1433	0.350	43.5 ± 2.0
1313	0.331	43.4 ± 1.5
1284	0.327	44.5 ± 1.0
1253		45.7 ± 1.2

Table 4. Specific heat of $Zr_{64}Ni_{36}$.

Finally, the melting enthalpy was determined from the length of the recalescence plateau as:

$$\Delta H_f = 14.7 \quad \mathrm{kJ/mol} \tag{8}$$

The investigations of Johnson used the same method, but he selected different alloys, namely NiNb of eutectic composition and a Zr24Ni-alloy, which is also eutectic and a good glass-former. For the specific heat of this latter alloy, they obtained data as given in Tab.5.

T(K)	c_p (J/(K mol))
1160	45.9
1248	46.0
1295	45.2

Table 5. Specific heat of Zr_{76} Ni_{24}

The experiment on NiNb went particularly well; it could be processed in the molten state for more than 3 hours, and an undercooling of ΔT = 50 °C could be obtained, giving a wealth of specific heat data on this alloy. They are shown in Tab. 6. In addition, when the sample finally solidified, it solidified into a metastable phase, the existence of which was presumed, but never proved before. This sample did not stick to the cage, so post-flight analysis will reveal the nature of the metastable phase.

T(K)	c_p (J/(K mol))
1466	48.8
1477	47.6
1498	47.9
1527	47.2
1559	47.3
1606	47.1

Table 6. Specific heat of $Ni_{60}Nb_{40}$

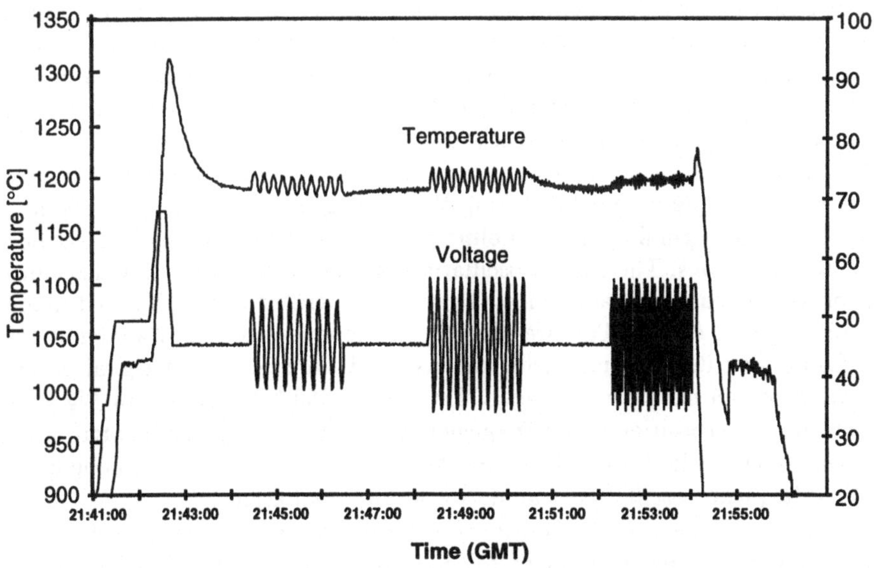

Fig. 4. A.C. calorimetry on ZrNi alloy: power modulation and temperature response

3.3 Class C: Surface Tension and Viscosity

These experiments use the oscillating drop technique [12] to measure surface tension and viscosity. In microgravity, the liquid samples perform oscillations around their spherical equlibrium shape. In that case, the simple formulae of Lords Rayleigh and Kelvin can be used to relate frequency ω and damping Γ of the oscillations to surface tension σ and viscosity η, respectively. Rayleigh's formula reads:

$$\omega_{\mathrm{R}}^2 = \frac{32\pi}{3}\frac{\sigma}{M} \tag{9}$$

while Kelvin derived the following expression:

$$\Gamma_{\mathrm{K}} = \frac{20\pi}{3}\frac{\mathrm{R}\eta}{M} \tag{10}$$

where M is the mass of the droplet and R its radius. Under terrestrial conditions, the above relations are not valid; corrections have to be made for the external forces, namely gravity and electromagnetic field. These corrections have been calculated for the Rayleigh formula by Blackburn and Cummings [13] and, only recently, by Bratz and Egry [14] for the Kelvin formula. The corrections are of the order of 5-10 % and difficult to quantify. In the case of viscosity, the formulae are derived under the assumption of purely laminar flow, an assumption which certainly cannot be made for earthbound levitation.

For TEMPUS IML-2, experiments on noble metals Au, Cu and its alloy AuCu were planned by Egry and Szekely in addition to an experiment on pure nickel. The experiments on Au and AuCu were performed successfully, no data points could be taken on the copper sample, and nickel was not processed. Instead, an additional run was performed on Fecht's ZrNi sample.

The heater is turned on for approximately 90 seconds to melt the sample and is run at a minimum value during the cooling phase which takes about 3 minutes. At the beginning of the cooling phase a short heating pulse is applied to excite oscillations. The sample oscillations are recorded with two video cameras, providing axial and radial view. They were operated in a high-speed-mode to avoid aliasing effects. The area of the visible cross section is analysed for each frame, and its variation yields the required time dependent signal of the oscillation. As an example, the signal for gold at 960 °C is shown in Fig.5. It consists of a superposition of low frequency oscillations, which can be attributed to translational oscillations of the entire sample in the restoring positioning field, and high frequency oscillations caused by surface oscillations.

This is more evident in the Fourier transform of the signal, which is shown in Fig.6. The peak at 2Hz corresponds to the translational oscillation, whereas the peak at 15Hz is the Rayleigh frequency of the surface oscillation. It is important to stress that the oscillation spectrum consists of a single peak, as predicted by theory for a spherical sample, in contrast to the multiple peak spectra found in terrestrial levitation experiments. This confirms once more that Rayleigh's formula is directly applicable to microgravity experiment without the correction needed for 1g-levitation experiments, as calculated by Cummings.

From the frequency of the peak, the surface tension could be derived using Rayleigh's formula, (9). This is shown for gold in Fig.7. The microgravity data

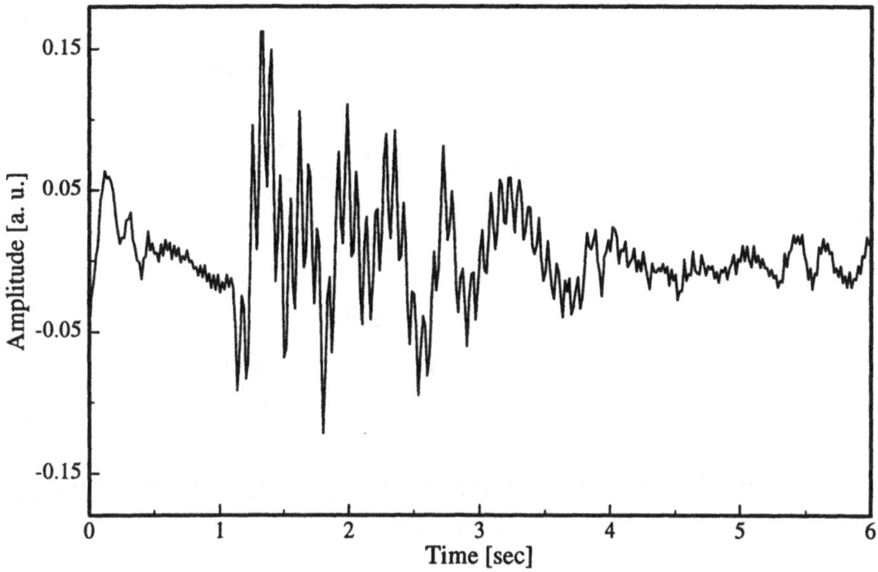

Fig. 5. Variation of the visible cross section for gold at 960 °C .

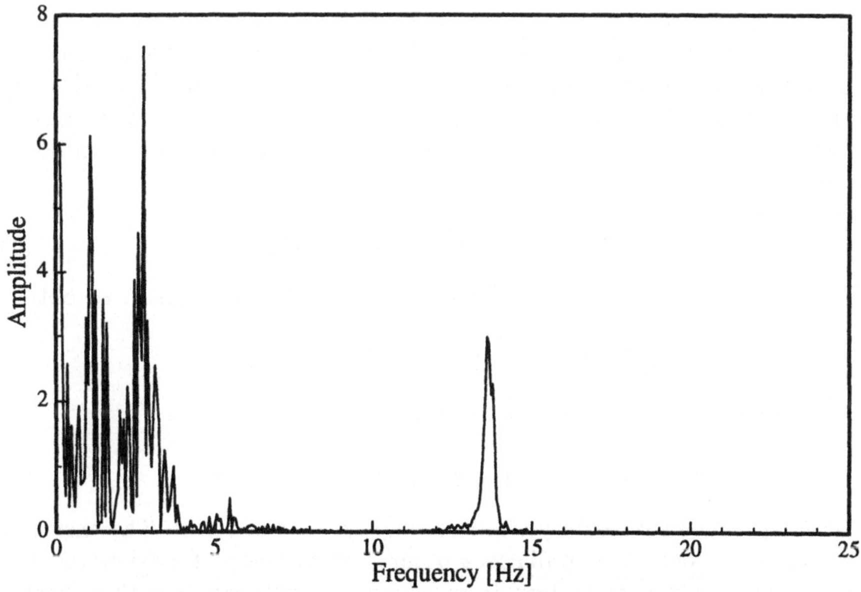

Fig. 6. Fourier spectrum of sample oscillations for gold at 960 °C .

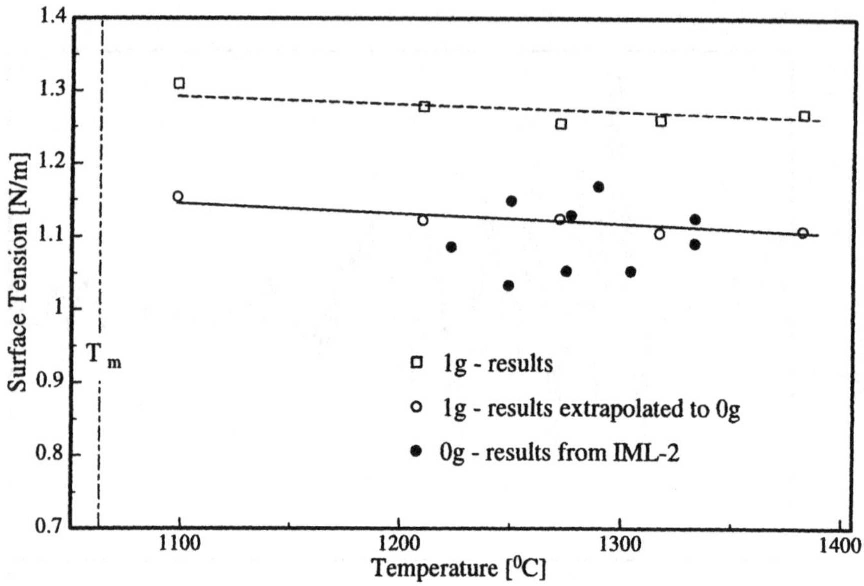

Fig. 7. Surface tension of gold; microgravity data are compared to 1-g data with and without correction

obtained during IML-2 are compared to earlier data obtained by the authors under 1-g conditions [15].

It can be clearly seen that there is excellent agreement between the two datasets, if the earthbound data are corrected according to Cummings. The same behaviour was also found for AuCu; no comparison to earthbound surface tension data could be made for ZrNi due to the lack of such data. The results of the surface tension measurements on IML-2 can be summarized by the following equations:

$$\sigma_{\mathrm{Au}}(T^{\circ}\mathrm{C}) = 1.149 - 0.14(T - 1064)\mathrm{N/m} \tag{11}$$

For $Au_{56}Cu_{44}$, a composition at which this alloy melts congruently, we obtain:

$$\sigma_{\mathrm{AuCu}}(T^{\circ}\mathrm{C}) = 1.205 - 0.15(T - 910)\mathrm{N/m} \tag{12}$$

For ZrNi at the eutectic composition $Zr_{36}Ni_{64}$, the result is:

$$\sigma_{\mathrm{ZrNi}}(T^{\circ}\mathrm{C}) = 1.545 + 0.08(T - 1010)\mathrm{N/m} \tag{13}$$

The experiments on ZrNi yielded an unexpected side result: The excitation pulses used to excite the oscillations occured in the case of ZrNi when there was already a slight undercooling. In each cycle, the appearance of these heating pulses and the onset of nucleation coincide indicating that the heating pulses may have triggered nucleation. Also, the undercooling with the excitation pulses was reproducibly $\Delta T = 40$ K , whereas during free cooling in the class B experiment on the same sample, an undercooling of $\Delta T = 70$ K was obtained. It is tempting to attribute this phenomenon to dynamic nucleation [16, 17].

It was also attempted to determine the viscosity from the damping of the oscillations. An elegant way of doing this is to apply a carefully designed filter to the Fourier spectra in order to eliminate the unwanted slow translational oscillations, and do an inverse Fourier transform obtaining a "purified" time signal. The result of such manipulation on the Fourier spectrum shown in Fig.6, is shown in Fig.8. By fitting an exponential to the envelope of the oscillations, a damping constant of $\Gamma = 0.74$ 1/s is easily determined. Inserting this into Kelvin's formula (10), using R=4 mm and M=5 g, we obtain $\eta = 46$ mPa s. This is nearly one order of magnitude larger than the accepted value for gold, namely $\eta = 5.0$ mPa s . The same discrepancy is also observed for AuCu. Therefore, we have to conclude that the damping of the oscillations cannot be attributed to viscosity alone, and other damping mechanisms must be present.

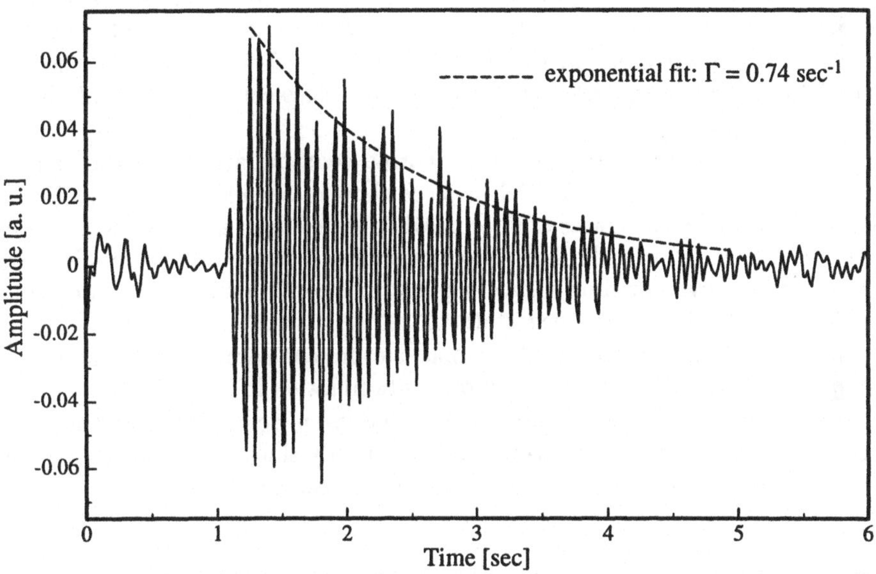

Fig. 8. Damping of surface oscillations for gold at 960 °C .

4. Summary and Outlook

4.1 Problem Analysis

Sample Contamination After the flight, it was decided to examine all samples for their contaminants by Auger, XPS and Electron Microscopy, in order to find out the source and reason for the contamination. This is of course essential to assure a succesful reflight of TEMPUS, with a perspective to finally obtain data in the deeply undercooled regime. In the following, a brief summary of these studies is given.

All samples investigated had foreign particles on the surfaces. These particles included alumina and silica in addition to pieces of many other sample materials loaded in TEMPUS. Each sample showed evidence of wear tracks on the surface except for those stuck to the cages. All samples had evidence of surface contamination by carbon, oxygen, and in many cases fluorine.

Particles of alumina most certainly originated from the sample pedestal. The pedestal is a sintered product with an alumina particle size range of 1-3 microns. EDS analysis of the pedestal showed Al, Si and in some places Ca. These pedestal particles were found imbedded on the surface of samples. Silica particles were evidenced primarily as small spheres. Some foreign particles on the surface of the pedestal were vapor coated with material from its own sample, demonstrating that this material was on the pedestal prior to processing.

It is certain that the samples and the sample holders were abraded by mutual contact. During transport, launch and landing this abrasion resulted in particulate material generation. In the TEMPUS facility these particulates floated around the chamber and resulted in subsequent cross-contamination of all the samples examined.

A source for carbon and possibly fluorine contamination was the sample shipping container insert. This delrin piece showed considerable outgassing of formaldehyde and some traces of fluorine when tested with an Residual Gas Analyzer. The pedestal was checked for outgassing and showed that formaldehyde was present in large amounts suggesting that this contamination was carried to the TEMPUS facility by the pedestals.

From the above we can conclude that the present sample holder and cage system was the primary cause of sample contamination in TEMPUS.

In the meantime, a new sample holder has been designed, consisting of a fused quartz glasss or sapphire tube, which not only shields the coils effectively of evaporation, but is also free of the problems discussed above.

Sample Stability As mentioned before, the sample stability of the samples during IML-2 was unsatisfactory leading to a loss of many samples. As was evident already during the mission, the equilibrium position of the samples was misaligned with the sample cage. This is due to the fact that the windings of a coil are not perfect circles: at the leads to the RF generator the circles do not close. Therefore, the minimum in the potential well is situated off-axis. This effect can be minimised through a careful design of the leads; the remaining effect can be compensated by introducing intentionally additional asymmetries which cancel the original ones.

In addition to the restricted space available for sample oscillations due to the misalignment, the coil design which was used in IML-2 had relatively weak restoring forces in radial direction, as compared to the axial force. This is a consequence of the design goal for this coil system, namely maximum heating efficiency. As it turned out, Zr could be melted and heated to 2000 °C with only 33 % of the available heating power. Therefore, a stronger radial force is possible at the expense of a smaller heating efficiency. Such a system has been already designed: It has essentially doubled the radial force and reduced the heating efficiency to approximately 40 % . It was already tested successfully on a parabolic flight.

After having solved the problems of sample contamination and sample stability, we are confident that we will obtain more scientific results, especially in the undercooled regime, on a reflight of TEMPUS.

4.2 Highlights

TEMPUS on IML-2 was the first successful attempt to establish electromagnetic levitation under microgravity conditions. Although all experiments suffered from sample contamination, which in most cases prevented undercooling, and from unstable sample positioning, which in many cases lead to a premature termination of the experiment, in total 48 hours of levitation time has been achieved and a number of interesting and promising results has been obtained. From a scientific point of view, the highlights are:

- Melting and heating of Zr up to 2000 °C and subsequent undercooling by $\Delta T = 160$ °C .
- Measurement of the specific heat of NiNb and ZrNi in the undercooled state.
- Surface tension measurements on Au, AuCu, ZrNi

Acknowledgements

In contrast to a laboratory experiment, experiments in space can only be carried out with the help of a large number of people. They all deserve credits for their excellent work during the IML-2 mission, but it is impossible to name them all. The TEMPUS team wishes to express explicitly its sincere gratitude to the cadre at the Huntsville Operations Support Centre (HOSC), and the astronaut team of IML-2, R. Cabana, J. Halsell Jr., C. Walz, R. Hieb, L. Chiao, D. Thomas, Ch. Mukai, and J. Favier. Financial support of the funding agencies DARA and NASA is also gratefully acknowledged.

References

1. R. Knauf, J. Piller, A. Seidel, M. Stauber, U. Zell, W. Dreier, in: 6th International Symposium on Experimental Methods for Microgravity Materials Science R. Schiffman ed., The TMS Society, Warrendale, 1994, 43-51
2. G. Lohöfer, P. Neuhaus, I. Egry, *High Temperature - High Pressure*, **23** (1991) 333-342
3. G. Lohöfer, *Q. Appl. Math.* **51** (1993) 495-518
4. V. Skripov, in: Current Topics in Materials Science, Crystal Growth and Materials, Vol.2, E. Kaldis, H.J. Scheel eds, North-Holland, Amsterdam, 1977, 328

5. B. Keene, *Int. Mat. Reviews* **38** (1993) 157-192
6. J. Lipton, W. Kurz, R. Trivedi, *Acta Metall.* **35** (1987) 957-964
7. J. Lipton, M.E. Glicksman, W. Kurz, *Mat. Sci. Engin.* **65** (1984) 57-63
8. P.J. Steinhardt, D.R. Nelson, and M. Ronchetti, *Phys. Rev. B* **28** (1984) 784
9. D.R. Nelson and F. Spaepen, *Solid State Phys.*, edited by H. Ehrenreich, F. Seitz, and D. Turnbull (Academic, New York, 1989), Vol.42, p. 1-90
10. D. Holland-Moritz, D.M. Herlach and K. Urban, *Phys. Rev. Lett.* **71** (1993) 1196-1199
11. H. Fecht and W. Johnson, *Rev. Sci. Instr.* **62** (1991) 1299-1303
12. S. Sauerland, K. Eckler, I. Egry, *J. Mat. Sci. Letters* **11** (1992) 330-333
13. D. Cummings and D. Blackburn, *J. Fluid Mech.* **224** (1991) 395-416
14. A. Bratz and I. Egry, Surface oscillations of electromagnetically levitated viscous metal droplet, submitted to J. Fluid Mech.
15. S. Sauerland, R. Brooks, I. Egry, K. Mills, *Containerless Processing: Techniques and Applications*, W. Hofmeister, R. Schiffman eds, TMS Warrendale, (1993), 65-70
16. B. Chalmers, Dynamic Nucleation, in: *Liquids: Structure, Properties, Solid Interactions*, T. Hughel Ed., Elsevier (1965), p. 308-325
17. J.J. Frawley, W.J. Childs, *Trans. AIME*, **242** (1968) 256-263

The Effect of Natural Convection on the Measurement of Mass Transport Coefficients in the Liquid State

J.P. Garandet[1], *C. Barat*[1], *J.P. Praizey*[1], *T. Duffar*[1], *S. Van Vaerenbergh*[2]

[1] Commissariat à l'Energie Atomique, DTA/CEREM/DEM/SES Centre d'Etudes Nucleaires de Grenoble, 38054 Grenoble Cedex 9 (F)
[2] Université Libre de Bruxelles, Service Chimie Physique, 50 Avenue F.D. Roosevelt, 1050 Bruxelles, Belgium.

Abstract

This paper focuses on the effect of natural convection on the accuracy of solute diffusion coefficient measurements in liquid alloys in microgravity. We consider the cases of both isothermal diffusion and thermodiffusion (Soret effect). Numerical simulations indicate that the error induced by additional convective transport scales with the square of the fluid velocity. Our main conclusion is that it is in principle possible to perform accurate measurements in space, but that the size of the capillaries used in the experiments should always be limited, especially in the case of highly concentrated alloys.

1. Introduction

The modelling and understanding of a variety of processes rely on the knowledge of transfer coefficients. However, an accurate measurement of these parameters is very difficult when the diffusivities are very low, e.g. for the case of liquid phase solute transport. Indeed, the unavoidable convective flows induced by the interaction of density gradients with gravity often lead to significant errors even in thin capillaries. A pendent question is thus to assess the possibility of realizing these measurements on earth, the alternative being to carry out the experiments aboard space vehicles where the intensity of gravity is reduced by a factor ranging from 10^3 to 10^6. Our purpose in this paper is to provide some new insights to this problem for the cases of both isothermal diffusion and thermodiffusion (Soret effect).

Since the axial component of the density (thermal and/or solutal) gradient is generally much larger than the lateral component, the experimental set up is maintained vertical for ground configurations. In space, the residual gravity transverse to the capillary axis provides the convective driving force. For modelling purposes [1] the cavity should be considered as being oriented horizontally. In this work, we rely on existing numerical simulations and scaling analyses [1, 2] of both the vertical and horizontal configurations to discuss the microgravity relevance of a given experiment.

2. Background

The most commonly used methods for liquid metals or semiconductors are the long capillary or shear cell techniques [3]. For isothermal diffusion coefficient measurements a one dimensional concentration step is followed over time in its development whereas for thermodiffusion one waits for the steady-state composition gradient to form from an initially homogeneous fluid. In both cases the sample is solidified at the end of the experiment and a concentration average is taken over slices normal to the capillary axis. The relevant transport coefficient is then deduced from a best fit of the resulting composition profile to a theoretical law, e.g. Gaussian error function for isothermal diffusion.

In our previous work [2] we used an idealized, two-dimensional planar geometry to model the transport phenomena during a simulated isothermal diffusion coefficient measurement. We imposed a fixed lateral temperature difference as a driving force for convections. The heat, momentum and mass transfer equations in the Boussinesq approximation

$$\frac{\partial T}{\partial t} + (\mathbf{V} \cdot \nabla) T = \kappa \nabla^2 T \tag{1}$$

$$\frac{\partial \mathbf{V}}{\partial t} + (\mathbf{V} \cdot \nabla) \mathbf{V} = -\nabla p / \rho_0 + \nu \nabla^2 \mathbf{V} + [1 - \beta_{\mathrm{T}}(T - T_0)] \mathbf{g} \tag{2}$$

$$\nabla \cdot \mathbf{V} = 0 \tag{3}$$

$$\frac{\partial C}{\partial t} + (\mathbf{V} \cdot \nabla) C = D \nabla^2 C \tag{4}$$

were solved using the FIDAP finite element code, along with appropriate boundary conditions (no slip on the walls, no solute flux out of the cavity and imposed lateral temperature difference). In the above equations, $T, \mathbf{V}, p, \mathbf{g}\,\mathrm{and} C$ respectively stand for the temperature, velocity, pressure, gravity and composition fields. The relevant thermophysical parameters are the thermal diffusivity κ, the kinematic viscosity ν, the solute diffusion coefficient D, the reference mass density ρ_0 and the thermal expansion coefficient β_{T}.

At the end of the simulated run we looked for the best fit of the composition profile using the error function. Such a procedure yields an effective diffusion coefficient D^*, that accounts for the additional convective transport. From a scaling analysis of the solute transport equation, carried out in parallel with the numerical simulations, we were able to conclude that the error induced by convection was proportional to the square of the average fluid flow velocity $\overline{W}$. We could thus write:

$$D^* = D[1 + \alpha \overline{W}^2 H^2 / D^2] \ . \tag{5}$$

H being a typical dimension of the cavity and $\alpha = 1/11$ a constant introduced to match the numerical data; incidentally the scaling analysis predicted a value $\alpha = 1/4$, in good order of magnitude agreement with the numerical result. We found that this "effective" diffusion coefficient could be used to capture the physics of the transport phenomena; indeed, except for intense convective flow, the simulated composition profiles keep a Gaussian error function appearance.

In practice this means that there is no way to determine *a posteriori* whether a given experiment has been carried out under favourable conditions. The effective diffusion coefficient may thus be significantly higher than the actual diffusion coefficient.

Only thermal convection and vertical cavities were considered in our previous work [2] and it is the purpose of the present text to discuss the effect of solutally driven convection in diffusion coefficient experiments on concentrated systems in space. However, we first have to assess the applicability of our effective diffusivity approach to solutally driven convection and to horizontal configurations characteristic of microgravity conditions.

On both points, we shall rely on an important paper by Henry and Roux [1] on the related topic of thermodiffusion. Incidentally it will be quite interesting to check that in Soret experiments the error induced by convection scales with the square of the fluid velocity. It will also allow us to discuss the space relevance of thermodiffusion measurements.

3. Effective diffusivity and thermotransport

Henry and Roux simulated numerically the heat, momentum and solute transport phenomena in a three dimensional horizontal cavity filled with a low Prandtl number fluid. The mass flux and the thermodiffusion coefficient are represented as $\mathbf{J}_\mathrm{M}$ and D_T. They used the following formulation [1]:

$$\mathbf{J}_\mathrm{M} = -\rho_0 D \nabla C - \rho_0 D_\mathrm{T} \nabla T \tag{6}$$

A key result was that the composition gradient in the central part of the cavity correlated univocally with the convective velocity, driven by either thermal or solutal density differences. This is very important for our present discussion, since it means that the temperature and composition gradients contribute in a similar manner to the flow. The fluid velocity, independently of the driving force, is thus the sole physically relevant parameter of the problem.

However, we still have to test the validity of the effective diffusivity concept against the numerical data of [1]. To do so, let us remark that under purely diffusive conditions at steady state, the thermodiffusion coefficient can be easily deduced from (6). Indeed, denoting the composition and temperature gradients G_D and G_T, we obtain:

$$D_\mathrm{T} G_\mathrm{T} + D G_\mathrm{D} = 0 \tag{7}$$

Let us now conjecture that, as in the isothermal case, the effect of convection can be included in an effective coefficient D^*. The total mass flux for the thermotransport problem is then given by:

$$\mathbf{J}^* = -\rho D^* \nabla C - \rho D_\mathrm{T} \nabla T \tag{8}$$

The steady state relation between the thermal and solutal gradients is thus:

$$D_\mathrm{T} G_\mathrm{T} + D^* G_\mathrm{D} = 0 \tag{9}$$

In the above expression, G stands for the composition gradient in the presence of convection. Comparing (7) and (9), we obtain:

$$DG_{\mathrm{D}} = D^{*}G \tag{10}$$

It can be easily seen that (10) is consistent with the expected inequalities $G_{\mathrm{D}} \geq G$ and $D^{*} \geq D$.

From (9) an analysis of the raw composition data assuming purely diffusive conditions, i.e. taking $D^{*} = D$, will lead to an effective value of the thermotransport coefficient:

$$D_{\mathrm{T}}^{*} = -\frac{DG}{G_{\mathrm{T}}} \tag{11}$$

From (7) and (10), we get $D^{*}G = DG_{\mathrm{D}} = -D_{\mathrm{T}}G_{\mathrm{T}}$, so that:

$$D_{\mathrm{T}}^{*} = \frac{D}{D^{*}}D_{\mathrm{T}} \tag{12}$$

Using (5), the above expression finally becomes:

$$D_{\mathrm{T}}^{*} = \frac{D_{\mathrm{T}}}{1 + \alpha \overline{W}^{2} H^{2}/D^{2}} \tag{13}$$

Since Henry and Roux relate the composition gradient (or equivalently the effective thermodiffusion coefficient) to the maximum fluid velocity, (13) has to be adapted for a comparison with the numerical data. Using the parallel flow analytical solution proposed in [4] for cylindrical horizontal geometries (i.e. the configuration studied in [1]), we get a relation between the average and maximum convective velocities:

$$\frac{|\overline{W}|}{|W_{\mathrm{M}}|} = \frac{12\sqrt{3}}{15\pi} \approx 0.441 \tag{14}$$

From $\overline{W} = (\nu/H)\overline{w}$, $\overline{W}^{2} H^{2}/D^{2}$ becomes $\overline{w}^{2}\mathrm{Sc}^{2}$. Taking into account the fact that all the numerical simulations were carried out with $\mathrm{Sc} = 60$ and noting that (14) holds in non-dimensional form, we get:

$$\frac{D_{\mathrm{T}}^{*}}{D_{\mathrm{T}}} = \frac{1}{1 + 700\alpha w_{\mathrm{M}}^{2}} \tag{15}$$

Shown in Fig.1 are the numerical data points (redrawn from figure 12 of [1]) and the full line is a curve of equation:

$$\frac{D_{\mathrm{T}}^{*}}{D_{\mathrm{T}}} = \frac{1}{1 + 290\alpha w_{\mathrm{M}}^{2}} \tag{16}$$

Equations (15) and (16) have similar forms which can be taken as a guarantee of the validity of our effective coefficient approach. To perfectly fit the numerical data, we would need to set $\alpha = 0.41$. This value is of the same order of magnitude as the $\alpha = 1/11$ of the 2D computations of [2] or the $\alpha = 1/4$ predicted by the scaling analysis. There is still a rather large discrepancy in terms of these proportionality factors, but we think that it can be ascribed to the different nature of the numerical simulations (2D versus 3D) and also to the fact that the modelled configurations are fairly different.

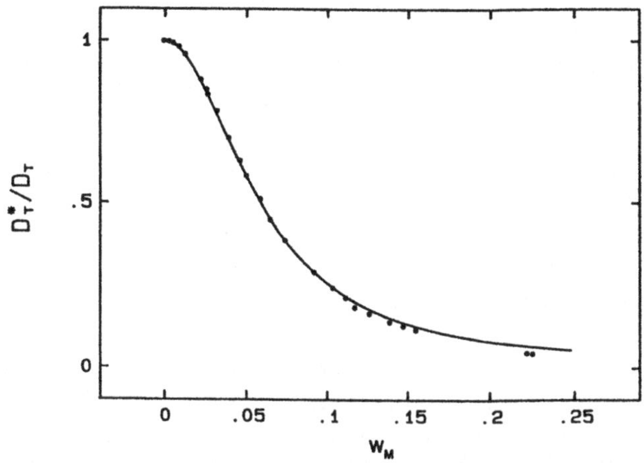

Fig. 1. Variation of the effective thermodiffusion coefficient with the convection velocity. Symbols: numerical data, full line after (16)

4. Thermosolutal convection in microgravity experiments

Having seen that the effective diffusivity approach was applicable to solutally driven convection and to horizontal configurations, we can now proceed to a discussion of the convective effect in microgravity. To relate the overall axial density gradient with the average flow velocity, the standard formula of [4] can be adapted to low Prandtl number fluids:

$$\overline{W} = \frac{1}{15\pi} \frac{\nabla\rho}{\rho} \frac{gH^3}{\nu} \tag{17}$$

With the average convective velocity once derived we can now estimate the nondimensional group $\overline{W}H/D$:

$$\frac{\overline{W}H}{D} = \frac{1}{15\pi} \frac{\nabla\rho}{\rho} \frac{gH^4}{\nu D} \tag{18}$$

For isothermal diffusion experiments the solutal gradients are the sole driving force. It should again be stated that no numerical simulations are available in this case, but we think that the arguments developed earlier are a sufficient guarantee. The density gradient thus reduces to $\nabla\rho/\rho = \beta_S G$.

Under the assumption of volume additivity, reasonable for liquid systems, the solutal expansion coefficient can be estimated from the formula giving the mass density of an A-B alloy :

$$\rho = \frac{\rho_A \rho_B}{x_A \rho_A + x_B \rho_B} \tag{19}$$

the x's and ρ's being the mass fractions and the mass densities of components A and B. In liquid metals, the difference between the mass densities of the constituents may be quite high, leading to solutal expansion coefficients values, expressed in inverse weight fraction ranging from 0.1 to 2.

The composition gradient G evolves with time from the initial step function distribution. For a discussion of the convective effects a relevant value is $G = \Delta C/H$, ΔC being the difference between the end compositions. As the error function best fit can only be performed if D is independent of concentration, ΔC is necessarily limited (to the order of 5 wt%).

For a gravity level characteristic of experiments in the NASA space shuttle $g = 10^{-3}\mathrm{ms}^{-2}$, a cell diameter $H = 10^{-3}$ m and thermophysical parameters typical of liquid metals ($\nu = 3 \cdot 10^{-7}\mathrm{m}^2\mathrm{s}^{-1}$, $D = 5 \cdot 10^{-9}\mathrm{m}^2\mathrm{s}^{-1}$), a nondimensional parameter $\beta_\mathrm{S}\Delta C$ of 0.01, we get $\overline{W}H/D = 0.14$.

It is quite difficult to accurately predict the resulting measurement error, since, as discussed earlier, the proportionality factor α depends on the simulated configuration. However, the above value $\overline{W}H/D = 0.14$ would probably not lead to observable effects. For capillary diameters in the mm range, when the solutal expansion coefficient is high, the composition difference should be reduced to ensure that $\beta_\mathrm{S}\Delta C$ does not exceed 0.01.

Let us now turn to a discussion of the possibility of realizing Soret coefficient measurements in space. The overall axial density gradient includes both a thermal and a solutal component:

$$\nabla\rho/\rho = -\beta_\mathrm{T}G_\mathrm{T} + \beta_\mathrm{S}G = -[\beta_\mathrm{T} + (D_\mathrm{T}/D)\beta_\mathrm{S}]G_\mathrm{T} \tag{20}$$

As noted in [1], depending on the signs of the solutal expansion and thermodiffusion coefficients, thermotransport will lead to an acceleration or a damping of the flow. In concentrated alloys the solutal and thermal contributions in (20) may be of the same order of magnitude. A typical value of the ratio D_T/D is 10^{-3} wt.fr./K [5].

The thermal expansion coefficient in liquid metals and the imposed temperature gradient are of the order of 10^{-4} K^{-1} and $10^4\mathrm{Km}^{-1}$, respectively. Using the same values as above ($H = 10^{-3}$ m, $g = 10^{-3}\mathrm{ms}^{-2}$, $\nu = 3 \cdot 10^{-7}\mathrm{m}^2\mathrm{s}^{-1}$, $D = 5 \cdot 10^{-9}\mathrm{m}^2\mathrm{s}^{-1}$) we get $\overline{W}H/D = 0.014$.

Such a low value of $\overline{W}H/D$ ensures that convective transport will not be a factor. However, the capillary diameter should always be limited to roughly 2 mm due to the H^4 dependence of $\overline{W}H/D$. Besides, in systems with exceptionnally high Soret coefficients, the effect of solutally driven flow may not be ruled out even at $H = 1$ mm.

5. Concluding remarks

Our purpose in this work was to discuss the possible effect of natural convection in liquid phase mass transport coefficients in microgravity for both isothermal and thermodiffusion experiments. Using existing numerical simulations we found that in both cases the measurement error scaled with the square of the fluid velocity and that an effective transport coefficient could be defined to account for the effect of convection. In practice this means that it is often impossible to check *a posteriori* whether an experiment has been carried out under favourable conditions, since the additional transport can be hidden under the pretence of accelerated diffusion.

In microgravity conditions the interaction of the axial density gradient with the transverse component of the residual gravity is the driving force for fluid flow. We propose a simple criterion (18) to estimate the possible convective effect. The key parameter of the problem appears to be the capillary diameter. Our results indicate tha, for experiments in capillaries of diameter smaller than 2 mm the convective effect should be limited, except may be in some extreme cases.

As a conclusion, it should be said that the arguments developed in this paper to study the microgravity relevance of transport coefficient measurements can be adapted to ground configurations. This was done in [2] for isothermal diffusion and we plan to extend this approach to the thermodiffusion problem in the near future.

Acknowledgments

The present work was carried out in the frame of the GRAMME agreement between the Centre National d'Etudes Spatiales and the Commissariat l'Energie Atomique. This text presents research results of the European Community Programme "Human Capital and Mobility", with the support of the Commission in the frame of the network CHRX-CT930106. Numerous fruitful discussions on the topic with Dr. D. Henry are also gratefully acknowledged.

References

1. D. Henry and B. Roux, *Phys. Fluids*, **29** (1986) 3562
2. J.P. Garandet, C. Barat and T. Duffar, *Int. J. Heat Mass Transfer* (1995)
3. G. Frohberg and Y. Malmejac, in "Fluid Sciences and Material Science in Space", Chap. 5, Ed. H.U. Walter, Springer Verlag (1987)
4. A. Bejean and C.L. Tien, *Int. J. Heat Mass Transfer*, **21** (1978) 701
5. J.P. Praizey, *Int. J. Heat Mass Transfer*, **32** (1989) 2385

Proboscis Container Shapes for the USML-2 Interface Configuration Experiment

P. Concus[1], *R. Finn*[2] *and M. Weislogel*[3]

[1] Lawrence Berkeley Laboratory and Department of Mathematics, University of California, Berkeley, CA 94720, U.S.A.
[2] Department of Mathematics, Stanford University, Stanford, CA 94305, U.S.A.
[3] Space Experiments Division, NASA Lewis Research Center, Cleveland, OH 44135, U.S.A.

Abstract

Cylindrical containers for the USML-2 Interface Configuration Experiment, having the form of a circular cylinder with two mathematically determined diametrically opposed "proboscis" protrusions, are discussed. Results of preliminary drop tower tests are shown, along with numerically computed fluid interface configurations.

1. Introduction

Small changes in container shape or in contact angle can give rise to large shifts of liquid in a microgravity environment. Such behavior suggests a means for managing fluids in microgravity and, as one specific possible application, for the accurate determination of contact angle. In connection with this application, we discuss certain containers designed for the forthcoming USML-2 Glovebox Interface Configuration Experiment (ICE) and depict their behavior in preliminary drop tower experiments. The containers are in the form of a circular cylinder with two diametrically opposed "proboscis" protrusions. These shapes are based on the canonical (single) proboscis containers introduced mathematically in [1], which have the properties in the absence of gravity that (i) fluid rises arbitrarily high over the entire proboscis for contact angles less than or equal to a critical value and (ii) the size of the proboscis can be made relatively as large a portion of the container cross section as desired. These properties allow overcoming some of the practical limitations of wedge containers; for the latter too little fluid may participate in the shift at a critical contact angle to be easily observable.

We include below some background material from [2], where computational results for the double proboscis containers are presented.

2. Governing equations

Consider a cylindrical container of general cross-section partly filled with liquid, as indicated in Fig. 1. According to the classical theory, an equilibrium interface

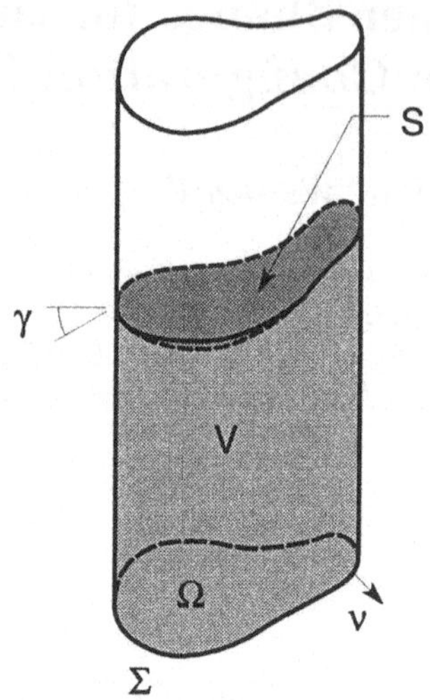

Fig. 1. Partly filled cylindrical container with base Ω.

in the absence of gravity between the liquid and gas (or between two immiscible liquids) is determined by the equations

$$\operatorname{div} Tu = \frac{1}{R_\gamma} \quad \text{in } \Omega \ , \tag{1}$$

$$\nu \cdot Tu = \cos\gamma \quad \text{on } \Sigma \ , \tag{2}$$

where

$$Tu \equiv \frac{\nabla u}{\sqrt{1 + |\nabla u|^2}} \ ;$$

see, e. g., [3], Chap. 1. In these equations Ω is the cross section (base) of the cylindrical container, Σ is the boundary of Ω, ν is the exterior unit normal on Σ, and

$$R_\gamma = \frac{|\Omega|}{|\Sigma| \cos\gamma} \ , \tag{3}$$

where $|\Omega|$ and $|\Sigma|$ denote respectively the area and length of Ω and Σ; $u(x, y)$ denotes the height (single-valued) of the interface S above a reference plane parallel to the base, and γ is the contact angle between the interface and the container wall, determined by the material properties. The volume V of liquid in contact with the base is assumed to be sufficient to cover the base entirely, and, for the mathematical results, the cylinder is assumed implicitly to be arbitrarily tall so that questions of behavior at a top do not arise. We restrict discussion to the case of a wetting liquid $0 \leq \gamma < \pi/2$ (the complementary non-wetting case can be easily transformed into this one). For $\gamma = \pi/2$, the solution surface is a horizontal plane for any cross-section.

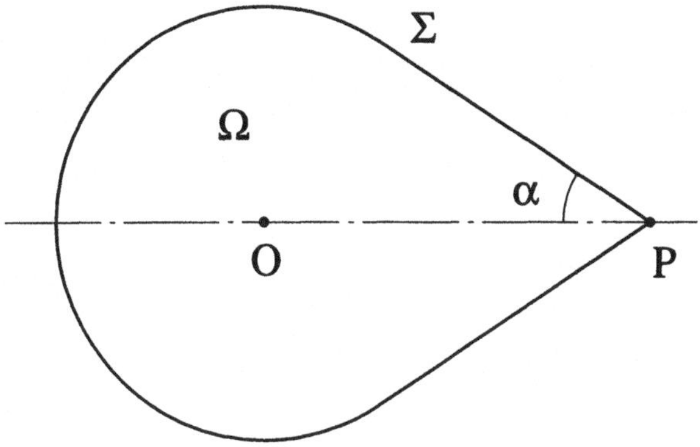

Fig. 2. Wedge container section.

3. Wedge container

For a cylindrical container whose section Ω contains a protruding corner with opening angle 2α, as in Fig. 2, the critical value of contact angle, at which behavior is discontinuous, is $\gamma_0 = \frac{\pi}{2} - \alpha$. For $\frac{\pi}{2} > \gamma \geq \gamma_0$ (and for fluid volume sufficient to cover the base) the height u can be given in closed form as the portion of the lower hemisphere with center at O meeting the walls with the prescribed contact angle γ. Thus the height is bounded uniformly in γ throughout this range. For $0 \leq \gamma < \gamma_0$, however, the fluid will necessarily move to the corner and rise arbitrarily high at the vertex, uncovering the base regardless of fluid volume. Details of this behavior can be found in [4], the initial study that revealed the discontinuous behavior, and in [3], [5], and [6]. Procedures for determining contact angle based on the phenomenon can give very good accuracy for larger values of γ (closer to $\pi/2$) but may be subject to experimental inaccuracy when γ is closer to zero, as the "singular" part of the section over which the fluid accumulates when the critical angle γ_0 is crossed then becomes very small and may be difficult to observe.

4. Canonical proboscis container

As a way to overcome the experimental difficulty, "canonical proboscis" sections were introduced in [1]. These domains consist of a circular arc attached symmetrically to a (symmetric) pair of curves described by

$$x = \sqrt{{R_0}^2 - y^2} + R_0 \sin\gamma_0 \ln \frac{\sqrt{{R_0}^2 - y^2}\cos\gamma_0 - y\sin\gamma_0}{R_0 + y\cos\gamma_0 + \sqrt{{R_0}^2 - y^2}\,\sin\gamma_0} + C\ , \tag{4}$$

and meeting at a point P on the x-axis, see Fig. 3. Here R_0, as well as the particular points of attachment, may be chosen arbitrarily. The (continuum of) circular arcs Γ_0, of which three are depicted by the dashed curves in Fig. 3, are

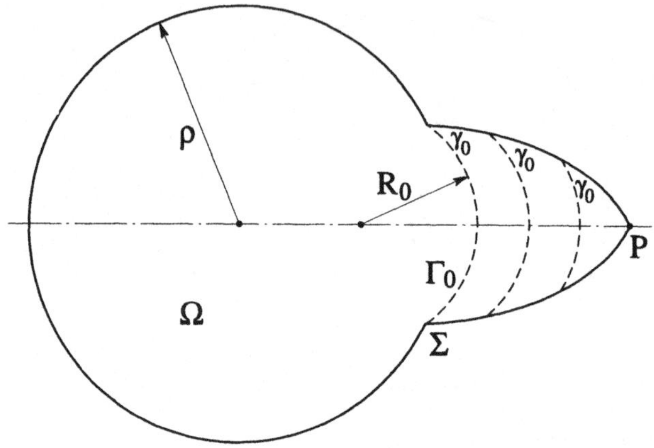

Fig. 3. Canonical proboscis container section showing three members of the continuum of extremal arcs.

all horizontal translates of one such arc, of radius R_0 and with center on the x-axis, and the curves (4) have the property that they meet all the arcs Γ_0 in the constant angle γ_0. If the radius ρ of the circular boundary arc can be chosen in such a way that R_0 is the value of R_γ from (3) for the value $\gamma = \gamma_0$, then the arcs Γ_0 become extremals for a "subsidiary" variational problem [7] (see also [3], Chap. 6, [6]) determined by the functional

$$\Phi \equiv |\Gamma| - |\Sigma^*| \cos\gamma + |\Omega^*|/R_\gamma \tag{5}$$

defined over piecewise smooth arcs Γ, where Σ^* and Ω^* are the portions cut off from Σ and Ω by the arcs. In the case of the section of Fig. 3, Σ^* and Ω^* lie to the right of the indicated arcs. It can be shown that every extremal for Φ is a subarc of a semicircle of radius R_0, with center on the side of Γ exterior to Ω^*, and that it meets Σ in angles $\geq \gamma_0$ on the side of Γ within Ω^*, and $\geq \pi - \gamma_0$ on the other side of Γ (and thus in angle γ_0 within Ω^* whenever the intersection point is a smooth point of Σ) [7],[3]. It is remarkable that whenever (3) holds, $\Phi = 0$ for every Ω^* that is cut off in the proboscis by one of the arcs Γ_0; see [1] and the references cited there.

In [1] a value for ρ was obtained empirically from (3) in a range of configurations, and it was conjectured that the angle γ_0 on which the construction is based would be critical for the geometry. That is, a solution of (1), (2), (3) should exist in Ω if and only if $\gamma > \gamma_0$. Additionally, the fluid height should rise unboundedly as γ decreases to γ_0, precisely in the region swept out by the arcs Γ_0 (the entire proboscis region to the right of the leftmost arc Γ_0 shown in Fig. 3). For these conjectures, which form the basis of our proposed procedure and for which the mathematical underpinnings were proved only partially in [1], complete mathematical proofs have been carried out [8]. Specifically, it has been established that a unique value of ρ can be obtained for any specified proboscis length and that the conjectured behavior of the fluid rise is the only one possible.

In [9] numerical solutions of (1), (2), (3) are depicted for canonical proboscis containers. Although the fluid rise in the corner is not discontinuous as occurs for a planar wedge, it can be "nearly discontinuous" in that the rise height in the proboscis is relatively modest until γ decreases to values close to γ_0, and then becomes very rapid at $\gamma = \gamma_0$. Furthermore, since the proboscis can be made relatively as large a portion of the section as desired, the shift can be easily observed for a broad range of γ_0. Through proper choice of the domain parameters for the cases considered, an effective balance can be obtained between conflicting requirements for contact angle measurement of a sharp near discontinuity (for accurate measurement) and a sizable volume of fluid rise (for ease of observation).

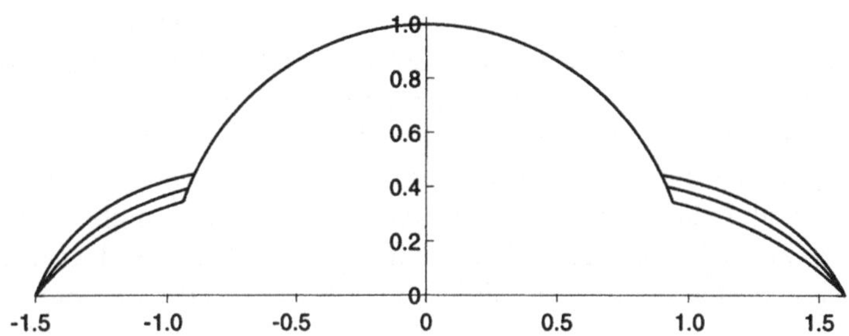

Fig. 4. Three superimposed double proboscis container sections. From uppermost to lowest, the pair of values of γ_0 for the left and right proboscides of each section are 20°/26°, 30°/34°, and 38°/44°.

5. Double proboscis container

For the USML-2 experiment, double proboscis containers will be used. These containers are similar to the single proboscis one of Fig. 3, except that there is a second proboscis diametrically opposite to the first, in effect combining two containers into one. The values of γ_0 in (4) generally differ for the left and right proboscides, whose values of γ_0 we denote by γ_L and γ_R, respectively. Similarly, we denote the values of R_0 for the left and right proboscides by R_L and R_R. In order for (3) to be satisfied for both proboscides, there holds

$$R_R \cos \gamma_R = R_L \cos \gamma_L .$$

Specifying the desired points of attachment and choosing ρ, the radius of the circular portion of the section, so that (3) is satisfied then yields the container section. (Such a ρ can be chosen for the cases discussed here, but a proof that such a choice is possible for any proboscis lengths has not yet been carried out for the double proboscis case.) The critical value for the container is the larger of γ_L and γ_R. For the containers considered here, we shall take $\gamma_R > \gamma_L$, so that the critical contact angle γ_0 for the container is equal to γ_R.

The upper half of the sections for the USML-2 experiment, superimposed on one another, are shown in Fig. 4. The sections have been scaled so that the circular portions all have radius unity. The meeting points of the vertices with the x-axis are, respectively, a distance 1.5 and 1.6 from the circle center. For the sections depicted in Fig. 4 the values of γ_L and γ_R are respectively 20° and 26° for the uppermost section, 30° and 34° for the middle section, and 38° and 44° for the lowest section.

For these containers the explicit behavior has not yet been determined mathematically in complete detail, as it has for the single proboscis containers. However, numerical computations in [2] and the known behavior of the single proboscis solution surfaces suggest that the behavior will be as follows: For contact angles $\gamma \geq \gamma_0$, as γ decreases to γ_0 the fluid will rise higher in the right than in the left proboscis, with the rise becoming unbounded in the right proboscis at γ_0. For contact angles between γ_L and γ_R the fluid will rise arbitrarily high in the right proboscis, but the height in the left will still be bounded. For smaller contact angles the fluid will rise up both proboscides arbitrarily high. By observing the liquid shift, one can then bracket the contact angle relative to the values of γ_L and γ_R. For a practical situation in which the container is of finite height with a lid on the top, the fluid will rise to the lid along one or both of the proboscides in the manner described above (providing the fluid volume is adequate).

The selected values of γ_L and γ_R for the three containers are based on the value of approximately 32° measured in a terrestrial environment for the contact angle between the experiment fluid and the acrylic plastic material of the container. The spread of values of contact angle covered by the three containers is intended to allow observation of possible effects of contact angle hysteresis, which is not included in the classical theory.

Typical behavior of the numerical solutions of (1), (2), (3) for the three double proboscis container sections in Fig. 4, for a range of contact angles γ, is illustrated in Fig. 5, which is taken from [2]. The numerically calculated solution surface $u(x, y)$ for (the upper half of) the 30°/34° domain is shown for four values of contact angle, 60°, 50°, 40°, and 35°. (The critical value for the domain is $\gamma_0 = 34°$.) Contour levels of the surfaces are indicated by the shading. As for the single proboscis containers, the computations indicate that rise heights are relatively modest until γ gets close to the critical value. At the container critical contact angle of 34°, the solution would rise arbitrarily high in the right proboscis. For contact angles less than or equal to 30°, the fluid would rise arbitrarily high in the left as well.

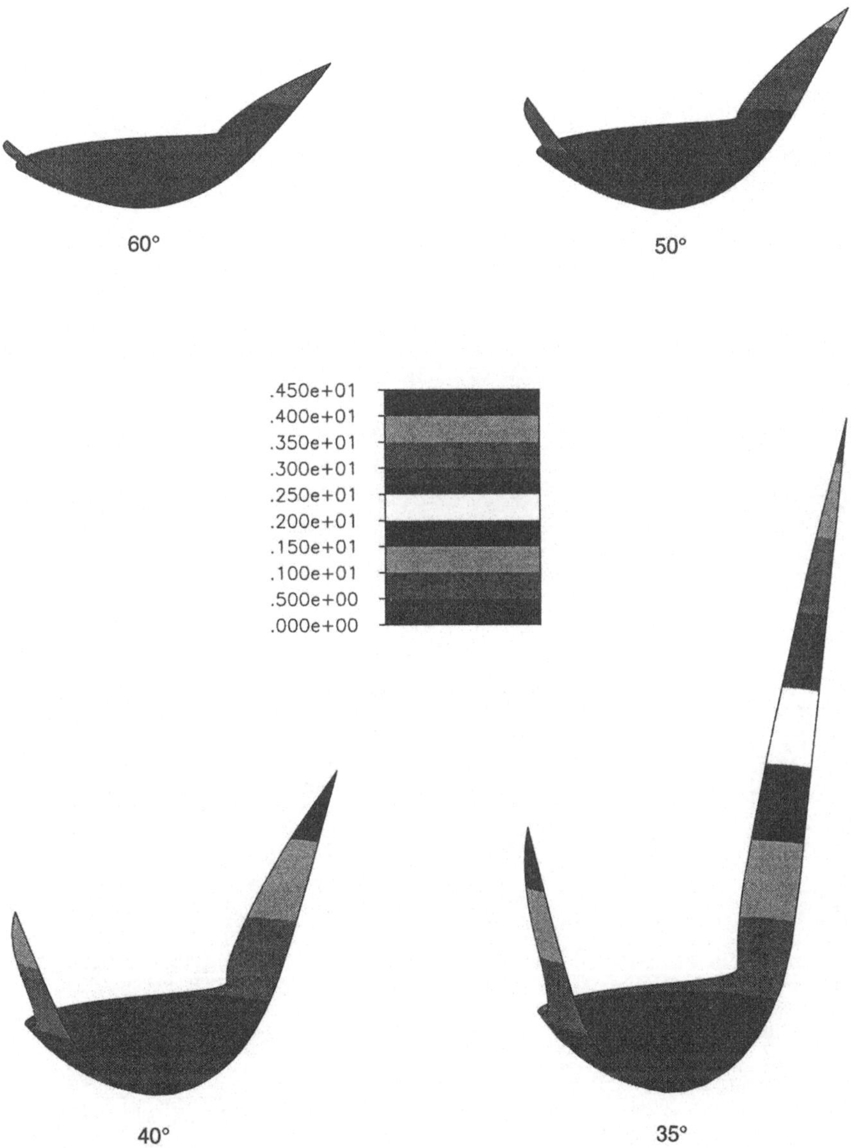

Fig. 5. Equilibrium interface for the 30°/34° (upper-half) double proboscis section for contact angles 60°, 50°, 40°, and 35°. $\gamma_0 = 34°$.

Fig. 6. (Transient) drop tower configurations after $\approx$ 5 Sec. of free fall. Upper row: 50% ethanol solution. Lower row: 60% ethanol solution. From left to right in each row are the 20°/26°, 30°/34°, and 38°/44° vessels, respectively.

6. Drop tower tests

The results of preliminary experiments carried out at the NASA Lewis Zero Gravity Facility 5.18-Sec. drop tower are shown in Fig. 6 for the three vessels of Fig. 4 and for two different liquids. The figure depicts the configurations after approximately five seconds of free fall. In the top row of Fig. 6, the liquid is a 50% ethanol solution (the liquid to be used for ICE), for which the equilibrium contact angle with the container wall was measured to be approximately 32°, with a measured receding/advancing hysteresis interval of approximately 18° to 43°. In the lower row the liquid is a 60% ethanol solution, for which the equilibrium contact angle is approximately 20°, with a 12°–30° receding/advancing hysteresis interval.

In the approximately five seconds of reorientation from an initial 1-g configuration, an indication could be observed of what might occur under the longer-term period of weightlessness of the orbiting Spacelab environment. For the 50% ethanol mixture, the fluid interfaces rise somewhat along the proboscis portions of the vessel, with a noticeably greater rise in the right proboscis than in the left for the 38°/44° vessel. For the 60% ethanol mixture the rise is more pronounced; in the 44° proboscis the substantial rise suggests that the fluid may be proceeding to the top of the container, in accordance with the mathematical results. In the other proboscides the less virgorous rise could be attributed to reorientation forces being smaller when the departure of contact angle from the critical value for the container is less. In a longer term low-g environment and with astronaut "tapping" of the vessels to encourage overcoming of contact line friction and hysteresis effects, we anticipate that information can be obtained as to what extent the mathematical and numerical predictions based on the classical Young-Laplace theory can be observed in practice and what the physical effects might be of factors not included in the classical theory.

7. ICE experiment

In addition to the three double proboscis containers depicted in Fig. 4, the USML-2 ICE experiment has also a wedge container. This container is constructed to allow the interior wedge angle 2α (see Fig. 2) to be varied, so as to permit observation of the wedge phenomenon for both the advancing and receding cases.

Acknowledgments

This work was supported in part by the National Aeronautics and Space Administration under Grant NCC3-329, by the National Science Foundation under Grant DMS91-06968, and by the Mathematical Sciences Subprogram of the Office of Energy Research, U. S. Department of Energy, under Contract Number DE-AC03-76SF00098.

References

1. B. Fischer and R. Finn, *Zeit. Anal. Anwend.* **12** (1993) 405–423.
2. A. Chen, P. Concus, and R. Finn, *On cylinder container sections for a capillary free surface experiment*, Paper AIAA 95-0271, 33rd AIAA Aerospace Sciences Meeting, Reno, NV, 1995.
3. R. Finn, *Equilibrium Capillary Surfaces*, Springer-Verlag, New York, 1986. Russian translation (with Appendix by H.C. Wente) Mir Publishers, 1988.
4. P. Concus and R. Finn, *Proc. Natl. Acad. Sci.*, **63** (1969) 292–299.
5. P. Concus and R. Finn, *Acta Math.*, **132** (1974) 177–198.
6. P. Concus and R. Finn, *Microgravity Sci. Technol.* **3** (1990) 87–92; Errata, **3** (1991) 230.
7. R. Finn, *J. reine angew. Math.* **353** (1984) 196–214.
8. R. Finn and T. Leise, *Zeit. Anal. Anwend.* **13** (1994) 443-462.
9. P. Concus, R. Finn, and F. Zabihi, in *"Fluid Mechanics Phenomena in Microgravity"*, AMD Vol. 154, Amer. Soc. Mech. Engineers, D. A. Siginer and M. M. Weislogel, eds., New York, 1992, pp. 125–131.

Response of a Liquid Bridge to an Acceleration Varying Sinusoidally with Time

I. Martínez, J.M. Peralez and J. Meseguer

CIDE-E.T.S.I. Aeronáticos, Universidad Politécnica, E-28040-Madrid

Abstract

The response of a long cylindrical liquid column subjected to an axial microgravity field has been experimentally studied on a TEXUS sounding rocket flight to check with theoretical predictions. The expected response of the liquid bridge was a quasi-static amphora-type deformation of the cylindrical shape. However, the experimental results showed a more complex behaviour. Nevertheless it has been possible to find out the reasons of this discrepancy except for a mysterious 0.5% uncertainty in the stimuli.

1. Introduction

This experiment belongs to a series (SL-1, TEXUS-12, SL-D1, TEXUS-29, SL-D2)[1] on the mechanics of long liquid columns [1]. The proposal was submitted (Feb-91) after conflicting results were obtained from a related Spacelab-D1 (Nov-85) analysis, aiming to elucidate the precise response to an axial acceleration in a presence of g-jitters before the experiments on Spacelab-D2 (Apr-93), although delays on TEXUS shifted it back to Nov-94.

The nominal configuration is a cylindrical liquid column of radius $R(z) = R_0$ and length L, anchored at the sharp edges of two coaxial solid discs (of radii R_0), held in ambient air by the interface tension σ. This configuration can only be studied in weightlessness. A silicone-oil column of R_0=15 mm, L=85 mm, (σ=0.02 N/m, density (ρ=920 kg/m^3 and kinematic viscosity (ν=10^{-5} m^2/s is used. The shape should be a perfect cylinder at equilibrium if no forces were applied. But the residual microgravity or g-jitters of the carrying platform (SL or TEXUS) is always present, and not very well known, introducing some uncertainty in the comparison with theoretical predictions. Besides, space experiments are so rare that one tries to explore many parameters, bridge stretching,

[1] TEXUS is the acronym of a German-European sounding rocket program providing up to 6 minutes reduced gravity during free parabolic fall of the order of $10^{-4}g_0$ with g_0 the gravity acceleration at the earth surface; SL is the acronym for the Spacelab, a laboratory in the cargo bay of the Space Shuttle; D1 and D2 are acronyms for the German spacelab missions 1 and 2, flown in 1985 and 1993 respectively. (Eds. note)

vibration, rotation, eccentric rotation, and so on without having time to make redundancy tests, small variations around a working set-point of parameters, and so on.

During SL-D1, one such a liquid column was undergoing shape oscillations in an amphora-type axisymmetric deformation, with an eigenfrequency and a damping as predicted by theory, but with a deformed mid-shape (averaged over the eigenperiod) corresponding to a theoretical acceleration of (70±10 $\cdot 10^{-5}$ m/s^2 (70±10 μg) constantly acting along the column axis. This high value could not be attributed to the Shuttle drag ($< 1\mu$g) and remains unexplained. A similar experiment was carried out in SL-D2, but then, applying the same analysis, only a value of 5±1 μg was found from the averaged deformation of the liquid column.

2. Scientific Objectives

The main goal of this experiment was to provide a calibrated microgravity acceleration (100±1 μg), well above the uncontrolled g-jitters of the platform ($< 1\mu$g), to check the static response of the liquid column acting as an accelerometer (microgravity accelerometry in the 10^{-3} Hz -10^{-1} Hz range is still controversial). As a constant controlled acceleration μg-platform is not available, a low varying sinusoidal motion was foreseen.

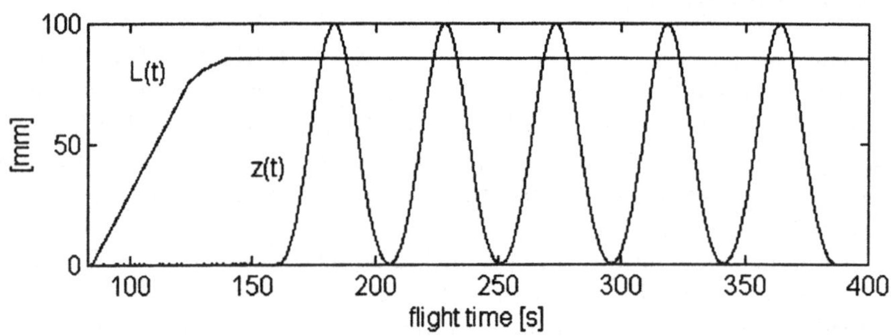

Fig. 1. Disc separation to form the bridge L(t), and cell oscillations z(t) in the experiment.

The acceleration was imposed by a slow sinusoidal translation of the liquid column cell (already used in previous TEXUS flights) along the common column-rocket axes. From the knowledge of the first natural period of the column (15 s) and optimisation of the microgravity time (6 minutes), a forcing period of T=45 s was chosen, and a corresponding amplitude of 100 mm peak-to-peak, in order to reach the acceleration quoted above. Six such cycles were allowed during the 5 minutes time available after the formation of the liquid column in orbit, although the first two were expected to be distorted by the initial transients (the half-damping time was known to be some 40 s). Lately, the analysis of Spacelab-D2 results prompted us to leave a full 1 minute to study the decay of the oscillations before re-entry and thus the forcing was reduced to 5 cycles, as shown in Fig.1. Observe that oscillations start without velocity jump but with a 100 μg acceleration jump. The nominal displacement law was $z(t) = z_m(1-\cos(2\pi t/T))$, with z_m =50 mm and T=45 s.

The expected response of the liquid column was a quasi-static amphora-type deformation of the cylindrical shape, best quantified by the radial deformation in a cross-section at 1/4 of the column length ($R_{1/4}$), as depicted in Fig. 2.

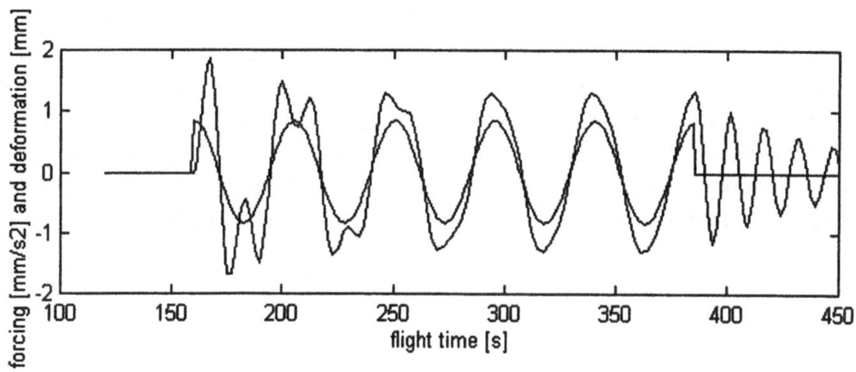

Fig. 2. Ideal excitation (the 5 perfect cycles of the forcing acceleration) and theoretical response of the liquid ($R_{1/4}$ deformation, with damping). After the transients and before the stopping there is a quasi-static amplification in phase.

Consider an initially cylindrical liquid column subjected to an axial acceleration that we name g to resemble gravity, although it is orders of magnitude smaller. The equation governing the equilibrium shape is the Young-Laplace non-linear second-order differential equation of capillary pressure (that accommodates the linear pressure profile inside the liquid to the constant value outside):

$$\sigma\left(\frac{R_{zz}}{1+R_z^{2\,3/2}} - \frac{1}{R(1+R_z^2)^{1/2}}\right) = p_0 + \rho g z$$

$$\text{with} \qquad R(0) = R(L) = R_0 \qquad \text{and} \qquad \pi\int_0^L R^2 dz = \pi R_0^2 L \qquad (1)$$

which can be numerically solved [2] for the parameters of this problem (R_0,L,ρ,σ) to find first a linear deformation for small g and then a limiting value of g beyond which the bridge breaks [2]. The pressure jump at the origin, p_0, is an

[2] $R_z = dR/dz; R_{zz} = d^2R/dR^2$.(Eds. note)

internal constant to be found from the three boundary conditions of the second order problem. In terms of the Bond number, Bo, numerical solving (1) yields $R_{1/4}/R_0 = 1 + 0.009 \cdot Bo$ and $R_{1/4}/R_{0\text{limit}} = 1.19$ (for $Bo = Bo_{\text{limit}}$=0.018). The linearised analytical solution is:

$$\frac{R_{1/4}}{R_0} = 1 + \frac{L}{4R_0}\left(1 - \frac{1}{\cos\frac{L}{4R_0}}\right) Bo = 1 + \frac{\rho R_0^2 L}{4\sigma}\left(1 - \frac{1}{\cos\frac{L}{4R_0}}\right) g \quad (2)$$

with the definition of the Bond number as $Bo = \rho g R_0^2/\sigma$. This experiment was not the first in the series. A good deal of work was spent not only analysing the nominal configuration but all conceivable possible departures: the influence of an error in the liquid volume injected (from the nominal cylinder), changes in liquid properties, residual spin of the TEXUS rocket (-3 rpm, and a value of 13 rpm would break the column), change in anchoring radii (in case of bad wetting), etc. Even ground simulation by mounting a Plateau tank in a centrifuge was considered, although abandoned for practical difficulties. All the lessons learnt from past experiments were also applied (redundancy of image recording, background grid to aid in image analysis, all information in the image, etc.).

3. Results

The results obtained were at first sight all right (large amphora-type deformations), but the detailed analysis was striking, as presented in Fig. 3, due to the fact that the sinusoidal motion was actually achieved by a crank-shaft mechanism, what was not realised by the investigators at the ground tests, where only pen-recorder plots were available (the data acquisition system was not operative).

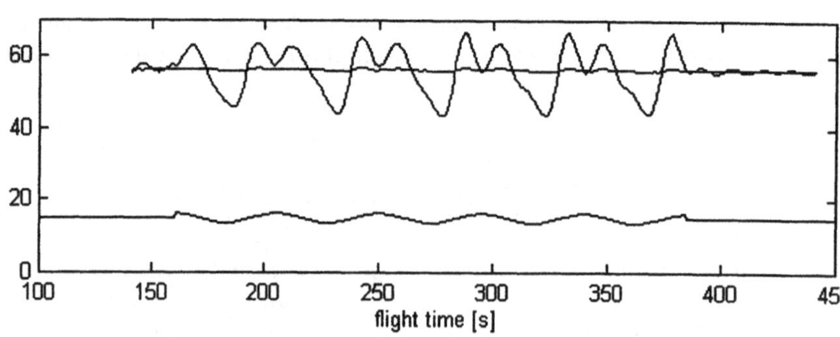

Fig. 3. Discrepancy found in the first analysis of video images. The lower part is the ideal expected $R_{1/4}$ deformation in mm, and the upper part the real $R_{1/4}$ deformation in pixels.

The analysis of this experiment is based on the recording of cell displacement by a linear potentiometer and the video recording of images of the liquid bridge against a background reference mask and a fixed reference bar to visualise cell displacement. The potentiometer data (used to draw Fig. 1) has a sampling rate of 0.22 s (4.6 Hz) and a resolution of 0.02 mm in 100 mm (12 bit), and from that one infers that the actual column length was L=85.15 ± 0.05 mm (nominally 85 mm), oscillations started at 160.4±0.1 s flight-time (nominally 160 s) and stopped at 386.6 ± 0.1 s (nominally 385 s), giving a period of 45.2±0.1 s (nominally 45 s), with an amplitude peak-to-peak of 99.8 ± 0.1 mm (nominally 100 mm). The frequency spectrum for the five cycle oscillation is shown in Fig.4, from two redundant sources, as well as that corresponding to the liquid deformation.

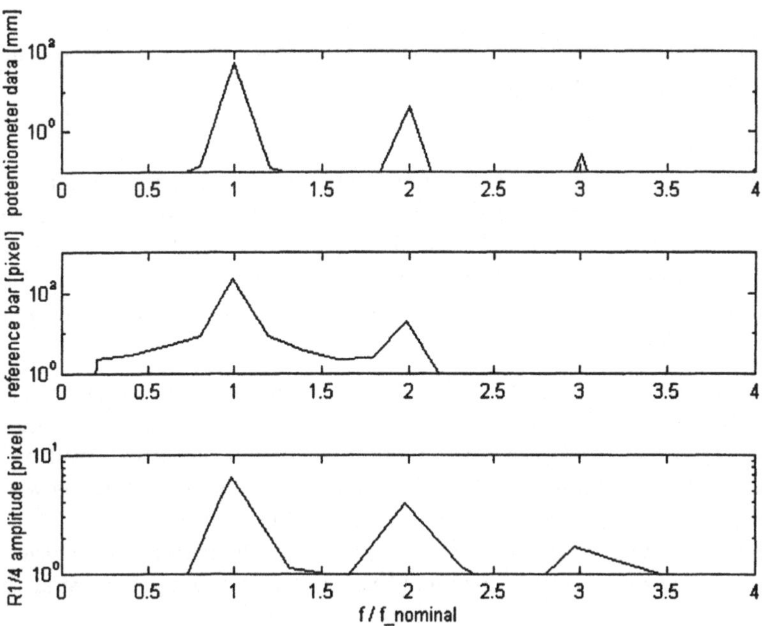

Fig. 4. Fourier analysis of the cell displacement measurement taken by the potentiometer transducer (top), by the reference bar in the image (mid) and of liquid response (deformation at $R_{1/4}$) (bottom).

Taking into account the uncertainty in the potentiometer data one should disregard the fourth harmonic contribution, but the third harmonic remains, with an amplitude of 0.2 ± 0.1 mm still unexplained. From the ratio of the second to the first harmonic, one may deduce the crank shaft length, what yields 158 ± 10 mm (150 mm according to design).

The analysis of the video images is not as straightforward as for the potentiometer data, mainly because of the distortions the analogue video signal suffers before digitization. For instance, for this experiment, an ideal quiescent cylindrical column once processed (CCD, videotape record, videotape playback, digitization) would present a noisy changing shape with uncertainties of 1 pixel in space and 0.5 pixels in time, although if one only considers odd or even videolines (but not both) and liquid column size (i.e. difference in position of the right and left liquid contours) the uncertainty can be reduced to ± 0.2 pixels in the image field, corresponding to ± 0.05 mm in the object field.

The video recording is classified according to the interest for further analysis as follows:

- The 53 s of column formation by liquid injection and disc separation (a small amphora-type deformation is visible, due to the jet injection dynamics).
- The 20 s of calm-down after filling. It turned out to be too short for a decay of the injection perturbations (the half damping time appears to be $t_{1/2} \approx 35 \pm 5$ s).
- The 90 s of the two first cycles, where transients from the start-up are distorting the periodic response.
- The 135 s of the last three cycles, practically periodic. Fig. 5 shows a sequence of frames.
- The 60 s of calm-down after stopping until re-entry. Astonishingly, the deformation after stopping the oscillation is found to be negligible, in contrast to the wide waving expected A perfect cylinder is seen by the naked eye. No residual acceleration is distinguishable in the finest analysis. Natural period appears to be 13 ± 1 s instead of the expected 15 s, and half-damping time $\approx 35 \pm 5$ s instead of 40 s expected.

The axisymmetry, judged by the position and evolution of the centre line (midway between liquid edges at every videoline) is perfect from the beginning up to re-entry of the rocket, where a C-mode deformations grows until breaking the bridge. The antisymmetry of the liquid shape to the mid-plane between the solid discs is also perfect after the long column is established.

Spectral analysis of the fix-reference bar motion digitised from the video images also shows the second harmonic (and a corresponding crank shaft length of 145 ± 5 mm, but not the third one because a resolution better than one pixel would be needed). The phase of the crank-shaft mechanism, with the flatten humps in acceleration at the upper part of the cell motion (at start-up and stop-down), is responsible for the dead of the oscillations after stopping the stimulus.

The compound oscillations of the whole TEXUS-rocket, the TEM-06-9 module [3] (fixed to the former by soft mounts), the moving fluid cell and the liquid column itself was analysed: $m_{\text{liquid}}=0.06$ kg, $m_{\text{cell}}=10$ kg, $m_{\text{module}}=60$ kg, $m_{\text{rocket}}=400$ kg. From the μg-accelerometers in the rocket, a train of pulses appear synchronised with our experiment timeline, with pulse amplitude precisely synchronised with the peaks in speed of our cell displacement, and not with acceleration, what might be explained by a series of beating of a loose joint at its bounds.

[3] TEM is the acronym for a TEXUS Experiment Module. The module 06-9 is a special module for fluid physics experiments. (Eds. note)

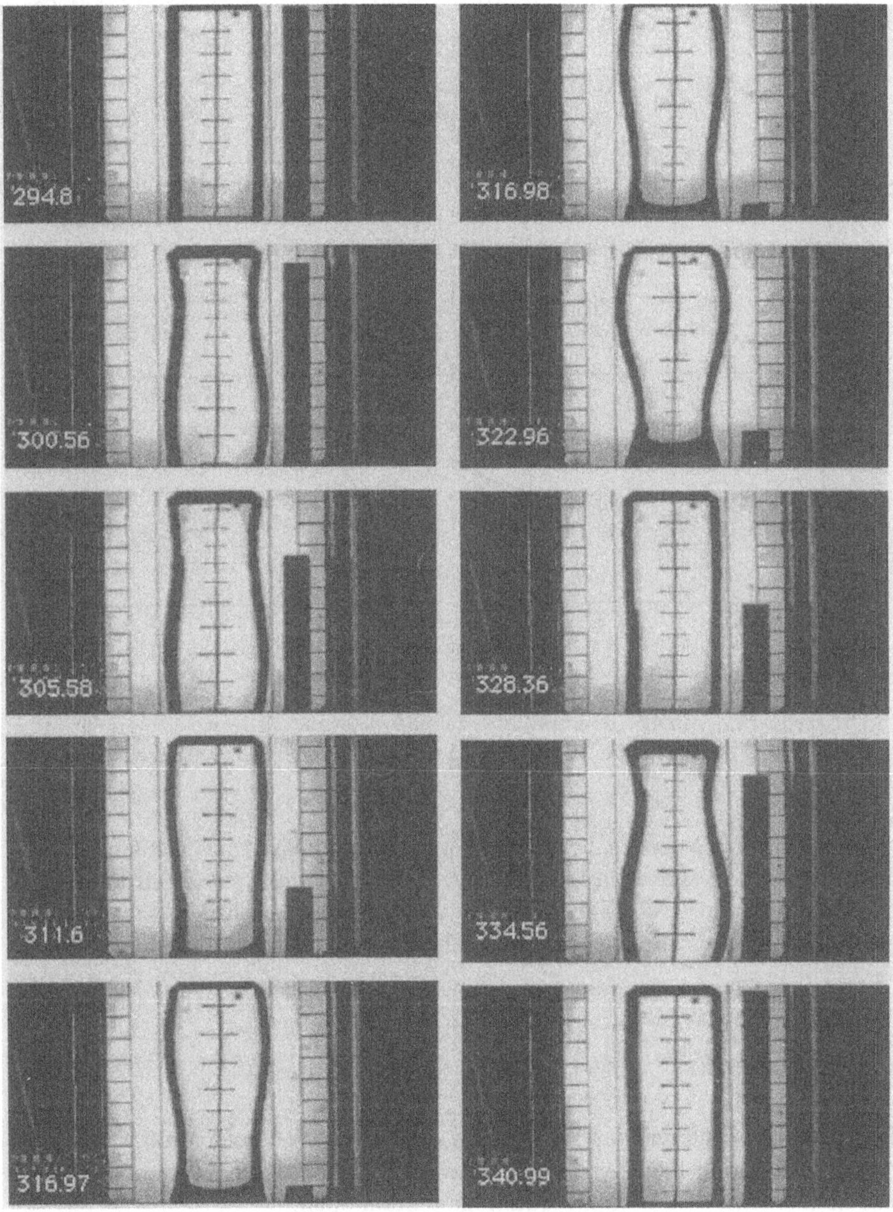

Fig. 5. A complete cycle of 45 s in the response of the liquid bridge starting at 294.2 s flight time.

A one-dimensional dumped-spring model is established to simulate the linear dynamics of the liquid column, once that available liquid bridge theory [3] shows that non-linear effects should be negligible and that the only relevant mode is the first (amphora-type oscillations):

$$\frac{d^2x(t)}{dt^2} + \frac{2\ln 2}{t_{1/2}}\frac{dx(t)}{dt} + \left(\frac{2\pi}{T}\right)^2 x(t) = K\left(\frac{2\pi}{T}\right)^2 \ddot{z}(t) \tag{3}$$

where x is the $R_{1/4}$ deformation ($R_{1/4} - R_0$), z the cell displacement, $t_{1/2}$=40 s, T=15.3 s and K=0.24 deduced from (1) for the static case. The gain in the frequency response of this linear system is shown in Fig.6.

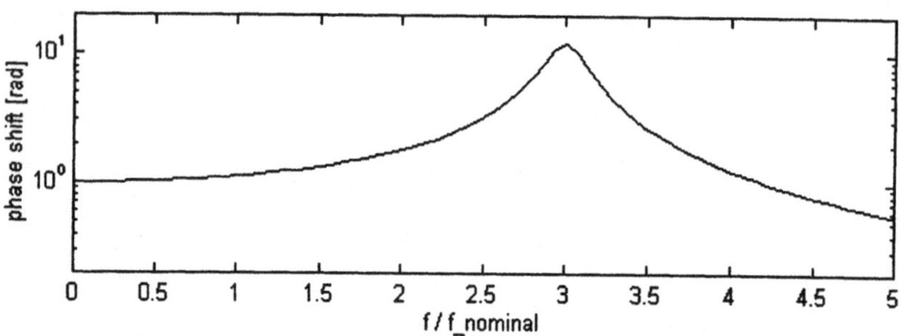

Fig. 6. Frequency response in amplitude of liquid deformation relative to amplitude of stimulus (f_{nominal} corresponds to the $T = 45$ s cycling at a third of the eigenfrequency).

Inclusion of the second harmonic from the crank-shaft mechanism explains the camel-shape response in Fig.3, but there still remains a small third harmonic contribution (some 0.5% of the intended displacement amplitude) that is pervasive to the analysis so far performed. The best matching achieved is shown in Fig.7.

4. Conclusions

Most aspects of the response of a highly sensitive long liquid column to a controlled acceleration have been understood in great detail with the theory developed so far. Only two puzzling questions remain open:

1. an artificial contribution of 0.5% in the applied stimuli to the third harmonic (and maybe of 0.4% to the forth), is found to better match the real liquid response than the mere crank-shaft motion.
2. a phase shift of precisely $\pi/2$ in the acceleration measured in the rocket body and the one applied.

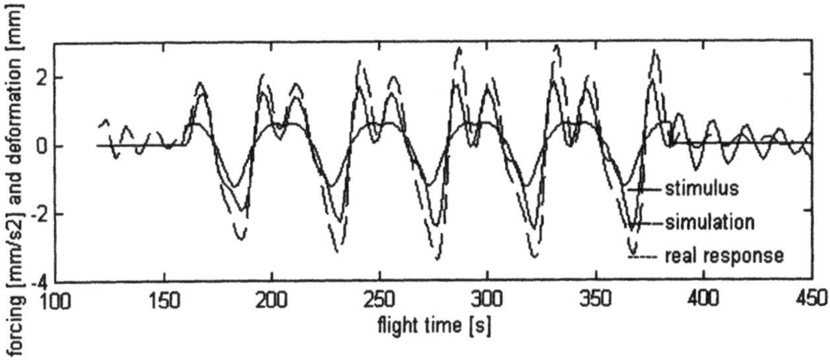

Fig. 7. Real excitation (the humped 5 cycles), real response of the liquid (from video) and expected response (from numerical simulation with (3) and third and forth harmonics).

Some lessons have been learnt for future work: 1) even at the high cost of space experiments, time periods to damp-out initial perturbations or to achieve periodic response must be generous, and 2) a negligible 0.5% noise in relative displacement at the eigenfrequency, when driving the system at one third of it, means a $3^2 \cdot 0.5 = 5\%$ noise in relative acceleration due to the centrifugal effect, and selectively amplified 12 times due to the weak damping, finally yields a $12 \cdot 3^2 \cdot 0.5 = 50\%$ noise in relative response of the liquid bridge.

Acknowledgement

The authors wish to acknowledge the invaluable help of the students E. Moreno, J. Paíno and M. Ramos, the ESA and ERNO TEXUS teams, and the financial support of the Spanish Grant CICYT ESP92-0001-CP.

References

1. I.Martinez, J.M. Perales and J. Meseguer, Stability of long liquid columns, in: Scientific Results of the German Spacelab Mission D2, P.R. Sahm, M.H. Keller, B.Schiewe (Eds.), DLR, Köln, Germany 1995, p.220
2. J.M. Perales, J. Meseguer and I. Martinez, Minimum Volume of Axisymmetric Liquid Bridges between Unequal Disks in an Axial Microgravity Field, J. Cryst.l Growth, **110** (1991) 855-861.
3. J. Meseguer and J.M.Perales, Non-steady Phenomena in the Vibration of Viscous Cylindrical Long Liquid Bridges, Microgravity Science and Technology, **5** (1992) 69-72.

Part V
Fluid Dynamics

Nonuniform Interfacial Tension-Driven Fluid Flows

M.G. Velarde

Instituto Pluridisciplinar, Universidad Complutense de Madrid 28040-Madrid, Spain

Abstract

The text provides a brief survey of the role of surface tension and surface tension gradients in thermocapillary flows. It discusses basic mechanisms for flow and flow instability, the role of boundary conditions, limiting approximations and dimensionless groups entering the usual problems dealt with in recent earth-based and low-g experimentation. A few comments about liquid layers are made followed by a more detailed account of drop and bubble dynamics.

1. Introduction

In this text I have tried to provide a brief survey (with no pretention of completeness) of the role of surface tension and surface tension gradients in thermocapillary flows. First, I discuss basic mechanisms for (forced) flow and flow instability, the role of boundary conditions, limiting approximations and dimensionless groups entering the usual problems dealt with in recent earth-based and low-g experimentation. Then, I give a few comments about liquid layers followed by a more detailed account of drop dynamics. The latter case is used to illustrate how the problem has been tackled since the pioneering work of Newton and Stokes.

2. The Role of Surface Tension and Its Gradient in Fluid Dynamics

Gravity in a fluid leads to stratification (density gradient) hence buoyancy. When stratification is unstable enough it provokes natural convection (if due to the presence of a thermal gradient, a threshold for instability is defined by a critical value of the Grashof, Gr, or Rayleigh number, Ra). On the other hand, a density gradient normal to the gravity vector provokes instantaneous (forced) convection. These are two modes of convection in the presence of body forces like gravity. Convection, one way or another is known to affect crystal growth, transport processes like the thermal diffusion, Soret effect, sedimentation, etc. However, convection is not always undesirable. It could be used to our advantage for, e.g. stirring liquids, mixing, or cooling processes or even to help maintaining concentration gradients. provided we know its extent and the factors on which it depends. In earth-based research it may not be easy to detect buoyancy-driven flows or if detected it may very well be quite difficult to eliminate them. Buoyancy-driven flows are, generally, laminar flows, with or without cellular patterns, and only at high enough Grashof or Rayleigh numbers convection becomes turbulent with high dissipation.

When there is an open surface or an interface exists between two liquids, the surface or interfacial tension accounts for the jump in normal stresses proportional to the surface curvature across the interface, hence affects its shape and stability. Gravity competes with the Laplace forces in accomodating equipotential levels with curvature. Their balance permits, for instance, the stable equilibrium of the spherical shape of drops and bubbles.

When surface tension varies with temperature or composition, and consequently, with position along an interface, its change takes care of a jump in the tangential stresses. Hence its gradient acts like a shear stress applied by the interface on the adjoining bulk liquid (Marangoni stress), and thereby generates flow or alters an existing one. Surface tension gradient-driven flows are known to also affect the evolution of growing fronts, and measurements of transport phenomena. The variation of surface tension along an interface may be due to the existence of a thermal gradient along the interface or perpendicular to it. In quite the same way as with gravity, in the former case we have instantaneous (forced) convection while in the latter, flow occurs past an instability threshold (defined by a critical value of the Marangoni number, Ma).

To qualitatively assess the role of gravity relative to forces induced by surface tension or its gradient, some dimensionless groups help. There is the (static) Bond number $\mathrm{Bo} = \mathrm{gl}^2 \Delta\rho/\sigma$, where $\Delta\rho$ is the density difference between two fluids (think of a drop of a fluid immersed in another liquid), σ is the interfacial tension and l is the space scale involved in the problem. In earth-based experiments, gravity overcomes surface tension effects when l is large enough. The natural extension of the Bond number is the modified (dynamic) Bond number $\mathrm{Bo}^* = \mathrm{gl}^2 \Delta\rho/\Delta\sigma$. Then to take into consideration the thermal gradients the Marangoni number is introduced, $\mathrm{Ma} = \mathrm{l}\Delta\sigma/\eta\kappa = \mathrm{PrRe}_\sigma$, where η is the dynamic viscosity ($\eta = \rho\nu$), κ is the thermal diffusivity, $\mathrm{Pr}=\nu/\kappa$ is the Prandtl number, and Re_σ denotes a surface Reynolds number, with the gradient-induced velocity scale $\mathrm{V}=\Delta\sigma/\eta$, and $\Delta\sigma = (d\sigma/dT)\Delta T$. Hence the Marangoni number is a Péclet number, $\mathrm{Pe}= \mathrm{V}\ \mathrm{l}/\kappa$.

Surface tension gradient-driven flows are, generally, laminar flows, with or without cellular patterns. For interface and bulk fluid moving tied together, only high enough Marangoni numbers lead to turbulent flows with high dissipation. For flows driven by capillary forces the Péclet/ Marangoni number appears in the heat equation. Then for high Prandtl number fluids the velocity field is slaved by the temperature while for low Prandtl number fluids the later slaves the former. Consequently, Reynolds (Kolmogorov) turbulence which is mostly inertial, appears as a different regime from turbulence in capillary-driven flows (with high Pr) where there is strong dissipation. Hence the interest in proceeding to high Péclet/Marangoni number flows to explore new features of turbulence with the numerical approach and benefiting from the opportunity offered by low-g experimentation.

For a clear assessment of the role of gravity we must proceed in every case to a detailed analysis. For illustration, following an argument developed long ago by S. Ostrach, I shall recall the role of gravity and how this influences the problem in a simple case. Consider Bénard convection, i.e., a liquid layer of horizontal extent d and vertical height l, open to air and subjected to a transverse thermal gradient, e.g. a temperature difference $\Delta T = T_{\mathrm{w}} - T_{\mathrm{c}}$, with T_{w} and T_{c} denoting warm and cold temperatures, respectively. Assume a steady convective motion, of components (u,v) along the (x,z) coordinates. To make dimensionless the thermohydrodynamic equations suitable scales are used. The following groups appear: Pr, $\mathrm{Re}_\sigma = \mathrm{V}^* \mathrm{l}/\nu$, Ma=Pr Re_σ, $\mathrm{Gr} = \mathrm{gl}^3 \Delta\rho/\nu^2$, and ar=d/l, where V^* is a reference velocity scale (an example has already been indicated). ar is an aspect-ratio whose inverse is called slenderness, particularly with liquid bridge, floating-zone configurations. $\Delta\rho/\rho_0 = \Delta\mathrm{T}$ (ρ_0 is a reference value). As the Rayleigh number is Ra=PrGr, minimization of buoyancy effects with common liquids is only realizable for layer depths below the millimeter scale. But then, wetting and contact-angle phenomena enter the problem in a non-negligible way.

At the open surface/interface the (simplest) dimensional tangential stress balance is

$$\eta \frac{\partial u}{\partial z} = -\frac{\partial \sigma}{\partial T} = -\frac{\partial \sigma}{\partial x} \frac{dT}{dx} \ .$$

It is here that the Marangoni number appears first as the driving mechanism for flow. If $[\mathrm{ar}^2 \mathrm{Re}_\sigma] \ll 1$ inertia can be neglected, and the effects of surface tension penetrate by viscosity all throughout the bulk of the fluid (viscous flow). With X=x/l, Z=z/d and $\mathrm{U}=\mathrm{u}/\mathrm{V}^*$ we have

$$\frac{\partial U}{\partial Z} = -\left[\frac{\partial \sigma}{\partial T} \frac{\Delta T}{\eta \mathrm{V}^*}\right] \frac{\mathrm{d}}{\mathrm{l}} \frac{d\Theta}{dX} \ ,$$

where

$$\mathrm{V}^* = \left[\frac{\partial \sigma}{\partial T} \frac{\Delta T}{\eta}\right] \frac{\mathrm{d}}{\mathrm{l}}$$

and $\Theta = (T - T_{\mathrm{c}})/\Delta T$. In the opposite situation, $[\mathrm{ar}^2 \mathrm{Re}_\sigma] \gg 1$, the effect of surface tension remains near the interface in a boundary layer, δ, much thinner than the liquid depth, $\delta \gg \mathrm{d}$ (boundary layer flow). Then δ rather than d is the scale for vertical displacements. $\delta/\mathrm{l} = \mathrm{ar}[\mathrm{Re}_\sigma]^{-1/2}$. Hence

$$\mathrm{V}^* = \left[\nu \left(\frac{\partial \sigma}{\partial T}\right)^2 \frac{(\Delta T)^2}{\eta^2 \mathrm{l}}\right]^{1/3} .$$

Note that $[\mathrm{ar}^2 \mathrm{Re}_\sigma] \ll 1$ does not imply that the Reynolds number is small. Indeed, Re_σ can be greater than unity provided the aspect-ratio is small enough.

$[ar^2 Re_\sigma] \gg 1$ does not preclude the flow from being laminar for large enough aspect ratio.

In the case $[ar^2 Re_\sigma] \ll 1$, the influence of buoyancy is provided by $Bo^* = ar\, Gr/Re_\sigma$. In the other case, for boundary layer flows, gravity appears through the parameter $\hat{Bo} = Bo^*/(ar^2 Re)^{1/3}$. In both cases the relevant Bond number is Bo^* and not Bo. Thus in a sufficiently low-g environment, surface tension gradient-driven flows will be significant. On the other hand, although buoyancy is negligible in low-g facilities, we not necessarily must go to space as clearly the expressions of Bo^* and $\hat{Bo}$ yield clues for possible reduction of gravity effects in earth-based experimentation.

Since the generation of surface tension inhomogeneity is related to the imposition of thermal gradients, in earth-based experiments buoyancy-induced flows can also occur simultaneously with surface tension gradient-driven flows. Then the relative magnitude of surface tension gradient and buoyancy forces is given by the ratio Ma/Ra which is the inverse of Bo^*.

Surface deformation and surface curvature appear in the complete balances of tangential and normal stresses, respectively. The former balance yields a capillary number $Ca = \eta V^*/\sigma = ar\,(Bo/Bo^*)$. For boundary layer flows the Weber number appears from the same normal balance, $We = l\, V^{*2} \delta\rho/\sigma$. Consequently, both Ca and We account, in their appropriate flow, for the relevance of surface deformation effects in a surface tension gradient-driven flow. Strictly speaking the only new parameter appearing is the contact angle at a boundary. Finally, the Galileo number $Ga = gl^3/\nu^2$, or GaPr, can be used to verify if in the pressence of gravity, buoyancy, and surface deformability, the Boussinesq approximation is respected ($GaPr \gg 1$, $Ra \ll GaPr$).

In flows with very low Re, inertia can be neglected in the flow (e.g. Stokes creeping flow) of very viscous fluids like the silicone oils, where however possible high values of Ma exist. Other conclusions are far reaching. For instance, in earth-based experiments, for sufficiently large Grashof numbers the flow would be induced by buoyancy and could be turbulent. If gravity were reduced several orders of magnitude the buoyancy might still predominate over the surface tension gradient-driven flow but the convection could be laminar. In these situations the free surface can be said to be passive with respect to the flow in the sense that the motion is generated by buoyancy. However, as gravity is further reduced the surface tension effects become dominant and will induce the primary motion, with the surface becoming an active factor in the convection process.

The specific dynamics of (passive or active) drops or bubbles immersed in another fluid is of our particular interest. Liquids used to power space vehicles or used in Shuttle cooling systems, upon evaporation lead to bubbles, hence the need of capillary forces, and eventually surface tension gradient-driven flows, to manage bubbles, isolated or in clusters following aggregation and possible coalescence in low-g. Boiling leads to formation of relatively small bubbles that in earth-based experiments float, move up but distribute all over the container while in low-g conditions bubbles tend to coalesce thus forming bigger ones that drastically affect the overall process. Sedimentation of particles or drops is also drastically altered by g.

When one phase is dispersed in another as drops or bubbles the approach of an eddy causing a surface deformation and a local reduction in surface tension can cause violent motion which in turn creates other eddies. The coalescence of drops with either different solutes or different solute concentrations, or the introduction of jets of solvent can cause similar sudden movements, altering mass transfer. Sedimentation generally affects these processes.

Instantaneous motion caused by purely capillary forces provokes that the surrounding fluid moves towards the cooler region while the drop or bubble (radius R) moves towards the warmer one. It is controlled by the Péclet/ Marangoni number, Ma=V^*R/η, where $V^* = [(\partial\sigma/\partial T)\beta_\infty/\eta]R$ depends on β_∞, the gradient at infinity. Note that when surface tension is not constant a spherical drop or bubble tends to become deformed due to the normal stress (im)balance at the interface. Then the Ca or We numbers control this effect and gravity has no role. Indeed at least in capillary-driven creeping flow, the curvature change is of order $\Delta\sigma/\sigma$, i.e. of order (Bo/Bo*), which is the capillary number. In such flows, when Ca is much smaller than unity, deformation is not really relevant.

High Péclet/Marangoni number flows, thus apparently predominant surface tension gradient-driven flows in earth-based experiments are different from analogous low-g flows. For instance, in an externally imposed thermal gradient, for the instantaneous surface tension gradient-driven flow of a drop or a bubble to be steady in the pressence of gravity, a non-zero hydrodynamic force is needed which is quite a different case from that in low-g conditions. Hence earth-based and low-g (high) Marangoni flows of drops or bubbles are different. Indeed, gravity-induced motions alter the temperature field in the fluid surrounding the drop or bubble thus changing the Marangoni forces acting on it.

An interesting point to note is that if a uniform velocity field, V_∞, is imposed at infinity in an otherwise homogeneneus medium in temperature or composition, drop activity like suitable surface reactivity with or without adsorption/desorption processes, internal heating, or high deformability, may produce a surface tension gradient-driven instability of the initial state, hence, self-propulsion enough to overcome viscous drag, and finally thrust with multiple steady velocity values for zero hydrodynamic force. Then, low level gravity values provide a way of discriminating among these different states. If the surrouding medium is not homogeneous and a thermal gradient exists on top of the uniform flow, V_∞, then multiplicity of steady states of flow is also possible for an active drop, and an interesting problem is the expected competition between such inhomogenity and any residual acceleration or g-jitter.

The surface tension variation acts dramatically on interphase transport processes, evaporation, adsorption/desorption processes, drop and bubble migration, etc. Indeed, when a highly surface-active substance is strongly adsorbed at an interface the resulting surface film tends to alter transport rates either hydrodynamically, by locally damping or enhancing eddies and ripples, or by causing stagnation over a considerable portion of the interface, or if the molecules form a highly packed film by imposing an (energy) barrier to the passage of other molecules across the interface.

Even minute traces of surfactant are known to produce dramatic effects at both quantitative and qualitative levels. For instance, theory and earth-based experiments indicate that transport processes, surface chemical reaction processes, catalysis phenomena,... can be altered by either buoyancy-driven or surface tension gradient-driven flows, or both appearing together, due to even mere ppm changes in composition. For instance, when a surfactant changes the tension, interfacial instability can appear during transfer out of the phase of higher kinematic viscosity or lower solute diffusivity. Amplification of interfacial convective motions can also be promoted by large $\Delta\sigma$ changes with composition, large thermal gradients, and large differences of kinematic viscosity and diffusivity between phases.

3. Liquid Layers

A natural candidate for space experimentation was the study of the relative strength of buoyancy- versus surface tension gradient-driven convection in heated liquid layers with a surface open to the ambient air, foreseing the suppression of the former. The first space mission carried indeed one such item albeit too a qualitative experiment. Since then a number of experiments have been carried out using different low-g facilities. S. Ostrach, J. Cl. Legros, J. Koster and others have explored in low-g facilities the onset and development of steady and time-dependent thermo/solutal, and double diffusion/thermosolutal, convective phenomena/convective instabilities occurring in single and multilayer systems (up to three liquid layers with two inner interfaces -hence possibility of thermal or mechanical coupling; liquids used had different viscosity and diffusivity ratios), with or without other constraints like varying aspect-ratio, rotation, electric or magnetic fields, etc. The multi-layer system is thought to mimmic features of an encapsulated melt in another immiscible liquid. The experimental set-ups and diagnostics have reached a high level of sophistication, comparable to similar lab methodologies in earth. Neededless to say, much has been learned about steady and oscillatory surface tension gradient-driven flows and instabilities with the related experiments carried out with liquid bridges subjected to thermal gradients.

A number of findings, both theoretical and experimental have been reported: various steady and oscillatory instabilities (viscous and boundary layer flows, with or without interface deformation, with varying liquid depths and aspect-ratios, with varying transport direction, with varying relative liquid properties and varying b.c.). On the other hand other recent discoveries also are worth noting: interfacial solitary waves and periodic wave trains with solitonic features excited and sustained by Marangoni stresses.

4. Drops and Bubbles

I shall discuss in some detail the case of drops or bubbles for it is illustrative of one and many problems and quite a number of recent earth-based and low-g experiments have been conducted. A liquid drop is a model for many natural systems as disparate as planets, nuclei, etc. It is also a g-detector. Drop dynamics provides the test ground for theoretical analysis (e.g. matched asymptotic expansions), experimental techniques and numerical methods (which on ocassion perform better than real experiments). The study of drops also allows to sequentially proceed from simple to complex. Clearly, drop dynamics is not just of mere fluid physics interest. It is of relevance in atmospheric sciences, chemical engineering, materials science, etc. Understanding some aspects of drop dynamics has led to the most spectacular albeit simple technological spin-off of microgravity space research. Let us see how the hydrodynamic force, generally drag on a drop has been estimated since the pioneering work of Newton and Stokes.

From Newton's experiments in 1710, and later observations, the magnitude of the drag force on solid drops/spheres in steady motion of a viscous fluid was given as

$$F_D = 0.22\pi R^2 \rho U^2 \tag{1}$$

where U is the relative velocity between particle and fluid, R is the particle radius, and ρ is the fluid density. This relation is for "large" values of U , for which inertial effects are important.

Stokes, in 1850, suggested that at very low relative velocities the inertial effects are so small that they can be omitted from the Navier-Stokes equations. Under these conditions, the total drag on a sphere is

$$F_D = 6\pi\eta RU \tag{2}$$

Oseen pointed out that at a great distance from the sphere the inertia terms become more important than the viscous terms, and suggested a possible improvement of the Stokes law (2) by taking inertial-force terms partly into consideration. His drag force is

$$F_D = 6\pi\eta RU\left(1 + \frac{3}{16}\mathrm{Re}\right) \tag{3}$$

where here $\mathrm{Re} = (2R)U/\nu$. Today we know that neither Stokes' nor Oseen's laws are uniformly valid, and the latter is not really an improvement of the former. Rather Stokes' analysis is valid in a small enough neighborhood of the sphere (creeping flow at zero Reynolds number) and Oseen's analysis though valid for matching at the flow velocity values far from the sphere is not valid when approaching it. The Oseen approximation although incorporates inertial terms is still a linear theory and as Stokes' is a steady state theory. Several authors did attempt, not always successfully, an improvement upon the Oseen and Stokes analyses.

The mathematical problems solved by Stokes and Oseen come from different approximations to the Navier-Stokes equations,

$$\frac{\partial \mathbf{v}}{\partial t} + \mathbf{v}\cdot\nabla\mathbf{v} = -\frac{1}{\rho}\nabla p + \nu\nabla^2\mathbf{v} \tag{4}$$

together with appropiate initial and boundary conditions. Using the vorticity $\boldsymbol{\omega} = \mathrm{rot}\mathbf{v}$, (4) becomes

$$\frac{\partial}{\partial t}\boldsymbol{\omega} + \mathbf{v} \cdot \nabla\boldsymbol{\omega} = \boldsymbol{\omega} \cdot \nabla\mathbf{v} + \nu\nabla^2\boldsymbol{\omega} \tag{5}$$

where $\boldsymbol{\omega} \cdot \nabla\mathbf{v}$ is the source of vorticity in the balance equation (yields stretching of vortex lines with consequent increase of vorticity), and ν is its diffusivity. For the axisymmetric motion of a sphere there always is a stream function, i.e. a velocity potential $\Psi(r, \Theta, \Phi)$ which is constant along a streamline (Φ is not a relevant variable in the axisymmetric case). The existence of Ψ is neither limited to steady motions nor to ideal, viscous-free flows. The existence of Ψ for axisymmetric motions depends entirely on the *kinematical* assumption of incompressibility, hence exists for both ideal and viscous flows for those two motions differ only in their *dynamical* properties.

In the axisymmetric case with (r, Θ, Φ) coordinates we have:

$$v_r = -\frac{1}{r^2 \sin\Theta}$$
$$v_\Theta = \frac{1}{r\sin\Theta}\frac{\partial\Psi}{\partial r}$$
$$v_\Phi = 0$$

Then (4) or (5) reduce to

$$\frac{\partial}{\partial t}(E^2\Psi) + \frac{1}{r^2\sin\Theta}\frac{\partial(\Psi, E^2\Psi)}{\partial(r,\Theta)} - 2\frac{E^2\Psi}{r^2\sin^2\Theta}\left(\frac{\partial\Psi}{\partial r}\cos\Theta - \frac{1}{r}\frac{\partial\Psi}{\partial\Theta}\sin\Theta\right) = \nu E^4\Psi \tag{6}$$

with

$$\frac{\partial(f,g)}{\partial(r,\Theta)} = \begin{pmatrix} \frac{\partial f}{\partial r} & \frac{\partial f}{\partial\Theta} \\ \frac{\partial g}{\partial r} & \frac{\partial g}{\partial\Theta} \end{pmatrix}$$

and

$$E^2 = \frac{\partial^2}{\partial r^2} + \frac{\sin\Theta}{r^2}\frac{\partial}{\partial\Theta}\left(\frac{1}{\sin\Theta}\frac{\partial}{\partial\Theta}\right)$$

Consequently, the Stokes creeping flow aproximation is the drastic reduction of (6) to

$$E^4\Psi = 0 \tag{7}$$

Equation (7) with the appropriate, non-penetration, no-slip/stick boundary condition and suitable asymptotic behavior for large r, yields

$$\Psi(r,\Theta) = (-\frac{1}{4}U\frac{R^3}{r} + \frac{3}{4}URr - \frac{1}{2}Ur^2)\sin^2\Theta \tag{8}$$

hence the hydrodynamic drag force (2). Note that from (8) follows that the disturbance of the sphere extends to infinity as $1/r$, and the presence of a boundary or another drop can modify the flow appreciably even when placed at a distance of many diameters from the drop.

Equation (7) [see also (5)] shows that vorticity obeys the Laplace equation $\nabla^2 \omega = 0$. The flow is due solely to steady diffusion of vorticity to infinity in all directions; the drop is a source of vorticity due to the no-slip boundary condition at its surface. Vorticity decreases as $1/r^2$ as originated from a dipolar source that produces for each component of ω equal positive and negative quantities at the surface of the sphere.

Oseen's apparent improvement of Stokes creeping flow approximation was a consideration of inertia terms linearized around the far field velocity hence a heuristic extension of the Stokes problem. It was not until the year 1957 that Stokes' and Oseen's results were properly put in context and generalized.

In 1911-12 another line of thought had developped. Rybczynski and Hadamard, independently, solved the Stokes problem for a liquid drop with flows outside and inside. Their extension of (2) is

$$F_D = 6\pi\eta_0 U R \left(1 + \frac{2\mu}{3(1+\mu)}\right) \tag{9}$$

with $\mu = \eta_i/\eta_o$ ($\eta_i = \eta_{\text{drop}}$). Clearly, the limit η_i going to infinity yields back Stokes law (2), while $\eta_o \gg \eta_i$ yields the corresponding law for a bubble, with the factor 6 being replaced by 4 in (2). In 1957 appeared a remarkable paper by Proudman and Pearson where earlier ideas proposed by Kaplun on matched asymptotic expansions were applied to the flow dynamics around the sphere. The basic concept was to consider Stokes solution as a local (or inner) solution of the problem and Oseen's as a regular (or outer) solution rather than considering Oseen's as an improvement upon Stokes'. The local (inner) solution was assumed to be valid in a spherical region of radius 1/Re around the sphere while the outer solution was valid from infinity down to the 1/Re neighborhood. In the overlapping zone both solutions were accepted as valid, hence the need for appropriately matching them. Proudman and Pearson found that for nonvanishing although low Reynolds number flows (Re≪1) the hydrodynamic drag on the sphere is

$$F_D = 6\pi\eta U R \left(1 + \frac{3}{16}\mathrm{Re} + \mathrm{Re}^2 + \frac{9}{160}\mathrm{Re}^2 \ln \mathrm{Re} + \ldots\right) \tag{10}$$

which shows the non-analytic form of the expansion. Extensions of the Proudman-Pearson paper were made by different people. For instance, Acrivos and Taylor using the Peclet number $2RU/\kappa$ considered the convective terms in the heat equation while using the Stokes flow in (6). They obtained a Peclet number expansion of the (convective) Nusselt number,

$$\mathrm{Nu} = 2 + \frac{1}{2}\mathrm{Pe} + 0.03404\mathrm{Pe}^2 + \frac{1}{4}\mathrm{Pe}^2 \ln \mathrm{Pe} + \frac{1}{16}\mathrm{Pe}^3 \ln \mathrm{Pe} + \ldots \tag{11}$$

which also shows a non-analytic form.

As already noted for flows driven by capillary forces in high Prandtl number fluids what indeed matters is the Peclet number and the Stokes approximation is quite a valid starting point (at Re=0, Pe_0 the velocity field is slaved by the temperature). Subramanian used matched asymptotic expansions to obtain the hydrodynamic force on a drop including convective terms in the heat equation while maintaining the Stokes approximation for the velocity field. His series expansion in terms of the Péclet/Marangoni number and subsequent improvement by his collaborator Merritt did not show any logarithmic term. This was due to the way the outer solution was treated, which was different from that of Acrivos and Taylor.

Taylor and Acrivos also showed that indeed for the deformation of the sphere (a bubble) to be noticeable the capillary number must be non-negligible. For a slightly deformable bubble they obtained the radius in term of the angle Θ:

$$R(\Theta) = R\left[1 - \frac{5}{96}\mathrm{ReCa}(3\cos^2\Theta - 1)\right] \tag{12}$$

for Re≪1 and Ca≪1. Note that at Re=0 (i.e. in the creeping flow approximation) a drop or a bubble remain spherical irrespective of the low or high value of the (constant) surface tension. Eq. (12) shows that a drop or a bubble will be deformed from the spherical shape when inertial effects are taken into acount. However, deformation may be relevant even when inertial effects are ignored if as earlier noted the surface tension σ is not constant. If the variation in interfacial tension $\Delta\sigma$ over the spherical surface is small compared to the value of the interfacial tension, the capillary number will be small, and the drop or bubble may be asummed spherical with little error.

Different authors have also extended the theory to account for various non-Newtonian fluid properties like first and second normal stress coefficients in viscoelastic fluids or multi-time scales (Oldroyd fluids) and power laws.

5. Drops and Bubbles. Low-G Experimentation, and Precursors

Young, Goldstein and Block were the first to realize the possibility of levitating a drop or a bubble by means of capillary forces (Marangoni stresses). They showed that a drop or a bubble placed in a temperature gradient instantaneously tends to move towards the hotter point. This is the motion of the drop relative to the flow induced along its surface by the lowering of surface tension at its leading pole (hotter than the rear pole). The Marangoni stresses not only help overcoming drag but even positive (bubble) or negative (drop) buoyancy, hence levitation for a sufficiently high temperature gradient. Using the Stokes-Rybczynski-Hadamard approximation they also computed the terminal velocity of a drop or a bubble in the field of gravity (Re≪1, Pe≪1), and experimentally checked the theoretical prediction within reasonable accuracy (within 20%) with an experiment using rising bubbles in a liquid layer heated from below (diameters $2R=10^{-3}-22\cdot10^{-3}$cm; dT/dz=10-90 K/cm). Later on Bratukhin, Evdokimova, Pschenichnikov, Briskman and Zuev did a similar experiment with rising bubbles in a laterally heated vertical liquid layer. They experimented using neutrally buoyant liquid water at 4°C . For the Young, Goldstein and Block problem the balance between capillary, buoyancy and hydrodynamic forces is

$$F_\sigma + F_g = AU \tag{13}$$

with the (Marangoni) capillary force

$$F_\sigma = \frac{4\pi R^2}{(1+\mu)(1+\delta)} \frac{d\sigma}{dT} (\nabla T)_\infty \tag{14}$$

and the buoyancy force

$$F_g = \frac{4}{3}\pi R^3 g(\rho_i - \rho_o) \tag{15}$$

As

$$A = 4\pi\eta_o \frac{1+\frac{3}{2}\mu}{1+\mu} R > 0 \tag{16}$$

the hydrodynamic force AU represents drag. $\delta = \lambda_i/\lambda_o$ is the ratio of thermal conductivities (drop to surrounding fluid). Clearly, AU is an extension of both the Stokes law ($\eta_i \to \infty, \mu \to \infty$) and the Rybzcynski-Hadamard law ($[\nabla T]_\infty = 0$).

In their experiment, Young et al. used a liquid bridge. An improvement eliminating possible capillary convection at the open sides was carried out by Hardy. He used a closed cavity with silicone oil and air bubbles ($2R$=5-25$\cdot 10^{-3}$ cm; dT/dz=40-140 K/cm). Hardy clearly noticed the role of contamination at the surface of the bubble. Further improvement came with an experiment by Merritt and Subramanian. Experimentalists started using drops rather than bubbles. Barton and Subramanian used neutrally buoyant drops (2R=20-600mm, dT/dz=2.4 K/mm). Recent earth-based and low-g (TEXUS, D2) work by Braun and colleagues on thermocapillary migration of drops provided the most accurate verification of the Young, Goldstein and Block prediction. They used flows with Péclet/ Marangoni numbers in the range $10^{-5} - 10^{-6}$, but with the non standard surface tension gradient behavior i.e. the surface tension increasing with the increase of temperature (2 butoxyethanol-water mixture with liquid/liquid phase separation at 61.14 °C on the lower branch of the closed miscibility gap; 2R=11mm, dT/dz=36.9K/m, $d\sigma/dT > 0$).

Worth mentioning are also experiments carried out by Neuhaus and Feuerbacher who found disagreement with Young, Goldstein and Block prediction but who invoked that the solution of the difficulty was in augmenting the theory with a surface dilational viscosity, in agreement with a conjecture already put forward long ago by Boussinesq and later on by Lucassen.

Another interesting experiment was made by Hähnel, Delitzsch and Eckelmann using a Plateau tank. They observed the motion of two water drops in butyl benzoate (2R=1.7 mm and 0.6 mm; dT/dz=2.07 K/cm). The curious thing about their experiment is that at first the bigger drop moved faster than the smaller one (first 479 s) and then the smaller drop overtook the larger which (at 689 s) lagged well behind the other. Their Péclet number was of order unity.

Worth mentioning are also the pionneering earth-based and low-g works by H. U. Walter and J. Siekman who addressed pertinent basic questions about single drop or bubble motions, drop-drop or bubble-bubble interactions, drop/bubble-wall interactions, drop/bubble interaction with melting or solidification fronts, etc. For instance, by focusing attention on the stability of multicomponent mixtures under low-g conditions and the possibilities of man-made composites by liquid phase sintering, etc. Walter was led to explore basic issues that, recently, other authors have come to address too. These and a plethora of related studies are still demanding further theoretical work and feedback from theory to go beyond 'weak' forces. Such is also the case of the following items:

- the early work on the motion of one or two interacting bubbles (as a by-product degassing) in liquids with the aid of a (Hele-Shaw cell) 'zero g simulator' and some other more recent experiments by J. Siekmann and collaborators (theory was took care of some of the experimental findings but certainly not of their high Marangoni cases),
- the work by J. Straub and colleagues on heat transfer with drops (including problems related to boiling e.g. nucleate boiling versus film boling),
- the drop tower experiments on thermocapillary bubble migration by H. J. Rath and colleagues (although some results exist for moderately high Marangoni and Reynolds number flows still the theory is far from complete),
- the recent space experiment on bubble and drop thermocapillary migration, with single items or a pair of apparently interacting items of unequal diameter, by Subramanian, demands theoretical understanding,
- the recent work by Viviani on thermocapillary migration of air bubbles in a liquid (n-heptanol aqueous solution) with a surface tension passing from decaying to growing with increasing temperature, hence the motion around the minimum. The bubbles failed to stop at the minimum. Does the location of the minimum of surface tension with temperature, obtained from equilibrium data, remain unchanged with the flow? Viviani's work follows the track of earlier studies by G. Petré, J. Cl. Legros and coleagues with the same aqueous alcohol solution,
- the seemingly break-through experiment on protein crystallisation inside a drop-shaped chamber by T. Molenkamp, L. P. B. M. Janssen and J. Drenth has shown the positive and negative role played by various convective motions (not all well quantified and understood) inside the drop-shaped chamber, etc.

In conclusion of this section I recall the great industrial spin-off of low-g work that came from the apparent failure of a low-g surface tension gradient-driven experiment in space. This was followed by a quick grasp of the reason for the failure that permitted a significant improvement of an earth-based technology. As L. Ratke and others have shown, surface tension gradient-driven drop migration can be used together with sedimentation and Stokes drag to take advantage of g for the earth-based continuous strip casting process formation of light-heavy monotectic alloys of otherwise immiscible materials (e.g. Al-Pb, Al-Bi, Al-Si-Pb,

Al-Si-Bi) during solidification. Recent low-g (first TEXUS then D2) experiments carried out by Prinz and Romero have definitely established this possibility.

6. Active Drops or Bubbles

The work on drops and bubbles so far recalled, refers to "passive" drops. Now I summarize some recent findings about "active" drops, obtained in collaboration with Yu. S. Ryazantsev, A. Ye. Rednikov and V. N. Kurdyumov.

By an "active" drop or bubble I mean a drop or a bubble with internal volume heat sources, with a surface where chemical reactions may occur, etc. Take e.g. a drop at rest in a homogeneous medium and assume that there is (uniform) internal heat generation or a surface chemical reaction. The state of rest can be unstable. Consider, for instance, the latter case with given uniform composition far off the drop. A composition fluctuation at the surface of the drop promotes the Marangoni effect which yields flows inside and outside the drop which can be sustained if the Marangoni effect is strong enough, i.e. past and instability threshold. Indeed, as the drop moves the flow brings to the leading pole the higher solute at the surface, it makes the concentration far off the drop larger than that in its vicinity; hence, the motion sustained past a certain threshold. Alternatively, if a velocity fluctuation tending to move the drop in certain direction spontaneously occurs, it breaks the initial spherical symmetry in composition, hence Marangoni stresses which in turn can help sustaining the velocity fluctuation past an instability threshold. If the initial state is that of a uniform, constant drop velocity or there is an externally imposed temperature or composition gradient, then instability is also possible leading to a different drop motion. These are not the only possible instabilities.

To illustrate how we have proceeded along the path set by earlier mentioned authors dealing with passive drops now I consider the case of a spherical drop moving with constant velocity in a temperature and composition homogeneous, infinitely extended surrounding fluid. Both the inner and outer fluids are taken immiscible. The surface tension is assumed to vary linearly with temperature. The outer fluid is assumed to have a uniform concentration of a solute which is allowed to react exo- or endothermally at the surface of the drop. Far off the drop both temperature and the concentration of solute are constants, hence, I insist, no external gradient. Stefan flow is negligible (convective flow of the reacting components in a direction normal to the surface where the reaction is taking place; it is generally a small effect for most chemical reactions and is normally important only in the presence of strong ablation or condensation).

In the low Reynolds and Péclet numbers approximation with however MaPe≈1, in dimensionless form the steady form of (6) becomes

$$\frac{\mathrm{Re}}{\nu^* r^2}\left[\frac{\partial(\psi_i, E^2\psi_i)}{\partial(r,\delta)} + 2E^2 L_r \psi_i\right] = E^4\psi_i \tag{17}$$

which is nonlinear and has to be considered together with the corresponding heat and mass diffusion equations, and appropriate boundary conditions. Here

Re is defined with the constant drop velocity, $\delta = \cos\Theta$, ν^* refers to (kinematic) viscosity (i=1 outer fluid,$\nu^* = 1$; i=2 drop, $\nu^* = \nu$), and L_r corresponds to the operator appearing in the third term of (6). The linear solution of (17) yields the hydrodynamic force

$$F = -4\pi\eta_1 rAU \tag{18}$$

If A is negative we have drag while if A is positive there is thrust, hence self-propulsion, and possible autonomous motion of the drop in a medium originally uniform. We have

$$\mathrm{A} = -[1 + (3/2)\mu + 3\mathrm{m}]/(1 + \mu + \mathrm{m}) \tag{19}$$

with $\mu = \eta_1/\eta_2$, and m a suitably scaled Marangoni number accounting for a balance between $d\sigma/dT$, the chemical reaction rate, and (viscous and heat) dissipation. We see that at $m = m_1 = -1/3 - \mu/2$ we have A=0, whereas at $m = m_2 = -1 - \mu$, A diverges to infinity. The study of both cases demands nonlinear analysis. Before referring to this we note that for A=3, the Marangoni effect combined with the chemical reaction yields the possibility of thrust, while for A=-1 we have drag, as well as for A= − 3/2, and even more, for A = − 3 there is enhanced drag due to the appearance of secondary backflow around the drop.

The weakly nonlinear result for ($m-m_1$) ($> 0, \ll 1$) provides the value of the hydrodynamic force

$$F = -4\pi\eta_1 R[A + B(RU/\nu_1)]U \tag{20}$$

with

$$B = -\frac{1}{4}A^2 + \frac{1}{2}\mathrm{m\,Pr}\frac{6 + 3L - AL}{1 + \mu + \mathrm{m}} \tag{21}$$

where here Pr= ν_1/κ_1 and L=κ_1/D are the Prandtl and the (inverse) Lewis number of the surrounding medium.

A representation of F versus U straightforwardly shows that for ($m-m_1$) $<$ 0 (A $>$ 0) there are three possible values of U for zero hydrodynamic force, hence possible autonomous motion of the drop. The addition of an external force field like buoyancy, or an externally imposed thermal gradient like in the Young et al. experiment, provides the possibility of three genuinely different non-zero velocities for a given force field strength. As already mentioned we have multiplicity of steady states of motion which are not all actually realizable. An estimate of the autonomous motion velocity yields about U≈0.5 R m/s with threshold at ΔT=0.02 K (respectively, ΔT=0.2K) for R=1 mm (respectively, R=0.1 mm). The other case yields that for $m < m_2$, there are three possible values of the hydrodynamic force for zero velocity, hence three possible coexisting levitation levels of which one is not stable. Levitation or motion is a consequence of the flow, which is a genuine nonlinear effect, and not a consequence of some external thermal gradient.

The most recent result obtained corresponds to the action of a time-varying gravity field as it occurs in space (g-jitter) and in some earth-based experiments. For some of the cases discussed earlier we have derived a time-dependent, weakly nonlinear vector-form equation for the velocity of the drop. This equation can be used not only to find the stationary regimes and analyze their stability, but also to consider time varying motions. In the simplified case of a small amplitude buoyancy force changing sinusoidally with time the result found is that an active drop capable of autonomous motion actually tends to move in a direction orthogonal to the direction of the time-varying force. Thus an active drop shows fascinating new features worth exploring with low-g levels.

Acknowledgments

The author is grateful to Prof. Yu. S. Ryazantsev, Prof. H. Linde, Prof. A. Sanfeld, Dr. H.U. Walter, Dr. A Ye. Rednikov and Dr. V.N. Kurdyumov for many enlightening discussions and fruitful collaboration over the past three years. This research has been sponsored by DGICYT (Spain) under Grant PB 93-81, by the European Union under Network Grant ERBCHRXCT 93-106 and by the Fundacin "Ramn Areces" (Spain).

On the next page follows a list of just a few references, needed/valuable for historical or technical reasons, with exclusion of ESA and related publications. We encourage the reader to get the ESA and ESA-coordinated publications as, presently, there is a wealth of earth-based studies and low-g results worth reading, several of which have been referred to in the present text [1].

References

1. A. Acrivos and T. D. Taylor, *Phys. Fluids* **5** (1962) 387-394.
2. K.D. Barton. and R.S. Subramanian, *J. Coll. Interf. Sci.* **133** (1989) 211-222.
3. Yu. Bratukhin, O.A. Evdokimova, A.F. Pschenichnikov, *Izvestia Akad. Nauk USSR* (1979), n. 5
4. Yu.K. Bratukhin, V.A. Briskman, A.L. Zuev, A.F. Pschenichnikov and V. Ya. Rivkind, in *Hydrodynamics and Heat-Mass Transfer in Weightlessness*, Nauka, Moscow (1982)
5. M. Hahnel, V. Delitzsch and H. Eckelmann, *Phys. Fluids* **A1** (1989) 1460-1466.
6. J. Happel and H. Brenner, *Low Reynolds Number Hydrodynamics*, Prentice-Hall, Englewood Cliffs, N.J. (1965).
7. S.C. Hardy, *J. Coll. Interf. Sci.* **69** (1979) 157-162.
8. S. Kaplun, *ZAMP* **5** (1954) 111-135.
9. S. Kaplun and P.A. Lagerstrom, *J. Math. Mech.* **6** (1957) 585-593.
10. P.A. Lagerstrom and J.D. Cole, *J. Rat. Mech. Anal.* **4** (1955) 817-882.
11. B.G. Levich, *Physicochemical Hydrodynamics*, Prentice-Hall, Englewood Cliffs, N.J. (1965).
12. B.G. Levich and V.S. Krylov, *Ann.Revs.Fluid Mech.* **1** (1969) 239-316 and references therein
13. R.M. Merritt and R.S. Subramanian, *J. Coll. Interf. Sci.* **125** (1988) 333-339.
14. S. Ostrach, *Ann. Revs. Fluid Mech.* **14** (1982) 313-345, and references therein.

[1] The interested reader may contact the editors for a list of books, conference proceedings or reports published by ESA on fluid sciences in space. Eds note.

15. I. Proudman and J. R. A. Pearson, *J. Fluid Mech.* **2** (1957) 237-262.
16. A.Ye. Rednikov, Yu.S. Ryazantsev and M.G. Velarde, *Phys. Fluids* **6** (1994) 451-468.
17. A.Ye. Rednikov, Yu.S. Ryazantsev and M.G. Velarde, *J. Coll. Interf. Sci.* **164** (1994) 168-180
18. A.Ye. Rednikov, Yu.S. Ryazantsev and M.G. Velarde, *Int. J. Heat. Mass Transf.* **37**, Supp. 1 (1994) 361-374
19. A.Ye. Rednikov, Yu. S. Ryazantsev and M.G. Velarde, *J. Non-Equilib. Thermodyn.* **19** (1994) 95-113
20. A.Ye. Rednikov, V.N. Kurdyumov, Yu. S. Ryazantsev and M.G. Velarde, *Phys. Fluids* **7** (1995), to appear
21. L.E. Scriven and C.V. Sternling, *Nature* **187** (1960) 186-188.
22. R.S. Subramanian, in *Transport Processes in Bubbles, Drops and Particles* (editors: R.P. Chhabra and D. de Kee), Hemisphere, N.Y. (1992), and references therein.
23. R.S. Subramanian, *AIChE. J*, **27** (1981) 646-654.
24. T.D. Taylor and A. Acrivos, *J. Fluid. Mech.* **18** (1964) 466-476.
25. M.G. Velarde and C. Normand, *Sci. Amer.* **243** (1980) 92-108.
26. G. Wozniak, J. Siekmann and J. Srulijes, *Z. Flugwiss. Weltraumforsch.* **12** (1988) 137-144, and references therein.
27. N.O. Young, J. S. Goldstein and M.J. Block, *J. Fluid Mech.* **6** (1959) 350-357.

Pure Thermocapillary Convection in a Multilayer System: First Results from the IML-2 Mission

P. Géoris and J.C. Legros

Microgravity Research Center - Université Libre de Bruxelles
Service de Chimie Physique EP CP 165 50 Av. F.D. Roosevelt, 1050 Brussels, Belgium.

Abstract

The first experimental results on pure thermocapillary convection in a multilayer system are presented. The Marangoni-Bénard instability in a symmetrical three layer system has been investigated during a six hours experiment which flew on the IML-2 mission June 1994. The system studied consisted of a silicone layer sandwiched between two identical Fluorinert layers. The thermal gradient applied perpendicular to the two free interfaces was up to five times the critical value for the onset of the Marangoni-Bénard instability in this configuration. The preliminary analysis of the experimental data, video recording and thermistor values, indicates that most of the scientific objectives have been met. The mechanical stability of a three layer system in microgravity has been achieved taking advantage of the differential wetting properties of the liquids. Valuable informations on the management of liquid/liquid interfaces in space have been gained. The critical temperature difference for the onset of convection has been estimated and is compatible with the results of a linear stability analysis. Digital image processing of some video sequences confirms the prediction of the numerical simulations showing that in this case the hot interface is driving the convective flow. The absolute amplitude of the velocity of the free interface is in good agreement with values obtained numerically.

1. Introduction

Thermocapillary and thermogravitational convection in multilayer systems has recently gained some interest since it is related to natural and industrial processes. For the gravitational case, most of the investigations were motivated by the modelling of earth mantle convection and the possibilities of thermal and mechanical coupling [1, 2, 3]. For pure thermocapillary flows, the most obvious application is the modelling of encapsulated floating zone. This technique has been suggested for the crystal growth of GaAs. Coating the melted GaAs with B_2O_3 should help to prevent the evaporation of the volatile As and thus to control the stoichiometry of the crystal [4].

Besides the crystal grower point of view, the fundamental features of convection in two or more superposed layers heated parallel or perpendicular to the free interfaces are still the subject of several recent papers [5][6].

The mechanism leading to the Marangoni-Bénard instability considering one layer as Pearson [7] or the two phases adjacent to the interfaces are different. In the first case, the instability results from the competition between thermal and viscous dissipation and surface tension forces along the interfaces. When both layers are taken into account, convective motion in the cold layer are stabilizing because they are damping the thermal gradients along the interfaces. In this case, the development of the instability will mainly depend on the relative efficiency of the heat transfer towards the interface in both layers.

This paper presents the very preliminary results of a microgravity experiment devoted to the Marangoni-Bénard instability in a three layer system. This experiment has flown in June 1994 on the IML2 mission of the American space shuttle.

2. Statement of the Problem

We consider the symmetrical three layer configuration shown on Fig.1. The thermal gradient is applied perpendicular to the interfaces. This problem is a natural extension of the single layer bounded by two free surfaces examined by Funada [8]. In that case, Funada and Géoris [9] have shown linearly and numerically that the hot interface exerts a strong stabilizing effect. This stabilizing effect exists also in the case one takes into account the existence of the bulk phase.

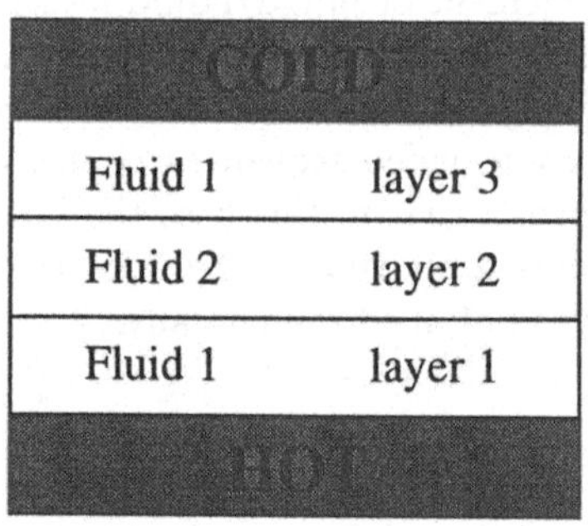

Fig. 1. Symmetrical three layer system.

3. Linear Stability Analysis

One assumes that the three layer systems has an infinite extension in the plane of the interfaces. Using the perturbation technique, the Navier Stokes and heat equations are linearized and can be solved for the marginal case. We define the Marangoni number with respect to the parameters of the layer 1.

$$\mathrm{Ma} = -\frac{\frac{dS}{dT}\Delta T_1 h_1}{\kappa_1 h_1} \tag{1}$$

where h_1 : height of the layer 1, κ_1 : heat diffusivity, μ_1 : dynamical viscosity, ΔT_1 : temperature difference across the layer 1 and dS/dT : interfacial tension dependence on temperature.

The linear differential system for the perturbation expressed in normal modes is the following:

$$\begin{aligned} (D^2 - k^2)(D^2 - k^2 - \omega)\, W_i &= 0 \quad , \qquad (i = 1,2,3) \\ (D^2 - k^2 - \mathrm{Pr}\omega)\, \Theta_i &= W_i \quad , \qquad (i = 1,2,3) \\ (D^2 - k^2 - \mathrm{Pr}\omega)\, \Theta_2 &= -\kappa\lambda W_2 \end{aligned} \tag{2}$$

The first equation reflects the bulk momentum balance written for each of the three incompressible layers (W: vertical velocity, $D = d/dy$, ω : time growth constant of the perturbation, k : spatial wave number and i: layer index). The two other ones result form the heat convecto-diffusive equations (Θ:temperature perturbation, $\lambda = \lambda_1/\lambda_2$: thermal conductivity ratio, $\kappa = \kappa_1/\kappa_2$: heat diffusivity ratio).

3.1 Boundary Conditions

Eighteen boundary conditions are needed to solve the linear system (2). They are found expressing: the rigidity and the high conductivity of the solid walls, that the interfaces are non deformable, the heat flux conservation, the temperature continuity between the layers and the Marangoni effect. More details on the boundary conditions can to be found in [10].

3.2 Results

In the simplified case where the layers have equal thicknesses and only monotonous instabilities - the imaginary part of ω vanishes - are examined, an analytical solution can be found for the marginal Marangoni number:

$$\mathrm{Ma} = \pm\frac{4}{\kappa\lambda k}\sqrt{\frac{((\lambda+1)\cosh^2 k - 1)((a_1 + \frac{a_1}{\mu})^2 - \frac{a_2^2}{\mu^2})}{(a_1^2 - a_2^2)(a_3^2 - a_4^2)^2(\frac{1}{\kappa^2} - \frac{2}{\kappa})}} \tag{3}$$

where

$$\begin{aligned} a_1 &= \frac{1}{\sinh k} - \frac{\cosh k}{k} \\ a_2 &= \frac{1}{k} - \frac{\cosh k}{\sinh k} \\ a_3 &= -\frac{1}{k\cosh k \sinh^2 k} - \frac{1}{2k^2 \sinh k} - \frac{1}{k\cosh k} + \frac{\cosh k}{2k^3} \\ a_4 &= \frac{\cosh^2 k}{k \sinh^2 k} - \frac{\cosh k}{2k^2 \sinh k} - \frac{1}{2k^3} - \frac{1}{2k} \end{aligned}$$

The influence of the parameters λ and μ on the critical Marangoni number are described at great length in [10].

The most striking features of the equation (3) is the influence of κ on the critical convective mode. For usual liquids whose interfacial tension decreases when the temperature increases, the flow is driven by the hot interface whereas the cold one is stabilizing, if $\kappa \ll 1$ and driven by the cold interface if $\kappa \gg 1$. If the diffusivities of the liquids are very similar, the convection is oscillatory at the onset.

4. Experimental Part

4.1 Liquid Selection

We have selected Fluorinert FC70 (3M) for the external layer and DC-200 10 cSt silicone oil. These fluids are transparent, non toxic, very stable and inert. The interface formed by these liquid is particularly unsensitive to contamination and thermocapillary flow can easily be observed in this couple of liquids without drastic cleanliness precautions.

The physical properties of these liquids are given in Tab. 1. The critical temperature difference depends on the surface tension variation with temperature. This parameter has been measured for the couple Fluorinert FC70 - silicone oil 50 cSt in the laboratory. The value obtained, $27 \cdot 10^{-5}\,\mathrm{Nm^{-1}K^{-1}}$ is similar to the value measured by Koster [11] for the couple FC70 - silicone oil 10 cSt $26 \cdot 10^{-5}\,\mathrm{Nm^{-1}K^{-1}}$.

	Fluorinet FC70 (1)	silicone 10 cSt (2)	1/2
density [$\mathrm{kgm^{-3}}$]	1940	935	2.07
kin. viscosity [$\mathrm{m^2s^{-1}}$]	$14 \cdot 10^{-6}$	$10 \cdot 10^{-6}$	1.4
dyn. viscosity [$\mathrm{kgm^{-1}s^{-1}}$]	$27.16 \cdot 10^{-3}$	$9.35 \cdot 10^{-3}$	2.90
heat diffusivity [$\mathrm{m^2s^{-1}}$]	$0.0348 \cdot 10^{-6}$	$0.095 \cdot 10^{-6}$	0.366
heat conductivity [$\mathrm{Wm^{-1}K^{-1}}$]	0.071	0.134	0.53
Prandtl	395	105	3.76
dS/dT [$\mathrm{Nm^{-1}K^{-1}}$]	$27 \cdot 10^{-5}$		

Table 1. Physical properties of the liquids used in the IML-2 experiment.

Since in our system, $\kappa < 1$ the convection will be driven by the hot interface only. It is expected that the layer 3 will be inactive. For these liquids the critical values derived from the equation (3) are: $\mathrm{Ma_c} = 499.58$, $k_c = 1.963$ and $\Delta T_c = 5.4\,^\circ\mathrm{C}$.

4.2 Hardware

The facility BDPU [1] is a multiuser facility designed for the study of capillary phenomena in microgravity. This facility offers to the investigator versatile advanced optical diagnostics including interferometry and thermography. BDPU is designed to accommodate dedicated test containers (TC) equipped with standard electrical and mechanical interfaces.

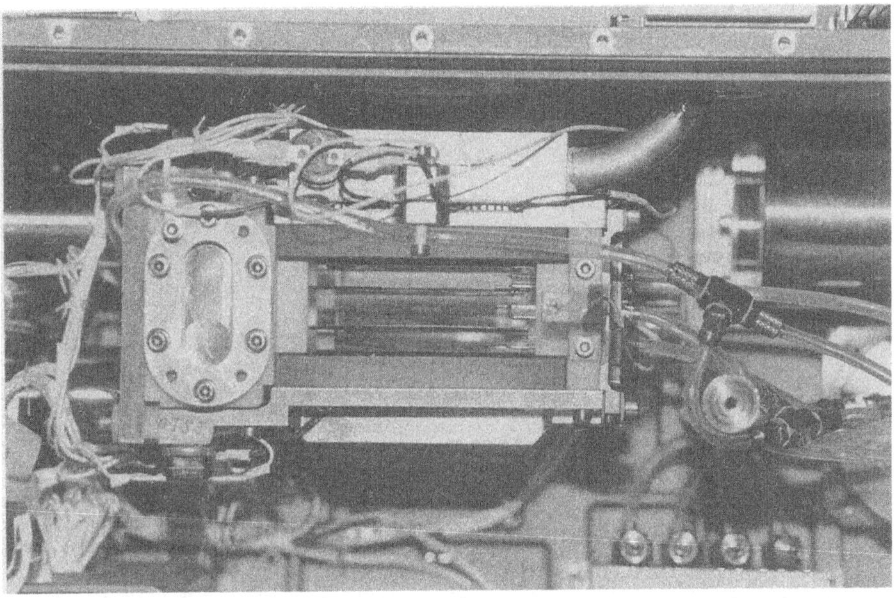

Fig. 2. Detailed view of the Test Container.

The test container: A TC has been developed by LABEN and Dornier. The cuvette containing the liquids consists in a one piece quartz frame closed by two sapphire windows. The internal dimensions of the cuvettes are: 24 x 50 x 35 mm^3. The size of the cuvette is limited by the Rayleigh Taylor instability which is always active in a symmetrical three layer system. The capillary length l of the layers has to be greater than their largest dimension L [12].

$$l = 2\pi\sqrt{\frac{S}{g(\rho_1 - \rho_2})} \quad > \quad L$$

[1] BDPU is the acronym of the Bubble Drop and Particle Unit. This is a special fluid physics experimentation facility developed under the direction of ESA for use in Spacelab missions. (Eds. note)

For our liquid combination silicone 10 cSt / Fluorinert FC70, the maximum level of acceleration allowed is:

$$g = \frac{4\pi^2 S}{l^2(\rho_{\mathrm{FC70}} - \rho_{\mathrm{si10cSt}})} = 1.1 \cdot 10^{-1}\mathrm{ms}^{-2} = 1.12 \cdot 10^{-2} \quad g_0$$

where $g_0 = 9.81\mathrm{ms}^{-2}$. This acceptable acceleration level is higher than the normal acceleration measured in the space shuttle. The restricted extension of the cuvette should stabilize the system, however, as it was shown by Dauby [13], this effect is much smaller than in the case of the Rayleigh-Bénard instability [14].

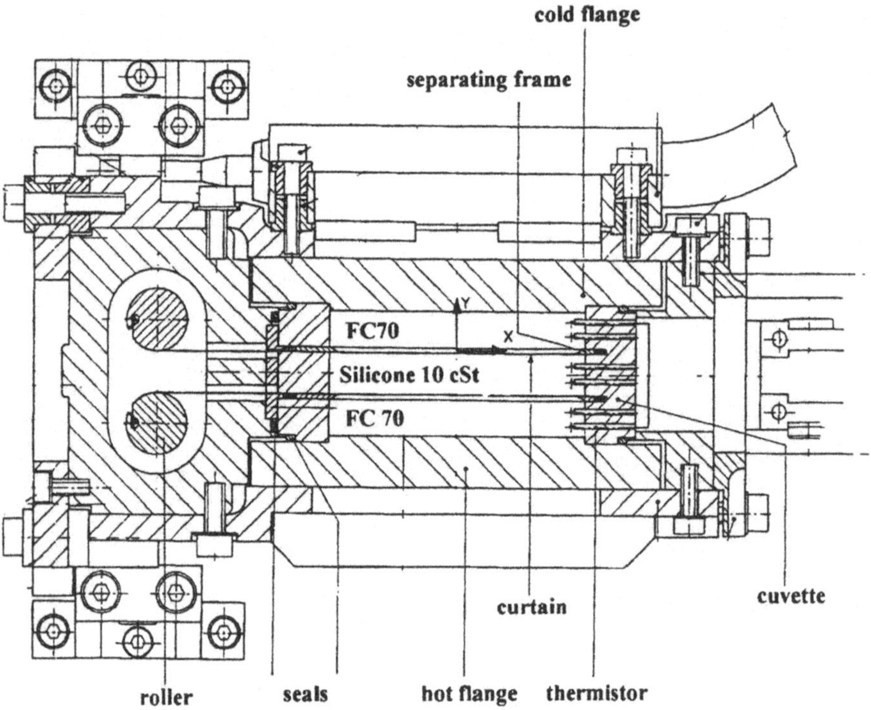

Fig. 3. Schematic drawing of the fluid cuvette. The cuvette is 50 mm wide and each layer is 8 mm thick.

The curtains: To avoid the mixing of the layers before the beginning of the experiment, the liquids are separated by two stainless steel foils. The tightness of the system is achieved with Vitton seals. At the beginning of the experiment, the curtains are pulled out and winded on two rollers.

The separating frames: Two stainless steel frames are inserted along the cuvette walls. They aim to keep the layers separated after the curtain retraction. The frames protrude the layers two millimetres. They are coated with Teflon on the Fluorinert side. The coating is absolutely necessary to prevent the creeping of the silicone oil around the Fluorinert layers after the curtain retraction phase.

Diagnostics

Light sheet illumination: BDPU provides a light sheet illumination device. This light sheet is one millimetre thick and oriented parallely to the longest side of the cuvette. As tracer particles, we have used silver coated glass spheres with diameters ranging from fifty to eighty micrometers. These spheres have been coated in the laboratory.

Infrared camera: An IR camera, AGEMA 900 type, permits to visualize the thermal field in the layer's plane through the sapphire flanges. This diagnostic has not given the expected results.

Thermistors: The cuvette is fitted with eight thermistors. Two of them are glued on the cold and hot flanges for the thermal regulation. The remaining six thermistor are in contact with the liquids since they protrude in the cuvette five millimetres.

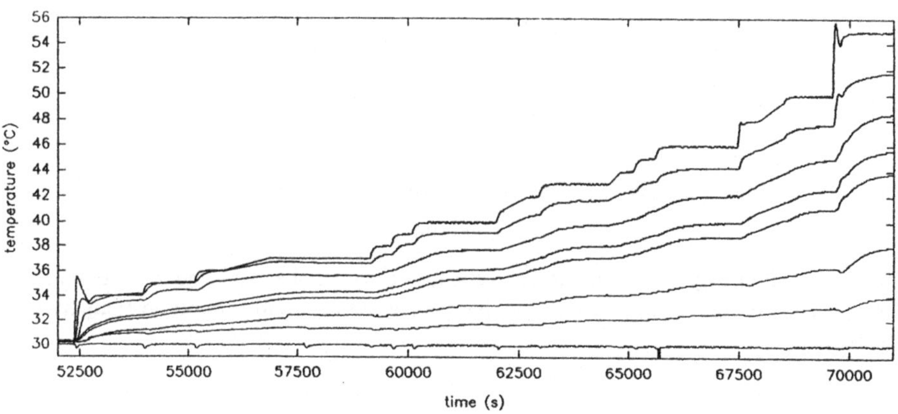

Fig. 4. Thermal data recorded during the experiment. The extreme temperature values correspond to the thermistors used for the thermal control. The six intermediate curves correspond to the thermistors in contact with the fluids.

5. Operations

Before the TC was inserted in the facility, the tracer particles were dispersed in the fluid layer by manual shaking. The homogeneity of the dispersion was checked out by the crew member and by the principal investigator (PI) thanks to the orbiter camera. The concept of manual stirring of the tracer particle has the advantage to simplify the TC and to increase its reliability.

After the cuvette has reached a steady isothermal state, the two curtains have been pulled out successively. The complete pull out of one curtain takes three minutes, no visible deformations of the interfaces have been observed during this phase. Then the temperature difference imposed between the flanges has been increased step by step to reach 25 °C at the end of the experimental run. The recording of the temperatures during the run is shown on Fig. 4 and 5.

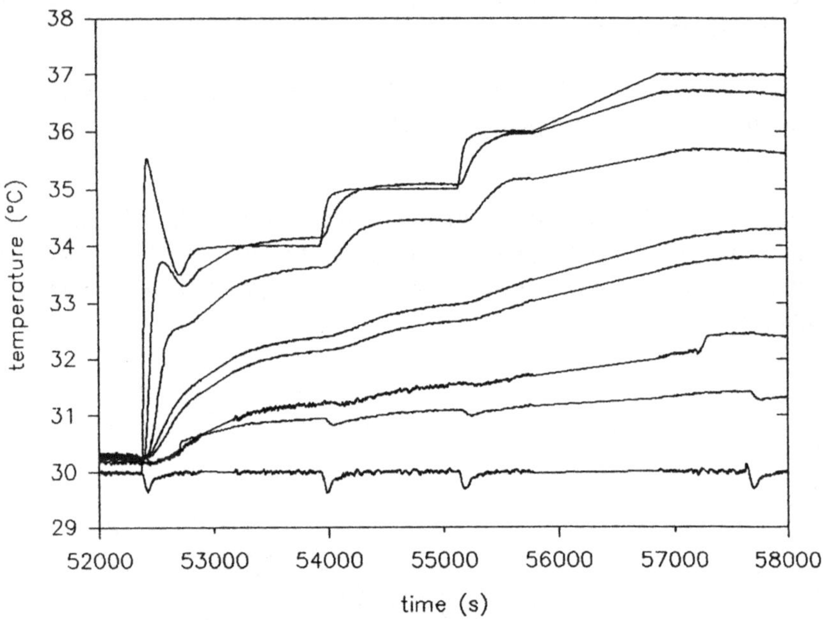

Fig. 5. Detail of Fig. 4 corresponding to the first stage of the experiment. The poor the thermal control loop is responsible for the overshoots and undershoots in temperature records.

6. Results

6.1 Onset of Convection

An experimental value for the critical temperature difference cannot be derived from the thermal data. The increase of thermal conductivity of the system due to convection is too small to be detected. Because of the lack of experimental time and the limitations of BDPU, the onset of convection could not be reached on a quasi steady manner. From the video pictures, we can however say that the convection starts between t=52700 s and t=55500 s corresponding to temperature differences of 4 °C and 6 °C . Live video pictures are not available

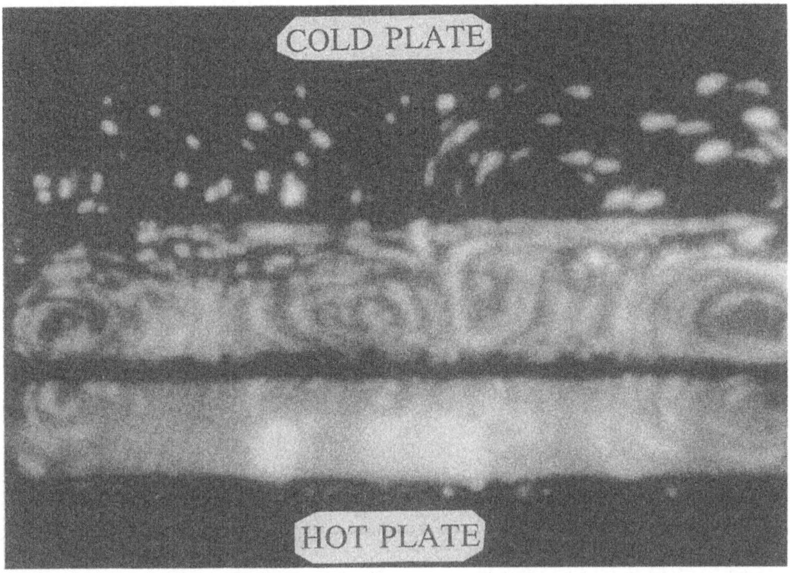

Fig. 6. Convective pattern obtained by long time exposure photography (150 s). The light sheet is located at 10 mm from the front window of the cuvette. The temperature difference across the layers is 10 °C .

between these two moments. Further analysis of low rate video data should permit to determine much more accurately the critical temperature difference. This experimental result is in good agreement with the linear stability analysis that predicts a critical temperature of 5.4 °C .

6.2 Convective Flow

A typical convective flow is shown Fig. 6. The interfaces are not visible because of the one millimetre Vitton seals inserted in the walls. The flow exhibits almost a perfect left-right symmetry. Four convective rolls are visible in the layers 1 and 2. Convection in the layer 3 is very weak, however some structure is also apparent. The hot interface is driving the flow. The two central rolls in the central layer are extending towards the lateral wall near the cold interface.

The video pictures are currently analyzed using a quasi real time image processing software provided by the European Space Agency. The horizontal velocity of the flow along the interfaces is plotted vs the hot interface coordinate and compared with the results of the numerical simulation on Fig. 7. The numerical program used is based on a purely explicit finite difference scheme. Each layer is represented by a regular rectangular 20 x 100 nodes grid. The top and bottom plates are considered to be rigid and conducting. The lateral walls are also rigid

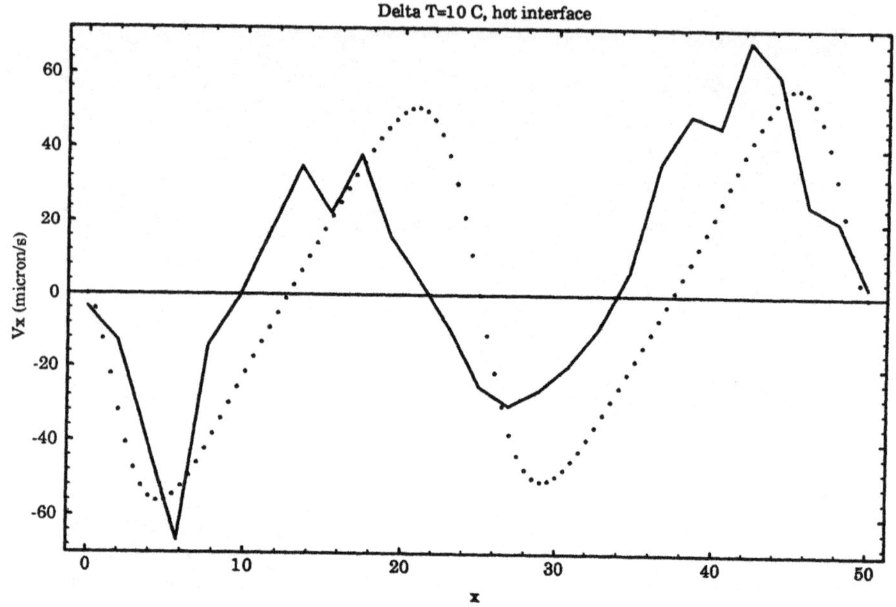

Fig. 7. Horizontal velocity along the hot interface at $\Delta T = 10$ °C . Solid line: experimental measurement, dotted line: numerical result.

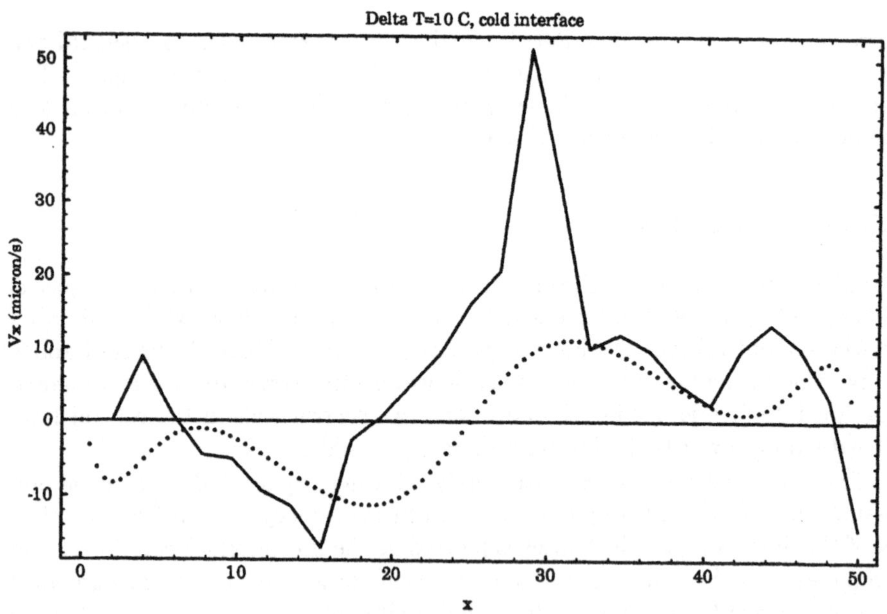

Fig. 8. As Fig. 7 but for the cold interface.

but we have assumed their thermal conductivity match the thermal conductivity of the fluids.

The agreement between numerical and experimental results on the amplitude of the velocity is excellent. The maximum velocity measured is 68 μms^{-1}. The experimental velocity profile is clearly not as symmetrical as the numerical one but the general shape is very similar. The flow in the central region is somewhat weaker in the experiment than in the numerical results.

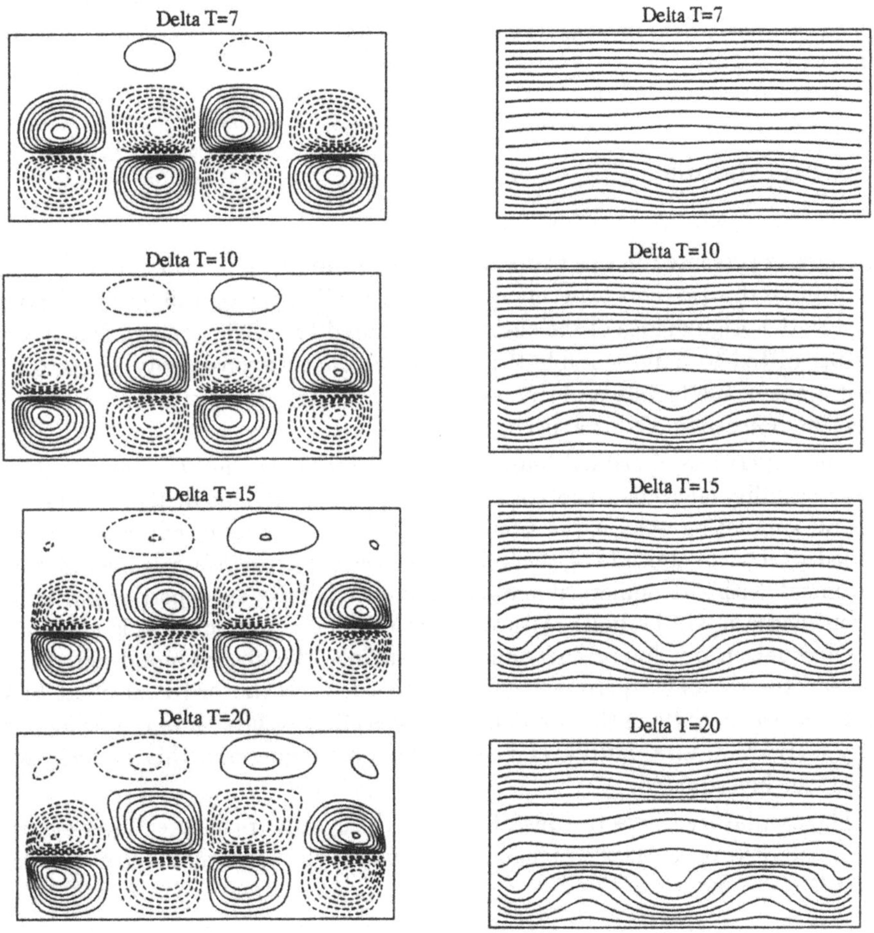

Fig. 9. Stationary 2-D convective patterns and isotherms for the FC70 - silicone oil 10 cSt - FC70 system. The dashed streamlines correspond to the clockwise convective rolls. ΔT=7 °C $\psi_{max} = 0.00259$, ΔT=10 °C $\psi_{max} = 0.00448$, ΔT=15 °C $\psi_{max} = 0.00675$, ΔT=20 °C $\psi_{max} = 0.00887$. The streamlines are plotted for equally spaced values of the stream function.

The same comparison is shown for the cold interface on Fig. 8. Clearly the numerical simulation is not reproducing correctly the experimental flow observed.

The maximum velocity of this interface is much greater than in the simulation. This discrepancy is probably due to the heat losses through the cuvette lateral walls. Indeed, at both interfaces, the thermocapillary flow of the side rolls is directed towards the lateral walls, this indicates that the temperature in the central part of the cuvette is greater than near the walls. We have also to remind that the numerical simulations are 2-D whereas the cuvette is 3-D. The 3-D structure of the flow has not been analyzed yet. The convective pattern evolution for increasing thermal temperature differences is given in Fig. 9. As in Fig. 6, the central rolls of the middle layer tend to prolongate along the cold interface towards the lateral walls but this effect is far less pronounced in the numerical results.

7. Conclusions

The results presented in this paper are very preliminary and no definitive conclusion can be drawn. This experiment was very challenging on the technical point of view, the original idea of the curtains associated to a Teflon coated frame edge has permitted to build a stable three layer system in microgravity. Despite the limitations of the facility and the poor quality of the video pictures, most of the scientific objectives of this experiment will be met.

The experimental critical temperature difference matches the results of the linear stability analysis. The convective pattern is well predicted by the numerical simulation even if the side wall effects are probably not yet correctly modelled. The magnitude of the velocity of the lower interface is also in good agreement with the numerical simulation and hopefully an experimental bifurcation diagram will be built. Although the interfacial deformation is not taken into account in our linear and numerical models, the general good agreement between theory and experiment confirms this is a second order effect at least when the aspect ratio of the layers is low. These first results on pure thermocapillary convection in a multilayer system are encouraging and this problem will be further investigated.

The oscillatory convective instability for the symmetrical three layer system will be studied on a Maxus flight in 1995 and on the MSL mission of spacelab in 1996.

Acknowledgments

This text presents results of the Belgian programme on Interuniversity Pole of Attraction initiated by the Belgian State, Federal Service of Scientific Technical and Cultural affairs. The scientific responsibility is assumed by its authors.

The authors wish to thank Dr T. Roesgen and R. Totaro of ESA/ESTEC for their invaluable help in the image processing of the video pictures.

References

1. L. Cserepes and M. Rabinowicz; *Earth and Planetary Science Letters* , **76** (1985) 193-207
2. S. Rasenat, F.H. Busse and I.Rehberg. *J. Fluid Mech.* **109** (1989) 519-540.
3. F.M. Richter and C. Johnson. *J. Geophysical Research* **79** (1974) 11.
4. E.S. Johnson. *J. Crystal Growth* **30** (1975) 249 - 256.
5. P. Colinet and J.C. Legros. *Phys. of Fluids* **6** (1994) 8.
6. Ph. Géoris, M. Hennenberg, I.B. Simanovskii, I.I. Wertgeim and J.C. Legros. *Micrograviity Science Tech.* **7** (1994) 90-97.
7. J.R.A. Pearson. *J. Fluid Mech.* **4** (1958) 4.
8. T. Funada. *J. Phys. Soc. Jap.* **55** (7) (1994) 2191-2202.
9. Ph. Géoris, PhD-thesis, Université Libre de Bruxelles, Belgium, 1994.
10. Ph. Géoris, M. Hennenberg, IB. Simanovskii, A Nepomniaschy I.I. Wertgeim and J.C. Legros. *Phys. Fluids. A.* **5** (1993) 7.
11. C.V. Burkersroda, A. Prakash and J.N. Koster. *Microgravity Q.* **4** (2) (1994) 93-99.
12. S. Chandrasekhar. *Hydrodynamic and Hydromagnetic Instability.* Dover Publication, Inc., New-York, 1981.
13. P.C. Dauby and G. Lebon. Submitted to Int. J. of Heat and Mass Transfer.
14. I. Catton. *Trans. ASME, J. Heat Transfer.* **C 92** (1970) 186-188.

On Vibrational Convective Instability of a Horizontal Binary Mixture Layer with Soret Effect

G.Z. Gershuni[1], *A.K. Kolesnikov*[2], *J.C. Legros*[3], *and B.I. Myznikova*[4]

[1] Department of Theoretical Physics, Perm State University, 614600 Perm, Russian Federation
[2] Department of Theoretical Physics, Perm Pedagogical University, 614600 Perm, Russian Federation
[3] Microgravity Research Center, Universite Libre de Bruxelles, B-1050 Brussels, Belgium
[4] Institute of Continuous Media Mechanics, Urals Branch of Russian Academy of Sciences, 614061 Perm, Russian Federation

Abstract

The stability of the mechanical equilibrium of a plane horizontal binary mixture layer with Soret effect in the presence of a static gravity field and high frequency vibration is studied theoretically. The horizontal boundaries of the layer are assumed to be rigid, isothermal and impermeable. A linear theory of stability is developed. The instability boundary and characteristics of the critical disturbances are determined.

1. Introduction

It is well known that vibration of a cavity filled with fluid in the presence of nonhomogeneity of temperature can provoce a mean flow even in the state of weightlessness. In some cases a state of quasi-equilibrium is possible and the problem of stability can be formulated. The study of stability of quasiequilibrium in weightlessness has been made in [1] - [4]. If static gravity exists, both mechanisms of instability are superimposed - the classic Rayleigh - Benard one and the specific vibrational one. In this paper we deal with the equilibrium and stability of a binary mixture in which the additional mechanism of stratification caused by concentration exists as well as that of dissipation connected with diffusion and thermodiffusion. It is demonstrated that the stability boundaries depend sharply on the thermodiffusion coefficient. The consideration is based on a system of equations for mean fields that is valid in the limiting case of very high (but not acoustic) frequencies and small amplitudes of vibration.

2. Description of the Problem and Basic System of Equations

We consider an infinite plane horizontal layer of binary mixture with Soret effect confined between two rigid, isothermal and impermeable planes $z = 0$ and

$z = h$. At $z = 0$ the temperature is maintained constant and equal to Θ , the temperature of the upper plane $z = h$ is chosen as the reference point. Both cases $\Theta > 0$ and $\Theta < 0$ will be considered. The layer is exposed to static gravity with the acceleration $\mathbf{g}(0, 0, -g)$ and to the high frequency vibration along an axis described by the unit vector $\mathbf{n}(n_x, 0, n_z)$ where $n_x = \cos\alpha$, $n_z = \sin\alpha$ and α is the angle between the axis of vibration and the horizontal $x-$axis.

To write down the system of equations in the proper (vibrating) system of coordinates it is necessary to change $\mathbf{g} \to \mathbf{g} + b\Omega^2 \cos\Omega t\mathbf{n}$; here b is the displacement amplitude, Ω is the angular frequency of vibration. Then we have a system of equations taking into account the thermodiffusional Soret effect:

$$\begin{aligned}
&\frac{\partial \mathbf{v}}{\partial t} + (\mathbf{v}\nabla)\,\mathbf{v} = -\frac{1}{\rho}\nabla p + \nu\Delta\mathbf{v} + g\,(\beta_1 T + \beta_2 C)\,\gamma + (\beta_1 T + \beta_2 C)\,b\Omega^2\cos\Omega t\,\mathbf{n}\\
&\frac{\partial T}{\partial t} + \mathbf{v}\nabla T = \chi\Delta T\\
&\frac{\partial C}{\partial t} + \mathbf{v}\nabla C = D\,(\Delta C + \alpha\Delta T)\\
&div\ \mathbf{v} = 0\ . \qquad (1)
\end{aligned}$$

Here β_1 is the coefficient of thermal expansion ($\beta_1 > 0$) , β_2 is the concentrational coefficient of density ($\beta_2 > 0$, C is the concentration of the lightest component) , D is the coefficient of diffusion, α is the thermodiffusional ratio; other notations are standard.

To describe the vibrational convection in the limiting case of asymptotically high frequencies the method of averaging can be applied successfully. It is necessary to subdivide each field into two parts: the slow varying with time (mean) part and the rapidly varying oscillating one, and then to obtain the closed system of equations for mean parts. We write it down in the nondimensional form using the following units : h for distance, h^2/ν for time, χ/h for velocity, Θ for temperature, $\beta_1\Theta/\beta_2$ for concentration and $\rho\nu\chi/h^2$ for pressure :

$$\begin{aligned}
&\frac{\partial \mathbf{v}}{\partial t} + \frac{1}{Pr}(\mathbf{v}\nabla)\,\mathbf{v} = -\nabla p + \Delta\mathbf{v} + Ra\,(T + C)\,\gamma + Ra_v\mathbf{w}\nabla\left[(T + C)\,\mathbf{n} - \mathbf{w}\right]\\
&Pr\,\frac{\partial T}{\partial t} + \mathbf{v}\nabla T = \Delta T\\
&Sc\,\frac{\partial C}{\partial t} + \frac{Sc}{Pr}\,\mathbf{v}\nabla C = \Delta(C - \varepsilon T)\\
&div\ \mathbf{v} = 0\\
&div\ \mathbf{w} = 0\\
&rot\ \mathbf{w} = \nabla(T + C)\times\mathbf{n}\ . \qquad (2)
\end{aligned}$$

Here $\mathbf{w}$ is the additional slow variable, the amplitude of the oscillating part of the velocity field.

Let us now formulate the boundary conditions:

$$z = 0: \qquad \mathbf{v} = 0, \quad \Theta = 1, \quad \frac{\partial C}{\partial z} - \varepsilon \frac{\partial T}{\partial z} = 0, \quad w_z = 0 \ ,$$

$$z = 1: \qquad \mathbf{v} = 0, \quad \Theta = 0, \quad \frac{\partial C}{\partial z} - \varepsilon \frac{\partial T}{\partial z} = 0, \quad w_z = 0 \ . \tag{3}$$

The formulation of the problem includes the following set of nondimensional parameters : the Rayleigh number Ra and its vibrational analog Ra_v, the Prandtl number Pr, the Schmidt number Sc, the parameter of thermodiffusion ε. The parameters are determined as

$$Ra = \frac{g\beta_1 \Theta h^3}{\nu\chi}, \quad Ra_v = \frac{(b\Omega\Theta h\beta_1)^2}{2\nu\chi}, \quad Pr = \frac{\nu}{\chi}, \quad Sc = \frac{\nu}{D}, \quad \varepsilon = -\frac{\beta_2 \alpha}{\beta_1} \ . \tag{4}$$

Finally let us write down the assumptions to be used in the averaging method: (i) the period of vibration must be small: $\tau \ll \min\left(h^2/\nu, h^2/\chi, h^2/D\right)$; (ii) the displacement amplitude must be small in the sense $b\beta_1\Theta \ll h$ and (iii) the relationship between the gravitational acceleration and the vibrational one must be as $g\beta_1\Theta \ll \Omega^2 h$.

3. Mechanical Quasi-Equilibrium

Under certain conditions the state of mechanical quasi-equilibrium is possible, i. e. the state when the mean velocity is equal to zero. For equilibrium fields T_0, C_0 and $\mathbf{w}_0$ one can obtain from the system (2):

$$\nabla (T_0 + C_0) \times [Ra\, \gamma - Ra_v\, \nabla (\mathbf{w}, \mathbf{n})] = 0 \ ,$$

$$\Delta T_0 = 0, \qquad \Delta C_0 = 0 \ ,$$

$$div\ \mathbf{w}_0 = 0, \qquad rot\ \mathbf{w}_0 = \nabla(T_0 + C_0) \times \mathbf{n}, \tag{5}$$

with the appropriate boundary conditions (3).

It is easy to see that in the case of a plane horizontal layer the mechanical quasiequilibrium is possible with linear profiles for T_0, C_0 and w_0:

$$T_0 = 1 - z, \quad \frac{dC_0}{dz} = -\varepsilon, \quad w_0 = -(1+\varepsilon)\cos\alpha\left(z - \frac{1}{2}\right) \ . \tag{6}$$

4. The Stability Problem

To study the stability of the state defined in (6) we introduce small disturbances $\mathbf{v}$, T', C', p', $\mathbf{w}'$. We consider 2D - disturbances because we expect that such kind of disturbances is the most dangerous by the analogy with the case of the one-component fluid [2]. Introducing the stream functions Ψ and F for the solenoidal vectors $\mathbf{v}$ and $\mathbf{w}'$ respectively, and considering the disturbances of the normal mode type

$$(\Psi, T', C', F') = (\varphi(z), \theta(z), \xi(z), f(z)) \ \exp(-\lambda t + ikx) \ , \tag{7}$$

we obtain the spectral amplitude problem. In the case of longitudinal vibrations ($n_x = 1$, $n_z = 0$) it has the form:

$$-\lambda \mathcal{D}\varphi = \mathcal{D}^2\varphi + ikRa(\theta + \xi) + ikRa_{\mathrm{v}}(1+\varepsilon)(\theta + \xi - f') \ , \tag{8}$$

$$-\lambda Pr\, \theta - ik\varphi = \mathcal{D}\theta \ , \tag{9}$$

$$-\lambda Sc\, \xi - ik\frac{Sc}{Pr}\, \varepsilon\varphi = \mathcal{D}(\xi - \varepsilon\theta) \ , \tag{10}$$

$$\mathcal{D}f = \theta' + \xi' \ , \tag{11}$$

$$\left(\mathcal{D} = \frac{d^2}{dz^2} - k^2\right) \ ,$$

$$z = 0, \quad z = 1: \qquad \varphi = \varphi' = 0, \quad \theta = 0, \quad f = 0, \quad \xi' - \varepsilon\theta' = 0 \ . \tag{12}$$

In the case of transversal vibrations ($n_x = 0, n_z = 1$) we have to change (8) and (11) for

$$-\lambda \mathcal{D}\varphi = \mathcal{D}^2\varphi + ikRa(\theta + \xi) + k^2 Ra_{\mathrm{v}}(1+\varepsilon) f \ , \tag{13}$$

$$\mathcal{D}f = ik(\theta + \xi) \ . \tag{14}$$

In the spectral problems φ, θ, ξ, f are the amplitudes of disturbances, k is the wave number and the eigenvalue λ is the decrement.

These two limiting cases are quite different physically. In the first case (the longitudinal vibration) there are two mechanisms of convective instability excitation - thermogravitational and thermovibrational, which are superimposed. In the second case (the transversal vibrations) the specific vibrational mechanism is not operative, and the vibration plays a purely stabilizing role [1].

5. Long Wave Disturbances

The conditions of impermeability enables us to expect that in some region of parameters the long wave disturbances (with $k = 0$) are responsible for instability. In this case one may develop the regular perturbation theory with k as a small parameter. It is possible to construct the solution of the spectral problem in the form of power expansions for all the amplitudes and the decrement,

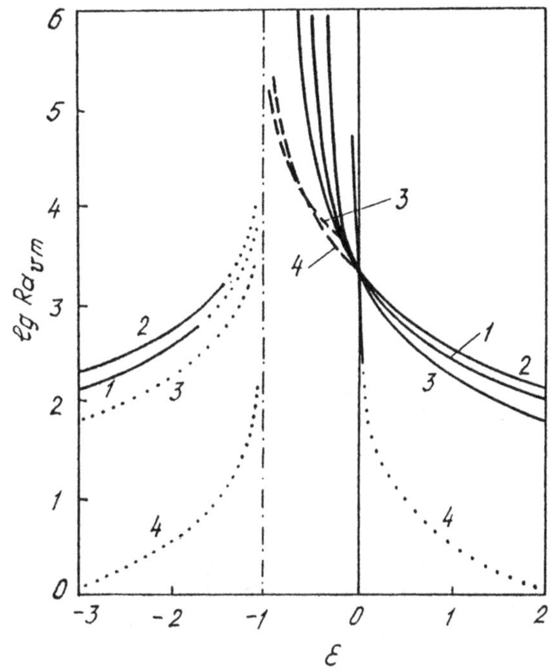

Fig. 1. Instability characteristics for the case of weightlessness (longitudinal orientation). Solid lines - monotonic cellular mode, dashed lines - oscillatory cellular mode, dotted lines - monotonic long wave mode; *1* – $Le = 1$; *2* – $Pr = 0.75, Sc = 0.5$; *3* – $Pr = 0.75, Sc = 1.5$; *4* – $Pr = 6.7, Sc = 677$.

and to obtain a system of successive approximations. The solvability condition for the system of the second order approximation allows us to determine λ_2, the first nonvanishing correction to the decrement. The condition $\lambda_2 = 0$ gives the boundary of the long wave instability. We have for the case of longitudinal vibrations

$$Ra + (1+\varepsilon)Ra_v = \frac{720Le}{\varepsilon}, \tag{15}$$

and for the case of transversal vibrations

$$Ra = \frac{720Le}{\varepsilon}. \tag{16}$$

Here $Le = Pr/Sc$ is the Lewis number.

To judge whether the long wave mode is the most dangerous or not it is necessary to determine the stability boundary for the cellular mode with the finite values of the critical wave number k_m. This requires the solution of the complete spectral problems (8) - (14).

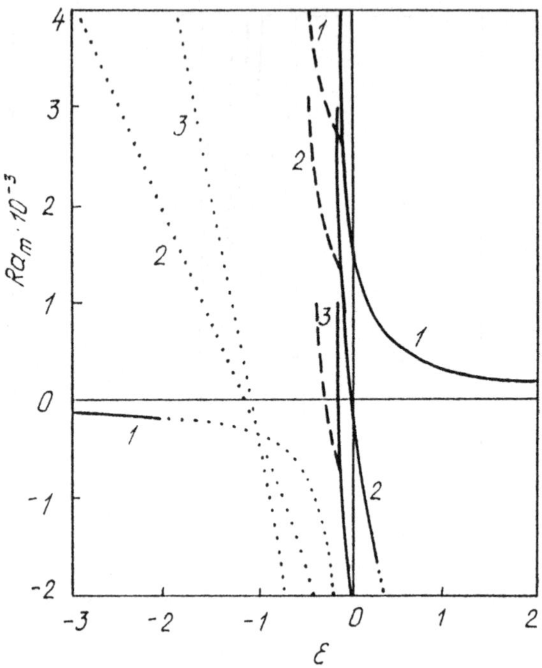

Fig. 2. Critical values of Rayleigh number versus nondimensional Soret parameter for gaseous mixture with $Pr = 0.75, Sc = 1.5$. *1* – $Ra_v = 0$; *2* – $Ra_v = 2129$; *3* – $Ra_v = 5000$.

6. Some Numerical Results

The numerical solution of the complete eigenvalue problem has been obtained by step - by - step numerical integrating using the Runge - Kutta - Merson method in combination with the shooting procedure. Thus we find the eigenvalue λ and afterwards we determine the boundary of stability from the condition $\lambda_r = 0$, where λ_r is the real part of λ. Some results of calculations are presented in Figs.1 - 4, where Figs. 1 - 3 refer to the case of longitudinal vibrations, while Fig.4 to the case of transversal ones. In all the figures the following notation is adopted: solid lines correspond to the monotonic cellular mode of instability with minimal critical values of the wave number $k_m \neq 0$; dashed lines correspond to the oscillatory cellular mode of instability, and dotted lines correspond to the long wave mode with $k_m = 0$.

First let us consider the case of weightlessness ($Ra = 0$). The vibrational Rayleigh number Ra_v is the regime parameter in this case, and its Ra_{vm} values minimized with respect to k are presented in Fig.1 as functions of the nondimensional parameter ε. It can be seen that very sharp dependency on ε takes place (the logarithmic scale for Ra_{vm} is used). The curves *1* - *3* correspond to the cases of gaseous mixtures; the curve *4* to the case of the typical liquid mixture:

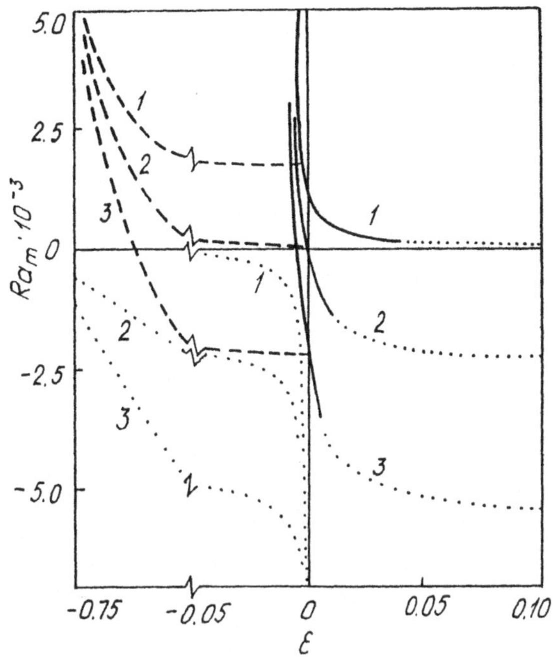

Fig. 3. Critical values of Rayleigh number versus nondimensional Soret parameter for liquid mixture with $Pr = 6.7, Sc = 677$. The nomenclature and numeration of lines are the same as in Fig. 2.

1 - $Le = 1$; *2* - $Le = 1.5$; *3* - $Pr = 0.75, Sc = 1.5$; *4* - $Pr = 6.7, Sc = 677$. In the region of the normal thermodiffusion effect ($\varepsilon > 0$) very strong destabilization takes place, while the region $\varepsilon < 0$ (anomalous Soret effect) corresponds to strong stabilization as $|\varepsilon|$ increases, and the oscillatory mode becomes more unstable (see curves *3* and *4*). In the region $\varepsilon < -1$ the monotonic instability exists.

Figure 2 corresponds to the case of a model gaseous mixture ($Pr = 0.75, Sc = 1.5$). The Lewis number is $Le = 0.5, Le < 1$, so the oscillatory mode of instability is possible in the region of the anomalous Soret effect ($\varepsilon < 0$). The family of instability curves in the plane ($\varepsilon - Ra_{\mathrm{m}}$) for several values of Ra_{v} is presented (*1* - $Ra_{\mathrm{v}} = 0$, *2* – $Ra_{\mathrm{v}} = 2129$, *3* – $Ra_{\mathrm{v}} = 5000$). The regions of stability are confined between two curves with the same numbers (*1* – *1*, *2* – *2*, *3* – *3*). As far as Ra_{v} increases the stability regions diminish. It is interesting to note that the oscillatory form of the instability may exist when the system is heated from above, and Ra_{v} is large enough (line *3*, $Ra_{\mathrm{v}} = 5000$).

The analogous family of stability curves for the case of the typical liquid binary mixture ($Pr = 6.7, Sc = 677$) is shown in Fig. 3 (*1* – $Ra_{\mathrm{v}} = 0$, *2* – $Ra_{\mathrm{v}} = 2129$, *3* – $Ra_{\mathrm{v}} = 5000$). It can be seen that only in the region of small positive or negative values of ε the instability is of the monotonic cellular

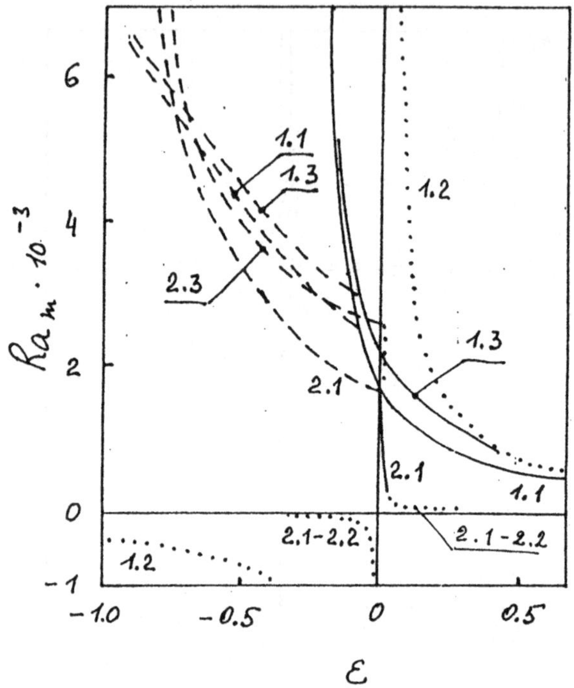

Fig. 4. Instability characteristics in the case of transversal orientation. *1* – gaseous mixture with $Pr = 0.75, Sc = 1.5$: *1.1* – $Ra_v = 0$, *1.2* – $k_m = 0$, *1.3* – $Ra_v = 1000$; *2* – liquid mixture with $Pr = 6.7, Sc = 677$: *2.1* – $Ra_v = 0$, *2.2* – $k_m = 0$, *2.3* – $Ra_v = 1000$.

character. The long wave instability plays the main role when either $\varepsilon > 0$, but not too small, or $\varepsilon < 0$ while heating from above.

Figure 4 corresponds to the case of transversal vibrations. As has been noted, the specific vibrational mechanism is not operative in this case, and the role of vibrations is purely stabilizing. The examples of instability curves are given for cases of the gaseous mixture ($1 - Pr = 0.75, Sc = 1.5$) and of the liquid mixture ($2 - Pr = 6.7, Sc = 677$). There are two limiting curves in the ($\varepsilon - Ra_m$) plane: the lowest line corresponds to the nonvibrating case $Ra_v = 0$, the highest one to the case of the long wave instability (16). All the other lines are confined between two limiting curves; the effect of the vibration is stabilizing. The numbers near the curves correspond to the following values of parameters: *1.1*, *2.1* – $Ra_v = 0$; *1.2*, *2.2* – $k_m = 0$; *1.3*, *2.3* – $Ra_v = 1000$.

Acknowledgements

This text presents results partly obtained in the framework of the Belgian programme on Interuniversity Poles of Attraction (PAI Nr.21) initiated by the Belgian State, Prime Minister's Office, Federal Office for Scientific, Technological and Cultural Affairs. The scientific responsibility is assumed by its authors.

The research described was made possible in part by Grant MF5300 from the International Science Foundation.

A stay of one author (G.G.) in Brussels was supported by the Human Capital and Mobility Programme of the European Community (contract CHRX CT 930106).

References

1. G.Z. Gershuni and E.M. Zhukhovitsky, *Soviet Physics Doklady*, **24** (1979) 894.
2. G.Z. Gershuni and E.M. Zhukhovitsky, *Fluid Dynamics*, **16** (1981) 498.
3. L.M. Braverman, *Fluid Dynamics*, **22** (1987) 657.
4. V.I. Chernatynsky, G.Z. Gershuni and R. Monti, *Micrograv. Quart.* **3**(1) (1993) 55 .

Thermocapillary Convection in Liquid Bridges with a Deformed Free Surface

V.M. Shevtsova, H.C. Kuhlmann and H.J. Rath

ZARM, University of Bremen, Am Fallturm, 28359 Bremen, Germany

Abstract

The two-dimensional thermocapillary flow in a liquid bridge of aspect ratio one with a curved free surface is investigated. The Stokes flow solution near the hot and cold corners becomes singular at the critical contact angle $\alpha_c = 128.7°$. Beyond this angle the flow should reverse locally. The length scale below which the flow reversal is expected cannot be predicted by the simple Stokes flow analysis. Numerical simulations indicate that the extention of the flow reversal region should be less than 10^{-4} in units of the height of the liquid bridge when $\alpha = 135°$.

1. Introduction

In the floating-zone crystal-growth technique the melt is suspended between the growing crystal and the feed rod. To mathematically modell the convective processes in the melt the so-called half zone model is frequently used. It consists of a cylindrical volume of liquid confined between two rigid surfaces, which are kept at different temperatures. In the absence of gravity, deviations from a cylindrical shape may be due to thermocapillary flow or a volume $V \neq \pi R_0^2 d$, where R_0 is the radius and d is the length of the liquid bridge. In the present work the influence of the volume constraint on the thermocapillary flow, driven by a linear variation of the surface tension $\sigma = \sigma_0 - \gamma(T - T_0)$, will be investigated. The first studies of this problem were done by [1]. Here, we extend the previous work and focus on the thermocapillary flow in the corners, where the free surface and the rigid wall meet at a contact angle α.

2. Mathematical Formulation

The geometry of the system is shown in Fig.1. We choose cylindrical coordinates (r, z). The rigid walls are at $z = \pm d/2$ and the free surface is located at $r = h(z)$, $h(\pm d/2) = R_0$. The imposed temperatures are $T_0 \pm \Delta T/2$ at $z = \pm d/2$. The aspect ratio is defined as $\Gamma = d/R_0$.

The governing Navier-Stokes and energy equations in dimensionless form for an incompressible fluid take the form

$$[\partial_t + \mathbf{U}\cdot\nabla]\,\mathbf{U} = -\nabla P + \nabla^2\mathbf{U}\ , \tag{1}$$

$$[\partial_t + \mathbf{U}\cdot\nabla]\,\Theta = -W + \frac{1}{Pr}\nabla^2\Theta\ , \tag{2}$$

$$\nabla\cdot\mathbf{U} = 0\ , \tag{3}$$

where $\mathbf{U} = U\mathbf{e}_r + W\mathbf{e}_z$ and d^2/ν, ν/d, and $\rho\nu^2/d^2$ are used as scales for time, velocity, and pressure, respectively. Θ is the deviation from the conducting temperature profile $\Theta = (T - T_0)/\Delta T - z$ of the ambient. On the rigid boundaries $z = \pm 0.5$, in units of d, we impose no-slip conditions $U = W = 0$ and constant temperatures $\Theta = 0$. On the axis ($r = 0$) the symmetry conditions $U = \partial_r W = \partial_r \Theta = 0$ are used.

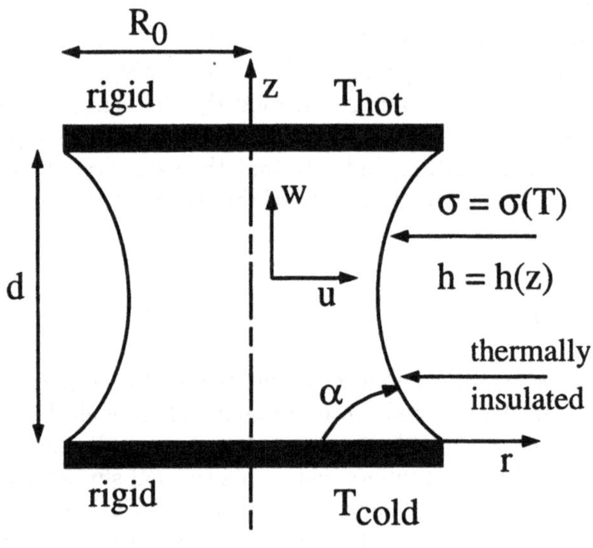

Fig. 1. Geometry of the system.

The normal and tangential stress balances between the viscous liquid and the inviscid gas on the free surface $r = h(z)$ are

$$\Delta P = \mathbf{n}S\mathbf{n} - \frac{1 - Ca(\Theta + z)}{\epsilon\, K}\ , \tag{4}$$

$$\boldsymbol{\tau}S\mathbf{n} + Re\frac{\partial\sigma}{\partial\tau} = 0\ , \tag{5}$$

where S is viscous stress tensor, $\mathbf{n}$ is the outward unit normal vector, $\boldsymbol{\tau}$ is the tangential unit vector, and K is the mean curvature. ΔP is the pressure deviation from the ambient. The dimensionless parameters arising are the Prandtl,

Reynolds, and Capillary number

$$Pr = \frac{\nu}{k}, \qquad Re = \frac{\gamma \Delta T d}{\rho_0 \nu^2}, \qquad Ca = \frac{\gamma \Delta T}{\sigma_0}$$

where k, ρ_0, and ν denote the thermal diffusivity, density, and kinematic viscosity, respectively. The Marangoni number is $Ma = Re \cdot Pr$. Equation (5) provides the driving force for the flow. The kinematic condition and the condition of thermal insulation on $r = h(z)$ complete the problem.

The interface location is determined by the hydrodynamic and the Laplace pressure (4). The relative importance of both is given by the ratio between the order of the magnitudes of the static $P_{\text{stat}} = \sigma_0/d$ and that of the dynamic pressure $P_{\text{dyn}} = \gamma \Delta T/d$,

$$P_{\text{stat}}/P_{\text{dyn}} = \gamma \Delta T/\sigma_0 = Ca.$$

In the case $Ca \ll 1$, (4) simplifies to the Young-Laplace equation

$$\Delta P = -\frac{1}{\epsilon}\left(\frac{h''}{\Gamma^2 \epsilon^2} - \frac{1}{h}\right) \qquad \text{where} \qquad \epsilon = \sqrt{1 + \frac{h'^2}{\Gamma^2}} \tag{6}$$

and dynamic surface deformations can be neglected.

Equation (6) was integrated by a shooting method to determine the free surface shape. To solve the Navier-Stokes equations (1-3) in the domain occupied by the liquid, the physical space was transformed to a rectangular numerical domain, using curvilinear body fitted coordinates defined by

$$(r, z) \rightarrow (\xi, \eta) \qquad \xi = \frac{r}{h(z)}, \quad \eta = z.$$

The transformed Navier-Stokes equations were then solved using a time-dependent finite-difference ADI technique in stream function, vorticity formulation ψ, ω, Θ. The time derivatives were discretized by forward-differences while a first order upwind scheme was used for the convective terms. The remaining terms were approximated by second order central differences. The resulting Poisson equation for ψ was solved by introducing an artificial time-derivative term [2]. For checks and comparisons central differences were also used occasionally for the convective terms.

3. Results

3.1 Global Solution Structure

The structure of the temperature field strongly depends on the contact angle, even in the heat conduction regime ($Ma \rightarrow 0$). The isotherm distributions for cylindrical volumes with different contact angles are shown in Fig.2 in the case of creeping flow ($Ma = 0.001, Pr = 1$). Unlike the rectangular cylinder with a linear temperature profile the isolines are more dense in the central part $z = 0$

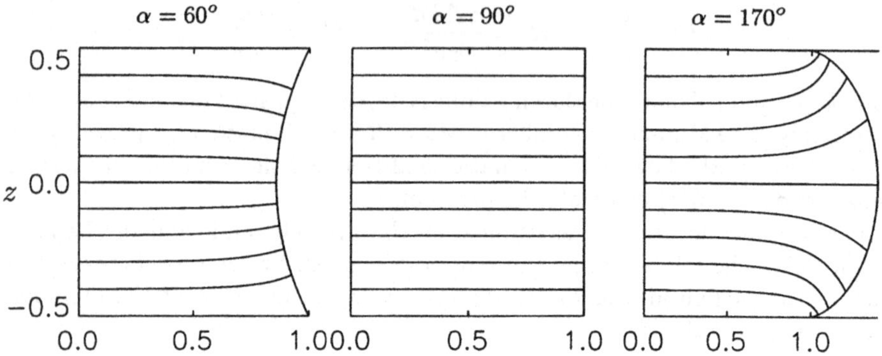

Fig. 2. Isotherms for cylindrical volumes with $Ma = 0.001$, $Pr = 1$ and varying contact angle α.

for concave volumes ($\alpha < \pi/2$). In the case of convex volumes ($\alpha > \pi/2$) the maximal temperature gradients are found near the corners $z = \pm 1/2$. This behaviour is due to the thermal insulation of the free surface.

It is well known that the surface velocity exhibits a spike near the cold wall due to high thermocapillary forces induced by a sharp temperature gradient when the Marangoni number becomes large. This feature is more pronounced for convex cylinders due to existence of even higher temperature gradients near the corners. Streamlines and isotherms for $Re = 1000$ and $Pr = 1$ are presented in Fig.3. Although the minimum value of the stream funcion $\psi_{\min}$ changes with α, the qualitative behaviour is similar for all contact angles. However, the flow structure near the corners is more complex.

3.2 Stokes Flow in a Thermocapillary Corner

For any Re and Pr there is a region near the corner where both inertia and thermal convection are negligible, since $\mathbf{U} = 0$ on the rigid boundary. The temperature on the free surface will then vary linearly with the distance from the corner and the flow is viscous. It is governed by the Stokes equation,

$$\nabla^4 \psi = 0 \ . \tag{7}$$

The solution of (7) according to the present boundary conditions, up to a constant determined by the local temperature gradient, is given by

$$\Psi = \frac{r^2}{4} \frac{(A \cos 2\theta + B \sin 2\theta + C\theta + D)}{\sin 2\alpha - 2\alpha \cos 2\alpha} \tag{8}$$

where

$$A = \sin 2\alpha - 2\alpha, \quad B = 1 - \cos 2\alpha, \quad C = -2B, \quad D = -2A \ , \tag{9}$$

r and θ are local cylindrical coordinates centered at the triple junction.

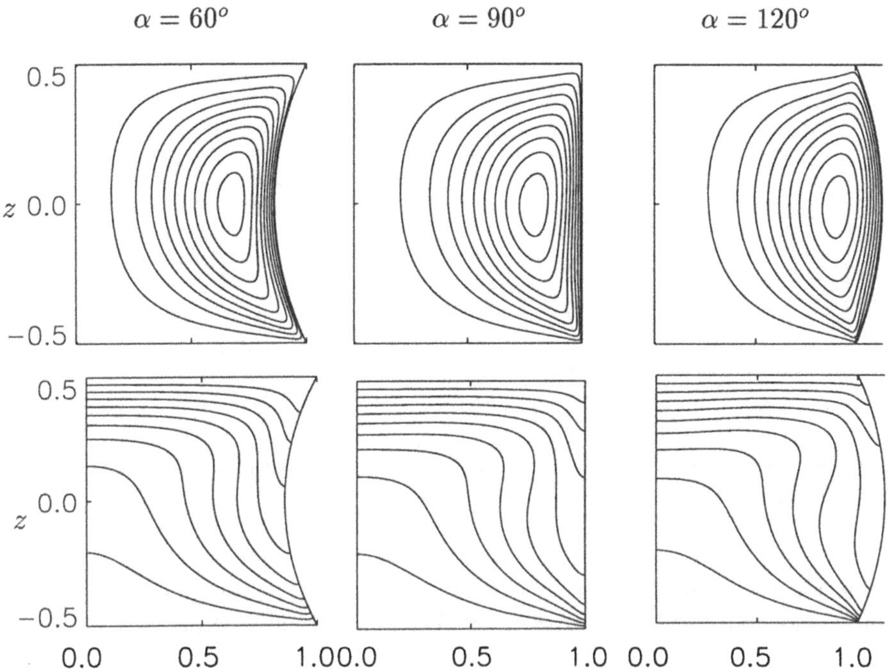

Fig. 3. Streamlines and isotherms for different contact angles α and $Re = 1000$, $Pr = 1$.

Apart from the well known vorticity singularity at the corner [3], ψ itself becomes singular everywhere when the contact angle is a zero of the denominator in (8). The only non-trivial root of

$$\sin 2\alpha - 2\alpha \cos 2\alpha = 0 \tag{10}$$

is $\alpha_c = 128.7°$. At this critical angle the stream function changes its sign. For $\alpha < \alpha_c$ the stream function is negative, while it is positive for $\alpha > \alpha_c$. Even though the Stokes flow approximation is not valid at $\alpha = \alpha_c$, there always exists a finite distance from the triple junction within which Stokes flow is a valid approximation, if $\alpha \neq \alpha_c$. The corresponding flow structures are shown in Fig.4 for a sub- and a supercritical contact angle. To accomodate for this local flow reversal when $\alpha > \alpha_c$ a flow separation on the free surface is expected.

3.3 Numerical Solution Near the Corner

To investigate the fine structure of the flow near the corner numerically, the computational area was successively subdivided. In a first step the steady-state solution in the whole cylindrical volume was calculated. Then we cut off a corner area as shown in Fig.5 and continue the calculation only in this sub-area. The grid resolution for each subdivision was twice that of the previous step. Linear interpolation was used to approximate the boundary values of ψ and ω along

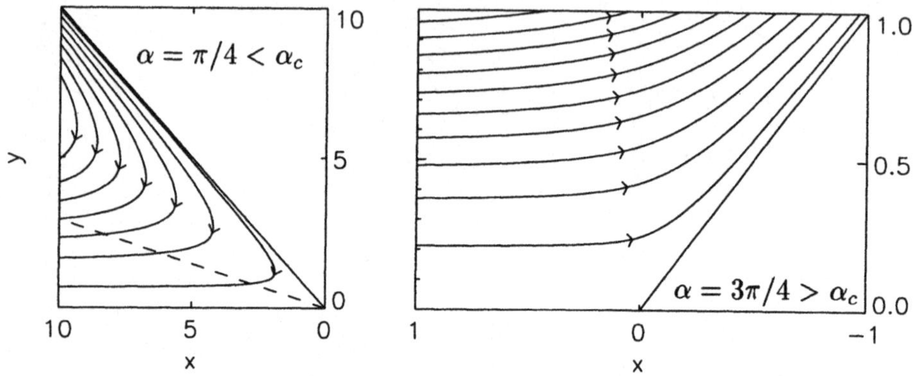

Fig. 4. Stream functions corresponding to Stokes flow solutions (8) near a corner for a subcritical (left) and supercritical angle (right).

the interior cut on the new grid. The numerical steady state solution on each sub-area was obtained by convergence of the transient calculation.

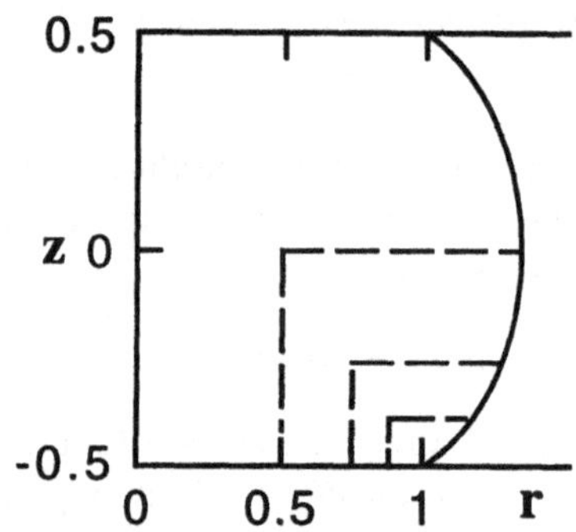

Fig. 5. Structure of successive subdivisions.

As the corner is approached, the flow will be dominated by Stokes flow. It is selfsimilar, i.e., the structure of the streamlines does not depend on the scale. As a test case, $\alpha = \pi/2$ was used. Selfsimilarity of the flow pattern was obtained within reasonable error bounds up to the seventh subdivision (height of sub-area is $4 \cdot 10^{-3}$) when terminating each transient calculation at $\tau = 2.5$. For further subdivisions, deviations from the exact solution occured. We attribute this to an error accumulation on the interior boundary. Further subdivisions are possible only if τ is increased. The same sequence of successive calculations was performed for the supercritical angle $\alpha = 135°$. The flow pattern was selfsimilar on every sub-area. However, the expected flow reversal could not be discovered up to the seventh subdivision of the computational domain. We conclude that a flow reversal, if present, must occur on even smaller length scales. In this context we note that the relative size of subsequent Moffat eddies [5] in rigid corners can be extremely small (depending on α).

4. Conclusions

Thermocapillary flows in liquid bridges were calculated for an extended range of contact angles. As α is increased from $\alpha = \pi/2$, the streamlines get crowded in the corners and the overall flow is enhanced. As $\alpha = \alpha_c$, the velocity field diverges in the corners and for $\alpha > \alpha_c$ local flow reversal is predicted. The length scale on which this flow reversal should occur was shown to be smaller than 10^{-4} for $\alpha = 135°$. Therefore, the numerical simulation on the original grid should not be sigificantly influenced by the presence of a small corner vortex for this contact angle. Further effort using different numerical techniques is required to settle this question.

In closing we note that the corner flow problem is of fundamental mathematical and numerical interest. The applicability of the results for large contact angles to real crystal growth applications is beyond the scope of the present investigation.

Acknowledgements

This work was supported by the Deutsche Agentur für Raumfahrtangelegenheiten under grant numbers 50 WM 1395 and 50 QV 8898.

References

1. Zh.Kozhoukharova and S.Slavchev, *J. Crystal Growth* **74** (1986) 236.
2. P. Roache, *Computational Fluid Dynamics*, Hermosa Publishers, Albuquerque, 1976.
3. A.Zebib, G.M.Homsy and E.Meiburg, *Phys. Fluids* **28** (1985) 3467.
4. F.Pan and A. Acrivos, *J. Fluid Mech.* **28** (1967) 643
5. H.K.Moffat, *J. Fluid Mech.* **18** (1963) 1.

Onset of Oscillatory Marangoni Convection in a Liquid Bridge

L. Carotenuto[1], *C. Albanese*[1], *D. Castagnolo*[1] *and R. Monti*[2]

[1] MARS Center - via Comunale Tavernola , 80144 Napoli, Italy
[2] DISIS, University of Napoli, P. Tecchio 80, 80125 Napoli, Italy

Abstract

It is well known that thermocapillary convection arises in liquid bridges when the isothermal support discs are at different temperatures. As the temperature difference increases the convective flow shows a transition from a steady axisymmetric to an oscillatory regime. Although a number of experiments have been performed and a considerable number of publications have appeared during the past ten years on oscillatory thermocapillary convection in liquid bridges, a definitive understanding of the physical mechanisms of the onset of instability and a coherent picture of the thermo-fluid-dynamic field has not yet emerged. On this subject the "Onset" experiment has been successfully performed during the D2 Spacelab mission using the Advanced Fluid Physics Module (AFPM). The experiment has investigated this transition for various bridge geometries and has provided quantitative measurements, which allow one to characterise the flow at the onset of the instability. The paper discusses the experimental observations, compars them with the available theoretical models of the instability. This comparison provides a validation of the oscillation description in terms of hydrothermal waves and also a new interpretation of the experimental results obtained previously. Finally a new scaling law for the oscillation frequencies at the onset is presented which agrees with all available experimental data.

1. Introduction

Thermocapillary convection, also called Marangoni convection, arises in liquid samples presenting a fluid-fluid interface with a nonuniform interface tension due to a temperature gradient along the interface itself. Liquid columns, held by solid circular discs, represent a typical configuration to study Marangoni convection. They model the floating-zone crystal growth technique and offer the ability to control the thermal stimuli imposed along the interface. Marangoni convection arises if a temperature difference (ΔT) is imposed between the two discs. If such a temperature difference exceeds a certain critical value ($\Delta \mathrm{T_c}$) a transition can be observed from steady axisymmetric (toroidal) flow to a time-dependent three-dimensional state, periodic in space and time. This regime is generally referred to as oscillatory Marangoni flow. Figures 1 show the flow pattern of the two regimes in a vertical section within the meridional plane that corresponds to the section illuminated by a light cut in the experiments. In the oscillatory regime the vortex in one half of the section appears smaller than the opposite vortex and the time dependence is observed as a periodic interchange of the vortices in the left and right sides of the zone.

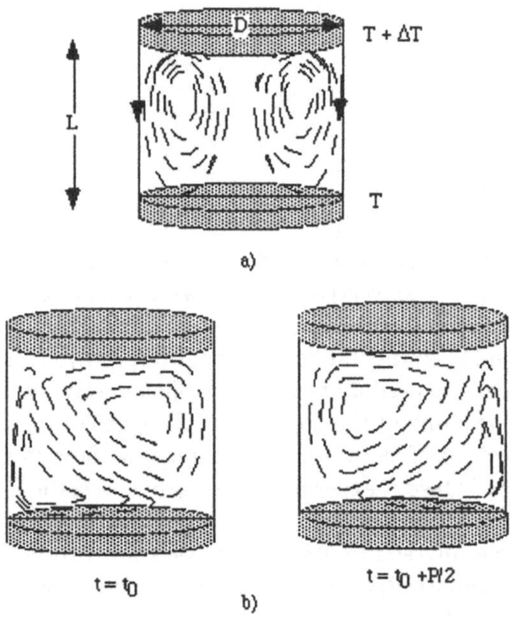

Fig. 1. Thermal Marangoni convection in a liquid bridge: a) axisymmetric regime; b) time dependent oscillatory regime at time t_0 and after a half period P/2.

A number of experiments have been performed by different investigators [1, 2, 3, 4, 5, 6, 7] who point out that the ΔT_c is a function of geometry (diameter D, length L, aspect ratio L/D), liquid properties (Prandtl number Pr) and thermal conditions at the free surface (Biot number Bi). These investigators also attempted to understand the "physical" mechanisms and to define critical characteristic parameters for the onset of instability, but the cause for the transition from a two-dimensional steady state to a time-dependent (oscillatory) flow is not yet completely understood. Ground experiments suggest that the flow organisation in the three-dimensional, time-dependent state could be characterised by an azimuthally rotating perturbation [1, 2, 3, 4, 5]. Microgravity experiments [6, 7] demonstrate that the onset of instability of thermocapillary convection is absolutely independent of natural convection or unsteady buoyant boundary layers related to gravity effects. They also provide qualitative and quantitative information about the dependence of the transition conditions on fluid properties and on the characteristic geometrical parametersof the bridge. However, due to the relatively small number of flight opportunities, to the limited available microgravity experimentation times and to the limits of the experimental facilities, the knowledge of the phenomenon is still incomplete. In particular there are some discrepancies between the present understanding of the flow organisation in the oscillatory regime and some predictions of theoretical studies as described below.

This paper illustrates the analysis of data provided by the Onset experiment performed in microgravity on board of Spacelab during the D2 mission. The results provide new insight regarding the flow structure in the oscillatory regime, and suggest a possible mechanism for the instability at large values of the Prandtl number.

2. Comparison between previous theoretical and experimental results

In recent years a number of theoretical and numerical studies have focused on the prediction of the critical conditions for the onset of oscillatory Marangoni flow. The most important contributions are those of Xu and Davis [8], Kuhlmann and Rath [9] and Neitzel et al. [10], who posed the problem in the framework of hydrodynamic stability theory to define the regions of nondimensional parameter space corresponding to sufficient conditions for stability and instability.

Linear stability analysis has been used in [8, 9] to investigate the stability of steady axisymmetric thermocapillary flow in a cylindrical liquid bridge. In [8] the basic state was obtained as an analytical solution for an infinite aspect ratio and in [9] as numerical solutions of the axisymmetric Navier-Stokes equations for unit aspect ratio together with the appropriate boundary and symmetry conditions. These authors solved the eigenvalue problem for the three-dimensional disturbances for various Prandtl numbers using spectral methods. On the other hand Neitzel et al. [10] used the energy method to determine the stability upper bound of the axisymmetric regime for different aspect ratios and unit Prandtl number. These results predict critical Marangoni numbers and the form of the most dangerous disturbances in the neighbourhood of the neutral stability point (i.e. close to the onset). The disturbed structure is characterised by the appropriate value of the critical azimuthal wave number (m). The eigenfunctions corresponding to each wave number are a couple of hydrothermal waves propagating in two opposite azimuthal directions. Because their relative amplitude is determined by non-linear effects these analyses cannot predict, if the oscillatory pattern is characterised by a pulsating disturbance (corresponding to a standing wave formed by the superposition of two equal amplitude counter-propagating waves) or by a rotating disturbance (corresponding to a single travelling wave travelling in the azimuthal direction).

In both cases, the instability appears as a temperature time oscillation at a fixed point. In the case of a single wave, the time-dependent temperature disturbance, periodic in space and time, is characterised by the rotation of hot (or cold) surface spots in the azimuthal direction. In the case of a standing wave, the spots oscillate in time, without any rotation.

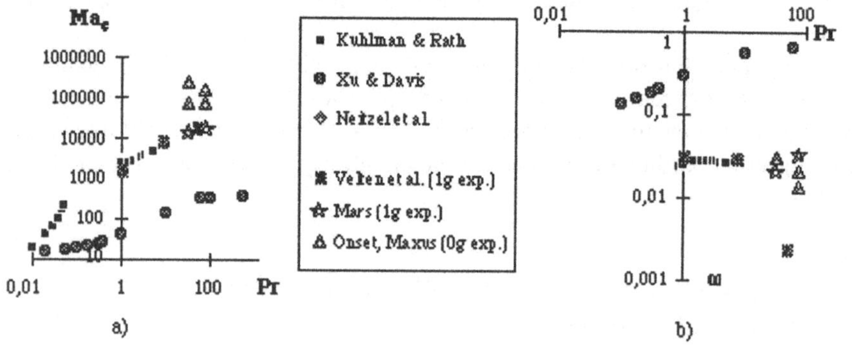

Fig. 2. Comparison between stability analysis and experimental results: a) critical Marangoni number Ma_c; b)angular frequency of the oscillations at onset.

According to the nature of thermal Marangoni convection it is obvious that the instability of steady laminar convection in high-Prandtl-number liquids is caused by the growth of temperature disturbances on the free surface of the liquid bridge. The disturbance amplification or damping on the free surface depends on the relative importance between convection and diffusion, as indicated by the magnitude of the Marangoni number

$$Ma = \frac{V_M L}{\alpha} = \frac{\Delta\sigma L}{\mu\alpha}$$

where V_M is the Marangoni velocity, $\Delta\sigma$ is the variation of the surface tension σ along the non-isothermal free surface, μ is the dynamic viscosity and α is the thermal diffusivity. In order to better position the present paper within the framework of numerical and experimental results obtained up to now, we summarize some data from numerical papers that consider the case of a liquid bridge in zero gravity and from experimental papers that report results obtained both on ground by means of micro-zone facilities and in space.

Figures 2 display the critical Marangoni number and angular velocity at the onset of instability as a function of the Prandtl number. The nondimensional experimental angular velocity is calculated using the measured frequency and scaled by V_M/L. The numerical data have been calculated using the same reference quantities. All the results have the same order of magnitude except those from [8], obviously due to the limiting hypothesis of an infinitely long liquid bridge.

It is important to point out that unsteady conditions induced by the time-dependent boundary conditions may affect the onset in space experiments. In

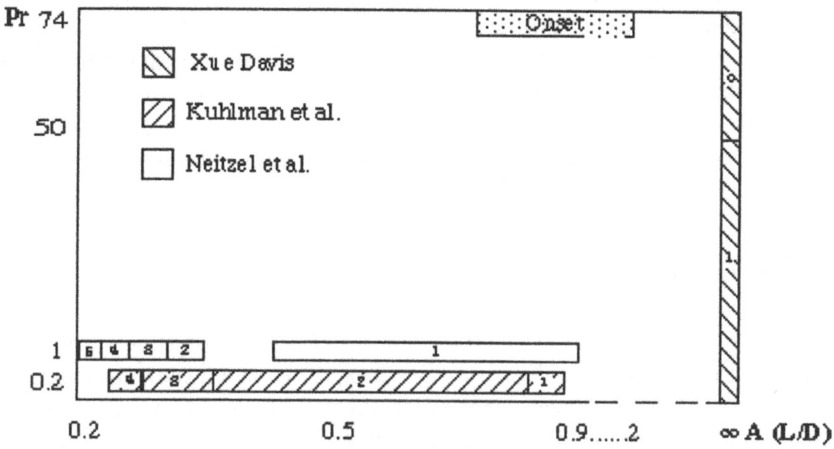

Fig. 3. Comparison between experiments and stability analysis: Oscillatory modes determined experimentally on ground in microzone columns and in microgravity.

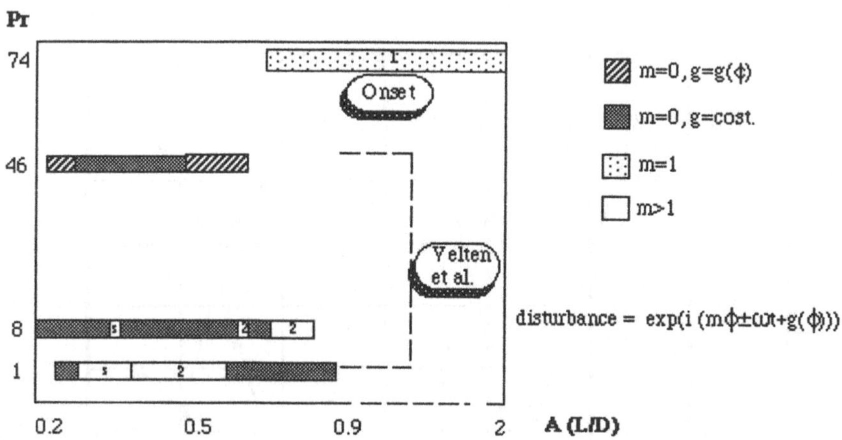

Fig. 4. Comparison between experiments and stability analysis: Oscillatory modes calculated by linear stability analysis in the case of zero gravity.

fact, short-duration experimental procedures (for example limited to 6-12 minutes in the case of sounding rockets) produced values of the critical Marangoni number larger than the expected values obtained from linear stability analysis of steady basic states as shown in Fig.2a.

However, the main discrepancy between numerical predictions and experimental results concerns the nature of the oscillatory mode. Figs. 3,4 show the m values calculated numerically and obtained by analysis of experimental data both as a function of aspect ratio and the Prandtl number Pr. Stability analysis predicts only non-axisymmetric modes with $m \geq 1$ depending on the aspect ratio considered. On the other hand previous ground experiments indicate $m = 0$, in particular, for high Pr numbers (Fig. 4). The Onset experiment due to the measurement of the complete surface temperature field yielded a value of $m = 1$ and also provided an explanation for the discrepancy with the previous ground results as described below.

3. Oscillatory regime

The Onset experiment, performed on board Spacelab during the D2 mission, allowed the investigation of the instability using a relatively slow, time-dependent flow and to use the advanced diagnostics of the Advanced Fluid Physics Module (AFPM). The description of this apparatus and the experiment is reported in [11]. The Onset experiment has provided the first complete experimental observation of the disturbance structure. The liquid velocity has been measured both at the surface and in the meridional plane of the bridge. These data have been correlated with the surface-temperature distribution measured with the AFPM thermocamera.

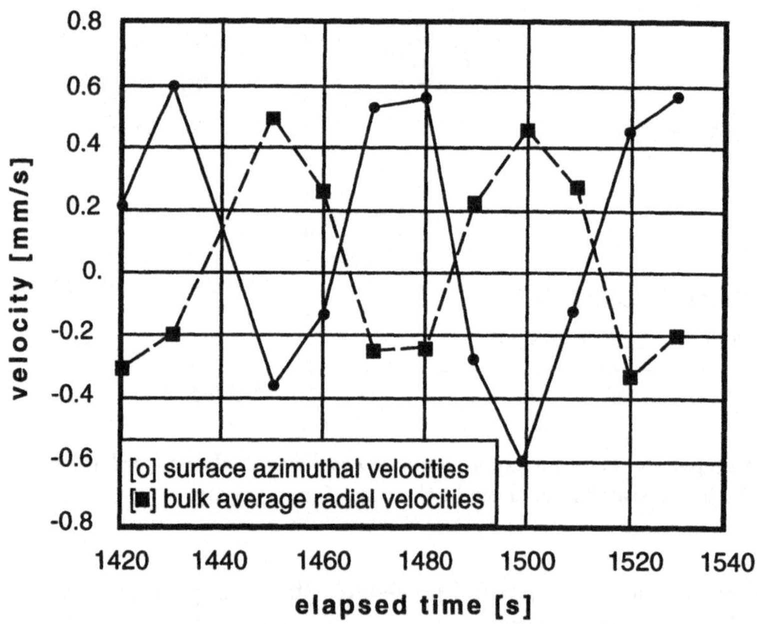

Fig. 5. Time evolution of the radial velocity in the meridian plane and azimuthal velocity at the surface.

In the axisymmetric regime the flow is characterised by the axial velocity (directed toward the hot disc in the bulk return flow and toward the cold disc at the surface) and the radial velocity (directed toward the surface close to the hot disc and toward the axis close to the cold disc). Due to symmetry the mean value of the radial velocity averaged in the meridional plane is zero. The azimuthal velocity is zero everywhere. With respect to this pattern the oscillatory regime shows the following features. The bulk flow has an average radial component in the meridian plane which changes its direction periodically with a zero mean value. The same behaviour also is observed for the azimuthal velocity component at the surface. Figure 5shows that they have opposite phases. Correlation with the thermocamera data confirms that the azimuthal velocity always is directed opposite to the azimuthal component of the surface temperature gradient as expected on the basis of the Marangoni effect. The corresponding flow organisation is shown in Fig.6. A possible mechanism for the oscillations is described below.

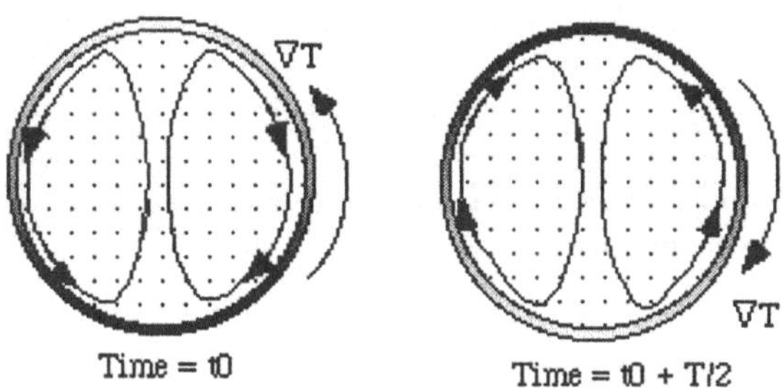

Fig. 6. Sketch of the flow pattern and surface temperature in a transversal section of the bridge with reference to Fig.5.

The thermocamera data show the development of a pulse disturbance at the onset. An example is shown in fig. 7 representing the time evolution of the surface-temperature distribution along the azimuthal direction measured at fixed axial position (z=L/10) close to the hot disc (the data refer to the run with 60 mm disc diameter and 45 mm bridge length). As shown the disturbance develops always in the same axial position without any rotation. Similar results have been obtained also from the other experimental runs with bridges of different diameter and length. The observed pattern is not axisymmetric ($m \neq 0$) and is compatible with a unit value (i.e. $m = 1$) of the azimuthal mode (only one maximum develops in the visible part of the free surface). This result is also compatible with theoretical predictions (see Fig. 3).

Further, evidence of a standing wave exists. In this pattern there is no phase shift between signals at different azimuthal positions. In this case the procedure used in [4] to obtain the oscillatory mode gives $m = 0$ even if the pattern is not axisymmetric. The present observations seem to indicate that a standing wave pattern with $m \neq 0$ could correspond to the $m = 0$ results of Fig. 4 thus explaining the difference between these and the predictions of stability analysis.

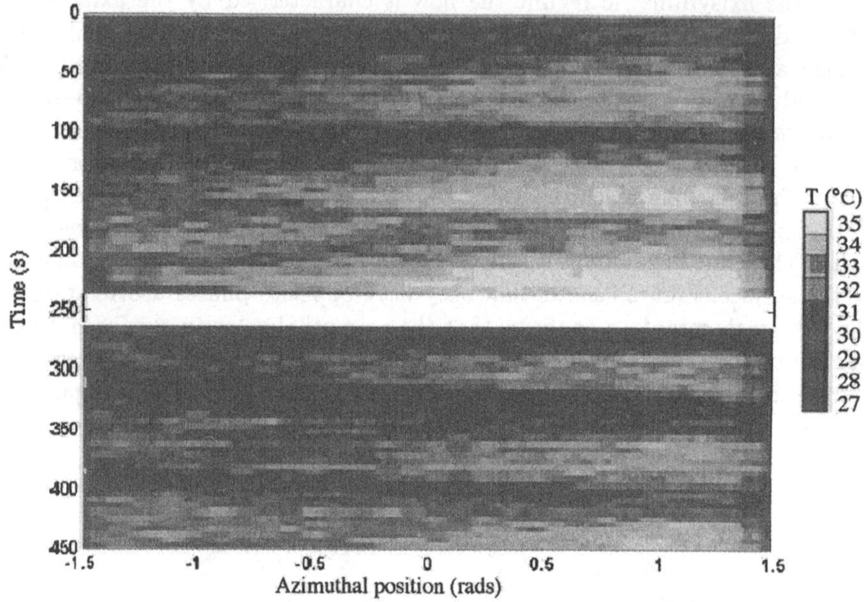

Fig. 7. Time evolution of the surface temperature measured close to the hot disc (axial position z=L/10; L=6 cm; D=6 cm; Δ T = 16 °C).

4. Critical oscillation frequency

Information on the transport mechanism of the disturbance can be obtained by the value of the oscillation frequency. A detailed numerical analysis of the propagation of a temperature disturbance is illustrated in [9]. According to [9] the thermal perturbation of the basic state appears as a spot in the bulk of the liquid bridge and then propagates to the surface by heat conduction. Once the spot approaches to the surface, it causes a convective thermocapillary motion in the azimuthal direction. The resulting convective surface flow, via continuity, causes a radial flow of cold liquid from the interior of the liquid bridge and of warmer liquid from the surface which creates, by immersing itself in the bulk, a new internal hot spot in an azimuthal position slightly rotated with respect to the previous one. By this way the disturbance propagates azimuthally. This mechanism becomes more effective as the temperature difference between the endwalls increases (i.e. as Ma increases), and leads to an oscillatory convective regime in which a hydrothermal wave rotates in the azimuthal direction on the surface of the liquid bridge.

Unfortunately this analysis does not apply to our case. In fact due to computational limits [9] considers only fluids with Pr$\leq$ 4. The Onset experiment, however, has been performed using silicone oil with Pr=74. In this case thermal diffusion should not have any significant effect on the propagation mechanism of the disturbance and its transport should be purely convective. Instead the

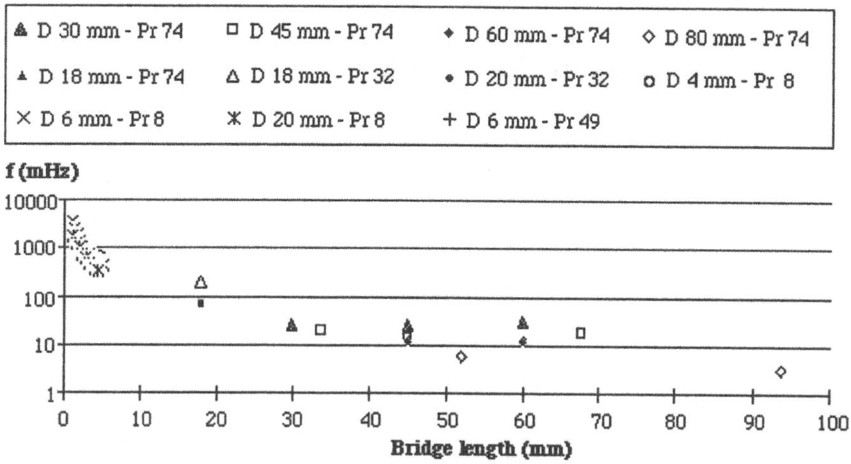

Fig. 8. Dependence of the oscillatory frequency measured at onset on the bridge length.

flow inversion may be explained as follows. Any temperature fluctuation which arises in the bulk is transported by the inner flow to the surface close to the hot disc. As soon as an azimuthal temperature gradient appears at the surface an azimuthal flow is established due to the Marangoni effect. Via continuity, the bulk return flow is modified: due to the increased surface flow departing from the warmer part of the surface, the return flow is mainly directed toward that part of the surface. Since the liquid of the return flow is colder than that at the surface it cools the surface below the average temperature at the same axial position. This overshooting induces the inversion of the azimuthal flow direction and then oscillations occur.

A preliminary qualitative validation of this hypothesis is provided experimentally by the measured values of the oscillation periods of the time-dependent convection which are certainly lower than the characteristic times $t_{\mathrm{d}} \approx \mathrm{L}^2/\alpha$ foreseen by diffusive transport over a distance L. The order of magnitude of the measured oscillation periods is about one minute, while the diffusive times, calculated for a distance of half a radius are $t_{\mathrm{d}} = R^2/4\alpha \approx$ 20-25 minutes. A second validation has been obtained by comparison with the results of other experiments performed on ground (with the microscale technique) [6, 4, 5] as well as in microgravity [3, 7] using liquids with a relatively high Prandtl number. All these experiments performed on liquid bridges of various lengths (ranging from 1 mm to 60 mm), yield values of the oscillation frequencies at onset which range from few Hz, for bridges of very small dimensions, down to few mHz for the larger liquid bridges. The measured values of the oscillation frequencies are shown in Fig. 8 as a function of the bridge length for various Prandtl numbers.

According to the hypothesis that the instability mechanism in high-Pr liquid bridges is provided by convective flows the oscillation frequencies should scale with a reference time related to the convective transport time instead of the

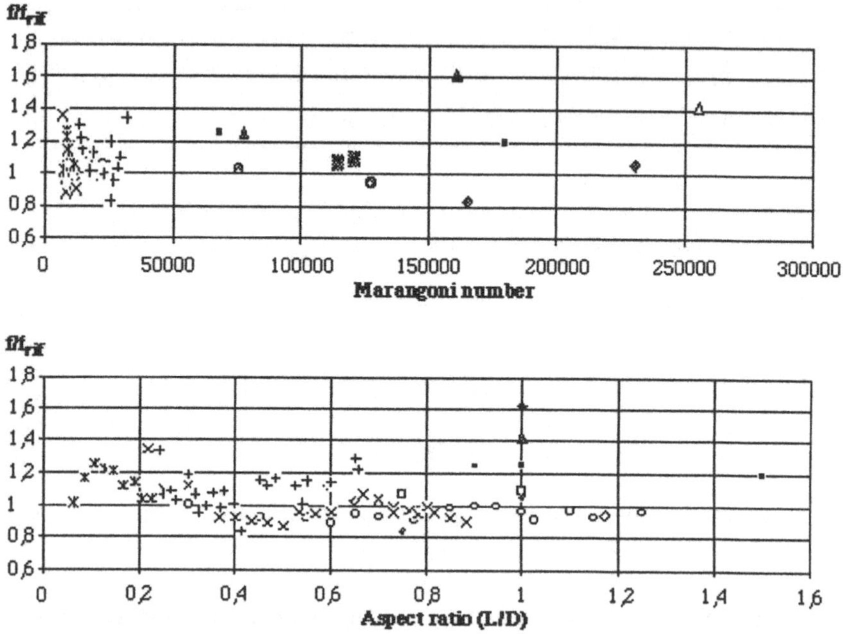

Fig. 9. Plots of the non-dimensional frequency f/f_r at onset versus Marangoni number and aspect ratio; symbols are reported in Fig. 8.

diffusive t_d defined above. Following an order of magnitude analysis [12] we define a reference velocity $V_\mathrm{r} = \omega L_\mathrm{r}$, where V_r and L_r are the reference velocity and length, respectively, and ω is the angular oscillation frequency. It has been found from the experimental data that an appropriate reference length is $L_\mathrm{r} = \sqrt{\mathrm{LD}}$, where L is the length of the bridge and D its diameter. According to the order of magnitude analysis, the reference velocity of the thermocapillary convection at the onset conditions is $V_\mathrm{r} = V_\mathrm{M} Ma^{-1/3}$ (V_M is the convective Marangoni velocity). This theoretical prediction for V_r has been demonstrated by the experimental measurements of surface velocity obtained by the Onset experiment [11]. Therefore, in a straightforward manner, the following expression is obtained as a reference for the oscillation frequency at onset:

$$f_\mathrm{r} = \frac{1}{2\pi}\frac{V_\mathrm{r}}{L_\mathrm{r}} = \frac{1}{2\pi}\frac{V_\mathrm{M}}{\sqrt{\mathrm{LD}}} Ma^{-1/3} \ .$$

This expression relates f_r to the geometrical parameters of the bridge, to the characteristic parameters of the fluid and to the critical conditions at onset, which are accounted for by using ΔT_c in the definition of V_M and Ma. The f_r values corresponding to each experimental measurement f of Fig. 8 have been calculated; the results are shown in Fig. 9 in terms of the nondimensional frequency f/f_r . The nondimensional frequencies at the onset of the time-dependent instability have been plotted as a function of the aspect ratio (L/D) and of the Marangoni number. All values are of O(1) and no trend emerges, even for the wide range of experimental conditions (the values of Prandtl number range from 8 to 74). These findings demonstrate that this scaling law is correct for Pr > 1. It is expected that this scaling, based on a purely convective transport of the thermal disturbance, should fail at lower values of Pr as a consequence of the increased importance of thermal-diffusion effects.

5. Conclusions

The microgravity Onset experiment has provided for the first time the structure of the oscillatory pattern of Marangoni convection in large liquid bridges due to the advanced diagnostics of AFPM (thermocamera and LED illumination). The flow and temperature fields seem compatible with a superposition of two counter-propagating hydrothermal waves, corresponding to a standing wave, which is also in agreement with recent 3D time-dependent numerical simulations and linear-stability analyses. A scaling law has been identified, which correlates the oscillation frequency at the onset to the critical conditions, the bridge geometry and liquid properties which holds for high Pr number liquids (Pr > 1).

Acknowledgements

The authors are grateful to Dr. E. Ceglia for his help in the data elaboration and to Prof. P. Neitzel for his suggestions and useful discussions. The authors also thank Dr. R. Savino for the comparison with results of the 3-dimensional numerical simulations performed at DISIS. This research has been performed with the financial support of the Italian Space Agency (ASI).

References

1. C.H.Chun, *Acta Astronautica*, **17** (1980) 479.
2. D.Schwabe, A.Scharmann, *J. Crys. Growth* **46** (1979) 125.
3. F.Preisser, D.Schwabe, A.Scharmann, *J. Fluid Mech.*, **126** (1983) 545.
4. R.Velten, D.Schwabe and A. Scharmann, *Phys. Fluid A* **2** (1991) 267.
5. R.Monti, D.Castagnolo, P.Dell'Aversana, G.Desiderio, S.Moreno, G.Evangelista, *An experimental and numerical analysis of thermocapillary flow of silicone oils in a micro-floating zone*, 43rd IAF Congress - Washington, USA (1992).
6. R.Monti, R.Fortezza, G.Mannara, *Results of the Texus 14b flight experiment on a floating zone-First approach towards Telesciences in Fluid Science*, 38th IAF Congress, Brighton (1987).
7. R.Monti, R.Fortezza and G.Desiderio, ESA SP-1132 (1992) 2.
8. J.J.Xu, S.H.Davis, *Phys. Fluids* **27** (1984) 1102.
9. M.Wanschura, V.M.Shevtsova, H.C.Kuhlmann,H.J. Rath, *Phys. Fluids* **7** (1995) 912.
10. G.P.Neitzel, C.C.Law, D.F.Jankowski, H.D.Mittelmann, *Phys. Fluids A*, **3** (1991) 2841.
11. C.Albanese, L.Carotenuto, D.Castagnolo, E.Ceglia, R.Monti, *Adv. Space Res.*, **16**, No.7, (1995).
12. L.G.Napolitano, *Acta Astronautica*, **9** (1982) 199.

Convection Visualization and Temperature Fluctuation Measurement in a Molten Silicon Column

S. Nakamura, K. Kakimoto and T. Hibiya

Fundamental Research Laboratories, NEC Corporation Miyukigaoka 34, Tsukuba 305, Japan

Abstract

Convection visualization and temperature fluctuation measurements in a molten silicon column were performed in a floating-zone configuration using a mirror furnace. The flow velocity of a tracer in a column of 10 mm in height and diameter was determined by flow visualization using X-ray radiography. The tracer velocity was found to be larger than 20 mm/sec. The fast fourier transform spectrum of the temperature fluctuations in the liquid bridge had a broad peak at a frequency of 0.5 Hz.

1. Introduction

In crystal growth processes of semiconductors thermocapillary convection (Marangoni convection) plays an important role in the formation of dopant striations in grown crystals. A crystal growth experiment with Si in a floating zone (FZ) system, which was carried out in microgravity, proved the existence of time-dependent Marangoni flow through dopant striations [1]. The critical Marangoni number for molten Si has been determined to be about 150 to 200, around which the flow changes from steady to time-dependent [2]. For crystal growth experiments with GaAs Rupp and et al. demonstrated that in a floating-zone with symmetrical thermal boundaries the flow in the melt was asymmetric [3].

In microgravity, since buoyancy driven convection is suppressed, many experiments focusing on Marangoni convection have been carried out with organic liquids and molten salts [4]-[6]. However, only a few model experiment have been done using molten semiconductors.

The purpose of this study was to investigate the structure of convection in a molten silicon column. X-ray radiography with tracer particles was used to observe the convective flow. (The same technique was used by Otto [7] to observe the phase separation in immiscible liquid metals.) Since the mode of convection is related to the mechanism of heat and mass transfer our experiments ultimately will led us to understand the mechanism of crystal growth in FZ processes.

2. Experimental Set Up

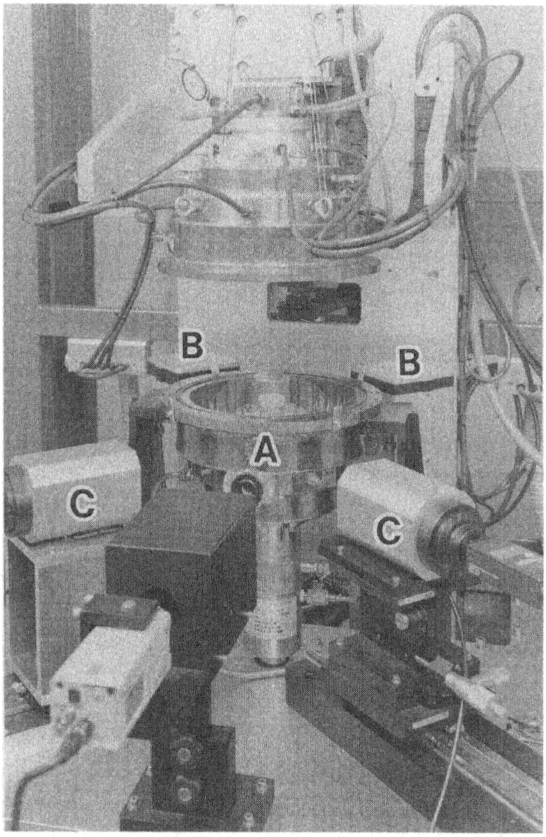

Fig. 1. Floating zone mirror furnace with X-ray visualization system; (A) mirror furnace; (B) X-ray generator;(C) X-ray imaging unit.

Figure 1 shows the ground-based experiment facility. A mono-ellipsoidal mirror furnace was used to heat the specimen. The maximum lamp power of the furnace was 1500 W. The highest temperature in a silicon volume of $10\times10\times10$ mm^3 is about 1800 K. Two sets of X-ray generators and detectors were used for the visualization of convection using tracer particles. The target size of the X-ray tube was 10 μm. The maximum voltage and current of the X-ray tube were 70 kV and 0.1 mA. The spatial resolution of the X-ray detector was about 50 μm in the observation of tracer particles in a molten silicon column of 10 mm diameter if the tracer particle core was a heavy metal.

Figure 2 (left) shows the system configuration of the experiment facility. The motion of tracers in the molten silicon column was simultaneously observed by two sets of X-ray systems and recorded by a video tape recorder (VTR).

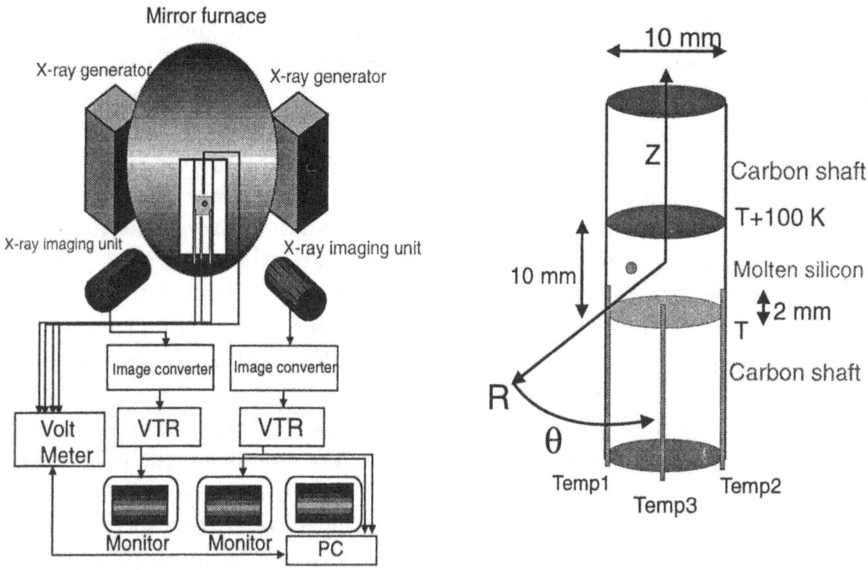

Fig. 2. *Left*:Experimental system configuration. *Right*:Specimen structure.

Since the original images were very faint, an image converter with an averaging function was used to improve the contrast of the image before recording. The three dimensional position of the tracers were calculated from the VTR images in which the position of the tracer was recorded every 1/30 second. The recording time error was considered to be negligible. The total uncertainty in determining the position, however, was estimated about 100 μm for our system.

To investigate thermocapillary convection in low-Prandtl-number liquids, such as molten semiconductors, we chose a liquid column system. This system had often been used for several kinds of liquids at near-room temperature. The merits of using the liquid column system in microgravity are not only that buoyancy convection can be avoided, but that a non-deformed melt surface can be obtained. The latter is more important for simplifying the shape of the molten silicon in computer simulations.

Figure 2 (right) shows schematically the structure of the molten silicon column. The molten silicon column was formed between carbon shafts of 10 mm in diameter. The atmosphere was 6N argon. To avoid spreading of the molten silicon on the carbon surface the surface of the shaft was coated with glassy carbon which is wet minimally by molten silicon (except for the interface between the carbon and the molten zone of the silicon) .

The structure of our tracer was as same as those with a tungsten core used for observing the flow of molten silicon in a crucible of the Czochralski (CZ) crystal growth process [8]. However, the size of our tracer in the present work was reduced using a zirconium oxide core, whose density is about one-third that of tungsten. The tracer, 0.9 mm in total diameter, consisted of a 350-μm ZrO_2 core coated with silica to adjust the density of the tracer to the density of molten silicon. Carbon was coated on the silica to improve the wettability by the molten silicon. Without the carbon layer the tracer could neither stay nor follow a flow inside the molten silicon column.

The temperature at the upper interface between the molten silicon and the carbon shaft was about 100 K higher than that of the lower interface in order to suppress buoyancy convection. Temperature fluctuations in the molten silicon column were measured using three Pt-Pt13%Rh thermocouples (Temp1 ~ Temp3) capped by a quartz tube. As shown in Fig. 2 (right) the tips of the thermocouples were positioned near the side surface of the silicon column 2 mm above the lower interface. The outer diameter of the quartz tube was about 0.95 mm. Its tip was coated with carbon.

This study was performed as a part of a feasibility study for experiments on board the TR-IA sounding rocket and the Japanese module for the space station. The first flight of the experiment is scheduled to perform temperature fluctuation measurements in a molten silicon column in August 1995. The specifications of the facility (Zone Melting Furnace) under development is shown in Tab. 1.

Item	Specifications for furnace
Temperature range	723 ~ 1823 K
Temperature stability	± 3 K at 1827 K for 1 hr
Temperature gradient	100 ~ 400 K/cm
Pulling rate	0.5 ~ 100 mm/hr
Sample dimension	ϕ 20 × 120 L
Atmosphere	Ar, N_2, 1×10^{-4} Torr ~ 1 atm.
Item	**Specifications for X-ray system**
X-ray tube voltage	70 KV
X-ray tube current	0.1 mA

Table 1. Specifications of the Zone Melting Furnace facility.

3. Results and Discussion

Figure 3(left) shows the trajectory of a tracer imposed into the radial-axial (r-z) plane in the polar coordinate. The rotation rate was 10 rpm. The Fig. 3(left) shows that the tracer moved downward near the melt surface and upward in the center of the column. This clearly suggests that the flow was dominated by the thermocapillary flow occurring at the melt surface. Under our experimental

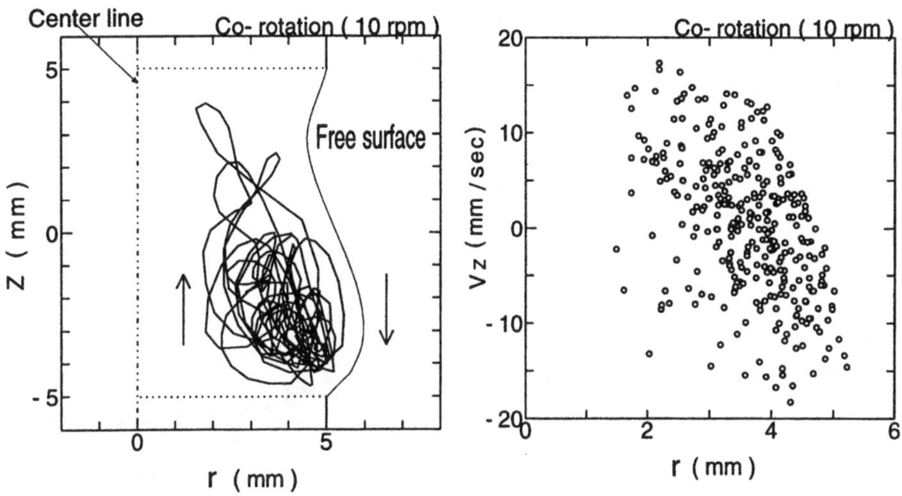

Fig. 3. *Left*: Tracer trajectory in the r (radius) - z (axial) plane. Marangoni convection on the surface influences the flow inside the molten silicon where the temperature at the upper interface was about 100K higher than at the lower interface.
Right: Tracer velocity of tracer particle in the r-z plane.

temperature conditions the Marangoni number Ma was estimated according to

$$\mathrm{Ma} = |\frac{\partial\sigma}{\partial T}| \cdot \frac{h\Delta T}{\nu\rho\kappa} \quad , \tag{1}$$

where ν is kinematic viscosity, ρ is density, κ is thermal diffusivity, ΔT is the temperature difference between the upper and lower interface, $\partial\sigma/\partial T$ is the temperature coefficient of the surface tension, and h is the height of molten zone. The temperature coefficient of the surface tension was assumed to be -0.2×10^{-3}N/mK since the temperature coefficient of surface tension for a carbon saturated silicon melt has not be measured. The temperature difference was 100 K. The Ma number for our system was 17070. Since the critical Marangoni number Ma_c from steady to time-dependent flow was considered to bearound 100-200 [2] the flow in the column was thought to be time-dependent. On the ground, the buoyancy force also must have an effect on the flow in the column. However, as shown in Fig. 3 (left), the buoyancy force for driving the convection was smaller than the thermocapillary force.

Figure 3 (right) shows the axial velocity V_z of the tracer as a function of radius R. At the surface, the tracer velocity was about 20 mm/s. Since the tracer diameter (~0.9 mm) was considered to be rather large compared to the thickness of the boundary layer of the Marangoni flow the tracer was not able to follow the streamline near the silicon surface.

Figure 4 (left) shows the temperature fluctuation in the silicon column. Temperature fluctuations indicates unsteady flow. It is, however, apparent that there is some phase correlation between the different temperature recordings. The fig-

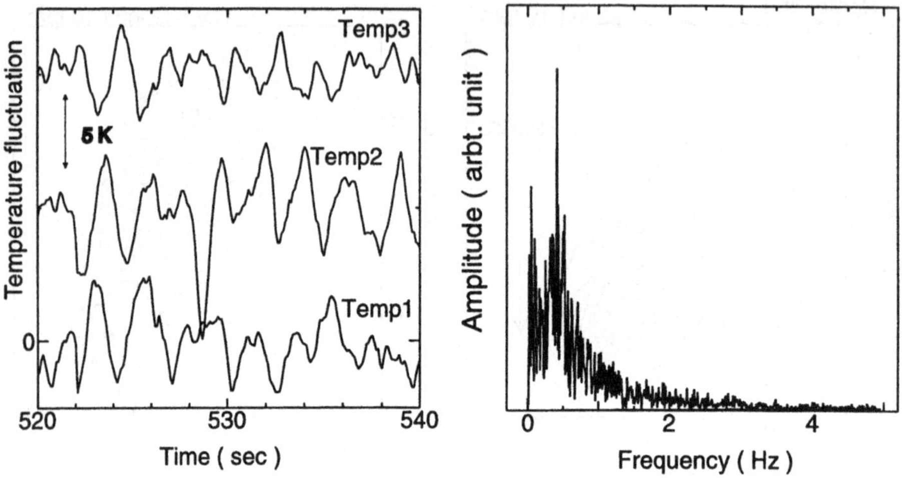

Fig. 4. *Left*: Temperature distribution in the molten silicon column. *Right*: FFT spectrum of the temperature fluctuations.

ure shows that temperatures at 90-degree angles (Temp1 and Temp3, Temp2 and Temp 3) seem to be out of phase. This relation may be explained by a two-fold rotating or asymmetrically fluctuating temperature. Since a temperature difference of about 30 K was observed between Temp1 and Temp2 it is, however, difficult to determine which intrinsic temperature asymmetry was observed or which asymmetrical temperature condition around the molten silicon caused this fluctuation.

Figure 4 (right) shows the fast Fourier transfrom (FFT) spectrum of the temperature fluctuations for the case of Temp2. There exists a broad peak about 0.5 Hz in the molten silicon. This frequency is due to large amplitude fluctuations in the temperature shown in the Fig. 4(left).

Acknowledgments

The authors thank Dr. Kyung-Woo Yi of the Fundamental Research Laboratories, NEC Corporation (present address: the Korea Institute of Science and Technology) for his fruitful discussion and M. Eguchi for his skillful technical assistance. We also thank team members for developing the zone melting furnace in the Space Development Department, NEC. This study was supported by the Frontier Joint Research Program of the Japan Space Utilization Promotion Center under the direction of National Agency of Space and Development in Japan.

References

1. A. Eyer, H. Leiste, R. Nitsche, Proc.Vth European Symp. on Materials and Fluid Sciences in Microgravity ESA SP-222 (1990) 173-181.
2. A. Cröll, W. Müller-Sebert, and R. Nitsche, Proc.VIIth European Symp. on Materials and Fluid Sciences in Microgravity ESA SP-295 (1990) 263-269.
3. R.Rupp, G. Müller, and G. Neumann, *J. Crystal Growth* **97** (1989) 34-41.
4. Y. Kamotani, S. Ostrach, M. Vergas, J.Crystal Growth, 66 (1984)83-90.
5. Ch-H. Chun, Proc.Vth European Symp. on Materials and Fluid Sciences in Microgravity ESA SP-222 (1984) 271-280.
6. D. Schwabe, R. Velten, A. Scharmann, *J.Crystal Growth* **99** (1990) 1258-1264.
7. G. H. Otto, Proc.Vth European Symp. on Materials and Fluid Sciences in Microgravity **ESA SP-222** (1990) 379-388.
8. K.Kakimoto, M. Eguchi, H. Watanabe, T. Hibiya, *J. Crystal Growth*, **88** (1988) 365-370.

The Micro Wedge Model: A Physical Description of Nucleate Boiling Without External Forces

J. Straub

Lehrstuhl A für Thermodynamik, Technische Universität München, Arcistraße 21, 80290 München

Abstract

Boiling experiments performed during the past years in microgravity environment obviously demonstrate that the existing theoretical models or semi-empirical correlation's extrapolated to lower gravity levels are in contradiction to the experimental findings. The overall heat transfer is hardly influenced by gravity while the bubble dynamics itself is strongly affected. Obviously external forces like gravity in pool boiling and shear forces in flow boiling only play a secondary role, while internal ones govern the process like interfacial forces. Based on these observations the microwedge model is proposed which can be regarded as an improvement of the microlayer model.

1. Introduction

In spite of the extensive technical application of nucleate boiling as the most efficient process of heat transfer from a heated solid surface to liquids the variety of phenomena and their interaction which contribute to the heat transfer are still inadequately understood [1, 2]. High power density systems on earth and in the new field of space technology demand comprehensive knowledge of the basic thermo- and hydrodynamic mechanisms and their limitations for their optimal use on earth and in space applications. Most equations of boiling heat transfer are characterized by a dominant gravity dependence of the complex mechanisms. Therefore it was doubted that boiling can be applied as an efficient process of heat transfer in microgravity at all, whereby the results of earlier works [3] have not been recognized because it was assumed that these results are affected by transient effects caused by the short duration of the microgravity time used in the early experiments.

Detailed investigations of the gravity dependency was realized by quasi steady state measurements within the TEXUS [1] program, in parabolic flights

[1] TEXUS is the acronym of a sounding rocket program, which provides during free parabolic fall of the rocket reduced gravity levels of the order $10^{-4} g_{\mathrm{Earth}}$ for approx-

using the KC 135 [2] aircraft [2], and recently in the BDPU during the IML-2 mission [4][3]. The statement of the results of these experiments is consistent: the heat transfer is hardly influenced by gravity and can not be described by any of the existing theories and correlations. Therefore the question arises what are the forces and the mechanisms which determine the heat transfer without buoyancy and maintain nucleate boiling in microgravity?

The heat flux Q between a heated or cooled solid wall and a fluid is generally described by Newton's law

$$Q = \alpha \cdot A(T_{\mathrm{w}} - T_{\mathrm{f}}) \tag{1}$$

where α is the heat transfer coefficient which determines the effectiveness of the transport process. A is the surface area of the heater, T_{w} is the temperature at the wall, and T_{f} is the fluid temperature which is in case of boiling the saturation temperature T_{sat} of the liquid. At boiling the heat transfer coefficient α is orders of magnitude larger compared to a single component liquid. Generally α can be described by the conservation equations for mass, momentum and energy if the physical process and the boundary conditions are known. However, the processes in two phase systems especially in boiling are too complex. At a certain temperature in the superheated liquid layer at the heater surface nucleation occurs, a bubble is formed, grows and detaches if the buoyancy force becomes larger than the holding forces. This process is repeated with high frequency and on many places - the nucleation sites - of the heater. The latent heat and the hot liquid in the wake of the departing bubble determine the overall heat transfer coefficient α. Heat transfer correlation based on that picture are gravity dependent. If all other parameters of the liquid and the heater are constant, the correlation can be analyzed with respect to their gravity effect by forming the ratio of the heat transfer coefficients:

$$\frac{\alpha_{\mu g}}{\alpha_{1\mathrm{g}}} = \left(\frac{a}{\mathrm{g}}\right)^{n} \tag{2}$$

where a/g is the actual system acceleration as fraction of the earth gravity g. The sign and the value of the exponent n indicates the change of the heat transfer ratio with gravity variation. In theoretical models based on the physical concept of bubble detachment by gravity n = 1/2. This means that the heat transfer coefficient is strongly reduced with decreasing acceleration. In Fig. 1 results obtained during KC 135 aircraft campaigns [2] are shown versus the

imately 6 minutes. The program is funded by the German Space Agency DARA and the ESA. TEXUS stands for **T**echnologische **EX**perimente **U**nter **S**chwerelosigkeit. (Eds. note)

[2] KC 135 is the abbreviation for an American parabolic flight program with a B707 airplane which provides quasi-periodic reduced gravity levels of the order of $10^{-2} g_{\mathrm{Earth}}$ for approximately 20 seconds, interrupted by gravity levels of two times g_{Earth}. (Eds. note)

[3] The acronym BDPU is explained in detail in the contribution of Géoris and Legros in this volume. (Eds. note)

heat flux density for various fluid states represented by the reduced pressures $p_r = p_{sat}/p$ which is the ratio of saturation pressure p_{sat} to the critical one. It may be recognize that the measured values under reduced gravity and at low heat flux values are even higher than the values at 1g. They are reduced only to 0.85 at higher heat fluxes.

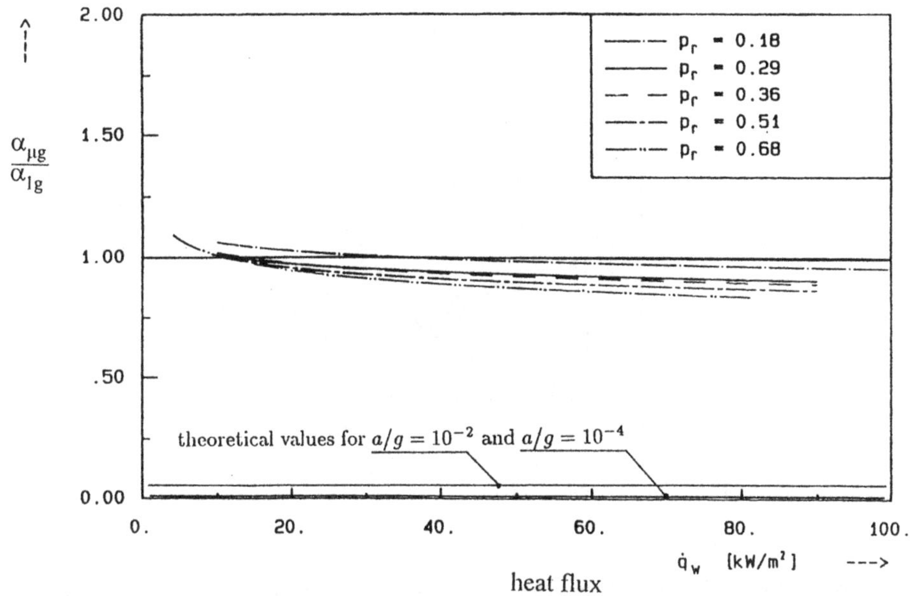

Fig. 1. Ratio of heat transfer coefficients versus the heat flux from KC 135 flights, results from flat plate 40 x 20 mm^2, parameter $p_r = p_{sat}/p$ and comparison to theoretical values.

These results lead to the conclusion that the primary mechanisms which determine the heat transfer are more or less independent of gravity. They are determined by effects which take place at the conduct line of the heater surface and the liquid. The secondary mechanisms are different at earth gravity and at microgravity. They determine the transport of the heat by the departing vapor bubbles to the bulk liquid. In microgravity these mechanisms are caused by the bubble dynamics itself, by coalescence processes, by surface tension and thermocapillary flow.

2. The Model

Stimulated by the publication of Wayner et al. [5] and Stephan [6] which treat the evaporation on the contact line of a thin liquid film and the evaporation in heat pipes we transferred these ideas to boiling bubbles [7]. The heat transfer is determined by the capillary wedge formed between the solid wall of the heater and the interface of the bubble (Figs. 2,3). Strong evaporation takes place at the liquid-vapor interface and the resulting momentum change across the interface is balanced by the liquid-vapor pressure difference, the surface tension and the adhesion pressure at the solid surface. This force balance can induce a liquid flow into the wedge.

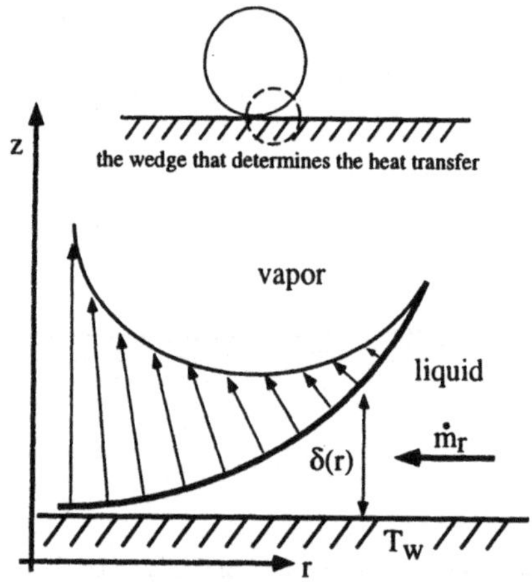

Fig. 2. Sketch fo the microwedge Model.

The evaporating mass flow $\dot{m}_{\mathrm{ev}}$ is caused by the difference between the interfacial temperature of the liquid T_{i} and the vapor (gas) temperature T_{g} and can be determined by the kinetic theory of Hertz-Knudsen:

$$\dot{m}_{\mathrm{ev}} = \beta \rho_g \sqrt{\frac{k}{2\pi m^*}} (\sqrt{T_{\mathrm{i}}} - \sqrt{T_{\mathrm{g}}}) \tag{3}$$

where k is the Boltzmann constant, m^* the mass per molecule, ρ_g the density of the vapor. β is the evaporation coefficient, which considers the real dynamic interaction of the molecules at the interface and is of the order of 10^{-3} according to recent measurements [8]. With (3) the interfacial heat transfer is determined as:

$$\dot{q}(r) = \dot{m}_{\mathrm{ev}} \Delta h = \alpha_{\mathrm{i}} (T_{\mathrm{i}} - T_{\mathrm{g}}) \tag{4}$$

where Δh is the latent heat of evaporation. From (3) and (4) the interfacial heat transfer coefficient can be derived with the assumption that the difference $T_{\mathrm{i}} - T_{\mathrm{g}}$ is small compared to T_{g}:

$$\alpha_{\mathrm{i}} = \Delta h \beta \rho_g \sqrt{\frac{k}{8\pi m^* T_{\mathrm{g}}}} \tag{5}$$

Between the heater and the liquid-vapor interface the liquid wedge of thickness $\delta(r)$ forms the thermal resistance for the heat flux to the interface. This liquid layer is very thin. Thus diffusive heat transport can be assumed:

$$\dot{q}(r) = \frac{\lambda}{\delta(r)}(T_{\mathrm{w}} - T_{\mathrm{i}}) \tag{6}$$

where λ is the thermal conductivity of the liquid, T_{w} is the surface temperature of the heater. It must be noticed that T_{w} and T_{i} cannot be assumed to be constant. $T_{\mathrm{w}} = T_{\mathrm{w}}(r)$ and $T_{\mathrm{i}} = T_{\mathrm{i}}(r)$ depend on $\dot{q}(r)$ respectively $\delta(r)$ and T_{w} depends on the system of heating and the thermal conductivity of the heater material respectivly.

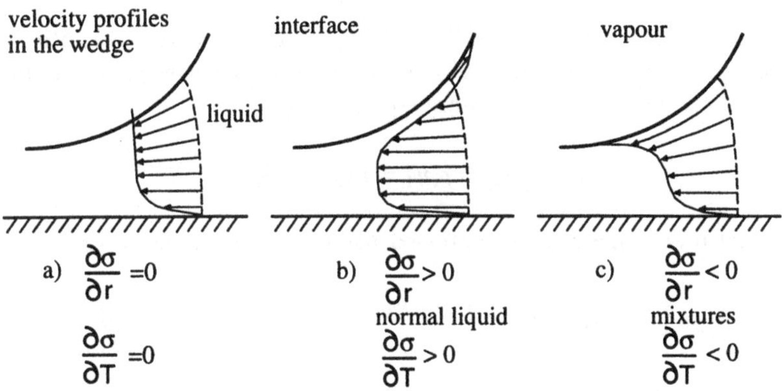

Fig. 3. Boundary conditions for the influence of thermocapillary flow at the liquid vapour interface.

At a large bubble curvature K the vapor temperature depends on the capillary pressure too, determined by the Clausius Clapyron equation:

$$T_g = T_{\mathrm{sat}}\left(1 + \frac{\sigma K}{\Delta h \rho_g}\right) \tag{7}$$

where σ is the liquid-vapor surface tension, T_{sat} the saturation temperature corresponding to the system pressure. Taking into consideration the resistance of conduction in the wedge (6) and the interfacial resistance (5) the heat flux can be written as:

$$\dot{q}(r) = \frac{T_{\mathrm{w}} - T_{\mathrm{g}}}{\frac{\delta(r)}{\lambda} + \frac{1}{\alpha_i}} \ . \tag{8}$$

The strong evaporation causes a momentum change and a pressure difference at the bubble interface which leads to a deformation of the bubble curvature and to a flow into the micro wedge. The liquid will be superheated along the heater surface and supports the evaporation. This flow explains the migration of smaller bubbles to the larger ones as observed in microgravity. They coalesce with them and this dynamic process generates secondary bubbles in the superheated boundary layer which grow, migrate and coalesce again. The bubble touches the solid surface. The adhesion force between the liquid and the solid surface is considered as the adhesion or disjoining pressure p_A. If the hydrodynamic pressure ($a/g \approx 0$) is not considered a force balance can be written as:

$$p_{li} = p_g - \sigma K + \rho_g w_g^2 - \rho_l w_l^2 + p_A \tag{9}$$

p_{li} is the pressure of the liquid at the interface in the non-equilibrium evaporating case, p_g is the vapor pressure inside the bubble assuming that it can be determined by the main bubble radius R in mechanical equilibrium with the bulk liquid:

$$p_g - p_l = \frac{2\sigma}{R} \tag{10}$$

with p_l the liquid pressure in the bulk. w_g and w_l are the velocities at the interface, ρ_g, ρ_l are the densities of gas and liquid, respectively. At the interface the gas and the liquid mass flows are equal. With (4:

$$\rho_g w_g = \rho_l w_l = \dot{m}_{ev} = \frac{\dot{q}}{\Delta h} \tag{11}$$

the momentum change can be expressed using (11) by the heat flux:

$$\rho_g w_g^2 - \rho_l w_l^2 = \rho_g w_g^2 \left(1 - \frac{\rho_g}{\rho_l}\right) = \frac{1}{\rho_g}\left(\frac{\dot{q}}{\Delta h}\right)^2 \left(1 - \frac{\rho_g}{\rho_l}\right) \tag{12}$$

The adhesion or disjoining pressure is defined [9] as:

$$p_A = -A\delta^{-3} \tag{13}$$

where A is the Hamaker or dispersion constant which accounts for the van der Waals intermolecular forces of the adsorbed liquid at the solid surface. The Hamaker constant has a positive or negative value depending on whether the liquid wets the surface or not. For the case of complete wetting A is about 10^{-20} J.

With (10), (12) and (13) equation (9) is written:

$$p_g - p_l = \frac{2\sigma}{R} - \sigma K + \frac{1}{\rho_g}\left(\frac{\dot{q}}{\Delta h}\right)^2 \left(1 - \frac{\rho_g}{\rho_l}\right) - A\delta^{-3} \tag{14}$$

This is the pressure difference for the mass flow into the wedge. The evaporation causes a recoil pressure which presses the bubble against the heater surface. It adds to the holding forces and flattens the curvature of the bubble at the contact with the heater.

Equation (14) is not only dependent on the radial distance r from the center of the bubble. After nucleation the bubble grows and therefore all parameters are time dependent too. The radial flow in the wedge of the bubble can be determined by solving the Navier Stokes equation for axial symmetry in two dimensions.

The boundary condition at the rigid wall is the non-slip condition. The second boundary condition at the liquid-vapor interface has a significant effect on the velocity distribution in the wedge. The temperature at the interface increases in the direction to the center. This temperature gradient (or in mixtures the resulting temperature-concentration gradient) causes interfacial tension driven flow which effects the flow profile in the wedge. In Fig.3 three cases are discussed:

1. No shear stress and no interfacial tension. The viscosity of the vapor is much smaller than the viscosity of the liquid and there shall be small temperature gradients at the interface.
2. Larger temperature gradients and a pure fluid. The surface tension decreases with an increase in temperature, thus temperature induced surface shear stress is opposite to the flow direction and impedes the flow.
3. In mixtures the concentration of the component with the lower vapor pressure increases in the flow direction. Normally this component has a higher surface tension. Therefore a surface tension gradient due to a concentration gradient can enhance the flow towards the wedge.

It could be that during the short growth time of a bubble under earth conditions the flow in the wedge cannot be fully developed. The fast growth moves the interface in opposite direction and forms the liquid wedge because the shear stress at the wall holds the liquid. This effect is known as the formation of the microlayer. Even in this case the heat transfer is dominated by the evaporation of this thin liquid layer formed in the wedge.

A rough onedimensional approximation can be given for steady state conditions by neglecting the inertia terms. This is an approach often used in "film theories". The momentum equation is written for the radial direction with the liquid velocity u:

$$\nu_{\mathrm{l}}\rho_{\mathrm{l}}\frac{d^2u}{dr2}=\frac{dp_{\mathrm{li}}}{dr} \tag{15}$$

Where ν_{l} is the kinematics viscosity of the liquid. With the boundary conditions:

$$\text{at}\quad z=0\quad u=0\qquad\text{and at}\quad z=\delta\quad \frac{du}{dz}=0$$

the local average velocity is obtained as:

$$\overline{u}(r)=-\frac{\delta^2}{3\nu_{\mathrm{l}}\rho_{\mathrm{l}}}\frac{dp_{\mathrm{li}}}{dr} \tag{16}$$

and the average mass flow rate as:

$$\dot{m}(r)=\rho_{\mathrm{l}}\overline{u}(r) \tag{17}$$

From the mass balance the evaporation mass flow rate follows:

$$\dot{m}_{\mathrm{ev}}=\frac{\dot{q}}{\Delta h}=-\frac{d}{dr}(\rho_{\mathrm{l}}\overline{u}(r)\delta)=-\frac{1}{3\nu_{\mathrm{l}}}\frac{d}{dr}\left(\frac{dp_{\mathrm{li}}}{dr}\delta^3\right) \tag{18}$$

The set of equations (8), (14) and (18) constitute a fourth-order nonlinear ordinary differential equation for the liquid layer thickness $\delta(r)$ in the wedge. The energy transfer per bubble Q_B is determined by integration of (8) over the bubble area and the growth time. To achieve the overall heat transfer it must be multiplied by the number N of all bubbles on the surface and the departure or growth frequency f.

$$\dot{Q} = Q_B N f \tag{19}$$

Recently solutions of a similar approach are given by Lay and Dhir [10] and by Stephan and Hammer [11].

3. Discussion

Here we will not present quantitative results from a numerical solution of the system of equations, but we like to discuss some qualitative conclusions following from this model. It explains observations made in microgravity experiments.

The first remarkable observation in microgravity is that the bubbles have a larger size but a lower departure frequency at the same overall heat flux. From geometrical consideration it follows that the total active heat transfer area of all bubbles is independent of the size of the bubbles if the active area below a bubble can be defined by the bubble radius R and a constant "wetting angle" ($\alpha : R_a = R\cos\alpha$. If the bubbles just touch each other the number of bubbles on the surface area A is $N = A/R^2\pi$ and the total active heat transfer area of all bubbles is $A_a = N(R\cos\alpha)^2$. Thus the ratio $A_a/A = \cos^2\alpha$ depends only on the wetting angle. In this case the active heat transfer area is generally independent of the bubble size.

A flow in the wedge (in our case in opposite direction to the radius r) can be maintained as long as $dp_{li}/dr > 0$. The recoil pressure presses a bubble in the center against the heater. There the curvature K is small. Therefore the adhesion pressure plays an essential role in this region.

With increasing heat flux the recoil pressure is increasing and stops the flow in the wedge. This will lead to a local dry-out and to a local instability between a non-wetting and a wetting situation below a bubble.

Dry-out, critical heat flux and the transition to film boiling will occur if at a certain radius r_c the mass flow into the wedge becomes smaller than the evaporating mass flow. This critical condition can be derived from (17)

$$\dot{\overline{m}}(r_c)\delta(r_c)2\pi r_c \leq \int_0^{r_c} \dot{q}(r)2\pi r dr$$

At higher temperatures resulting from high heat fluxes even the adhesion layer will evaporate. This accelerates the transition to film boiling. The whole bubble base will evaporate and by coalecence with neighboring bubbles a vapor film can spread over the entire area. The high heat flux below a bubble can consume the stored energy in the material of the heater surface and reduce locally the wall temperature. This depends on the heating system and the thermal conductivity of the heater material. Thus the evaporation is stopped and with it the recoil pressure. The surface tension transforms the bubble to a spherical shape. This induces a momentum force and a liquid flow such that the bubble can depart from the surface.

Other mechanisms for bubble departure are horizontal and vertical coalescence and lift-off by bubbles growing below larger ones. The influence of Marangoni convection and the heat pipe effect at boiling in subcooled liquids is discussed in [7].

Acknowledgements

The author would like to acknowledge gratefully the support of the microgravity research by DARA and ESA.

References

1. V.K. Dhir, *Int. J. Heat and Fluid Flow*, **12** (1991) 290-314.
2. J. Straub, M. Zell, and B. Vogel, in: Proc. of the 9th IHTC, Jerusalem, Ed. G. Hetsrony Vol. 1, pp. 91-112; and M. Zell, Untersuchung des Siedevorgangs unter reduzierter Schwerkraft, Dissertation T.U. München, 1991.
3. R. Siegel, in: Effects of Reduced Gravity on Heat Transfer in Advances in Heat Transfer, J. P. Hartnett and T. F. Irvine Jr.; Eds., Academic Press, New York.
4. J. Straub, J. Winter, G. Picker and M. Zell, Boiling on a miniature heater under microgravity - a simulation for cooling of electronic devices. ASME National Heat Transfer Conference, Portland August (1995) USA.
5. P.C. Wayner, C.Y. Tung, M. Tirumala, and J.H. Yang, *J. of Heat Transfer*, **107** (1988) 182-189.
6. P. Stephan, *Wärmedurchgang bei Verdampfung aus Kapillarrillen in Wärmerohren*, Fortschritt Berichte-VDI Reihe 19, Nr.59.
7. J. Straub, in: Proc. of the Third World Conference on Experimental Heat Transfer, Fluid Mechanics, and Thermodynamics held in Honolulu Hawaii Oct. 1993 and Experimental Thermal and Fluid Science 1994: pp. 253-273.
8. J. Straub, J. Winter, G. Picker, and M. Zell, Study of Vapor Bubble Growth in Supersaturated Liquid, National Heat Transfer Conference, USA, Portland, August 1995.
9. B.V. Derjaguin and Z.M. Zorin,in: Proc. 2nd Conf. on Surface Activity (London) Vol. 2 pp. 145-152.
10. J.H. Lay, V.K. Dhir, in: Heat Transfer 1994, Proc. of the 10th IHTC, Brighton, UK, 10PB 17, pp. 105, 110.
11. P. Stephan and J. Hammer, *Wärme und Stoffübertragung* **30** (1994) 119, 125.

Theoretical Models for Boiling at Microgravity

L.G. Badratinova[1], *P. Colinet*[2], *M. Hennenberg*[2] *and J.C. Legros*[2]

[1] Lavrentyev Institute of Hydrodynamics, 630090 Novosibirsk, Russia
[2] Chimie Physique E.P., Université Libre de Bruxelles, B-1050 Brussels, Belgium

Abstract

In this work initial results are presented on the theoretical modelling of the following aspects related to microgravity boiling: the rate of phase change at moving interfaces, the thermocapillary effect at the phase separating boundary, the influence of thermal effects on the Rayleigh – Taylor instability of a heated static vapour layer below its liquid and the formation of the liquid macrolayer beneath vapour bubbles.

1. Introduction

Despite many years of experimental investigation the predictability of microgravity boiling is yet impossible [1]. The classical formulas for peak heat flux [2, 3, 4] for example, can no longer define its actual values at reduced gravity levels. Indeed, the existing theoretical models have been specifically developed for ground-based applications. The critical wavelength of the Rayleigh – Taylor instability [5] is used [3, 4] for the estimation of the distance between vapour masses on a plane heater in the stage of developed nucleate boiling. The critical wavelength has a small value at earth conditions and becomes very large as the gravity level decreases (for the water/air system under gravity $10^{-4}g_0$, $g_0 = 981\text{cm/s}^2$) its value is equal to 160 cm). When the horizontal dimensions of a system are less then the value of the critical wavelength, the Rayleigh – Taylor instability does not occur since all perturbations are stabilized by surface tension. Because of the small heater areas used in the microgravity experiments performed up to now the existing theoretical models cannot be applied.

The focus of this work is related to interactions between the mechanisms of the thermocapillary and the Rayleigh – Taylor instability at the liquid-vapour interface in the case of low Bond numbers. The nonequilibrium rate of phase change is described using the irreversible thermodynamics. The influence of the nonequilibrium effect on the thermocapillary one is described in terms of the phenomenological mass transport coefficient. It is proven that at high heat flux in a very thin vapour layer the Rayleigh – Taylor instability is replaced by the thermocapillary one if the mass transport coefficient is sufficiently small. A possible mechanism of the liquid macrolayer formation is discussed which is due to the thermocapillary instability of a residual vapour layer remaining on a heater after bubble break off.

2. Rate of phase change

To describe theoretically boiling in microgravity it is necessary to know precisely the thermal conditions at the interfaces. In studies on condensation and evaporation the rate of phase change is generally defined by the famous Hertz – Knudsen equation. For a flat interface, along which the liquid and the vapour temperatures are assumed to be equal, the Hertz – Knudsen equation [6] is

$$J = \beta(M/2\pi RT)^{1/2}[p_s(T) - p_0(T)] \tag{1}$$

where J is the nonequilibrium mass flux across the interface, M is the molecular weight of the vapour, $p_s(T)$ is the saturation pressure at surface temperature T, $p_0(T)$ is the vapour pressure just beyond the interface, R is the universal gas constant and β is the constant accommodation coefficient. According to the theoretical considerations of the kinetic theory the value of β should be close to unity. However, reported experimental values are of order $10^{-3} - 10^{-1}$ and there also exist discrepancies among the experimental data of various investigations.

One can derive an equation for the kinetics of phase transition in the framework of irreversible thermodynamics. Assuming continuity of the temperature along the interface $T_v = T_\ell \quad (= T)$ Onsager's law suggests [7] that

$$J = K[\mu_\ell(p_\ell, T) - \mu_v(p_v, T)] \ . \tag{2}$$

Here K is the positive phenomenological coefficient, ρ_v and ρ_ℓ are the densities of vapour and liquid respectively. J is the rate of phase change. By definition $J = \mathbf{J} \cdot \mathbf{n}$, where $\mathbf{J} = \rho_\ell(\mathbf{v}_\ell - \mathbf{v}_\Sigma) = \rho_v(\mathbf{v}_v - \mathbf{v}_\Sigma)$ is the mass flux vector at the interface, $\mathbf{v}_\ell$ and $\mathbf{v}_v$ are the velocities of the liquid and vapour at the interface, $\mathbf{v}_\Sigma$ is the interface velocity and $\mathbf{n}$ is the surface normal unit vector directed into the vapour phase. Equation (2) together with the definition of J show that the mass will leave the phase with the highest chemical potential toward that with the lowest one. Let us remark that if the temperature jump across the liquid-vapour interface is taken into account, then other transport coefficients (see [8]) appear in the phenomenological equation for the mass and energy fluxes because there exists then a new thermodynamic force $\delta T = T_\ell - T_v$. In accordance with the requirements of irreversible thermodynamics the phenomenological coefficient K depends on the surface temperature T and the vapour pressure p_v. Equation (2) can be interpreted as an approximation to the general mass flux equation where the thermodynamic force δT is neglected.

In (2) the liquid and vapour pressures p_ℓ and p_v and the temperature T vary along the moving interface. Introduce now a function $p_s(T)$ which relates the value of T to the corresponding saturation vapour pressure for a flat isothermal interface at a constant temperature T . Assume further that for the liquid and vapour phases the equations of state are respectively $\rho_\ell = \rho_\ell^0(p,T)$, $\rho_v = \rho_v^0(p,T)$ and that everywhere along the liquid-vapour interface one has the inequalities $|p_\ell - p_s(T)|/p_s(T) \ll 1$, $|p_v - p_s(T)|/p_s(T) \ll 1$. One then can linearize the function K and the chemical potentials in the vicinity of the point $(p_s(T),T)$ using a Taylor expansion. With the assumption of local thermodynamic equilibrium inside each phase near the interface one can use the classical thermodynamical relation for the pressure derivative of the chemical potential [7]. The one obtains a linearized form of (2)

$$J = K(T)\left[\frac{p_s(T) - p_v}{\rho_v(T)} - \frac{p_s(T) - p_\ell}{\rho_\ell(T)}\right], \tag{3}$$

where $\rho_v(T)$, $\rho_\ell(T)$ and $K(T)$ stand for $\rho_v^0(p_s(T),T)$, $\rho_\ell^0(p_s(T),T)$ and $K(p_s(T),T)$ respectively. Far from the critical boiling point the vapour density is much lower than the density of liquid; assuming also $p_v \sim p_\ell$ and neglecting the last term in (3), one obtains $J = K(T)\,[p_s(T) - p_v]\,/\rho_v(T)$. This equation presents a formal analogy with the Hertz–Knudsen equation (1) since they are both proportional to the difference $p_s(T) - p_v$. But it has been deduced in a different way, without the ideal gas approximation for the vapour phase. In this equation one can use the correct state equation, experimentally evaluated for the vapour; assuming the interface is flat and isothermal, one can compare this equation with (1) and obtains the expression of the phenomenological coefficient K through the accommodation coefficient β :

$$K = \beta\rho_v(T)(M/2\pi RT)^{1/2}. \tag{4}$$

This expression is rough because it is obtained with the help of the linearization of the original (2). The nonlinear terms, neglected in (3), remain yet to be estimated. Probably, the investigation of the nonlinear approximations to (2) and its extension to the case where the temperature jump δT is taken into account will help to understand why so large discrepancies exist among the reported experimental data on the accommodation coefficient values. Note that in the nonlinear approximations the dependence of the phenomenological coefficient on the vapour pressure cannot be neglected.

The last term in (3) can be important for strongly curved interfaces. For this case, using the interfacial balance of normal momentum, we obtained from (3) the following approximate formula (H - is the mean curvature of moving interface) :

$$J = \frac{\rho_\ell(T) - \rho_v(T)}{\rho_\ell(T)\rho_v(T)} K(T)\left[p_s(T) - p_v - \sigma\frac{\rho_v(T)}{\rho_\ell(T) - \rho_v(T)}H\right]. \tag{5}$$

Equation (5) takes into account the influence of the surface tension on the rate of phase change.

3. The degree of nonequilibrium number

The rate of phase change depends on the interfacial temperature (2). Therefore, the thermocapillary force and the surface gradient of the nonequilibrium mass flux are interrelated. The inverse coefficient K^{-1} measures the deviation from the equilibrium state of the interface. Indeed, in the limiting case $K^{-1} \to 0$ it follows from (2) that the chemical potentials are equal: the moving interface is in a state of quasi–equilibrium. The influence of nonequilibrium mass transfer on the interfacial balances of momentum and energy is increased with increasing values of K^{-1}. In the case where $K^{-1} \to \infty$ the equality $J = 0$ obtained from (2) is the non–leak condition at the interface. Consequently, the thermocapillary problem with no phase change corresponds to this latter case.

The behavior of a liquid–vapour system can be quite different in the opposite cases "no phase change" and "quasi-equilibrium". We shall demonstrate this by examining the linear stability of a motionless liquid-vapour system (Fig. 1). A horizontal vapour layer lies between a wall being laterally infinite and which is heated at a constant temperature T_w and a liquid phase which extends infinitely. There is no mass transfer across the interface in the basic state, but it may occur in the perturbed state. We consider the heat conduction and dynamics of both phases, assuming them to be incompressible viscous fluids. Equation (3) is linearized at the saturation temperature T_0 of the unperturbed interface and is used as an interfacial boundary condition. This physical set–up allows one to investigate the interplay between the thermocapillary, the nonequilibrium phase exchange and the Rayleigh – Taylor mechanisms.

Our particular problem shows that the influence of the thermocapillary effect on the Rayleigh – Taylor instability depends essentially on the degree of nonequilibrium of the perturbed interface. The variation of the nonequilibrium mass flux along the interface gives a contribution to the total thermocapillary force characterized by the following dimensionless parameter,

$$s_n = \sigma_T T_0 / (K_0 L \nu_v).$$

Here $\sigma_T > 0$ is the temperature coefficient of surface tension, ν_v is the kinematic viscosity of the vapour, L is the latent heat of vaporization and $K_0 = K(T_0)$. Since s_n is inversely proportional to the phenomenological coefficient, it is a measure of the degree of interface nonequilibrium. Thermocapillary instability may exist only for high values of this parameter.

4. Results of the stability analysis

a) When s_n is infinite, the phase change effect is completely ignored. The problem is reduced to the study of the coupling between the pure Marangoni and the Rayleigh – Taylor instability mechanisms. The relative contribution of the destabilizing gravity mechanism to the total surface deformation effect is characterized by the dimensionless group

$$G = k_* N_g / (24 N_{cr}) = (\rho_\ell - \rho_v) g d_v^3 / (24 \rho_v \nu_v k_\ell)$$

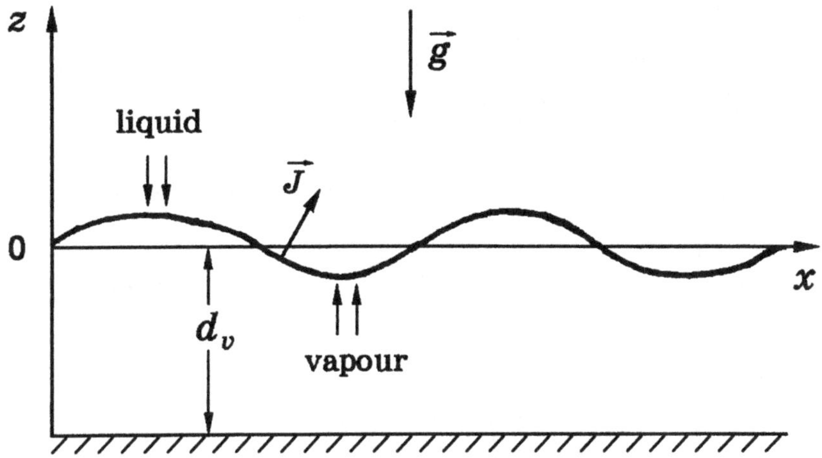

Fig. 1. Physical configuration of the liquid – vapour system

where $k_* = k_v/k_\ell$ is the thermal diffusivity ratio, d_v is the depth of the vapor layer. The parameters $N_g = (\rho_\ell - \rho_v)gd_v^2/\sigma_0, N_{cr} = \rho_v\nu_v k_v/(\sigma_0 d_v)$ are the Bond number and the crispation group. σ_0 is the value of the surface tension at $T = T_0$. If $G \ll 1$ the Rayleigh – Taylor instability is replaced by the thermocapillary surface-wave instability, in the case of a common liquid - vapour systems sufficiently heated from below,. This is demonstrated in Fig. 2. The marginal non–oscillatory instability curves are presented for the water system at $T_0 = 373K$. Note that in this "no phase–change" case the Marangoni number can be used to characterize the instability. However, the result is presented with respect to the number $Q = \lambda_v(T_w - T_0)/(\rho_v k_v L)$ –modified Jakob number, because it is more convenient for comparison with the "quasi-equilibrium" case, λ_v is the vapour thermal conductivity. For $G = 2.44 \times 10^{-6}$ the number Q is plotted versus the dimensionless wave number ω (the length scale is d_v). The curve 1 intersects the axis $Q = 0$ at $\omega = \sqrt{N_g} = 1.98 \times 10^{-5}$, the crossing point is not seen because it is very close to the coordinate origin. In the case $G \ll 1$ for large positive Q the condition of instability is given by the inequality $\omega < \omega_a$, where $\omega_a = \sqrt{24N_{cr}/k_*}$ is the position of the asymptote. The ratio between the actual critical wavelength $(2\pi d_v/\omega_a)$ and the critical wavelength $(2\pi d_v/\sqrt{N_g})$ of the pure Rayleigh – Taylor instability (which pertains to isothermal systems) is equal to the square root of the small parameter G. Note that the condition $G \ll 1$ is equivalent to the following inequality for the vapor depth $d_v \ll d^* = (24\rho_v\nu_v k_\ell)^{1/3}/([\rho_\ell - \rho_v)g]^{1/3}$. For the water system at earth gravity acceleration, $d^* = 1,7 \times 10^{-3}$ cm.

b) An essential difference of the "quasi-equilibrium" case ($s_n = 0$) with the previous one is that the thermocapillary instability is now absent and that the Marangoni number cannot be introduced. Here the thermocapillary number

$$s_e = \sigma_T T_0/(\rho_v L d_v)$$

characterizes the thermocapillary effect. For $s_e \ll 1$ the influence of the ther-

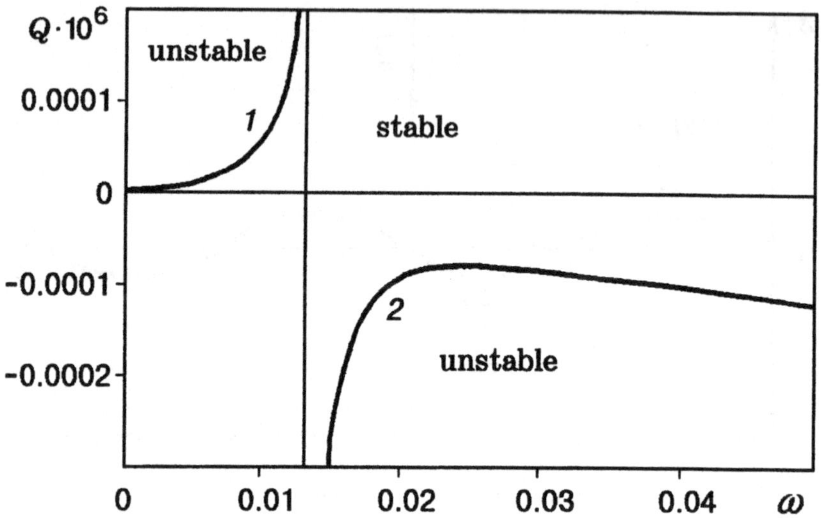

Fig. 2. Marginal stability curves for the water system at $T_0 = 373K$ in the case of no phase-change, $g = 981 \times 10^{-4} cm/s^2$, $d_v = 5 \times 10^{-4}$ cm

mocapillary effect on the neutral stability condition is shown to be negligible. The condition $s_e \ll 1$ is equivalent to the following inequality $d_v \gg d_* = \sigma_T T_0/\rho_v L$. The value of d_* is very small. Indeed for the water system at $373K$, $d_* = 5.5 \times 10^{-6}$ cm.

Our analytical investigations show that for the system heated from below ($Q > 0$) the phase change mechanism stabilizes the long-wave perturbations. In contrast with the previous case, the positive number Q now characterizes the stabilizing influence of the change of phase mechanism. With small G the Rayleigh – Taylor instability is prevented for sufficiently large positive Q. Such a situation is shown in Fig. 3 for the case when $G = 0.195$.

There is one single marginal stability curve which intersects the axis $Q = 0$ at $\omega = \sqrt{N_g}$ and at $\omega = 0$. The maximal value Q_m on this curve is equal to 7.2×10^{-9}. It is reached for $\omega = 0.00078$. For $Q > Q_m$ the motionless state of system is stable with respect to all monotonical perturbations.

c) Consider the case where the value of s_n is finite and positive. It is noteworthy that the thermocapillary effect is now characterized by two dimensionless parameters: s_n and s_e. Assume $G \ll 1$, $Q > 0$. Then owing to the disappearance of gravity waves, instability can be generated only by the thermocapillary mechanism. Instead of Q, the Marangoni number may be formed from dimensionless groups by the thermocapillary number s_e. However, the possible existence of thermocapillary instability depends on the value of the dimensionless group

$$\alpha = k_*/(N_{cr} s_n) = \sigma_0 K_0 L d_v/(\rho_v k_\ell \sigma_T T_0).$$

For high values of Q the instability criteria is formulated only in terms of the number α as follows. For $s_e \ll 1$ the thermocapillary instability exists when the inequality $\alpha < 4/3$ holds and it is non–existent when this inequality is violated. For $s_e = 1$ the static state is unstable at $\alpha < 0.692$ and stable if $\alpha > 0.692$. For the water system we calculated K_0 from (4) at $T = 373K$. Then for $d_v = d_* = 5.5 \times 10^{-6}$cm ($s_e = 1$) we get the instability criteria in terms of the accommodation coefficient: $\beta < 1.2 \times 10^{-3}$. In the light of the existing experimental data one can assume that this inequality is satisfied for real water systems where β is strongly influenced by gaseous contamination and impurity adsorption at the interface.

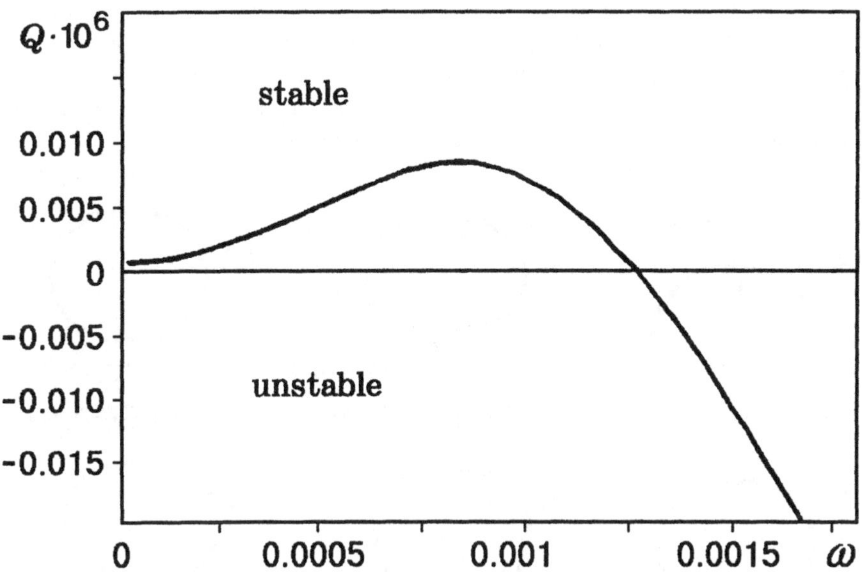

Fig. 3. Marginal stability curve for the water system at $T_0 = 373K$ in the case of a quasi–equilibrium interface, $g = 981 \times 10^{-3} cm/s^2$, $d_v = 0.05$ cm

d) In the case where it is the heat flux, rather than the temperature, that is kept constant at the heated wall, we were not able to give the explicit formulation of the thermocapillary instability criteria applicable for common liquid/vapour systems. Here the gravity influence dissapears when the parameter $G' = k_*^2 N_g/(144 N_{cr}^2)$ is much less than unity. Our initial results predict that the instability occurs at small values of the dimensionless group $\alpha' = \alpha k^*/N_{cr}$.

Note finally that all results presented in this section have been derived under the assumption of small Bond numbers. When the heated vapour layer lies above the liquid, the thermocapillary instability described above also exists and it is independent of gravity if the dimensionless parameter G (or G') is small as compared with unity.

5. On the mechanism of macrolayer formation

The model of the critical heat flux, proposed in [4] attributes this phenomenon to the evaporation of the macrolayer on a heater. In this model the macrolayer is formed on a heater after the vapour bubble detachment. The critical heat flux is attained when the heat flux is high enough to evaporate the macrolayer before the bubble detachment. The Kelvin – Helmholtz instability which is at the dividing boundary between the vapour and the superheated liquid serves to estimate the height of the macrolayer through the critical wavelength of this instability. The Kelvin – Helmholtz instability appears on the interface of two non-viscous fluids which are in relative motion with respect to each other [5]. The inviscid model for fluids excludes the possibility for the examination of the thermocapillary and phase change effects in the macrolayer. It cannot explain [9] many experimentally proven facts such as the dependence of the critical heat flux on the thickness and contamination of the heated wall.

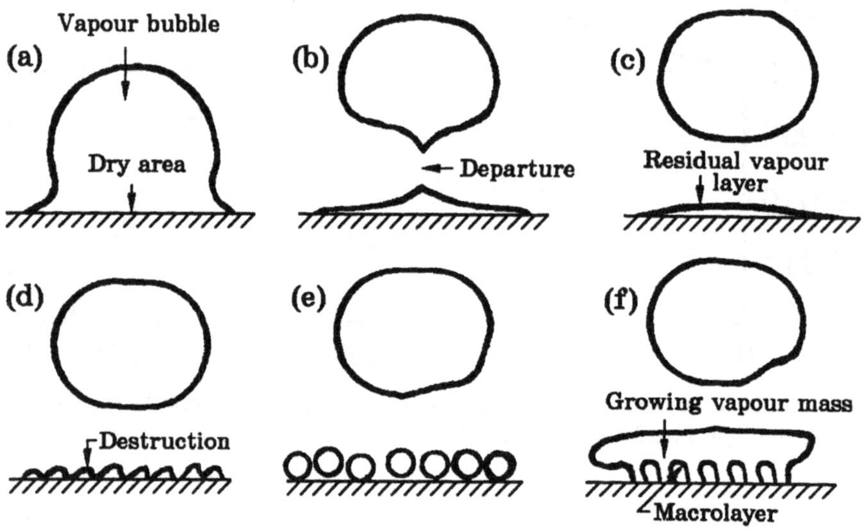

Fig. 4. Bubble departure process: (a) formation of a neck (b) break off (c) condensation of a residual vapour mass (d) destruction of a condensing vapour film (e) possible formation of small bubbles (f) mechanism of growth of the next bubble.

In a recent paper [10] the experimental observations on the macrolayer formation process are described and a new model of macrolayer formation is proposed. At high heat fluxes the observations show that when a bubble detaches and the liquid rushing onto the heater surface comes into contact with its dry area a large amount of nucleation occurs simultaneously. In the proposed model nucleate bubbles quickly grow and coalesce simultaneously producing oblate vapour film above the heated surface. The macrolayer is formed on the surface of the heater by coalescence of primary bubbles or of secondary (coalesced) bubbles.

We assume that simultaneous nucleation occurring after the liquid contact with a dry area can be explained in terms of the thermocapillary instability of a thin residual vapour layer on the heated wall condensing after the bubble break off. A possible existence of such an instability has been shown in the previous sections for sufficiently small vapour layer thickness. The departure process such as shown in Fig. 4a–c may take place (see [11]) for the vapour mass covering a dry area of a heater. When the thickness of the thinning residual vapour layer reaches a critical value the surface thermocapillary waves appear at the interface between the liquid and the residual vapour layer (Fig. 4d). These surface waves lead to the destruction of the vapour layer with the subsequent formation of a liquid macrolayer from which the vapour is supplied to the above growing vapour mass (Fig. 4f) by liquid evaporation. After complete evaporation of the macrolayer the vapour mass will grow on a dry area as shown in Fig. 4a. After the bubble detachment the macrolayer formation process can be repeated again. After the destruction of the residual vapour layer a large amount of small vapour volumes (Fig. 4d) covers the heated surface. The behavior of these small vapour bubbles (nucleate sites) cannot be predicted from the linear stability analysis performed here. They grow and then can take the spherical shape (Fig. 4e) forming the macrolayer by the coalescence. In our stability analysis the heater roughness effect is not taken into account. Thus the thermocapillary instability conception predicts that the macrolayer can be formed also on the perfectly smooth heated surfaces. This conception can be verified by experimental investigations with very smooth heater surfaces. Also, the instability analysis gives new dimensionless groups $(G, G', Q, s_{\mathrm{n}}, s_{\mathrm{e}}, \alpha, \alpha')$ which predict, in particular, the important role of viscous forces in the vapour phase and of the nonequilibrium phase transition inside the macrolayer. The instability conception predicts the space periodicity of the destruction process (Fig. 4d) and, as a consequence, no liquid supply into the macrolayer during its evaporation. This agrees with the existing experimental data [10]. The described instability is independent of gravity. Hence it is reasonable to assume that it also pertains to residual vapour layers on an inclined and vertical heated surfaces.

6. Conclusions

Applying an irreversible thermodynamics approach we have analyzed the possibile ways for the extension of the Hertz – Knudsen equation to moving interfaces. Knowledge of the correct rate equation is necessary to describe the thermocapillary effect in boiling where it is influenced by the mass transfer across the interface. Thermocapillary instability of a thin static vapour layer confined between a heated plate and the liquid bulk have been proven for which the critical wavelength is much less than the critical wavelength of the Rayleigh –Taylor instability. The thermocapillary instability may cause the destruction of a residual vapour layer on the heated surface leading to the formation of macrolayer beneath vapour masses.

Acknowledgements

One of us (L.B.) has enjoyed financial support from the Région de Bruxelles through a Research in Brussels Grant. The text presents research results obtained in the framework of the Belgian program on Interuniversity Pole of Attraction (P.A.I. n^0 21) initiated by the Belgian State, Prime Minister's Belgian Federal Office for Scientific, Technological and Cultural Affairs. The scientific responsibility is assumed by its authors.

References

1. J. Straub, M. Zell and B. Vogel, *Microgravity Sci. Technol.*, **6** (1993) 239-247
2. S.S. Kutateladze, *Zhurn. Techn. Fiz.* **20**(11) (1950) 11 1389-1392
3. N. Zuber, *Hydrodynamic aspects of boiling heat transfer.* U.S.AEC REP. AECU 4439. Tech. Inf. Ser. V. Oak Redge, Tenn. (1959)
4. Y. Haramura, Y. Katto, *Int. J. Heat Mass Transfer* **26** (1983) 389-398
5. S. Chandrasekhar, *Hydrodynamic and Hydromagnetic Stability.* Clarendon Press, Oxford (1961)
6. M. Knudsen, *The kinetic theory of gases.* Methuen's Monographs on Physical Subjects, London (1953)
7. R. Haase, *Thermodynamics of Irreversible Processes*, Dover Publications, 1990
8. W.J. Bornhorst, G.N. Hatsopolous, *Transactions of the ASME, J. of Applied Mechanics* (1967) 840-846
9. Y. Katto, Proc. of the Engineering Foundation Conference on Pool and External Flow Boiling, Santa Barbara, California, March 22-27,1992, Published on Behalve of the Engineering Foundation by ASME, Eds. V.K.Dhir, A.E.Bergles, pp.151-164
10. T. Kumada, H. Sakashita, *Int. J. Heat Mass Transfer* **38** (1995) 979-987
11. J. Mitrovic, *Int. J. Heat Mass Transfer* **26** (1983) 955-963

Chemically Driven Convection in the Belousov-Zhabotinsky Reaction

K. Matthiessen and S.C. Müller

Max-Planck-Institut für molekulare Physiologie, Rheinlanddamm 201
D-44139 Dortmund, Germany

Abstract

The Belousov-Zhabotinsky reaction is a frequently used example for an excitable medium showing propagating oxidation waves. When traveling in a liquid solution without a gel, the chemical waves induce a convective flow in the reaction medium. We present an overview of the different types of flow behaviour in a thin layer of a the Belousov-Zhabotinsky solution. Large chemical wavelength lead to individual flow patterns, each associated with a single wave front, the socalled discrete flow. Short chemical wavelengths induce a cooperative phenomenon in the form of large global flow waves traveling through the fluid layer. The chemical reaction can induce this convection by producing local density changes as well as by generating gradients of surface tension. These coupling mechanisms are a consequence of local changes in the chemical composition of the medium.

1. Introduction

In the research on nonlinear system dynamics there have emerged, among others, two main groups of pattern forming systems with spatial degrees of freedom. One group comes from the more physical side of this scientific field and deals with hydrodynamic instabilities due to which spatio-temporal patterns occur in the form of specific flow fields. One of the most established phenomena belonging to this group is the Rayleigh-Bénard convection [1, 2]. In a solution layer between two horizontal plates, where the lower one has a higher temperature, characteristic convection rolls appear if the temperature difference exceeds a critical threshold. In a system with a free surface hexagonal convection cells form due to the Marangoni instability [3]. In this case local changes in the surface tension destabilize the former roll pattern. In nature Rayleigh-Bénard type convection occurs, for example, in the atmosphere, forming large convection rolls or in the earth's core, largely influencing weather phenomena and the continental drift, respectively.

The other main group, which addresses pattern formation from the more chemical side, is concerned with reaction-diffusion systems. In this case, traveling excitation waves can be observed [4] as well as stationary periodic patterns

Fig. 1. An undisturbed pair of spiral waves in a liquid layer of BZ solution, observed at 490 nm with a 2D spectrophotometer (Image side length: 1 cm)

of the Turing type [5, 6] and, as shown very recently, self replicating spots [7]. Representative systems of this group are the Belousov-Zhabotinsky (BZ) reaction [4] and the Chloride-Iodate-Malonic Acid (CIMA) reaction [8]. If the BZ reaction is prepared in a continuously stirred tank reactor (CSTR), the spatially homogeneous case is realized, where characteristic temporal oscillations between the oxidized and the reduced state occur. In a spatially extended system the reaction can be prepared as an excitable medium which supports excitation waves traveling by diffusion of reagents [9]. In a two-dimensional system circular as well as spiral geometries of the waves are observed. The BZ reaction has become a model system for excitable media because it can be prepared and observed easily. Media of this type play an important role, especially in biology. One example is the propagation of action potentials on neurons [10] or, in systems with higher dimensionality, excitation waves on neural tissue like in spreading depression [11] or on the heart muscle [12]. On the other hand, the CIMA reaction is the first chemical example of a system that shows stationary Turing patterns [6]. In nature this mechanism is, for example, supposed to be responsible for pattern formation on the skin of animals [13].

A system which combines effects arising both from hydrodynamic instability and reaction-diffusion coupling, is a BZ reaction proceeding under certain conditions in a thin layer of liquid solution. In the case of a uncovered reaction dish, one can observe propagation of excitation waves as well as convection of the Marangoni type due to the evaporative cooling of the surface. The convection cells lead to a visible deformation of the chemical wave pattern [14, 15]

Fig. 2. A pair of spiral waves in a liquid BZ layer deformed by natural Marangoni convection. A description of the measurement is given in [15].

(see Fig. 2). If evaporation is suppressed by covering the reaction dish, leaving just a small air gap, convection induced by another driving force can occur: now the chemical waves themselves force the reaction medium to move. This flow is called chemically driven convection [16]. From the hydrodynamical point of view this type of convection represents a fascinating type of intrinsic forcing of the hydrodynamic system, leading to interesting new types of flow patterns. For the reaction-diffusion system the convection acts as a second transport mechanism in addition to the diffusion, accelerating the front velocity and deforming the shape of the wave fronts.

2. The Belousov-Zhabotinsky Reaction

In the following section we describe the preparation and observation of the Belousov-Zhabotinsky reaction in more detail. For our experiments the reaction solution is prepared as a mixture of the components malonic acid, bromate, bromide, sulfuric acid, and, as a catalyst and redox indicator, ferroin. The malonic acid acts as the substrate. During the reaction it is consumed and degrades to other organic products, CO_2 and water. This reaction may follow different reaction paths, indicated by different intermediate concentrations in the reaction solution. During the reaction the system can switch between these paths, which leads to oscillations in concentrations of the intermediates. Due to the ferroin these oscillations are indicated by a clearly visible color change. As substrate is consumed, the whole system is on the way to the thermodynamic equilibrium and only the intermediates are oscillating. This is also the reason why batch systems have only a limited life time of several hours, depending on the chemical recipe used.

For the observation of chemical oscillations the reaction solution is filled into a stirred vessel to obtain spatial homogeneity. The solution exhibits periodic or, in special cases, aperiodic oscillations. For a detailed description of these oscillation types, the ageing of the system is a great problem. To solve it, the reaction is performed in a continuously fed stirred tank reactor (CSTR), where the system receives permanently new reactants. In a CSTR the behavior of the chemical reaction is examined as a function of the flow rates of the different reactants. As a result, for example, period doubling and chaotic oscillations are obtained [17].

Also unstirred spatially inhomogeneous systems may exhibit chemical oscillations with perhaps a spatial phase difference in the oscillation state, forming so-called phase waves. Of more interest are systems which remain in only one state for a long time and undertake a single oscillation if they are externally disturbed. Systems of this type are called excitable media. In the spatially inhomogeneous case a local excitation of the system influences the solution in the surroundings via diffusion of chemical components, turning it also to the excited state. As a result, this effect leads to the propagation of a chemical wave front. Behind this front the solution returns to the reduced state, remaining unexcitable for some time. This phase is called the refractory period and is responsible for the annihilation of colliding wave fronts in the system. Exciting the system in a single point leads to a circular wave which propagates away from this point with the same velocity in each direction. A spatial inhomogeneity in the system which generates chemical wave fronts with a specific frequency is called a pacemaker. In a two-dimensional system a pacemaker produces a so-called target pattern (Fig. 3).

In addition to the circular waves, a two dimensional BZ system may also show spiral waves. Due to the propagation laws of chemical waves an open end of a wave front curls up into a spiral [18]. As the breakup of a wavefront leads to two open ends, the spirals are frequently generated as a pair with opposite sense of rotation. This pair of spiral waves generates chemical wave fronts with a very short wavelength (Fig. 4). A spiral predominates a pacemaker in the system. Both are generating wave fronts, but the spiral has the higher frequency. As a consequence the point between them, where the fronts collide and annihilate, is moving towards the pacemaker until it is consumed. This is one reason why spiral waves are often called defects. Once they are produced, they occupy the whole system and it is not so easy to remove them. Spiral waves are typical characteristics of excitable media. They are found in a large number of systems besides the BZ reaction [11, 12, 19, 20].

Nowadays the spatially inhomogeneous BZ reaction is usually prepared in a gel matrix to avoid mechanical disturbances of the system and the influence of hydrodynamic flow [21]. Especially in continuously fed systems a gel matrix is necessary to avoid disturbances due to the inflow of the reactants. However, for

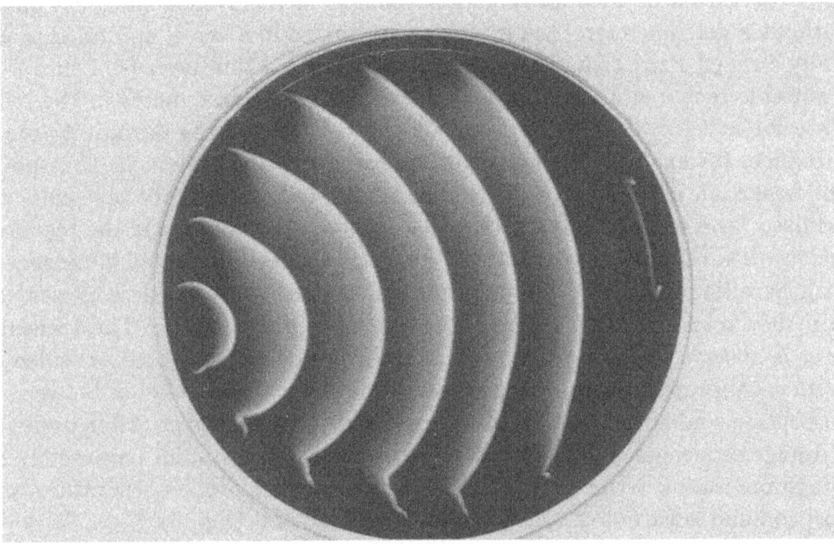

Fig. 3. A target pattern in a liquid Belousov-Zhabotinsky system, consisting of 4 ml BZ solution, filled in a Petri dish with a diameter of about 7 cm. The figure shows the pattern about 20 min after the preparation of a system with a pacemaker at the left boundary.

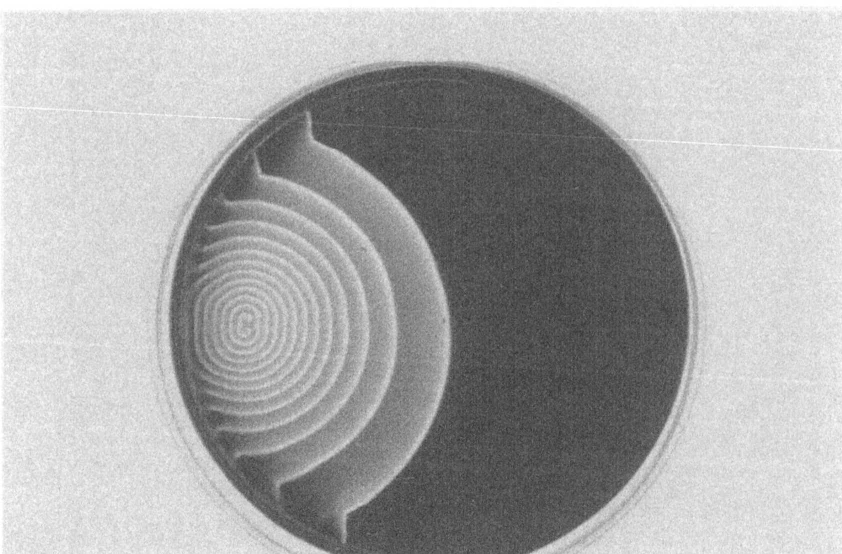

Fig. 4. A spiral wave in a liquid BZ system with parameters like in Fig. 3 To get the pair of spirals, initially a circular wave was induced. After a while its wave front was broken with a gentle blast of an air stream. The figure shows the pattern about 5 min after the induction of the spirals.

the observation of chemically driven convection the system must be prepared without a gel. Therefore, the reactants are mixed in a vessel and filled in a specially cleaned Petri dish, where the solution forms a thin layer (< 1 mm). Using a suitable recipe and a well prepared dish, the solution remains in the reduced state for at least 20 min, indicated by the red color of the ferroin. A local disturbance, for example with a silver wire, induces the system at that point to undertake one oscillation to the oxidized state, indicated by the blue color of the oxidized form of the catalyst ferriin. To observe color changes in the reaction solution with a specific wave pattern, the Petri dish is illuminated homogeneously with parallel blue light (490 nm) from below. This wavelength is chosen to obtain the largest possible contrast between ferroin and ferriin. The transmitted light is observed with a video camera and the images are stored on video tape. With a calibrated set-up it is possible to obtain a direct relation between light absorption and concentration of the oxidized catalyst ferriin. After digitization of image sequences, it is possible to estimate various system parameters using image processing techniques [22], for example, wave profiles, curvature, or the position and traces of the tip of spiral waves [23, 24] (Fig. 5).

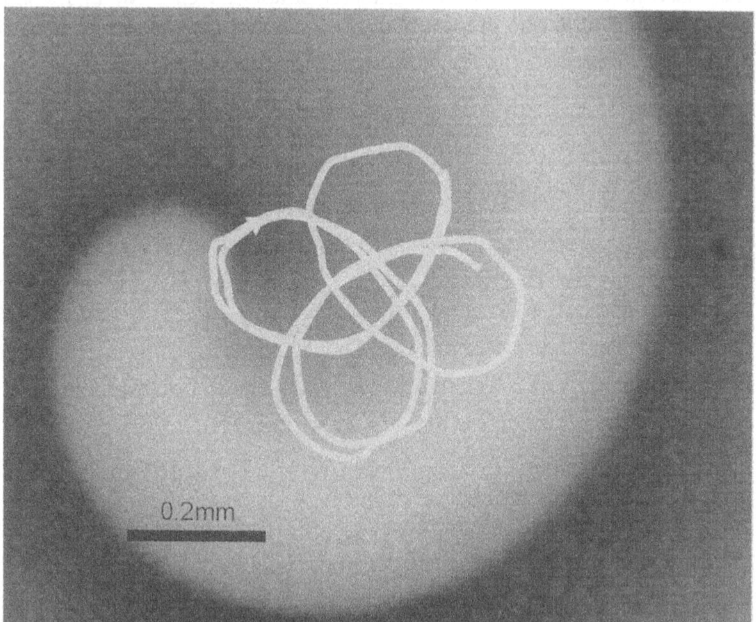

Fig. 5. Digitized video image of the tip of a spiral wave. In a superposition the trace of the tip during the last five rotations is displayed. The observed type of composed motion of the spiral tip is called meandering [22].

3. The Observed Flow Behavior

We now turn to the description of effects related to the occurence of flow fields in aqueous BZ layers without a gel. The local hydrodynamic state of the system can be detected by observing suspended small tracer particles (standard dow latex, 0.5 μm) with a microscope (Fig. 6). Due to the large magnification the depth of focus is short enough to obtain a resolution up to 30 μm in vertical direction. On the other hand, the high magnification restricts the horizontal region for velocity measurements to nearly a single point as the microscope image has a side length of only 0.4 mm. With special software for 2D velocimetry it is possible to calculate the local flow velocity from the recorded motion of the particles automatically [25]. Thus this set-up allows to measure the flow velocity at one point of the system as a function of time.

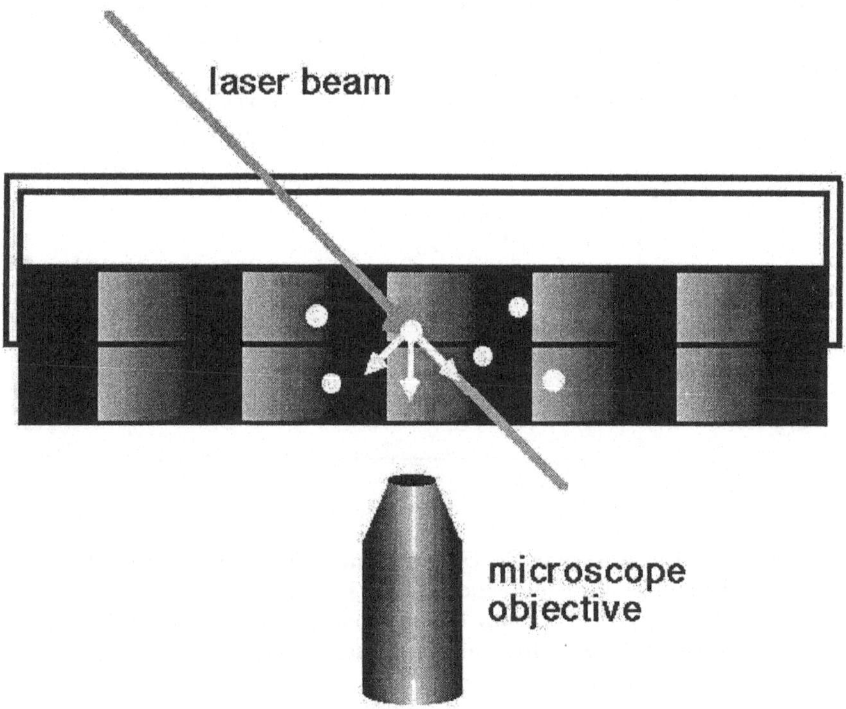

Fig. 6. Principle of the flow measurements. The Petri dish is filled with the BZ-solution with suspended latex particles. These particles are illuminated by a He-Ne-laser and the scattered light is observed with an inverse microscope.

The observed flow behavior in chemically driven convection depends highly on the underlying chemical wave pattern, in particular on the wavelength. Thus, the flow in a system with circular waves is quite different from that in a system with spiral waves. In the case of circular front geometries the wavelength is usually comparatively long and therefore each single wave front creates its individual flow field [16]. The preparation of a system with circular wave propagation for the flow measurements contains only one center of wave generation. Usually this pacemaker is induced externally near the boundary of the dish by use of a silver wire (Fig. 3) and flow measurements are carried out in the center of the dish. At the surface of the fluid layer a strong flow towards the approaching front is measured which exceeds in some measurements a velocity of 100 μm/s. This is close to the velocity of chemical wave propagation. As the front passes by there is a rapid change in flow direction. Near the bottom the behavior is in principle the same, but the flow velocity is smaller and its direction is opposite. After the wave has passed the measuring point, there is a smooth decrease in the flow velocity (Fig. 7). Including velocity data in the vertical direction these measurements reveal a strong descending flow in the chemical wave front and a moderate ascending flow in the outer regions. Hence each chemical wave front generates a pair of convection rolls and carries them through the solution. Figure 8 shows a reconstruction for the case of a circular wave pattern like in Fig. 3. The strong descending flow in the front indicates that this is the region where the main coupling between reaction and convection takes place.

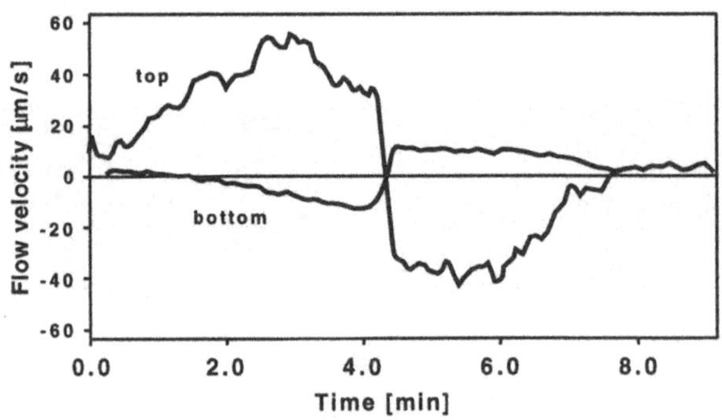

Fig. 7. Observed flow velocity in the case of a single circular wave front. The wave is generated in the middle of the dish and the flow is measured at the top and near the bottom of the fluid layer half the way to the boundary to get rotationally symmetric conditions. Passage of the wave front at the measuring point t=4.1 min.

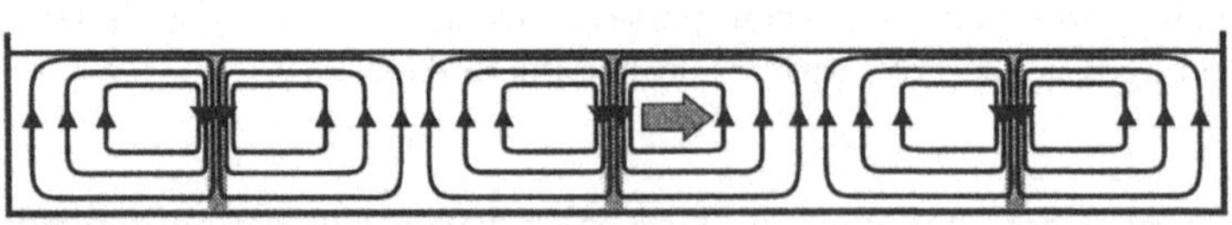

Fig. 8. Reconstruction of the convection pattern in a vertical cut through the Petri dish. The gray bars show the chemical wave fronts and the black arrows the hydrodynamical flow direction. Waves propagate in the direction of the gray arrow.

In a system with spiral wave propagation the flow behavior is different. For the flow measurements the preparation of the system contains just one pair of spirals with a distance of about 0.5 cm between their centers. Usually the spiral centers are induced about 1 cm from the boundary of the Petri dish by breaking a wave front of a circular wave with a gentle blast of air (Fig. 4). Flow measurements are carried out at the surface of the fluid layer during the evolution of the spiral waves after the initial wave break. The first few wave trains generated by the spirals show a rather long wavelength (> 3 mm), but, as a consequence of the dispersion relation of chemical waves, the wave trains move closer and closer together until they reach a stationary distance which is slightly less than 1 mm [18]. After about 15 min the whole Petri dish is filled densely with chemical waves.

During this evolution process the flow behavior changes as well and passes through several regimes [24] (Fig. 9). In the early stage of the experiments, when the wavelength is long, the flow is similar to that found for circular wave propagation. But when the wavelength gets shorter, this kind of flow stops and a second phase with almost no surface flow starts. In a third phase, after 15-20 min, when the whole dish is filled with wavetrains, a completely new flow behavior occurs spontaneously. Now the flow direction at the surface changes periodically, about once with every passing wave front. This so-called oscillatory flow is the typical flow behavior in a fully developed spiral wave. After about 40 min the system dies due to the aging of the solution and the oscillatory flow stops.

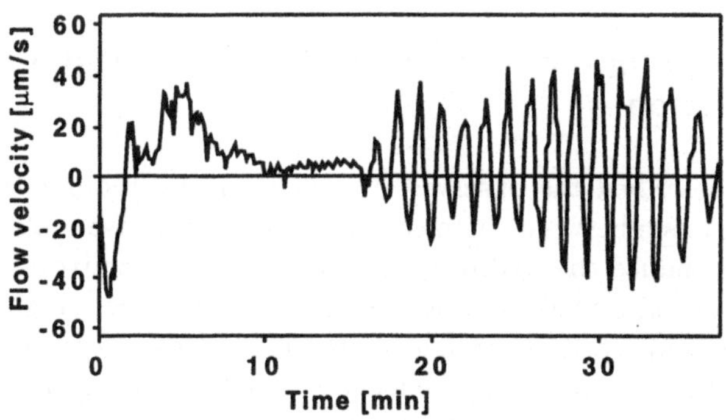

Fig. 9. Surface flow velocity in the middle of the Petri dish during the evolution of a spiral wave. After a discrete phase is observed with circular wave propagation and a quiet phase the flow reaches an oscillatory behavior, when the whole dish is filled with very close wave trains.

In the case of circular waves one period of the chemical oscillation corresponds exactly to one period of the convection rolls. Hence the ratio of about 2:1 in the case of spiral waves is remarkable. Here two chemical oscillations are necessary to induce one complete hydrodynamic period. However, it is important to note that this correlation is not perfectly fulfilled. The period ratio is only close to 2:1 which indicates that there is no more a precise correlation between single wave fronts and the local convection. Consequently, the observed flow appears to be a cooperative phenomenon of a larger group of wave fronts in the dish.

For this complex flow behavior it is impossible to reconstruct the three-dimensional flow field just from the measurements at a single point. Unfortunately the presented method of direct flow measurement does not allow to obtain simultaneously the velocity distribution in the whole dish. We found, however, that it is possible to acquire more information about the global features of the flow, if one takes into account the influence of the hydrodynamics on the shape of the chemical wave fronts. In fact, it has been shown that the convection deforms the wave fronts in a characteristic manner, depending on the flow direction relative to the direction of wave front propagation [26]. The deformation of the wave fronts changes the local light transmission in the system, as can be visualized by image processing. Regions where the surface flow is directed towards the direction of wave propagation appear about 5% darker compared to the case without convection. The flow in the opposite direction results in a higher light transmission [27]. This way one can obtain information about the global flow behavior just by observing the transmitted light over the whole dish. The oscillatory flow turns out to feature large global flow waves extending over about ten chemical wavelengths and traveling through the dish towards the center of wave generation (Fig. 10). This very complex interplay between a large group of dense chemical waves and global flow in the whole Petri dish is a unique phenomenon and theoretically not yet understood.

4. The Coupling Mechanism Between Reaction and Convection

After having described the observed flow phenomena, the question about the coupling mechanism between chemical waves and convective motion arises. How can the waves induce the hydrodynamic flow in the reactive medium? There are two main regions, where such a coupling can take place (Fig. 11): the bulk, where the chemical reaction can change the local density r(x) and the surface, where the reaction may influence the local surface tension s(x). Density gradients as well as surface tension gradients may be caused by changes in temperature or directly by the alterations of the local chemical composition of the medium. From calorimetric measurements it is well known, that the Belousov-Zhabotinsky reaction is exothermic and most of the heat is produced in the oxidized state [28]. Hence temperature gradients must be expected in the medium. In addition, the existence of different types of ions, especially the ferroin and ferriin ions, should lead to surface tension gradients and, via different hydration, to gradients in density [29].

To clarify the role these different types of coupling mechanisms play in chemically driven convection, it is necessary to measure the magnitude of the effects. The temperature changes between the oxidized and the reduced state in the BZ-

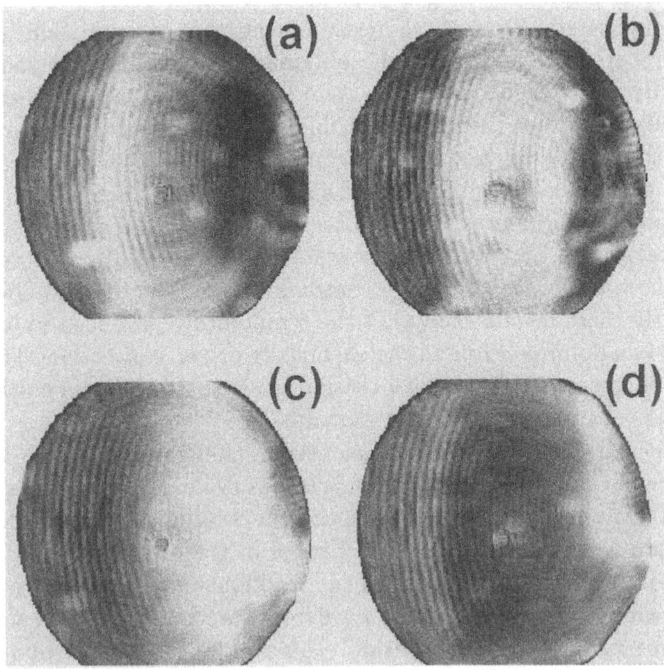

Fig. 10. Global flow waves in the BZ reaction shown in four snapshots of the Petri dish during the oscillatory flow. Time interval between the images: 16.8 s. In the dish a spiral wave is induced near the right boundary. The images are processed by background subtraction, moving averaging to eliminate the structure of the chemical waves, and contrast enhancement. The bright and dark stripes represent a surface flow in and against the direction of wave propagation respectively.

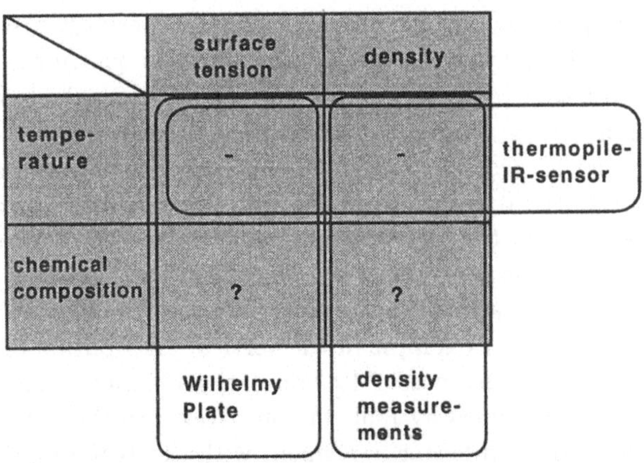

Fig. 11. Table of the possible coupling mechanisms between chemical wave propagation and hydrodynamical flow. The rounded rectangles show the regions, where special measurements can clarify the role, these effects may play in the phenomenon.

reaction are estimated by Ungvarai et al. in a calorimeter, i. e. for a spatially homogeneous system. From these measurements, a maximum value of 0.1 K for the temperature difference could be obtained [28]. Preliminary measurements in a spatially inhomogeneous system were done recently using a thermopile-IR-detector [30]. As a consequence of heat diffusion the measured values are smaller and show that the wave fronts are accompanied by a local temperature maximum with a temperature difference of about 50 mK. This maximum results in a local density minimum as well as in a minimum of the surface tension near the chemical wave front. Both effects would force the solution to an ascending flow near the wave front, which is in contradiction to the observed flow behavior. Consequently these results show that the temperature produced by the reaction does not play a dominant role in the chemically driven convection. The observed flow must be generated by surface tension or density gradients caused directly by the chemical composition of the medium.

Kusumi et al. measured the surface tension changes in the BZ-reaction [31]. The authors also used a spatially homogeneous system by preparing the BZ reaction in a stirred vessel and used for their measurements the Wilhelmy plate method. As a result, surface tension changes in the order of 2 mN/m between the oxidized and the reduced state of the reaction were obtained. In a spatially inhomogeneous system the chemical wave fronts are expected to be accompanied by a local maximum in surface tension. This will lead to a descending flow near the wave front, which is consistent with our observations. Thus, the good agreement of the measured surface tension changes with the flow behavior supports the view that the chemically driven convection is of the Marangoni type.

For the density changes there exist no reliable measurements yet. Pojman et al. calculated values only for the different hydration of ferroin and ferriin [29]. With this value Wu et al. were able to obtain in principle the correct flow behavior in numerical simulations [32]. Consequently, there is no decision yet whether the observed convection phenomena are density or surface tension driven or a superposition of both effects. Further experiments, especially direct measurements of the density changes will answer this question.

In the literature an increasing number of publications deals with the chemically driven convection. Numerical simulations are now carried out, using the Oregonator model to describe the chemical part. For the hydrodynamic part the stationary Navier-Stokes equations for incompressible fluids are added [33]. In these studies usually the hydrodynamics are linked to the chemical pattern assuming a variable density which is proportional to one of the Oregonator variables [32, 34]. The simulations lead to flow patterns which look quite similar to the measurements but, for example in the vertical distribution of the horizontal flow velocity, there are differences. Also simulations taking into account a change in surface tension are underway [35]. In comparison to the density simulations they seem to be in an even better agreement with the experiments, but for more information, one has to await further experiments. First steps have also been undertaken to address the problem on an analytical basis by applying amplitude equations of the Ginzburg-Landau type [36].More detailed work is in progress along these lines.

Acknowledgement

This work was supported in part by the European Space Agency, Paris.

References

1. H. Bénard, *Rev. Gen. Sci. Pure Appl.* **11** (1900) 1261 and 1309
2. E. L. Koschmieder, *Adv. Chem. Phys.* **26** (1974) 177
3. M. J. Block, *Nature* **178** (1956) 650
4. A. N. Zaikin, A. M. Zhabotinsky, *Nature*, **225** (1970) 535
5. A. M. Turing, *Trans. R. Soc. Lond.* **B 237** (1952) 37
6. V. Castets, E. Dulos, J. Boissonade, P. DeKepper, *Phys. Rev. Lett.* **64** (1990) 2953
7. K.-J. Lee, W. D. McCormick, J. E. Pearson, H. L. Swinney, *Nature* **369** (1994) 215
8. Q. Ouyang, J. Boissonade, J. C. Roux, P. DeKepper, *Phys. Lett.* **A 124** (1989) 282
9. *Oscillations and Travelling Waves in Chemical Systems*, edited by R. J. Field and M. Burger (Wiley, New York, 1971)
10. R. Fitzhugh, *Biophys. J.* **1** (1961) 445
11. 11. N. A. Gorelova, J. Bures, *J. Neurobiol.* **14** (1983) 353
12. J. M. Davidenko, P. Kent and J. Jalife, *Physica* **D 49** (1991) 182.
13. J. D. Murray, *Mathematical Biology*, Springer-Verlag, Berlin, Heidelberg, New York, p. 435 (1989)
14. K. I. Agladze, V. I. Krinsky, A. M. Pertsov, *Nature* **308** (1984) 834
15. S. C. Müller, Th. Plesser, B. Hess, *Chemical Waves and Natural Convection* in M. G. Valerde (Ed.) Physical Hydrodynamics: Interfacial Phenomena, p. 423, Plenum Press, New York (1988)
16. H. Miike, S. C. Müller, B. Hess, *Phys. Lett.* **A 141** (1989) 25
17. 17. V. Petrov, V. Gáspár, J. Masere, K. Showalter, *Nature* **361** (1993) 240
18. J. P. Keener, J. J. Tyson, *Physica* **D21** (1986) 307
19. P. Foerster, S. C. Müller, B. Hess, *Development* **109** (1990) 11
20. H. H. Rotermund, W. Engel, M. Kordesch, G. Ertl, *Nature* **343** (1990) 355
21. T. Yamaguchi, L. Kuhnert, Zs. Nagy-Ungvarai, S. C. Müller, *J. Phys. Chem.* **95** (1991) 5831
22. S. C. Müller, Th. Plesser, B. Hess, *Physica* **D 24** (1987) 71
23. W. Jahnke, W. E. Scaggs, A. T. Winfree, *J. Phys. Chem.* **93** (1989) 740
24. G. S. Skinner, H. L. Swinney, *Physica* **D 48** (1991) 1
25. 25. Y. Aizu, T. Asakura, *Appl. Phys.* **B 43** (1987) 209
26. H. Miike, S. C. Müller, *Chaos* **3** (1993) 21
27. K. Matthiessen, S. C. Müller, *Global Flow Waves in Chemically Induced Convection*, Phys. Rev. E, in the press (1995)
28. E. Körös, M. Orbán, Zs. Nagy, *Acta Chim. Acad. Sci. Hung.* **100** (1979) 29
29. J. A. Pojman, I. R. Epstein, *J. Phys. Chem.* **94** (1990) 4966
30. M. Böckmann, S. C. Müller, to be published

31. K. Yoshikawa, T. Kusumi, *Chem. Phys. Lett.* **211** (1993) 211
32. Y. Wu, D. A. Vasquez, B. F. Edwards, J. W. Wilder, *Phys. Rev.* **E 51** (1995) 1119
33. Th. Plesser, H. Wilke, K. H. Winters, *Chem. Phys. Lett.* **200** (1992) 158
34. H. Wilke, *Interaction of Traveling Chemical Waves with Density Driven Hydrodynamic Flow*, Physica D, in the press (1995)
35. H. Wilke, private communication (1995)
36. M. Diewald, H. R. Brand, *Chem. Phys. Lett.* **216** (1993) 566

Part VI

Combustion

Combustion Processes Under Microgravity Conditions

F.A. Williams

Center for Energy and Combustion Research, Department of Applied Mechanics and Engineering Sciences, University of California, San Diego La Jolla, CA 92093-0411, USA

Abstract

A classification of combustion processes is presented which can be used for discussing influences of microgravity. Profound effects of gravity on combustion arise from the large fractional density changes that generally are associated with combustion. Thus it seems to be rather the rule than the exception that combustion experiments under microgravity conditions reveal previously unanticipated phenomena. A few of these phenomena and their theoretical explanations are presented here. Particular attention is devoted to flame balls and to alcohol droplet combustion. Some accomplishments of these and other microgravity combustion studies are summarized. Finally, combustion topics which appear to hold greatest promise for benefiting from future microgravity experiments are indicated.

1. Introduction

Combustion processes are exothermic transformations of chemical energy into thermal energy and as such involve elements of thermodynamics, fluid mechanics, transport processes and chemical kinetics. Combustion generally involves large temperature increases that typically result in substantial fractional density decreases, by a factor between two and ten in representative situations. As a consequence of this density change, earthbound combustion processes generally are influenced strongly by buoyancy. Under microgravity conditions buoyancy is nearly removed, and consequently the characteristics of combustion processes can be modified profoundly. The nature of these modifications is the subject of this article.

The relevance of gravity to commonly experienced combustion phenomena has long been known. For example, a standard method for testing flammability of gas mixtures is to fill with the mixture a glass tube, 1 m long and 25 mm in diameter, to discharge a sufficiently intense spark at one end of the tube and to observe whether the flame initiated by the spark succeeds in propagating to the opposite end. The mixture is judged flammable if this propagation is successful, which occurs only if the mole fraction of fuel in the mixture exceeds a minimum value, termed the lean flammability limit and which is less than a maximum value (the rich flammability limit), above which there is insufficient oxidizer to sustain propagation. It is well established that the flammability limits determined in this manner differ for upward and downward flame propagation, [1] thereby demonstrating a significant role of gravity in this type of combustion process.

Efforts to achieve microgravity conditions for purposes of scientific study of combustion phenomena date back to the pioneering work of Kumagai [2], who in the 1950's studied droplets burning for about 1 s of free fall in a chamber dropped from the ceiling to the floor of the laboratory. Work at NASA in the 1960's recognized some of the unique flammability phenomena under weightlessness that could potentially affect fire hazards in spacecraft [3]. Beginning in that decade, scientific studies of influences of gravity on structures of gaseous jet diffusion flames were performed in NASA drop towers [4, 5]. More then twenty years ago an evaluation was conducted of the scientific advances that might come from investigations of combustion processes employing microgravity conditions [6] At the beginning of the 1980's, an AIAA volume reported in much greater depth the promise of experiments at reduced gravity for contributing to the progress of combustion science [7]. The past decade and a half has now seen a number of discoveries in fundamentals of combustion which are due to to microgravity research. Some of these discoveries will be discussed here.

First, a classification of combustion processes is presented, within which microgravity combustion studies can be categorized. Next, a few particular combustion phenomena relevant to microgravity conditions are addressed in greater detail. Finally, prospects for future advances of combustion knowledge through microgravity investigations are assessed.

2. Classification of Combustion Processes

It is convenient to divide combustion processes into four subareas, namely, gaseous premixed flames, in which the gaseous reactants are well mixed prior to initiation of the flow and combustion process, gaseous diffusion flames, in which the gaseous fuel and oxidizer are initially unmixed and must diffuse towards each other to react, condensed-fuel combustion, in which one or more of the reactants are in a condensed phase, rather than the gaseous phase, at least initially, and detonation, in which combustion is initiated by a sufficiently strong shock wave. Combustion processes in the last of these categories generally are dominated by inertial effects of gasdynamics that overpower gravity and thereby render microgravity conditions of little relevance. By contrast, microgravity conditions typically exert strong influences on combustion processes in the other three categories.

2.1 Premixed Flames

The most spectacular advances in combustion science through microgravity research have occurred in the area of premixed flames. This is because there are a number of mechanisms for stabilizing or destabilizing premixed flames that operate at extended time scales so that buoyancy interferes with their observation. Flame balls in fuel-lean hydrogen-air mixtures, in which individual balls of flame develop and then remain stationary, were first observed in microgravity [8]. Self-extinguishing flames, which propagate outward from an ignition source for a long time, and then suddenly extinguish, also were first observed in microgravity [9]. Thus, a number of unique premixed-flame phenomena are encountered under microgravity conditions.

2.2 Diffusion Flames

Diffusion flames also behave quite differently under microgravity, again mainly because of the longer residence times available with the greatly reduced influences of buoyancy. Unlike premixed flames, diffusion flames are nonpropagating and possess fewer mechanisms of instability. Microgravity observations of the combustion of gaseous hydrocarbon fuels injected into air have revealed these flames to be generally rounder, thicker, sootier, more stable and cooler than corresponding flames under normal gravity [10, 11, 12]. There is increasing understanding of the reasons for these differences in flame structures.

2.3 Condensed-Fuel Combustion

Many processes in the combustion of condensed fuels are substantially affected by microgravity. These include smoldering, spray combustion, droplet combustion, metal combustion, dust-cloud ignition and combustion, and flame spread. The first combustion experiments performed in the Space Shuttle concerned flame spread along solid fuel surfaces and demonstrated spread mechanisms of a different character than observed at normal gravity [13]. Theoretical analyses of flame spread along solid rods demonstrated the significant role of microgravity [14]. Space experiments planned for droplet combustion are also expected to reveal new combustion phenomena [15]. There are many different physical phenomena that can influence condensed-fuel combustion, such as surface-tension-driven flow, radiation, condensed-phase mixing and heat transfer, condensed-phase and gas-phase pyrolysis, and surface chemistry. Intricate interactions among these phenomena and buoyancy can be affected strongly by a microgravity environment. Especially noteworthy here are processes such as smoldering that require long periods of time to develop and that therefore demand space experiments for proper scientific investigation of microgravity influences. Another significant aspect of this extensive area of investigation is that, in microgravity, material flammability may differ substantially from flammability at normal gravity, through buoyancy effects.

3. Flame Balls

Consider an infinite, quiescent space at a temperature T equal to T_∞, containing a uniform fuel-oxidizer mixture with a great excess of oxidizer and a mass fraction Y of fuel equal to Y_∞. If there is no convection at all, then the steady-state, spherical symmetrical equations for conservation of fuel and of energy can be written, respectively as

$$\rho D \frac{1}{r^2}\frac{d}{dr}\left(r^2 \frac{dY}{dr}\right) = \omega \tag{1}$$

and

$$\lambda \frac{1}{r^2}\frac{d}{dr}\left(r^2 \frac{dT}{dr}\right) = -q\omega \tag{2}$$

Here r is the radial coordinate, ρ is the gas density, D denotes the diffusion coefficient of the fuel, ω is the rate of consumption of mass of fuel per unit volume by chemical reactions, λ represents the thermal conductivity of the mixture, and q stands for the energy released per unit mass of fuel consumed. These equations express balances between molecular diffusion and chemical reaction and between heat conduction and chemical heat release, respectively. Many simplifying assumptions have been introduced in writing these equations, such as constancy of D and λ. The rate ω is a function of Y and T which is typically proportional to Y and increases rapidly with increasing T.

Zel'dovich [17] was the first to observe that there exists a simple solution to these equations which describes flame balls. This solution maintains $Y = 0$ and $T = T_0$, a constant value, inside a sphere of radius R, while $Y_\infty - Y$ and $T - T_\infty$ decrease with increasing r, in proportion to $1/r$, outside the sphere. In this solution, advantage is taken of the strong temperature dependence of w to replace that source function by a delta function located at $r = R$. Corresponding steady solutions in infinite media do not exist in cylindrical or planar geometry because in those cases $1/r$ would become lnr or r, respectively, both of which diverge at infinity. Thus, flame balls may exist, but not flame rods or flame slabs.

The Zel'dovich flame-ball solution is readily derived from equations (1) and (2). Multiplication of the first by q and then addition to the second gives

$$\frac{\rho D q}{\lambda}\frac{d}{dr}\left(r^2 \frac{dY}{dr}\right) + \frac{d}{dr}\left(r^2 \frac{dT}{dr}\right) = 0 \tag{3}$$

which, with the further assumption that $\rho D q/\lambda$ is constant, can readily be integrated twice to show that

$$T + \frac{\rho D q}{\lambda} Y = \frac{c_1}{r} + c_2 \tag{4}$$

where c_1 and c_1 are constants of integration. Boundedness at $r = 0$ requires that $c_1 = 0$, and the boundary conditions at infinity then give

$$T = T_\infty + \frac{\rho D q}{\lambda}(Y_\infty - Y) \tag{5}$$

In particular, $T_0 = T_\infty + \rho D q/\lambda Y_\infty$ is the constant flame temperature through the interior of the flame ball. In the delta-function approximation, outside the flame ball $\omega = 0$, and equations (1) and (2) are then readily integrated twice to give

$$T = T_\infty + (T_0 - T_\infty)\frac{R}{r} \qquad Y = Y_\infty(1 - \frac{R}{r}) \tag{6}$$

The value of R is determined from the reaction rate by integrating across the thin flame sheet. Where $\omega \neq 0$ (2), for example, is very nearly

$$-\frac{q\omega}{\lambda} = \frac{d^2T}{dr^2} = \frac{1}{2}\frac{d}{dT}\left[\left(\frac{dT}{dr}\right)^2\right] \tag{7}$$

so that, since dT/dR inside the flame ball, while $dT/dR = -(T_0 - T_\infty)R$ just outside according to equation (6), integration gives

$$R^2(T_0 - T_\infty)^2 = \int \frac{q}{\lambda}\omega dT \tag{8}$$

in which equation (5) can be employed to express ω explicitly in terms of T and thereby enable the integral across the reaction sheet, essentially the same as the integral from $T = T_\infty$ to $T = (T_0$ to be evaluated.

Zel'dovich also considered the time-dependent conservation equations and investigated the stability of these simple flame-ball solutions, finding them to be unstable. Stable flame balls therefore were not anticipated to exist, until microgravity experiments in a drop tower revealed apparently stable flame balls in fuel-lean hydrogen-air mixtures [8]. As can be imagined, this sparked a flurry of activity by theoreticians. It was found that if a radiant energy loss term is added to equation (2), then there are two solution branches, one unstable, corresponding to the Zel'dovich type of solution, and another, with a larger radius R , being stable when the Lewis number L is less than unity. Here

$$L = \frac{\lambda}{\rho D c_p} \tag{9}$$

where c_p denotes the specific heat at constant pressure of the gas. Flame balls of large enough radii can thus be stabilized by their radiant energy losses [18].

Has this theoretical prediction been verified experimentally? In earth gravity buoyancy distorts flame balls and causes them to rise, severely limiting observation times. Test times in drop towers are too short for proof of long-time stability of flame balls. Experiments were performed in aircraft flying parabolic trajectories, but the gravity quality is too poor, and the flame balls are observed to bounce around as the gravity vector varies. Experiments by P.D. Ronney are planned for the April 1997 launch of MSL-1 with the Space Shuttle. These experiments should provide sufficient time, at sufficiently low gravity levels, for definitive testing of flame-ball theories.

4. Alcohol Droplet Combustion

Drop-tower experiments have been performed on the microgravity combustion of droplets of pure alcohols and of binary mixtures of alcohols [19, 20]. The purpose of microgravity here is to facilitate achievement of spherical symmetry, enabling accurate comparison with predictions of correspondingly simplified theories to be made. Sufficiently small free droplets can achieve a good degree of spherical symmetrical combustion at normal gravity,21 but for larger droplets natural convection distorts the flame substantially. Experiments with initially pure methanol droplets showed the flame sheet which surrounds these droplets to extinguish before the droplet disappeared. [19, 20, 21]. Extinction occurs abruptly when the droplet diameter has decreased to a critical value, the extinction diameter. It is of interest to calculate the extinction diameter theoretically and to compare the theoretical prediction with experiment. This entails consideration of the gas-phase chemical kinetics for combustion of the fuel vapor with the oxygen in the atmosphere, because extinction occurs when the time needed for combustion to occur exceeds the time available in the diffusion flame which surrounds the droplet, the latter decreasing with decreasing droplet diameter d_l being proportional to d_l^2/D by dimensional analysis, where D is an appropriate gas-phase diffusion coefficient.

4.1 Alcohol Combustion Chemistry

Through recent clarifications of rates of elementary steps, accurate knowledge of the gas-phase chemical kinetics of methanol combustion now exists. Table 1 contains a minimal list of the elementary steps and their rate parameters; each reaction rate is the product of the concentrations of the reactants with the specific reaction-rate constant k. This fourteen-step mechanism can be simplified further by taking advantage of the fact that most of the reaction intermediaries are so reactive that their production and consumption rates nearly balance, that is, they achieve chemical-kinetic steady states. For example, for an alcohol containing n carbon atoms, for which the overall combustion process is $\mathrm{C_nH_{2n+1}OH + 3/2nO_2 \rightarrow nCO_2 + (n+1)H_2O}$ steady-state approximations may be introduced for all radicals except the important H atom and result in a four-step reduced mechanism for the alcohols that can be written as

$$\begin{aligned}
&\mathrm{C_nH_{2n+1}} + \gamma\mathrm{H} + \left[\beta - (\mathrm{n}+1+\tfrac{1}{2})\right]\mathrm{H_2O} + \tfrac{1}{2}\left[3\mathrm{n} + \tfrac{1}{2}\gamma - \beta(1+\alpha)\right]\mathrm{O_2} \\
&\qquad \rightarrow \beta(\mathrm{H_2} + \alpha\mathrm{CO}) - (\mathrm{n} - \alpha\beta)\mathrm{CO_2} && (I) \\
&\qquad \mathrm{CO + H_2O \leftrightarrows CO_2 + H_2} && (II) \\
&\qquad \mathrm{2H + M \rightarrow H_2 + M} && (III) \\
&\qquad \mathrm{3H_2 + O_2 \leftrightarrows 2H + 2H_2O} && (IV)
\end{aligned}$$

noindent where α is the number of CO molecules produced per H_2 molecule produced, β is the number of H_2 molecules produced per fuel molecule consumed, and γ, a quantity of critical importance, is the number of radicals destroyed for each fuel molecule consumed.

No.	Reactions	A [cm³/smol]]	n	E [kcal/mol]
	Hydrogen-Oxygen Chain			
1f	$H + O_2 \rightarrow OH + O$	$3.52 \cdot 10^{16}$	-0.7	17070
1b	$OH + O \rightarrow H + O_2$	$1.15 \cdot 10^{14}$	-0.324	175
2f	$H_2 + O \rightarrow OH + H$	$5.06 \cdot 10^{4}$	2.67	6290
2b	$OH + H \rightarrow H_2 + O$	$2.22 \cdot 10^{4}$	2.67	4398
3f	$H_2 + OH \rightarrow H_2O + H$	$1.17 \cdot 10^{9}$	1.30	3626
3b	$H_2 + H \rightarrow H_2 + OH$	$6.65 \cdot 10^{9}$	1.30	4352
4f	$OH + OH \rightarrow H_2O + O$	$k = 5.46 \cdot 10^{11} \exp(0.00149T)$		
4b	$H_2O + O \rightarrow OH + OH$	$k = 4.19 \cdot 10^{7} T^{.58} \exp(0.003T)$		
	Hydroperoxyl Formation and Consumption			
5[a]	$H + O_2 + M \rightarrow HO_2 + M$	$6.76 \cdot 10^{19}$	-1.4	0
6	$HO_2 + H \rightarrow OH + OH$	$1.70 \cdot 10^{14}$	0.0	874
7	$HO_2 + H \rightarrow H_2 + O_2$	$4.28 \cdot 10^{13}$	0.0	1411
8	$HO_2 + OH \rightarrow H_2O + O_2$	$2.89 \cdot 10^{13}$	0.0	-497
	Water-Gas Shift			
9f	$CO + OH \rightarrow CO_2 + H$	$4.40 \cdot 10^{6}$	1.5	-741
9b	$CO_2 + H \rightarrow CO + OH$	$4.97 \cdot 10^{8}$	1.5	21446
	Formyl Formation and Consumption			
10	$CHO + H \rightarrow CO + H_2$	$1.00 \cdot 10^{14}$	0.0	0
11[c]	$CHO + M \rightarrow CO + H + M$	$2.85 \cdot 10^{14}$	0.0	16795
12	$CH_2O + H \rightarrow CHO + H_2$	$2.26 \cdot 10^{10}$	1.05	3278
	Fuel Consumption and Formaldehyde Formation			
13	$CH_2OH + H \rightarrow CH_2O + H_2$	$3.00 \cdot 10^{13}$	0.0	0
14	$CH_3OH + H \rightarrow CH_2OH + H_2$	$4.00 \cdot 10^{13}$	0.0	6092

[a] Chaperon Efficiencies: N_2,O_2 : 1.0, CO : 1.0, CO_2 : 3.8, H_2 : 2.5, H_2O : 12.0
[b] Chaperon Efficiencies: same as *a* except H_2O: 16.3

Table 1. The chemical-kinetic mechanism and the associated rate parameters in the form $k = AT^n \exp(-E/R_gT)$.

For methanol, this four-step description involves introducing steady-state approximations for CH_2OH, CH_2O, CHO, HO_2, OH and O in Tab. 1 and results in $\gamma = 2$, giving a reduced mechanism that can be written as

$$\begin{array}{lcl} CH_3OH + 2H \rightarrow 3H_2 + CO & & CO + H_2O \leftrightarrows CO_2 + H_2 \\ 2H + M \quad \rightarrow \quad H_2 + M & \text{and} & 3H_2 + O_2 \leftrightarrows 2H + 2H_2O \end{array}$$

with corresponding rates

$$k_{14} C_F C_H \qquad \frac{k_{9f}}{K_3} C_H \cdot \left(\frac{C_{CO} C_{H_2O}}{C_{H_2}} - C_{CO_2} \frac{K_3}{K_9}\right)$$

$$k_5 C_M C_H C_{O_2} \qquad \text{and} \qquad k_{1f} C_H \left[C_{O_2} - \frac{C_H^2 C_{H_2O}^2}{C_{H_2}^3 K_1 K_2 K_3^2}\right]$$

Here C_i denotes the concentration of species i, subscript F representing the fuel, CH_3OH, and the K's are equilibrium constants. The four overall steps are, respectively, fuel consumption, the water-gas shift, radical recombination, and the oxygen-consumption step, which always produces radicals through the branched hydrogen-oxygen chain. An asymptotic analysis of the flame structure has been performed with this four-step mechanism to predict extinction diameters of methanol droplets [22]. The predicted extinction diameters are in good agreement with results of full numerical integrations employing more complicated sets of conservation equations that include more elementary steps than shown in Table 1. In other words, the descriptions of the gas-phase transport processes and chemical kinetics are good enough so that one can be reasonably confident about the predicted extinction diameters.

4.2 Methanol Flame Extinction

Table 2 lists the three experiments in which extinction diameters were measured for initially pure methanol droplets. The table also gives the initial droplet diameter in the second column and the observed extinction diameter in the third. The fourth column is the predicted extinction diameter, which clearly is much too small. The difficulty with the prediction is that it neglects influences of absorption of water by the liquid fuel during combustion. Water is one of the two main products of combustion and is produced in the gas-phase flame, diffuses back to the methanol droplet, and is adsorbed there during the first part of the combustion history. Water then gradually builds up inside the liquid and, during the second part of the combustion process, evaporates along with the methanol. What is needed, therefore, is not a theory for the combustion and flame extinction of a pure methanol droplet, but rather a theory for the combustion and flame extinction of a binary liquid droplet mixture of methanol and water. Since water is inert, the chemical kinetics are the same as for the pure fuel, and the gas-phase analysis can be done equally accurately. Such theories have now been completed [22]. As is expected, since water is inert, the binary liquid droplet extinguishes much more easily, that is, its extinction diameter is larger.

Predicted extinction diameters for theories in which account is taken of adsorption of water in initially pure methanol depend strongly on how the binary liquid phase is treated. There are two limiting liquid-phase behaviors. In one limit, there is no convection in the liquid phase, and the adsorbed water diffuses into the stationary liquid droplet from its surface. The liquid-phase diffusion coefficient is so small that, in this limit, a thin water boundary layer forms quickly in the liquid at its surface, and the water soon begins vaporizing from this boundary layer, along with the fuel. In the opposite limit, liquid-phase convection is so strong that the liquid rapidly becomes well mixed, and the water concentration always is spatially nearly uniform throughout the droplet. In the second limit, much more water is adsorbed by the droplet, and the adsorption period lasts longer (about half of the combustion lifetime), but the percentage of water which evaporates with the fuel during the second stage is much larger. The first limit will be called the diffusion limit because of the dominant role played by spherical symmetrical liquid-phase diffusion, and the second will be called the convection limit because of the essential role of liquid-phase convection.

Atmosphere and reference	Initial droplet diameter [mm]	Experimental extinction diameter [mm]	Extinction diameter with no water absorption [mm]	Extinction diameter in the diffusion limit [mm]	Extinction diameter in the convection limit [mm]	Final water mass fraction in the convection limit
50% O_2, 50% He [19]	1.0	0.32	0.018	0.034	0.29	90%
air [20]	0.43	0.17	0.026	0.040	0.14	60%
air [21]	0.24	0.07	0.026	0.040	0.075	40%

Table 2. Comparison between theoretical and experimental conditions

The fifth and sixth columns in Table 2 give the predicted extinction diameters for the diffusion limit and the convection limit, respectively [22]. It is seen that in the diffusion limit the extinction diameters are less than twice those obtained with no water adsorption and they are much smaller than the experimental extinction diameters. Therefore there must be some convection in the liquid in these experiments. The predictions of the convection limit are actually seen to be in excellent agreement with experiments. One must therefore conclude that the convection must be strong enough to produce nearly complete mixing, leading to a droplet of nearly uniform composition. The last column gives the predicted percentage of the liquid that has become water at the time of extinction, and it is seen that a substantial amount of the initially pure methanol droplet has become water by the time when combustion ceases, which is in agreement with measurements [21] made on collected final droplets. Figure 1 shows the predicted extinction diameter d_{le} and water mass fraction in the liquid at extinction $Y_{H_2Ol_e}$ as functions of the initial droplet diameter d_{l0}, for initially pure methanol droplets burning in three different room-temperature atmospheres, according to the convection limit. The nearly linear dependence of the extinction diameter on the initial diameter is in good agreement with the few available experiments and is a consequence of the mass-transfer dynamics that result in water absorption during the first half of the combustion history and vaporization during the second half.

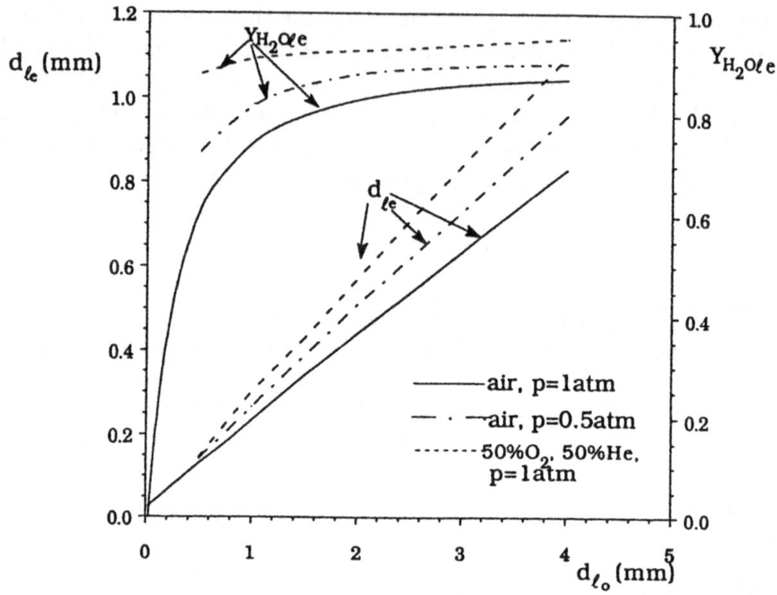

Fig. 1. Dependence of the extinction diameter and of the final water fraction on the initial diameter, predicted in the convection limit for methanol droplet combustion in three different room-temperature atmospheres.

4.3 Potential Resolution of a Paradox

The need for liquid-phase convection in these explanations poses a paradox. In one of the microgravity experiments [20], not only were initially pure methanol droplets burned, but also droplets which were initially binary mixtures of methanol and dodecanol have been studied. Over a wide range of mixture fraction these binary mixtures are known from reasonably extensive previous work to exhibit disruptive burning, sometimes termed "microexplosion". That is, at some stage in the combustion history, a bubble nucleates in the interior of the liquid and rapidly expands, turning the liquid into a large, thin shell that then ruptures into a large number of very small droplets that very rapidly burn to completion. The nucleation and growth for disruption of these alcohol mixtures are observed to be quite spherically symmetrical. The disruptive burning mechanisms of these binary mixtures is well understood physically. As combustion proceeds more volatile fuel vaporizes preferentially, leaving the less volatile component to build up in concentration in a thin liquid boundary layer at the droplet surface. As the surface concentration of this less volatile fuel increases, so does the surface temperature, eventually approaching the equilibrium vaporization temperature of this less volatile constituent. For binary mixtures of fuels of sufficiently different volatility, such as methanol and dodecanol, this final equilibrium vaporization temperature can exceed the nucleation temperature of the original mixture. Heat conduction in the liquid, from its high-temperature surface to the interior where the original mixtures is presumed to still exist, then causes the nucleation of that mixture, which leads to the bubble growth and disruption. Detailed theoretical calculations based on this mechanism have produced good agreement with measured disruption conditions. The paradox is that this mechanism precludes thorough liquid-phase convection. With strong liquid convection, there is no disruption mechanism. How can any one experimental sequence [20] show, side by side, extinction for initially pure methanol and disruption for the mixture?

The explanation suggested here is the solutal Marangoni effect. Water adsorbs to a much lesser extent in the dodecanol, which rapidly builds up at the surface of the binary liquid, than in the pure methanol. It is therefore suggested that water adsorption plays no significant role for the binary fuel. On the other hand, for the water-methanol mixture, a strong solutal-Marangoni instability is to be anticipated. The higher surface tension of the water-enriched mixture will draw the surface towards any spot that momentarily acquires excess water in a random fluctuation. But since the initially adsorbed water is mainly at the surface, this motion further enriches the spot in water, because the material drawn towards the spot is replenished from within by methanol-rich liquid, reducing the surface water concentration away from the spot. No thorough solutal Marangoni stability analysis has been completed for this configuration, involving diffusive mass transfer of both constituents across the evaporating interface in opposite directions. Therefore, of course, the bifurcation analysis needed to define the consequent convection patterns has not been addressed either. Theoretical work on these questions deserves to be done, and relevant normal-gravity experiments can also be envisioned. Since the corresponding surface-tension variations of the binary alcohol mixtures are smaller, the effect is not expected for the methanol-dodecanol droplets. Also, similar qualitative reasoning indicates that the ordinary thermal Marangoni effect is not destabilizing in these configurations. Marangoni flow, which can be important in flame spread across liquid surfaces, for example, seems in general to arise less often than one might anticipate in liquid-fuel combustion problems, but nevertheless may be significant in explaining the alcohol paradox.

5. Other Combustion Phenomena Learned from Experiments Under Microgravity Conditions

A number of other combustion phenomena have been found in microgravity studies, for example in premixed flames [23] and in diffusion flames [24]. Since there is too much information of this kind to be covered properly here, the reader is referred to recent conference proceedings [25] for a thorough presentation. It may be stated in general that the rate of progress has been increasing exponentially, especially in recent years.

6. Combustion Topics of Future Promise for Microgravity Investigation

Seven specific topics are listed here as especially warranting further investigation under microgravity conditions. These areas have been identified from considerations of what has been accomplished in the past and of what more is needed in the future.

6.1 Premixed-Flame Instabilities

Premixed flames exhibit a number of instability phenomena, many of which are stabilized in earthbound laboratories mainly by gravity [26]. Microgravity studies could be designed to provide careful tests of theoretical predictions of these instabilities, possibly revealing specific new influences of particular chemical-kinetic mechanisms.

6.2 Flammability Limits

As indicated in the introduction, flammability limits are known to depend on gravity through influences of natural convection on heat loss. Microgravity experiments can therefore aid in measuring "true" limits, controlled only by finite-rate chemical kinetics and radiant energy loss. Special phenomena, such as flame balls, may nevertheless extend limits appreciably under microgravity for highly diffusive reactants. Thus, there appears to be much to be learned about flammability limits through microgravity investigations.

6.3 Ignition

Autoignition and stimulated ignition processes, too, are often affected by natural convection, except at very short time scales. This applies to ignition of both gaseous and condensed-phase fuels. A very limited number of studies of ignition under microgravity conditions have been performed [27], and much more work can be done in this area.

6.4 Flame Extinction

Section 4.2 identified one of many types of extinction studies that benefit from investigation in microgravity. There is good potential for obtaining further clean extinction results from microgravity studies.

6.5 Soot Production

Soot production occurs through pyrolysis of carbon-containing fuels over fairly long time scales. The associated chemistry is complex and is influenced strongly by temperature-time histories. Observed increased sooting tendencies of diffusion flames under microgravity are a consequence of the increased residence times found in fuel-rich elevated-temperature regions when natural convection is reduced. Investigations under microgravity conditions therefore afford the potential for revealing altered chemical pathways to soot.

6.6 Flame Spread Along Fuel Rods

This topic is one of a number of flame-spread processes that can be quite different under microgravity conditions [14]. Further studies of flame spread in microgravity for this and other configurations seem warranted not only for scientific reasons but also to help in assessing fire hazards in spacecraft environments.

6.7 Pollutant Production in Combustion

Pollutants emitted from flames include oxides of nitrogen, carbon monoxide, unburnt hydrocarbons, particulate matter such as soot, and oxides of sulfur, for example. The chemical and physical processes by which these emissions occur differ greatly for the different pollutants and are affected by the combustion environment. Experiments under microgravity offer appreciably revised environments so that different pollutant production routes may occur. Little or no microgravity research has been performed on this topic, which certainly deserves consideration for future investigations.

7. Conclusions

Although, in the field on combustion, fewer microgravity experiments have been performed than in a number of other fields, nevertheless significant discoveries in combustion already have come from microgravity investigations. There are many areas of combustion that could benefit substantially from future microgravity studies. Carefully planned microgravity experiments would appreciably advance fundamental understanding in combustion science.

References

1. H.F. Coward and G.W. Jones, "Limits of Flammability of Gases and Vapors," Bulletin 503, U.S. Bureau of Mines, U.S. Government Printing Office, Washington, D.C., 1952.
2. S. Kumagai, *Jet Propulsion*, **26** (1956) 786.
3. J.H. Kimzey, "Flammability During Weightlessness," NASA TMX-58001, 1966.
4. T.H. Cochran and W.J. Masica, Thirteenth Symposium (International) on Combustion, The Combustion Institute, Pittsburgh, PA, 1970, p. 821-829.
5. J.B. Haggard and T.H. Cochran, *Combustion Science and Technology*, **5**, (1972) 291-298.
6. A.L. Berlad, C. Huggett, F. Kaufman, G.H. Markstein, H.B. Palmer and C.H. Yang, "Study of Combustion Experiments in Space," NASA CR-134744, PSRI TR-4034, Contract NAS3-17809, NASA Lewis Research Center, Cleveland, Ohio, November 1974.

7. T.H. Cochran, ed., *Combustion Experiments in a Zero-Gravity Laboratory*, Progress in Astronautics and Aeronautics, Vol. 73, American Institute of Aeronautics and Astronautics, New York 1981.
8. P.D. Ronney, *Combustion and Flame*, **82** (1990) 1-14.
9. P.D. Ronney, *Combustion and Flame*, **62** (1985) 121-133.
10. M.Y. Bahadori,D.P. Stocker, D.F. Vaughan, L. Zhou and R.B. Edelman, in: *Modern Developments in Energy, Combustion and Spectroscopy*, F.A. Williams, A.K. Oppenheim, D.B. Olfe and M. Lapp, eds., Pergamon Press, Oxford, 1993, pp. 49-66.
11. L. Bonneau, J.M. Most, P. Joulain and A.C.Fernandez-Pello, "Flat-Plate-Flow Diffusion Flames in Microgravity," AIAA Paper No. 93-0826, January 1993.
12. J.L. Torero, L. Bonneau, J.M. Most, and P. Joulain, "The Effect of Gravity on a Laminar Diffusion Flame Established over a Horizontal Flat Plate," 30th COSPAR Scientific Assembly, Hamburg, Germany, July 1994.
13. S. Bhattacharjee, R.A. Altenkirch, and K. Sacksteder, *Combustion Science and Technology*, **91** (1993) 225-242.
14. C.S. Tarifa, A. Liñan, J.J. Salvá, G. Corchero, G.L. Juste, and F. Esteban, Final Report, ESA Contract No. 6934/86/F/Fl. Departamento de Motopropulsion y Termofluidodinamica, Escula Tecnica Superior de Ingenieros Aeronauticos, Universidad Polytecnica de Madrid, Spain, February 1988.
15. F.A. Williams, in: Third International Microgravity Combustion Workshop, H.D. Ross, editor, NASA Conf. Publ., to appear, 1995.
16. J.L. Torero, A.C. Fernandez-Pello, and D. Urban, *AIAA Journal*, **32** (1994) 991-996.
17. Y.B. Zel'dovich, Theory of Combustion and Detonation of Gases, Izd. Akad. Nauk SSSR, Moscow, 1944.
18. J.D. Buckmaster, G. Joulin, and P.D. Ronney, *Combustion and Flame*, **84** (1991) 411-422.
19. S.Y.Cho, M.Y. Choi, and F.L.Dryer, in: *Twenty-Third Symposium (International on Combustion*, The Combustion Institute, Pittsburgh, PA, 1991, pp. 1611-1617.
20. J.C. Yang, G.S. Jackson, and C.T. Avedisian, in: *Twenty-Third Symposium on Combustion*, The Combustion Institute, Pittsburgh, PA, 1991, pp. 1619-1625.
21. A. Lee, and C.K. Law, *Combustion Science and Technology*, **86** (1992) 253-265.
22. 22. Zhang, B.L., Card, J.M. and Williams, F.A., *Combustion and Flame*, submitted, 1995.
23. D. Durox, F. Baillot, P. Scouflaire, and R.Prud'homme, *Combustion and Flame*, **82** (1990) 66-74.
24. M.Y. Choi, F.L. Dryer, and J.B. Haggard, in: *Twenty-Third Symposium (International on Combustion*, The Combustion Institute, Pittsburgh, PA 1991, pp. 1597-1604.
25. H.D. Ross, editor, Third International Microgravity Combustion Workshop, NASA Conf. Publ., to appear, 1995.
26. P. Clavin, *Progress in Energy and Combustion Science*, **11** (1985) 1-59.
27. M. Tanabe, M. Kono, J. Sato, J. Knig, C. Eigenbrod, and H.J.Rath, in: *Twenty-Fifth Symposium on Combustion*, The Combustion Institute, Pittsburgh, PA, to appear, 1995.

Flat Plate Diffusion Flames: Numerical Simulation and Experimental Validation for Different Gravity Levels

J.L. Torero, H-Y Wang, P. Joulain and J.M. Most*

Laboratoire de Combustion et de Dtonique Universit é de Poitiers - UPR 9028 au CNRS - ENSMA Site du Futuroscope - BP 109 Chasseneuil-du-Poitou 86960 FUTUROSCOPE CEDEX France

Abstract

A laminar diffusion flame is established over a horizontal flat plate burner when ethane is injected into a stream of air flowing parallel to the burner. A three dimensional stationary numerical solution is obtained for the temperature and velocity fields with the aim of better understanding the effect of buoyancy. The chemical reaction is described by the flame sheet approximation, therefore, fuel, oxidiser, inert gases and products are assumed to go through a single global reaction step with an infinitely fast reaction rate. The results obtained from the numerical simulation are compared with experiments conducted under different gravity levels. The parameters studied are velocity and temperature profiles as well as the flame geometry. The numerical simulation describes very well the geometry of the flame in the presence of gravity, in its absence, the stationary model is accurate only close to the leading edge. As the distance from the leading edge increases the numerical solution under predicts the stand-off distance.

1. Introduction

Condensed combustible materials often burn such that a diffusion flame is established over its surface. This form of fuel burning is commonly encountered in fires and, thus, is of particular interest in the fire safety field.

Emmons [1] showed that the evaporation rate of the fuel could be expressed as a function of the mass transfer number only. Numerous studies have been conducted on the diffusion flame established over a flat condensed fuel. The works of Williams [2] and Drysdale [3] provide extensive reviews on the subject.

Limited attention has been given, however, to the aerodynamical structure of the flow and the effect of gravity. Hirano and co-workers [4, 5] used a porous plate through which a gaseous fuel was uniformly injected. Ramachandra and Raghunandan [6], on a similar burner, studied the effect of injection and free stream velocities on the stability of vaporised n-heptane flames. A gas burner

* Corresponding Author, Present Address: Department of Fire Protection Engineering, University of Maryland at College Park, College Park, Maryland 20742-3031, USA

was also used to study the effect of the burner geometry on separation occurring at the leading edge for air forced velocities of the order of 1 m/s [7]. By substituting the combustible fuel by a porous gas burner the problem is simplified significantly, fuel supply is no longer coupled to heat transfer and therefore fuel and oxidiser supply to the reaction can be varied independently. Although the similarity solution, obtained by Emmons [1], is no longer possible, this configuration allows for a better study of phenomena such as entrainment due to thermal gradients.

There are very few studies dealing with the effect of gravity on a chemically reacting boundary layer. Fernandez-Pello and Pagni [8] introduced a single non-dimensional parameter to study forced, mixed or free combustion of a vertical fuel surface. Pagni [9] pointed out the importance of buoyancy in the determination of the flame length. Kodama et al. [10] and Mao et al. [11] included a buoyancy term in numerical solutions of a burning vertical wall. Using a porous burner, Bonneau et al. [12] obtained some preliminary experimental results on a vertical flame in micro-gravity. In a horizontal configuration, a buoyantly induced pressure gradient parallel to the wall was discussed by Lavid and Berlad [13]. Torero et al. experimentally studied the effect of buoyancy on extinction limits and flame shape [14, 15].

In this work, a numerical study is conducted to further explore the role of buoyancy on a diffusion flame established over a horizontal porous gas burner. The results are compared with experiments conducted at different gravity levels. The parameters studied are the oxidiser flow velocity, oxygen concentration, fuel injection velocity and gravity level.

2. Theoretical Analysis and Numerical Features

2.1 Mathematical Formulation

The conservation expressions for all of the variables are written in the elliptic form by using Cartesian co-ordinates for the geometrical configuration presented in Fig 1.

For mass conservation:

$$\frac{\partial(ru)}{\partial x}+\frac{\partial(rv)}{\partial y}+\frac{\partial(rw)}{\partial z}=0$$

For momentum and all scalar quantities can be written as :

$$\frac{\partial(\rho u\phi)}{\partial x}+\frac{\partial(\rho v\phi)}{\partial y}+\frac{\partial(\rho w\phi)}{\partial z}=\frac{\partial}{\partial x}\left(\Gamma_\phi\frac{\partial\phi}{\partial x}\right)+\frac{\partial}{\partial y}\left(\Gamma_\phi\frac{\partial\phi}{\partial y}\right)+\frac{\partial}{\partial z}\left(\Gamma_\phi\frac{\partial\phi}{\partial z}\right)S_\phi$$

The details of the variables are shown in Table 1.

The flame sheet model is used to describe the chemical reaction in which fuel (C_2H_6), oxidiser (O_2), products (H_2O, CO_2) and inert gas (N_2) are assumed to go through a single global reaction step with an infinitely fast reaction rate, resulting in an infinitely thin reaction zone that can be described by:

$$\nu_F C_2H_6+\nu_O O_2+\nu_I N_2\Rightarrow\nu_P(H_2O+CO_2)+\nu_I N_2+\Delta H_F$$

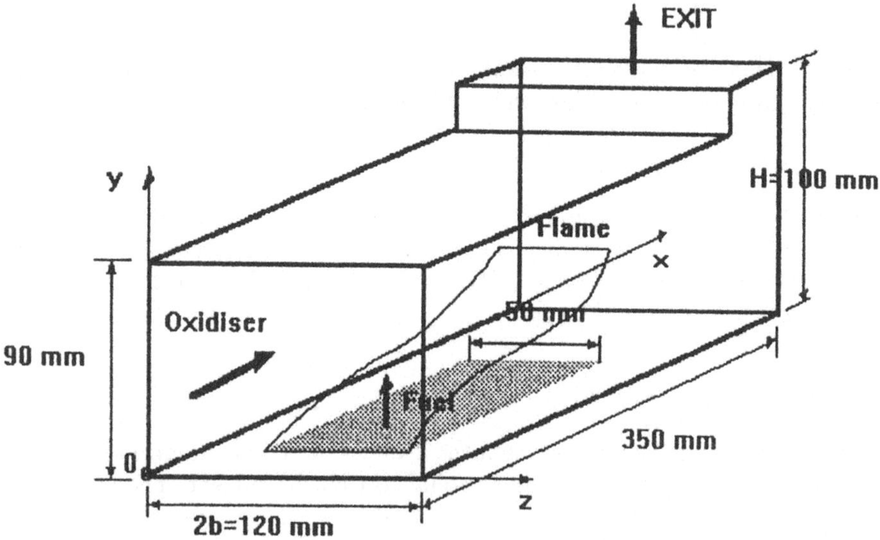

Fig. 1. Schematic of the experimental test section

ϕ	Γ_ϕ	S_ϕ
u	μ	$-\frac{\partial p_d}{\partial x} + \frac{\partial}{\partial x}(\mu \frac{\partial u}{\partial x}) + \frac{\partial}{\partial y}(\mu \frac{\partial v}{\partial x}) + \frac{\partial}{\partial z}(\mu \frac{\partial w}{\partial x})$
v	μ	$-\frac{\partial p_d}{\partial y} + (\rho_\infty - \rho)g + \frac{\partial}{\partial x}(\mu \frac{\partial u}{\partial y}) + \frac{\partial}{\partial y}(\mu \frac{\partial v}{\partial y}) + \frac{\partial}{\partial z}(\mu \frac{\partial w}{\partial y})$
w	μ	$-\frac{\partial p_d}{\partial z} + \frac{\partial}{\partial x}(\mu \frac{\partial u}{\partial z}) + \frac{\partial}{\partial y}(\mu \frac{\partial v}{\partial z}) + \frac{\partial}{\partial z}(\mu \frac{\partial w}{\partial z})$
f	μ/Sc	0

Table 1. Summary of the general variable ϕ, the transport coefficient Γ_ϕ and the source terms S_ϕ

The coupling of the equations is done by introducing the Shvab-Zeldovich formulation that results in the mixture fraction equation without a source term. The heat-release parameter ΔH_F is reduced to:

$$H_\mathrm{C} = (1-\chi)\Delta H_\mathrm{F} \quad \text{where} \quad \chi = \chi_\mathrm{A} - \chi_\mathrm{R} \quad \text{and} \quad \Delta H_\mathrm{F} = 4.745 \cdot 10^7 \mathrm{J/kg}$$

Although the flame location is obtained by means of an infinitely fast reaction rate, temperatures have to be corrected by decreasing the heat release by a factor χ_A (χ_A= 0.85 for all conditions). A second corrective factor,χ_R, has to be introduced to correct for radiative heat losses from the flame to the environment (χ_R=0.35 for this work). This "constant-χ" approach was previously employed by Markstein [16], who determined a value for based on experiments conducted at normal gravity. Flame colour and temperature are similar for normal and microgravity experiments [12], therefore, the same value of χ_R is used for all calculations. The Schmidt number (Sc) is set equal to 0.7.

For adiabatic flames, the enthalpy equation is identical to the mixture-fraction equation (f), so the mean temperature is obtained without solving the energy conservation equation, thus, obtaining

$$T = [f(H_{\mathrm{F}} + H_{\mathrm{C}} - H_{\infty}) - Y_{\mathrm{F}} H_{\mathrm{C}} + H_{\infty})]/C_{\mathrm{P}}$$

where H is the enthalpy and the mean density is evaluated from the ideal gas law ($\overline{\rho T} = \rho_{\infty} T_{\infty} =$ constant). The variation of the molecular viscosity with temperature is taken to be $\mu = (T/T_{\infty})^{3/4} \sum_i (Y_i \mu_{0,i})$ [17] and the mean specific heat is given by $C_{\mathrm{P}} = \sum C_{\mathrm{p},i} Y_i$.

2.2 Boundary Conditions

It is assumed that the flow enters in the x-direction with a uniform axial velocity and at a temperature and pressure equal to that of ambient. Therefore, at the inlet (x=0)

$$u = U_{\infty}, v = 0, w = 0, p = 0, f = 0, T = T_0$$

at the outlet (y=H), zero gradient conditions are imposed thus:

$$\frac{\partial \phi}{\partial y} = 0 \qquad \text{where } \phi = u,v,w \text{ and } f$$

The boundary conditions at all solid walls are stated as

$$u = 0, \quad v = 0, \quad w = 0, \quad \left.\frac{\partial f}{\partial n}\right|_w = 0$$

The last condition states that the wall is adiabatic, where n means normal to the wall. For the fuel surface (y=0) we set

$$u = 0, \quad v = \dot{m}_S \rho, \quad \dot{m}_S (1 - f_0) = \frac{\mu}{\mathrm{Sc}} \left.\frac{\partial f}{\partial y}\right|_w$$

Because of symmetry with respect to the z-axis, the domain to be studied will be reduced to half the channel $0 < z < b$. At $z = b$, the symmetry conditions are

$$\frac{\partial \phi}{\partial z} = 0 \quad \text{where} \quad \phi = u,v,f \text{ and } w = 0$$

2.3 Numerical Procedure

The flow equations are solved numerically by employing the SIMPLEC algorithm [18], which is based on a finite volume method developed by Patankar [19]. An under-relaxing factor, by which the convergence can be assured, of 0.6 - 0.7 was used in the equations, as well as in updating the density and the viscosity.

A grid with 35(x) x 26(y) x 11(z) interior nodes is generated using a series distribution. A run time of about 20 minutes was necessary to compute accurately the reacting flow field on a DEC3100/Alfa. The (absolute) residuals for the u,v,w, and equations were summed for all cells and scaled with an appropriate value (here, the scaling value was taken as the greatest residual in the course of iterations). When the scaled residuals for all variables were all smaller than 0.02, the solution was considered to be converged.

Limited computer memory did not allow for an absolute grid independence but the variation observed in the temperature and velocity distributions throughout consecutive grid refinements was at least two orders of magnitude smaller both in value and distribution.

3. Experimental Apparatus

The experiments are conducted in a small scale, horizontally oriented, combustion tunnel along with the supporting instrumentation (Fig. 1). Three of the walls are Pyrex windows to enable visual observation. Ethane, 99.4% pure, is uniformly injected through a porous burner mounted on the floor of the combustion chamber. The oxidiser and fuel velocities are governed with controlled mass flow meters.

Information on the flame is obtained from a CCD video camera and six Chromel-Alumel thermocouples. The video cameras provide a top and a side view of the flame. The thermocouples are placed on a vertical line with their junction located on the plane of symmetry. The thermocouples generate significant perturbations on the flow, therefore, they are placed far away from the leading edge (120 mm). It is important to clarify that the measurements are not representative of the flame temperature but only give a characteristic temperature of the combustion products.

Experiments under reduced gravity conditions were conducted in a drop tower (ZARM) and by using the Caravelle aircraft of CNES. Up to 5 sec. of micro-gravity ($10^{-5}g_0$) were obtained at the drop tower and up to 25 sec. ($10^{-3}g_0$) with the Caravelle. A more detailed description of the experimental hardware and procedures can be found in the work of Bonneau et al. [12].

4. Experimental Results

A series of images of flames for different gravity levels are presented in Figs. 2 and 3. The oxidiser velocity is for all cases 0.115 m/s, the fuel injection velocity is 0.0052 m/s and the oxygen concentration 22%. Several experiments with other velocities were conducted but will not be presented here since these flames show well the fundamental characteristics and physical aspects of the flame.

In normal gravity the flame consisted of two distinctive zones, a first zone, close to the leading edge, where a boundary layer flame can be observed and a second zone, further upstream, where a buoyantly induced plume is present. The first zone is characteristically blue and the second zone yellow. For low oxygen concentrations the flame is pale and unstable at the leading edge. As the oxygen concentration increases the flame becomes more stable and brighter. The first zone remains blue and the second yellow, but both colours become more intense.

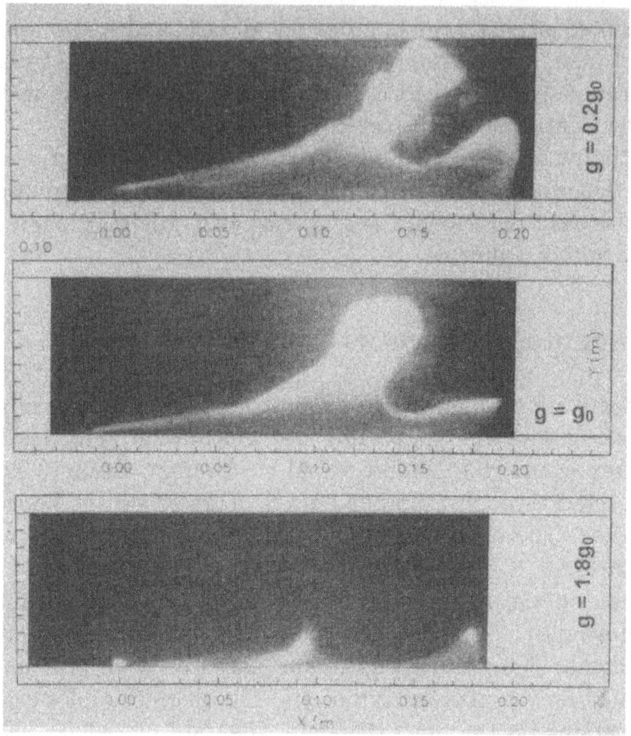

Fig. 2. Photographic images of flames for different gravity levels

The distance from the leading edge at which the plume appears remains constant with the oxygen concentration, but the size of the plume decreases significantly.

Experiments were also conducted at different gravity levels (0.2 and $1.8g_0$). Images of these flames are presented in Fig.2. Increasing the gravity level turns the flame bluer, smaller and more unstable. Decreasing the gravity level has the opposite effect, the boundary layer zone increases in size and the stand-off distance increases.

In micro-gravity the flame is curved when close to the leading edge, turning linear as the distance from the leading edge increases. As the injection velocity increases so does the region where the flame is linear. For high injection velocities the flame is almost entirely linear and a curved flame is only present very close to the leading edge. The geometry of the flame is also affected by the oxygen concentration (Fig.3), for low oxygen concentrations the flame is almost linear through out the plate, as the oxygen concentration increases the parabolic zone increases in size and the stand-off distance increases. For all gravity levels, the brightness of the flame increases with the oxygen concentration.

In microgravity all flames are blue on the oxidiser side and yellow on the fuel side and there is a significant effect of the ceiling on the flame shape. The ceiling reflects the flame, when it gets too close, generating a point of maximum stand-off distance. The geometrical limitation of the chamber has a significant

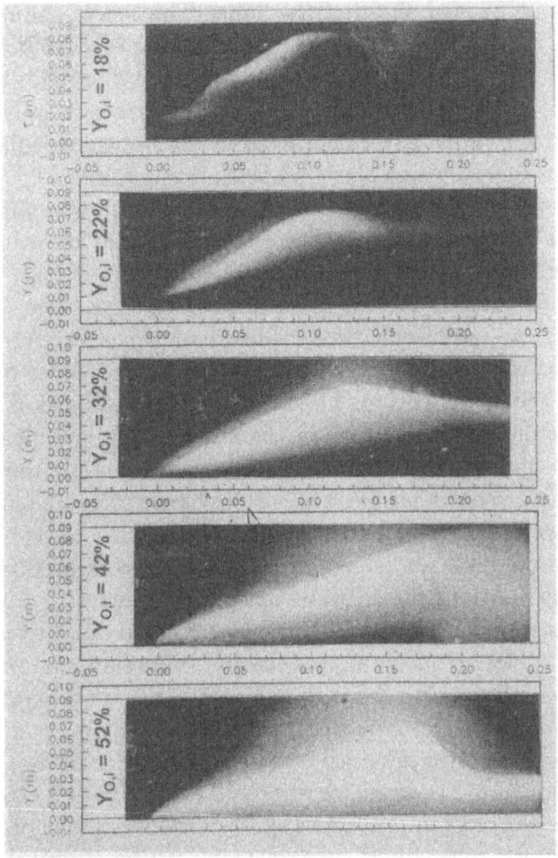

Fig. 3. Photographic images of micro-gravity flames for different oxygen concentrations

effect downstream but no observable effect upstream of the point of maximum stand-off distance. For low oxygen concentrations significant three dimensional effects could be observed. These effects along with the influence of the lateral walls are extensively discussed by Torero et al. [14] and will not be treated here.

5. Numerical Solution

A numerical simulation was conducted for all experimental cases but only the results obtained for the conditions previously described will be presented. The observations presented in this section are applicable to all other conditions.

The velocity and temperature distribution along the chamber was obtained for 22% oxygen concentration and simulations were conducted for different gravity levels. The centreline temperature and velocity distributions are presented in Figs. 4 and 5, respectively, and correspond to the images presented in Fig.2.

It can be observed that in the presence of any gravitational acceleration a recirculation zone is formed ahead of the flame. The size of the recirculation zone increases with the gravity level but the magnitude of the velocity vectors is not significantly influenced by gravity. A characteristic velocity for this recirculation flow has been previously determined [14] to be approximately 0.1 m/s. These velocities are in very good agreement with the numerical calculations. A smaller recirculation zone can be also observed behind the flame.

The temperature distributions show that the maximum temperature remains almost constant for different gravity levels (Fig.4). In the presence of a significant gravitational acceleration a plume becomes evident ($g > 0.2g_0$). The plume splits in two as gravity increases and the flame height decreases. The zone of maximum temperature corresponds well to the visible flame zone observed in Fig.2.

The results corresponding to experiments conducted under microgravity conditions deserve special attention. The temperature distribution of Fig.4 and the velocity profiles of Fig.5 show that in the absence of a significant gravitational acceleration the maximum temperatures are found outside the zone where the flow is affected by viscous forces. The visible flame presented in Fig.3, for 22% oxygen concentration, does not correspond well with the maximum temperature zone, the stand-off distance is significantly under estimated. Fig.6 shows the temperature distributions for different oxygen concentrations. All flames are in micro-gravity and the temperature distributions correspond to those flames presented in Fig.3. It can be observed that the zone of maximum temperature differs significantly from the visible flame. For high oxygen concentrations, maximum temperature and visible flame correspond well close to the leading edge, in the region where the flame was observed to be parabolic. For 52% oxygen concentration good agreement is observed for approximately 100 mm downstream of the leading edge. As the oxygen concentration decreases the zone of good agreement decreases in size, for 18% almost no agreement can be observed.

6. Concluding Remarks

By comparing numerical solutions and experimental results, the limitations of the numerical model are identified. The numerical model properly describes the structure of the flow where the effect of buoyancy is overwhelming. Near the leading edge the numerical calculations agree well with the experimental results.

Buoyancy is the controlling parameter determining the geometry of the flame in the presence of gravity. In the absence of gravity the numerical model seems to need a time criteria which reflect better the nature of the interaction between a finite chemical reaction and mass transport to the reaction zone.

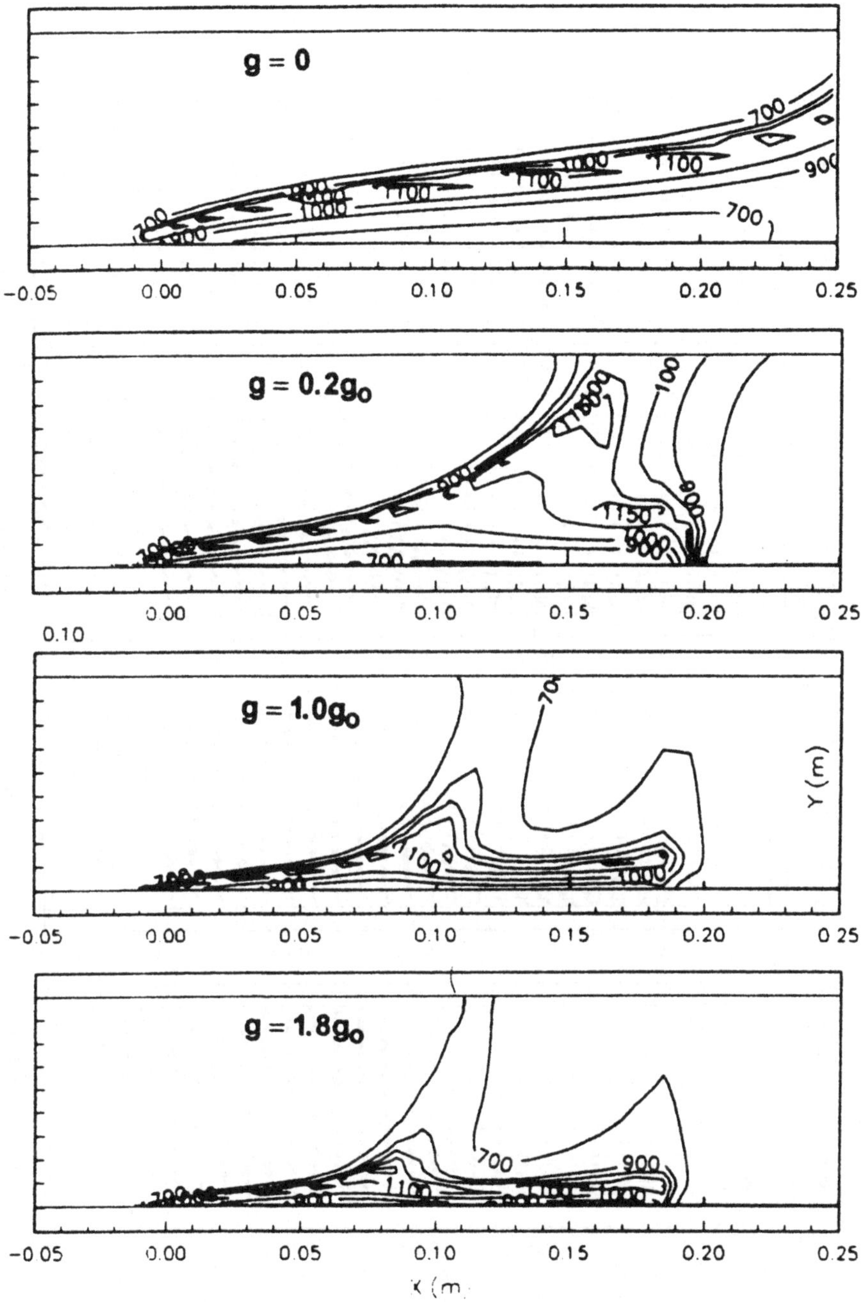

Fig. 4. Numerically obtained temperature distributions for various gravity levels

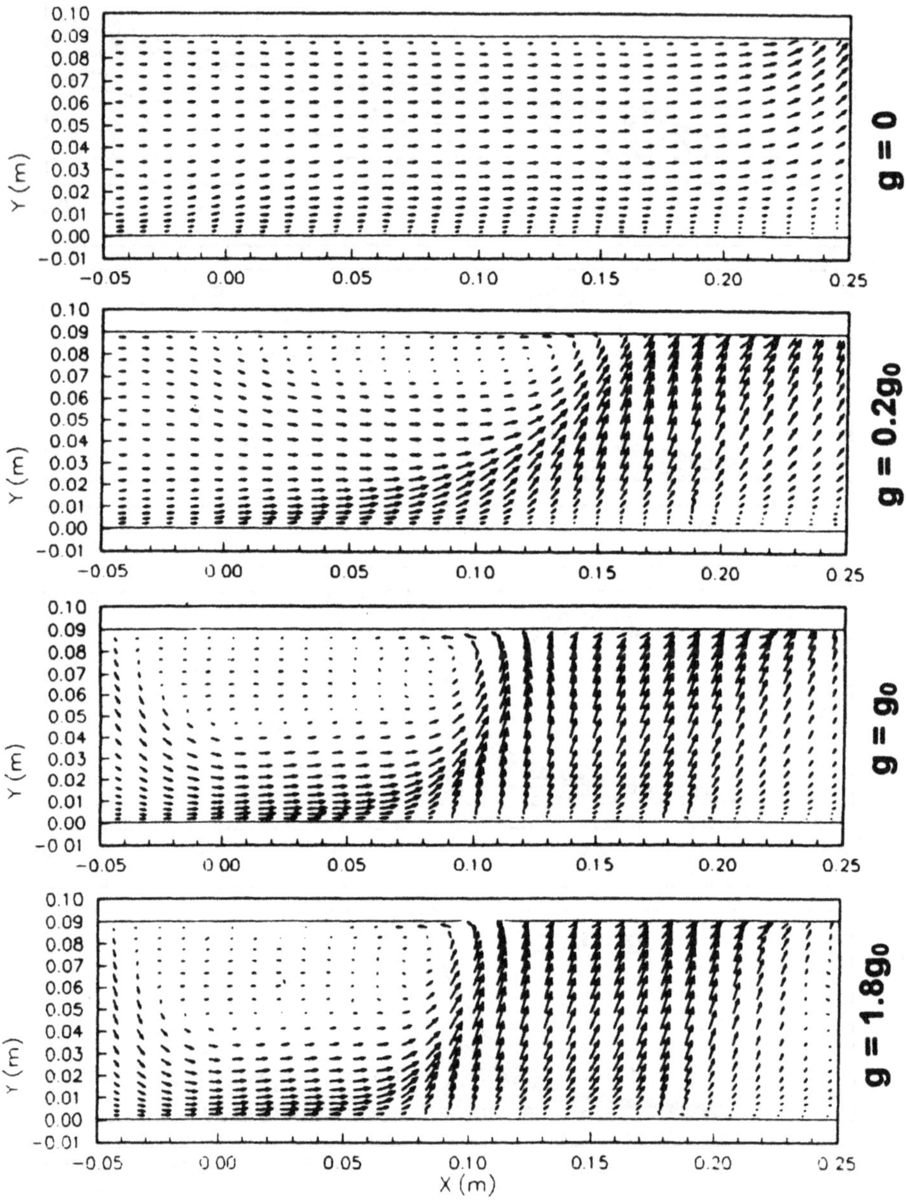

Fig. 5. Numerically obtained velocity distributions for various gravity levels

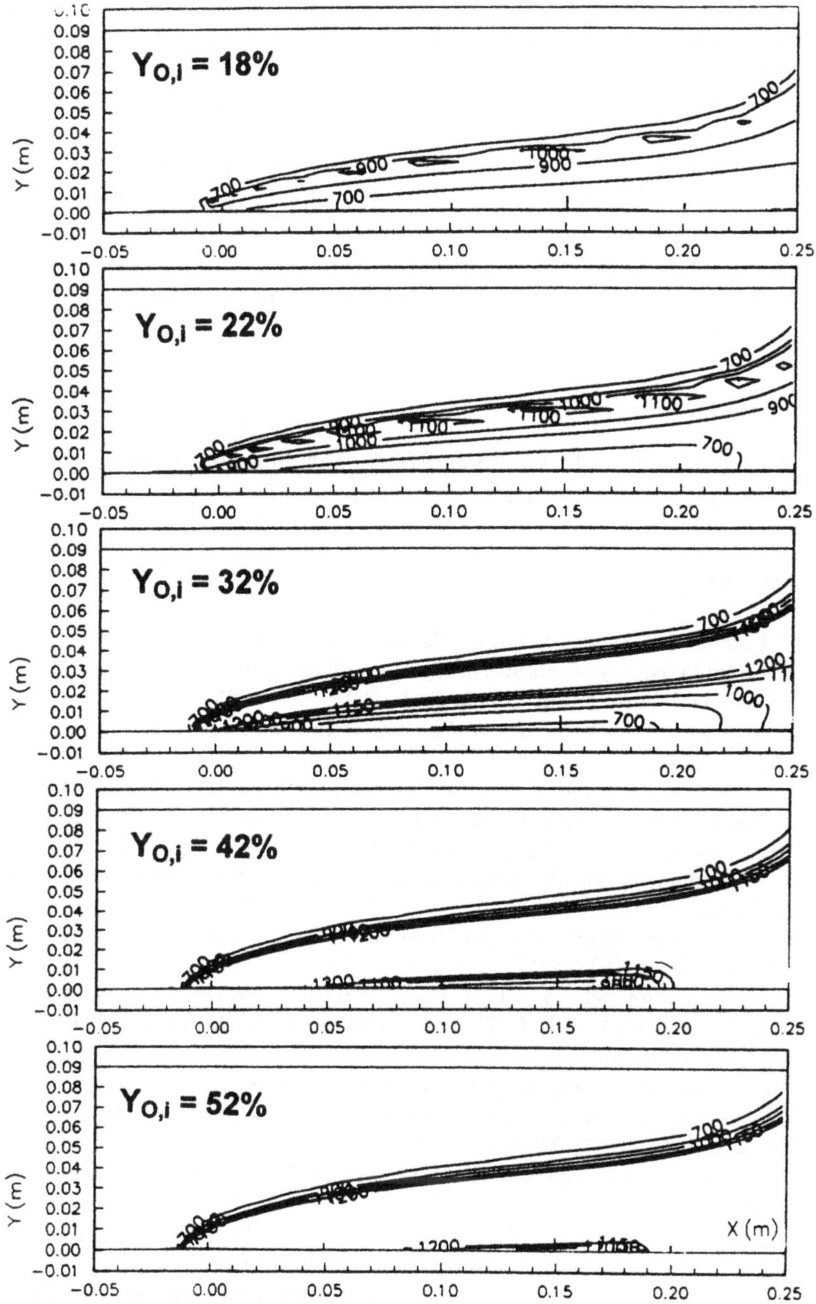

Fig. 6. Numerically obtained temperature distributions for microgravity flames

Comparison between the experimental results and the numerical and solutions is limited to direct observation of video images. In normal gravity, the visible flame boundary and the location of the buoyant plume are used to characterise the flame location. In microgravity, the stand-off distance is defined by the boundary between the blue and yellow regions. Qualitative comparison with the velocity profiles and temperature distributions, obtained numerically, and with the flame stand-off distance, obtained analytically, have provided further insight on the role of buoyancy on this type of flame. Due to the lack of experimental diagnostics and the numerical and analytical approximations used throughout this work, the agreement between predictions and experiments should be taken as approximate.

Acknowledgements

This work is sponsored by CNES and ESA. The numerous technical discussions with Prof. A.C.Fernandez-Pello are much appreciated.

References

1. H. Emmons, *Z.Angew. Math. Mech.*, **36** (1956) 60-71.
2. F.A. Williams, *Combustion Theory*, Second Edition, The Benjamin/Cummings Publishing Company Inc., Menlo Park, California, 1985, pp.485-519.
3. D. Drysdale, *An Introduction to Fire Dynamics*, John Wiley and Sons, New York, 1985, pp.226-252.
4. T. Hirano, K. Iwai and Y. Kanno, *Astronautica Acta*, **17** (1972) 811-818.
5. T. Hirano and Y. Kanno, Fourteenth Symposium (International) on Combustion, The Combustion Institute, Pittsburgh, pp.391-398, 1973.
6. A. Ramachandra and B. N. Raghunandan, *Combustion Science and Technology*, **36** (1984) 109-121.
7. Ji Soo Ha, Sung Hoon Shim and Hyun Dong Shin, *Combustion Science and Technology*, 75 (1991) 241-260.
8. A. C. Fernandez-Pello and P. J. Pagni, Mixed Convective Burning of a Vertical Fuel Surface, ASME-JSME Thermal Engineering Joint Conference Proceedings, Vol.4, Honolulu, 1983.
9. P. J. Pagni, *Fire Safety Journal*, **3** (1981) 273-285 .
10. H. Kodama, K. Miyasaka and A. C. Fernandez-Pello, *Combustion Science and Technology*, 53 (1987) 37-50.
11. C. P. Mao, H. Kodama and A. C. Fernandez-Pello, *Combustion and Flame*, 57 (1984) 209-236.
12. L. Bonneau, P. Joulain, J. M. Most and A. C. Fernandez-Pello, Flat Plate Diffusion Flame Combustion in Microgravity, AIAA-93-0826, 1993.
13. M. Lavid and A. L. Berlad, Sixteenth Symposium (International) on Combustion, The Combustion Institute, Pittsburgh, 1977. pp.1557-1568.
14. J. L. Torero, L. Bonneau, J. M. Most and P. Joulain, Twentyfifth Symposium (International) on Combustion, The Combustion Institute, Pittsburgh, 1994 (in press).
15. J. L. Torero, L. Bonneau, J. M. Most and P. Joulain, *Advances in Space Research*, **16**(7) (1994) 149-152.

16. G. H. Markstein, Sixteenth Symposium (International) on Combustion, The Combustion Institute, Pittsburgh, 1977, pp. 1407-1419.
17. F. Tamanini, *Combustion and Flame*, **30** (1977) 85.
18. J. P. Van Doormaal and G. D. Raithby, *Numerical Heat Transfer*, **7** (1984) 143-163.
19. S. V. Patankar, *Numerical heat Transfer and Fluid Flow*, McGraw Hill, New York, 1980.

High Pressure Droplet Burning Experiments in Reduced Gravity

C. Chauveau, X. Chesneau, B. Vieille, A. Odeide and I. Gökalp

Laboratoire de Combustion et Systèmes Réactifs Centre National de la Recherche Scientifique 45071 Orléans cedex 2, France

Abstract

The paper summarizes research on high pressure burning of single fuel droplets and describes recent results obtained under normal and reduced gravity conditions with suspended droplets. Parabolic flights were used to create a reduced gravity environment. Droplet burning experiments using methanol, n-heptane, n-hexane and n-octane were performed at ambient air temperature and pressures in the range of 0.1 MPa to 12 MPa. The combination of high pressure droplet burning experiments with reduced gravity is crucial in order to reduce the pressure enhanced natural convection effects and also to extend the applicability of the fibre suspended droplet technique when the surface tension decreases due to the closeness of thermodynamic critical conditions. The experimental results presented in this paper show a decrease of the droplet burning time with pressure in the subcritical domain. The minimum burning time observed by other investigators around the critical pressure was not found in this study for the four fuels investigated. This may be attributed to the pressure enhanced natural convection effects due to the residual gravity during parabolic flight experiments.

1. Introduction

Depending on the surrounding flow and thermodynamic conditions, a single droplet may experience several gasification regimes, ranging from the envelope flame regime to pure vaporization [1]. The characteristic times of droplet gasification are largely influenced by the surrounding pressure and temperature. Gasification of liquid droplets under high pressure conditions is indeed one of the main phenomena characterizing liquid fueled energy systems. It is also conjectured that droplets initially injected in the high pressure combustion chamber under subcritical temperatures may heat up and attain the supercritical thermodynamic state before their total gasification. Many theoretical and computational studies have addressed sub- and supercritical droplet vaporization phenomena under spherical symmetry [2]-[12] or axisymmetric conditions [13, 15], by solving the conservation equations associated with the problem. Most of the contributions take into account the transitory effects, the thermodynamic non-idealities and the ambient gas solubility in the liquid phase due to high pressures. Some new approaches based on molecular dynamics are also emerging [16].

A parametric investigation of single droplet gasification regimes is therefore helpful in providing the necessary physical ideas for sub-grid models used in spray combustion numerical prediction codes. A research program has been initiated at the LCSR to explore the vaporization regimes of single and interacting hydrocarbon [1] and liquid oxygen droplets [17] under high pressure conditions. The research program also includes the investigation of single and interacting droplets vaporizing in laminar or turbulent, isothermal or heated flows [18]. In parallel to pure liquid droplet studies, bicomponent hydrocarbon droplets are also investigated [19]. Droplet interaction effects are explored for the configuration of three droplets in tandem vaporizing in a laminar heated flow [20]. Reduced gravity studies are conducted to reproduce spherical symmetry conditions essentially for burning droplet studies [1] and [21].

This paper summarizes the status of the LCSR program on the high pressure burning of single fuel droplets; recent results obtained under normal and reduced gravity conditions with suspended droplets are presented. In the work described here parabolic flights of the CNES Caravelle were used to create a reduced gravity environment of the order of $10^{-2}g_0$. For all the droplet burning experiments reported here, the suspended droplet initial diameters are scattered around 1.5 mm ; and the ambient air temperature is 300 K. The ambient pressure is varied between 0.1 MPa and 12 MPa. Four fuels are investigated : methanol (Pc = 7.9 MPa), n-heptane (Pc = 2.74 MPa), n-hexane (Pc = 3.01 MPa) and n-octane (Pc = 2.48 Mpa).

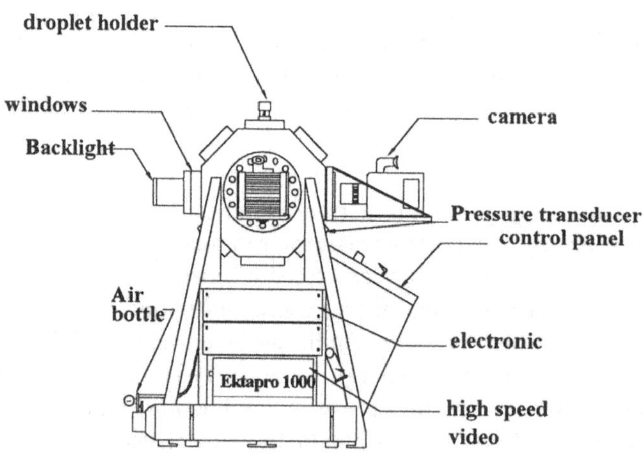

Fig. 1. Droplet Gasification Facility (DGF-HP)

2. Description of the Experimental Techniques

2.1 High Pressure Droplet Gasification Facility

Droplet gasification experiments under stagnant and variable pressure conditions are performed in a specially designed High Pressure Droplet Gasification Facility (HP-DGF). The HP-DGF is designed to investigate the gasification regimes of suspended or free-floating droplets under variable pressure conditions, up to 12 MPa. Low pressure droplet burning experiments can also be performed. This facility can be used during the parabolic flights of an aircraft. This apparatus has been fully described previously [1]. For high pressure droplet burning experiments an improved version of the injection system and a heated coil for ignition are used. The coil is located about 15 droplet diameters below the suspending fiber and is used to heat the air in the vicinity of the droplet. The coil heating current is shut down when droplet autoignition is attained. The sequence of injection and ignition operations is computer controlled. The results presented in this paper are obtained with droplets suspended on a quartz fiber of 0.2 mm diameter with an enlarged extremity of 0.4 mm diameter.

2.2 Diagnostics

The principle diagnostic systems we are using are based on the visualization of the droplet burning phenomena. Suspended droplets are illumninated from the back; this provides the necessary contrast for direct threshold definition. The recording equipment is based on the high speed video camera, Kodak-Ektapro 1000, which can record up to 1000 full frames per second, and allows detailed analysis of droplet burning. The Ektapro 1000 Motion Analyzer's real-time viewing makes possible to follow the investigated phenomena frame per frame. A second camera has been added to the Kodak-Ektapro 1000 system, to be able to observe simultaneously the droplet and the flame with appropriate magnification rates. The digitized images from the Ektapro system are transferred to a microcomputer where the image analyses are performed. For the details of the image analysis [1] and [17] can be consulted.

3. Results and Discussion

Burning droplet experiments have been conducted both under normal gravity conditions and under reduced gravity conditions. An example of the residual gravity level recorded during a parabolic flight experiment is given on Fig. 2. The figure shows the time variation of the gravity level synchronized with the camera acquisition, during the 20 seconds of the microgravity period. In this sequence the average gravity level corresponding to the actual droplet burning time (0.69 s for a n-octane droplet burning at 8 MPa) is $2 \cdot 10^{-2} g_0$.

Figure 3 presents the time histories of the normalized squared droplet diameter versus normalized time for methanol droplets burning under normal gravity. For the pressure range explored (Pr = 0.9), the droplet projected surface area

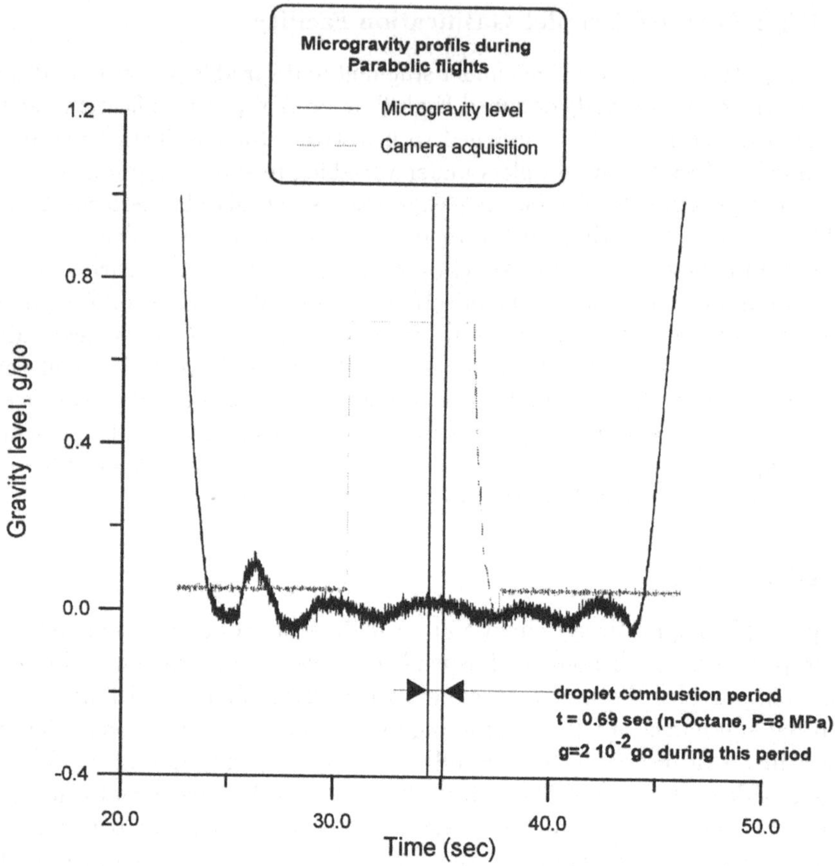

Fig. 2. Typical microgravity level versus time during a parabolic flight

regresses quasi-linearly with time, which allows for the determination of an average burning rate K. A similar behaviour is also observed for the experiments conducted under reduced gravity.

The average burning rates determined under normal and reduced gravity are compared in Fig. 4 for a pressure range extending up to Pr = 1.15. As expected, the reduced gravity K values are lower than the normal gravity values; both increase continuously with pressure. The strong increase of K with pressure above Pr = 0.8 under normal gravity remains unexplained. The results obtained with n-heptane droplets are shown on Fig. 5. As the critical pressure of n-heptane is lower, Pr values larger than 4 are obtained in the experiments in microgravity. The lowering of the pressure effect above the critical pressure is clearly observed.

The results obtained in microgravity with the four fuels tested are shown in Fig. 6. In this figure the burning rate at a given pressure is normalized by

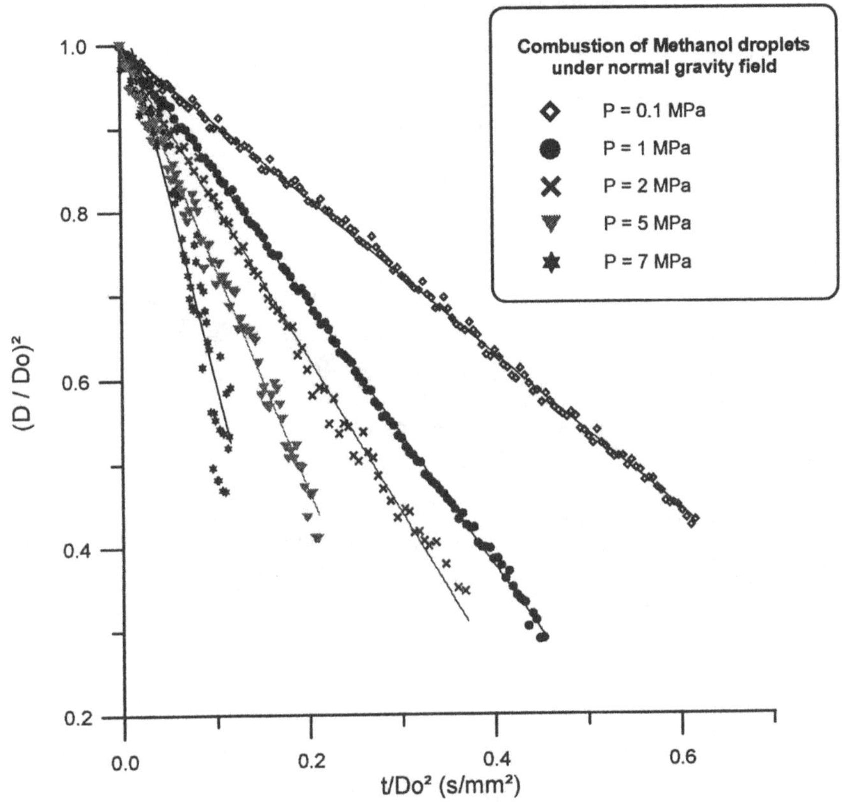

Fig. 3. Evolution of the normalized squared droplet diameter versus normalized time

its value at the critical pressure. The scattering of the experimental data become significant for high pressures, while a single behaviour characterizes the subcritical domain.

An interesting feature of these results is that we did not observe a maximum burning rate around the critical pressure, as was observed by Sato in [22]. Figure 7 shows the corresponding burning times for a droplet of 1 mm initial diameter versus the reduced pressure. The subcritical domain is characterized by a strong decrease of the burning time, which seems to be stabilized during the whole supercritical domain. In [22], a minimum of the burning time near the critical pressure was clearly observed. The main differences between the two experiments are related to the reduced gravity levels; the experiments of [22] were conducted by using drop towers, where the residual gravity level is much lower. The increase of the burning time observed in [22] in the supercritical domain can

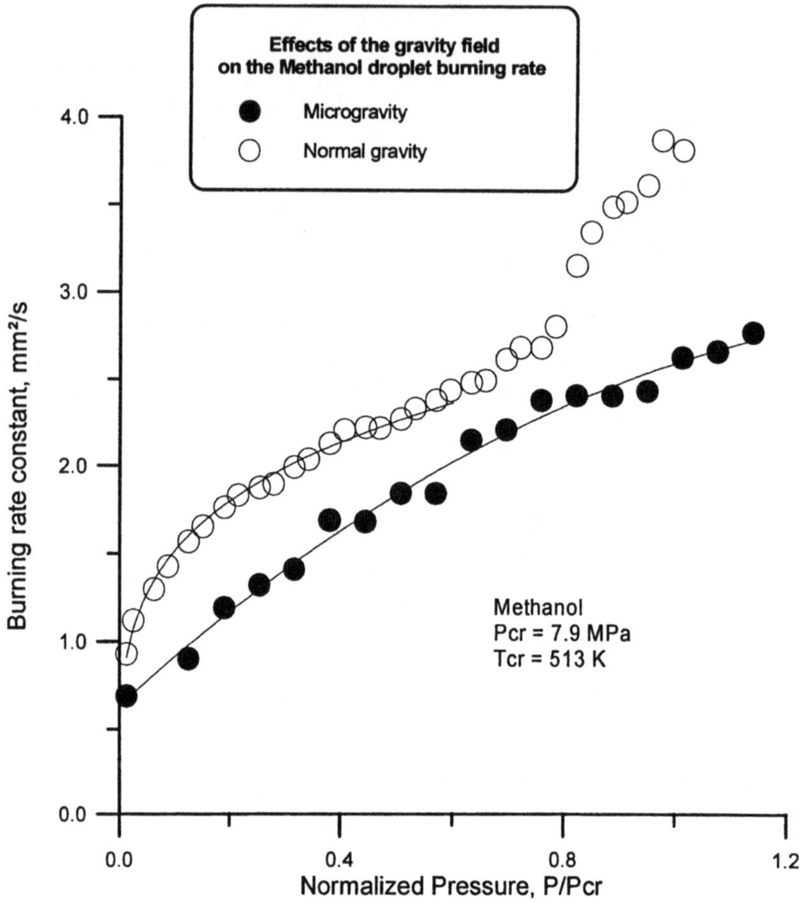

Fig. 4. Evolution of the burning rate constant versus normalized pressure.

be explained by the onset of a diffusion controlled regime after the vanishing of the gas-liquid interface. In the present experiments the residual gravity induced natural convection is enhanced with increasing pressure; also in the supercritical domain, the reduction of the surface tension makes the dense gas phase much more sensible to convection effects. All together these two effects may therefore explain the present experimental observations. In order to quantify these predictions, an attempt to introduce a Grashof number correction is presently under development.

4. Summary and Future Work

In the work described here, emphasis is put on recent results obtained in high pressure droplet burning experiments conducted both under normal gravity and

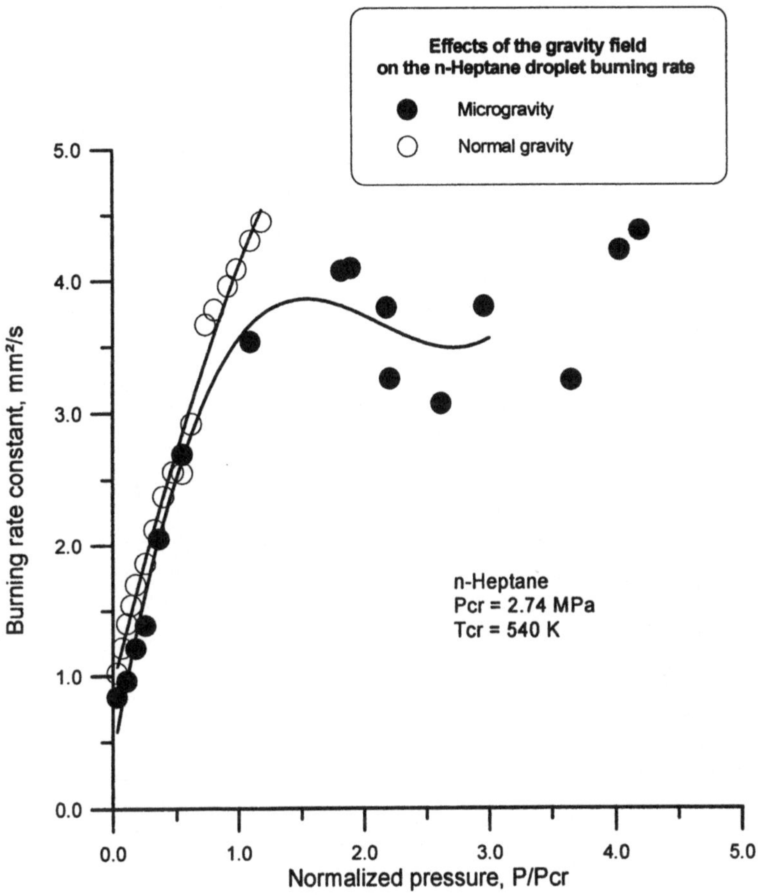

Fig. 5. Evolution of the burning rate constant versus normalized pressure.

during the parabolic flights of the CNES Caravelle, which creates a reduced gravity environment of the order of $10^{-2}g_0$. The combination of high pressure droplet burning experiments with reduced gravity is crucial in order to reduce the pressure enhanced natural convection effects and also to extend the applicability of the fibre suspended droplet technique when the surface tension decreases due to the closeness of thermodynamic critical conditions. The experimental results presented in this paper show a decrease of the droplet burning time with pressure in the subcritical domain. The minimum burning time observed by other investigators around the critical pressure was not found in this study for the four fuels investigated. This may be attributed to the pressure enhanced natural convection effects due to the residual gravity during parabolic flight experiments.

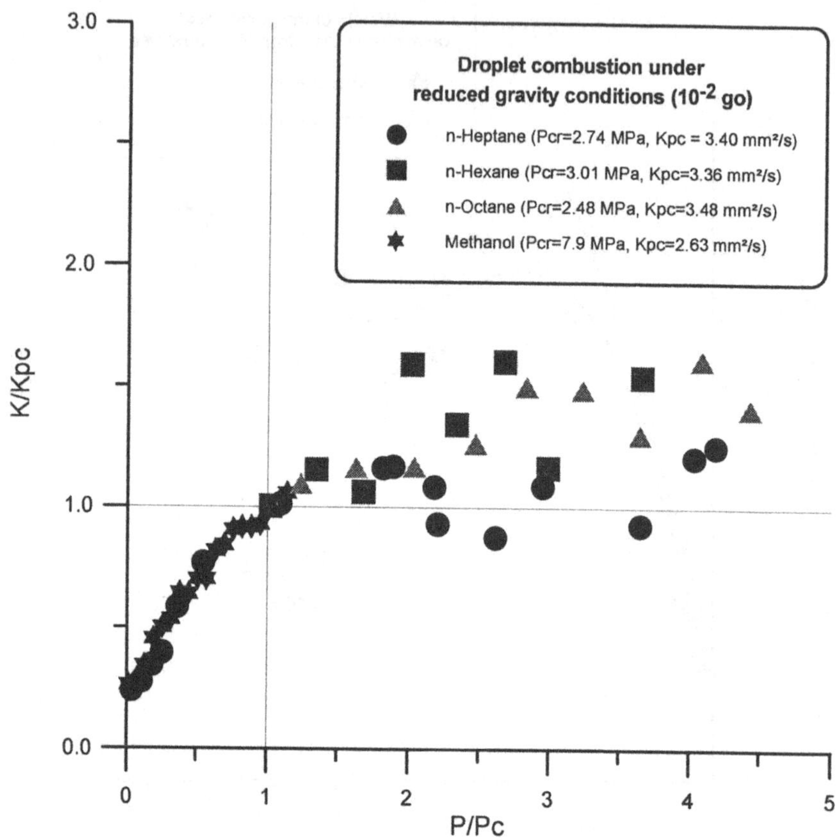

Fig. 6. Evolution of the normalized burning rate constant versus normalized pressure.

The investigation of high pressure effects will continue to be the main focus of the future LCSR work on droplet gasification regimes. The supercritical regime will be approached in the pure vaporization experiments with the help of a new version of the HP-DGF which is under construction; it will allow to increase the temperature as well as the pressure of the ambient medium. This apparatus will be used to investigate the vaporization and burning of mono- and multi-component fuel droplets at high pressure and high temperature conditions and also to compare normal gravity and reduced gravity experiments to infer natural convection effects. An electrodynamic balance is developed and will be added to the apparatus. It will allow to conduct supercritical vaporization experiments under normal gravity conditions. Parabolic flights will be continued to be used for suspended droplet experiments. A new high pressure droplet burning module will be developed to be used during sounding rocket experiments with ESA.

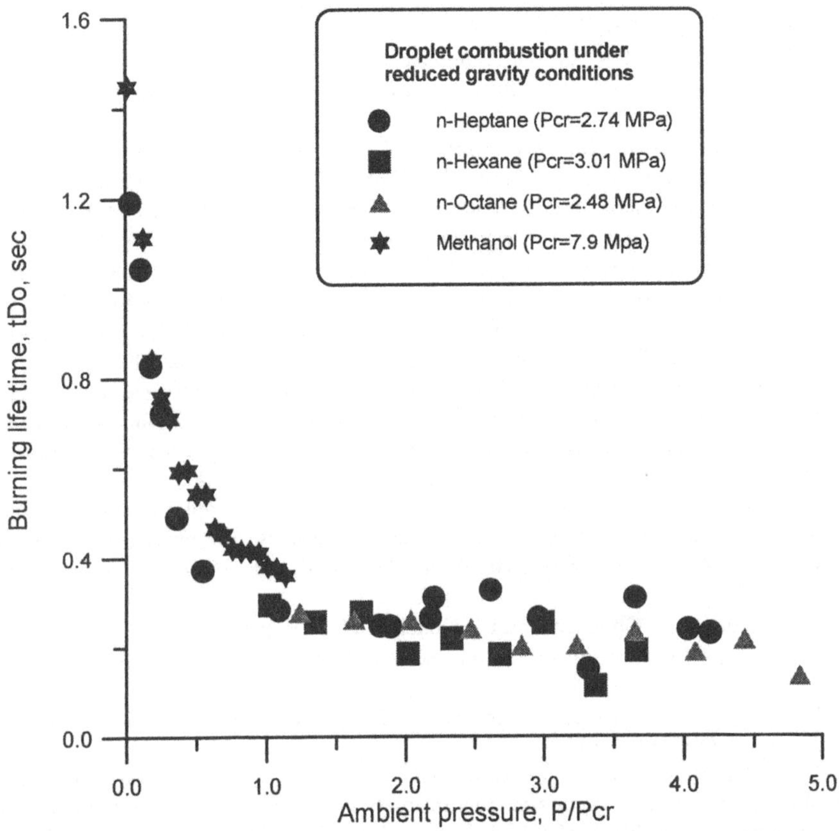

Fig. 7. Evolution of the normalized burning life time versus normalized pressure.

Acknowledgments

This work is supported by the CNRS/CNES joint research program on "Transport phenomena in reduced gravity" and by the European Space Agency Microgravity Program. The continuous support of CNES (program monitor Dr. B. Zappoli) and of ESA (program monitor Dr. H.U. Walter) is greatly appreciated.

References

1. C. Chauveau, X. Chesneau and Gökalp, I., *AIAA* Paper No. 93-0824, 1993
2. D.B. Spalding, *ARS J.*, **29** (1959) 828-835
3. P.R. Wieber, *AIAA J.* **1** (1963) 2764-2770.
4. J.A. Manrique and G.L. Borman, *Int. J. Heat and Mass Transfer* **12** (1969) 1081-1095.

5. D.E. Rosner and W.S. Chang, *Combust. Sci. Technol.* **7** (1973) 145-158.
6. T. Kadota and H. Hiroyasu, *Bull. JSME* **19** (1976) 1515-1521.
7. R.J. Litchford and S.M. Jeng, *AIAA* Paper No. 90-2191, 1990.
8. K.C. Hsieh, J.S. Shuen and V. Yang, *Combust. Sci. Technol.* **76** (1991) 111-132
9. E.W. Curtis and P.V. Farrell, *Combust. Flame* **90** (1992) 85-102.
10. J.P. Delplanque and W.A. Sirignano, *Int. J. Heat Mass Transfer* **36** (1993) 303-314.
11. V. Yang and N.N. Lin, "Vaporization of liquid oxygen droplets in supercritical hydrogen environments", submitted to *Combust. Sci. Technol.*, 1993
12. H. Jia and G. Gogos, *Int. J. Heat Mass Transfer* **36** (1993) 4419-4431.
13. J.P. Delplanque and W.A. Sirignano, *Atomization and Sprays*, **4** (1994) 325-349.
14. W.A. Sirignano, J.P. Delplanque, J.H. Chiang and R. Bahtia, "Liquid propellant droplet vaporization : a driving mechanism for rocket combustion instability", paper presented at the *First International Symposium on Liquid Rocket Combustion Instability*, The Pennsylvania State University, January 18-20, 1993
15. C.C. Hsia and V. Yang, "Lox droplet vaporization in a supercritical forced convective environment", paper presented at the *Fifth Annual Symposium of the Propulsion Engineering Research Center*, The Pennsylvania State University, University Park, PA, September 8-9, 1993
16. L.N. Long, M.M.Micci and B.C. Wong, "supercritical droplet evaporation modelled using molecular dynamics on parallel processors", paper submitted for presentation at the 30th Joint Propulsion Conference, Indianapolis, IN, June 27-29, 1994
17. X. Chesneau, C. Chauveau and I. Gökalp, *AIAA* Paper No. 94-0688, 1994
18. I. Gökalp, C. Chauveau, O. Simon and X. Chesneau, *Combust. Flame* **89** (1992) 286-298.
19. I. Gökalp, C. Chauveau, H. Berrekam and N.A. Ramos-Arroyo, *Atomization and Sprays*, **4** (1994) 661-676.
20. N.A. Ramos-Arroyo, C. Chauveau, C., and I. Gökalp, "An experimental study of three interacting vaporizing fuel droplets in a forced convective laminar flow", presented at the Sixth ICLASS, Rouen, July 18-22, 1994, also submitted to Atomization and Sprays
21. C. Chauveau, X. Chesneau and I. Gökalp, *Adv. Space Res.* **16** (7) (1995) 157-160.
22. J. Sato, J., Studies on droplet evaporation and combustion in high pressures, *AIAA* paper 93-0813, 1993

Springer-Verlag und Umwelt

Als internationaler wissenschaftlicher Verlag sind wir uns unserer besonderen Verpflichtung der Umwelt gegenüber bewußt und beziehen umweltorientierte Grundsätze in Unternehmensentscheidungen mit ein.

Von unseren Geschäftspartnern (Druckereien, Papierfabriken, Verpackungsherstellern usw.) verlangen wir, daß sie sowohl beim Herstellungsprozeß selbst als auch beim Einsatz der zur Verwendung kommenden Materialien ökologische Gesichtspunkte berücksichtigen.

Das für dieses Buch verwendete Papier ist aus chlorfrei bzw. chlorarm hergestelltem Zellstoff gefertigt und im pH-Wert neutral.

Lecture Notes in Physics

For information about Vols. 1–429
please contact your bookseller or Springer-Verlag

Vol. 430: V. G. Gurzadyan, D. Pfenniger (Eds.), Ergodic Concepts in Stellar Dynamics. Proceedings, 1993. XVI, 302 pages. 1994.

Vol. 431: T. P. Ray, S. Beckwith (Eds.), Star Formation and Techniques in Infrared and mm-Wave Astronomy. Proceedings, 1992. XIV, 314 pages. 1994.

Vol. 432: G. Belvedere, M. Rodonò, G. M. Simnett (Eds.), Advances in Solar Physics. Proceedings, 1993. XVII, 335 pages. 1994.

Vol. 433: G. Contopoulos, N. Spyrou, L. Vlahos (Eds.), Galactic Dynamics and N-Body Simulations. Proceedings, 1993. XIV, 417 pages. 1994.

Vol. 434: J. Ehlers, H. Friedrich (Eds.), Canonical Gravity: From Classical to Quantum. Proceedings, 1993. X, 267 pages. 1994.

Vol. 435: E. Maruyama, H. Watanabe (Eds.), Physics and Industry. Proceedings, 1993. VII, 108 pages. 1994.

Vol. 436: A. Alekseev, A. Hietamäki, K. Huitu, A. Morozov, A. Niemi (Eds.), Integrable Models and Strings. Proceedings, 1993. VII, 280 pages. 1994.

Vol. 437: K. K. Bardhan, B. K. Chakrabarti, A. Hansen (Eds.), Non-Linearity and Breakdown in Soft Condensed Matter. Proceedings, 1993. XI, 340 pages. 1994.

Vol. 438: A. Pękalski (Ed.), Diffusion Processes: Experiment, Theory, Simulations. Proceedings, 1994. VIII, 312 pages. 1994.

Vol. 439: T. L. Wilson, K. J. Johnston (Eds.), The Structure and Content of Molecular Clouds. 25 Years of Molecular Radioastronomy. Proceedings, 1993. XIII, 308 pages. 1994.

Vol. 440: H. Latal, W. Schweiger (Eds.), Matter Under Extreme Conditions. Proceedings, 1994. IX, 243 pages. 1994.

Vol. 441: J. M. Arias, M. I. Gallardo, M. Lozano (Eds.), Response of the Nuclear System to External Forces. Proceedings, 1994, VIII. 293 pages. 1995.

Vol. 442: P. A. Bois, E. Dériat, R. Gatignol, A. Rigolot (Eds.), Asymptotic Modelling in Fluid Mechanics. Proceedings, 1994. XII, 307 pages. 1995.

Vol. 443: D. Koester, K. Werner (Eds.), White Dwarfs. Proceedings, 1994. XII, 348 pages. 1995.

Vol. 444: A. O. Benz, A. Krüger (Eds.), Coronal Magnetic Energy Releases. Proceedings, 1994. X, 293 pages. 1995.

Vol. 445: J. Brey, J. Marro, J. M. Rubí, M. San Miguel (Eds.), 25 Years of Non-Equilibrium Statistical Mechanics. Proceedings, 1994. XVII, 387 pages. 1995.

Vol. 446: V. Rivasseau (Ed.), Constructive Physics. Results in Field Theory, Statistical Mechanics and Condensed Matter Physics. Proceedings, 1994. X, 337 pages. 1995.

Vol. 447: G. Aktaş, C. Saçlıoğlu, M. Serdaroğlu (Eds.), Strings and Symmetries. Proceedings, 1994. XIV, 389 pages. 1995.

Vol. 448: P. L. Garrido, J. Marro (Eds.), Third Granada Lectures in Computational Physics. Proceedings, 1994. XIV, 346 pages. 1995.

Vol. 449: J. Buckmaster, T. Takeno (Eds.), Modeling in Combustion Science. Proceedings, 1994. X, 369 pages. 1995.

Vol. 450: M. F. Shlesinger, G. M. Zaslavsky, U. Frisch (Eds.), Lévy Flights and Related Topics in Physics. Proceedigs, 1994. XIV, 347 pages. 1995.

Vol. 451: P. Krée, W. Wedig (Eds.), Probabilistic Methods in Applied Physics. IX, 393 pages. 1995.

Vol. 452: A. M. Bernstein, B. R. Holstein (Eds.), Chiral Dynamics: Theory and Experiment. Proceedings, 1994. VIII, 351 pages. 1995.

Vol. 453: S. M. Deshpande, S. S. Desai, R. Narasimha (Eds.), Fourteenth International Conference on Numerical Methods in Fluid Dynamics. Proceedings, 1994. XIII, 589 pages. 1995.

Vol. 454: J. Greiner, H. W. Duerbeck, R. E. Gershberg (Eds.), Flares and Flashes, Germany 1994. XXII, 477 pages. 1995.

Vol. 455: F. Occhionero (Ed.), Birth of the Universe and Fundamental Physics. Proceedings, 1994. XV, 387 pages. 1995.

Vol. 456: H. B. Geyer (Ed.), Field Theory, Topology and Condensed Matter Physics. Proceedings, 1994. XII, 206 pages. 1995.

Vol. 457: P. Garbaczewski, M. Wolf, A. Weron (Eds.), Chaos – The Interplay Between Stochastic and Deterministic Behaviour. Proceedings, 1995. XII, 573 pages. 1995.

Vol. 458: I. W. Roxburgh, J.-L. Masnou (Eds.), Physical Processes in Astrophysics. Proceedings, 1993. XII, 249 pages. 1995.

Vol. 459: G. Winnewisser, G. C. Pelz (Eds.), The Physics and Chemistry of Interstellar Molecular Clouds. Proceedings, 1993. XV, 393 pages. 1995.

Vol. 461: R. López-Peña, R. Capovilla, R. García-Pelayo, H. Waelbroeck, F. Zertuche, (Eds.), Complex Systems and Binary Networks. Lectures, México 1995. X, 223 pages. 1995.

Vol. 462: M. Meneguzzi, A. Pouquet, P.-L. Sulem (Eds.), Small-Scale Structures in Three-Dimensional Hydrodynamic and Magnetohydrodynamic Turbulence. Proceedings, 1995. IX, 421 pages. 1995.

Vol. 463: H. Hippelein, K. Meisenheimer, H.-J. Röser (Eds.), Galaxies in the Young Universe. Proceedings, 1994. XV, 314 pages. 1995.

Vol. 464: L. Ratke, H. U. Walter, B. Feuerbach (Eds.), Materials and Fluids Under Low Gravity. Proceedings, 1994. XVIII, 424 pages, 1996.

New Series m: Monographs

Vol. m 1: H. Hora, Plasmas at High Temperature and Density. VIII, 442 pages. 1991.

Vol. m 2: P. Busch, P. J. Lahti, P. Mittelstaedt, The Quantum Theory of Measurement. XIII, 165 pages. 1991.

Vol. m 3: A. Heck, J. M. Perdang (Eds.), Applying Fractals in Astronomy. IX, 210 pages. 1991.

Vol. m 4: R. K. Zeytounian, Mécanique des fluides fondamentale. XV, 615 pages, 1991.

Vol. m 5: R. K. Zeytounian, Meteorological Fluid Dynamics. XI, 346 pages. 1991.

Vol. m 6: N. M. J. Woodhouse, Special Relativity. VIII, 86 pages. 1992.

Vol. m 7: G. Morandi, The Role of Topology in Classical and Quantum Physics. XIII, 239 pages. 1992.

Vol. m 8: D. Funaro, Polynomial Approximation of Differential Equations. X, 305 pages. 1992.

Vol. m 9: M. Namiki, Stochastic Quantization. X, 217 pages. 1992.

Vol. m 10: J. Hoppe, Lectures on Integrable Systems. VII, 111 pages. 1992.

Vol. m 11: A. D. Yaghjian, Relativistic Dynamics of a Charged Sphere. XII, 115 pages. 1992.

Vol. m 12: G. Esposito, Quantum Gravity, Quantum Cosmology and Lorentzian Geometries. Second Corrected and Enlarged Edition. XVIII, 349 pages. 1994.

Vol. m 13: M. Klein, A. Knauf, Classical Planar Scattering by Coulombic Potentials. V, 142 pages. 1992.

Vol. m 14: A. Lerda, Anyons. XI, 138 pages. 1992.

Vol. m 15: N. Peters, B. Rogg (Eds.), Reduced Kinetic Mechanisms for Applications in Combustion Systems. X, 360 pages. 1993.

Vol. m 16: P. Christe, M. Henkel, Introduction to Conformal Invariance and Its Applications to Critical Phenomena. XV, 260 pages. 1993.

Vol. m 17: M. Schoen, Computer Simulation of Condensed Phases in Complex Geometries. X, 136 pages. 1993.

Vol. m 18: H. Carmichael, An Open Systems Approach to Quantum Optics. X, 179 pages. 1993.

Vol. m 19: S. D. Bogan, M. K. Hinders, Interface Effects in Elastic Wave Scattering. XII, 182 pages. 1994.

Vol. m 20: E. Abdalla, M. C. B. Abdalla, D. Dalmazi, A. Zadra, 2D-Gravity in Non-Critical Strings. IX, 319 pages. 1994.

Vol. m 21: G. P. Berman, E. N. Bulgakov, D. D. Holm, Crossover-Time in Quantum Boson and Spin Systems. XI, 268 pages. 1994.

Vol. m 22: M.-O. Hongler, Chaotic and Stochastic Behaviour in Automatic Production Lines. V, 85 pages. 1994.

Vol. m 23: V. S. Viswanath, G. Müller, The Recursion Method. X, 259 pages. 1994.

Vol. m 24: A. Ern, V. Giovangigli, Multicomponent Transport Algorithms. XIV, 427 pages. 1994.

Vol. m 25: A. V. Bogdanov, G. V. Dubrovskiy, M. P. Krutikov, D. V. Kulginov, V. M. Strelchenya, Interaction of Gases with Surfaces. XIV, 132 pages. 1995.

Vol. m 26: M. Dineykhan, G. V. Efimov, G. Ganbold, S. N. Nedelko, Oscillator Representation in Quantum Physics. IX, 279 pages. 1995.

Vol. m 27: J. T. Ottesen, Infinite Dimensional Groups and Algebras in Quantum Physics. IX, 218 pages. 1995.

Vol. m 28: O. Piguet, S. P. Sorella, Algebraic Renormalization. IX, 134 pages. 1995.

Vol. m 29: C. Bendjaballah, Introduction to Photon Communication. VII, 193 pages. 1995.

Vol. m 30: A. J. Greer, W. J. Kossler, Low Magnetic Fields in Anisotropic Superconductors. VII, 161 pages. 1995.

Vol. m 31: P. Busch, M. Grabowski, P. J. Lahti, Operational Quantum Physics. XI, 230 pages. 1995.

Vol. m 32: L. de Broglie, Diverses questions de mécanique et de thermodynamique classiques et relativistes. XII, 198 pages. 1995.

Vol. m 33: R. Alkofer, H. Reinhardt, Chiral Quark Dynamics. VIII, 115 pages. 1995.

Vol. m 34: R. Jost, Das Märchen vom Elfenbeinernen Turm. VIII, 286 pages. 1995.

Vol. m 35: E. Elizalde, Ten Physical Applications of Spectral Zeta Functions. XIV, 228 pages. 1995.

Vol. m 36: G. Dunne, Self-Dual Chern-Simons Theories. X, 217 pages. 1995.

Vol. m 37: S. Childress, A.D. Gilbert, Stretch, Twist, Fold: The Fast Dynamo. XI, 410 pages. 1995.

Vol. m 38: J. González, M. A. Martín-Delgado, G. Sierra, A. H. Vozmediano, Quantum Electron Liquids and High-T_c Superconductivity. X, 299 pages. 1995.

Vol. m 39: L. Pittner, Algebraic Foundations of Non-Commutative Differential Geometry and Quantum Groups. XII, 469 pages. 1996.

GPSR Compliance

The European Union's (EU) General Product Safety Regulation (GPSR) is a set of rules that requires consumer products to be safe and our obligations to ensure this.

If you have any concerns about our products, you can contact us on ProductSafety@springernature.com

In case Publisher is established outside the EU, the EU authorized representative is:

Springer Nature Customer Service Center GmbH
Europaplatz 3
69115 Heidelberg, Germany

Batch number: 10650754

Printed by Printforce, the Netherlands